Nanomaterials Chemistry

Edited by
C. N. R. Rao, A. Müller,
and A. K. Cheetham

1807–2007 Knowledge for Generations

Each generation has its unique needs and aspirations. When Charles Wiley first opened his small printing shop in lower Manhattan in 1807, it was a generation of boundless potential searching for an identity. And we were there, helping to define a new American literary tradition. Over half a century later, in the midst of the Second Industrial Revolution, it was a generation focused on building the future. Once again, we were there, supplying the critical scientific, technical, and engineering knowledge that helped frame the world. Throughout the 20th Century, and into the new millennium, nations began to reach out beyond their own borders and a new international community was born. Wiley was there, expanding its operations around the world to enable a global exchange of ideas, opinions, and know-how.

For 200 years, Wiley has been an integral part of each generation's journey, enabling the flow of information and understanding necessary to meet their needs and fulfill their aspirations. Today, bold new technologies are changing the way we live and learn. Wiley will be there, providing you the must-have knowledge you need to imagine new worlds, new possibilities, and new opportunities.

Generations come and go, but you can always count on Wiley to provide you the knowledge you need, when and where you need it!

William J. Pesce
President and Chief Executive Officer

Peter Booth Wiley
Chairman of the Board

Nanomaterials Chemistry

Recent Developments and New Directions

Edited by
C. N. R. Rao, A. Müller, and A. K. Cheetham

WILEY-VCH Verlag GmbH & Co. KGaA

The Editors

Prof. Dr. C. N. R. Rao
Jawaharlal Nehru Centre for Advanced
Scientific Research
CSIR Centre of Excellence in Chemistry and
Chemistry and Physics of Materials Unit
Jakkur P.O.
Bangalore 560 064
India

Prof. Dr. Dr. h.c. mult. A. Müller
University of Bielefeld
Faculty of Chemistry
P. O. Box 10 01 31
33501 Bielefeld
Germany

Prof. Dr. A.K. Cheetham
University of California
Materials Research Laboratory
Santa Barbara, CA 93106-5130
USA

■ All books published by Wiley-VCH are carefully produced. Nevertheless, authors, editors, and publisher do not warrant the information contained in these books, including this book, to be free of errors. Readers are advised to keep in mind that statements, data, illustrations, procedural details or other items may inadvertently be inaccurate.

Library of Congress Card No.: applied for

British Library Cataloguing-in-Publication Data
A catalogue record for this book is available from the British Library.

Bibliographic information published by the Deutsche Nationalbibliothek
Die Deutsche Nationalbibliothek lists this publication in the Deutsche Nationalbibliografie; detailed bibliographic data are available in the Internet at ⟨http://dnb.d-nb.de⟩.

© 2007 WILEY-VCH Verlag GmbH & Co. KGaA, Weinheim

All rights reserved (including those of translation into other languages). No part of this book may be reproduced in any form – by photoprinting, microfilm, or any other means – nor transmitted or translated into a machine language without written permission from the publishers. Registered names, trademarks, etc. used in this book, even when not specifically marked as such, are not to be considered unprotected by law.

Printed in the Federal Republic of Germany
Printed on acid-free paper

Typesetting Asco Typesetter, Hong Kong
Printing betz-druck GmbH, Darmstadt
Binding Litges & Dopf GmbH, Heppenheim
Wiley Bicentennial Logo Richard J. Pacifico

ISBN 978-3-527-31664-9

Contents

Preface *XI*

List of Contributors *XIII*

1   Recent Developments in the Synthesis, Properties and Assemblies of Nanocrystals *1*
P.J. Thomas and P. O'Brien

1.1   Introduction *1*
1.2   Spherical Nanocrystals *1*
1.2.1 Semiconductor Nanocrystals *1*
1.2.2 Metal Nanocrystals *4*
1.2.3 Nanocrystals of Metal Oxides *6*
1.3   Nanocrystals of Different Shapes *7*
1.3.1 Anisotropic Growth of Semiconductor and Oxide Nanocrystals *7*
1.3.2 Anisotropic Growth of Metal Nanocrystals *14*
1.4   Selective Growth on Nanocrystals *17*
1.5   Properties of Nanocrystals *18*
1.5.1 Electronic and Optical Properties *18*
1.5.2 Magnetic Properties *21*
1.6   Ordered Assemblies of Nanocrystals *22*
1.6.1 One- and Low-dimensional Arrangements *22*
1.6.2 Two-dimensional Arrays *24*
1.6.3 Three-dimensional Superlattices *26*
1.6.4 Colloidal Crystals *29*
1.7   Applications *30*
1.7.1 Optical and Electro-optical Devices *30*
1.7.2 Nanocrystal-based Optical Detection and Related Devices *31*
1.7.3 Nanocrystals as Fluorescent Tags *33*
1.7.4 Biomedical Applications of Oxide Nanoparticles *33*
1.7.5 Nanoelectronics and Nanoscalar Electronic Devices *34*
1.8   Conclusions *35*
       References *36*

2 Nanotubes and Nanowires: Recent Developments 45
S. R. C. Vivekchand, A. Govindaraj, and C. N. R. Rao

2.1 Introduction 45
2.2 Carbon Nanotubes 45
2.2.1 Synthesis 45
2.2.2 Purification 50
2.2.3 Functionalization and Solubilization 54
2.2.4 Properties and Applications 60
2.2.4.1 Optical, Electrical and Other Properties 60
2.2.4.2 Phase Transitions, Mechanical Properties, and Fluid Mechanics 66
2.2.4.3 Energy Storage and Conversion 68
2.2.4.4 Chemical Sensors 68
2.2.5 Biochemical and Biomedical Aspects 69
2.2.6 Nanocomposites 71
2.2.7 Transistors and Devices 72
2.3 Inorganic Nanotubes 75
2.3.1 Synthesis 75
2.3.2 Solubilization and Functionalization 77
2.3.3 Properties and Applications 79
2.4 Inorganic Nanowires 79
2.4.1 Synthesis 79
2.4.2 Self Assembly and Functionalization 90
2.4.3 Coaxial Nanowires and Coatings on Nanowires 92
2.4.4 Optical Properties 92
2.4.5 Electrical and Magnetic Properties 97
2.4.6 Some Chemical Aspects and Sensor Applications 100
2.4.7 Mechanical Properties 101
2.4.8 Transistors and Devices 102
2.4.9 Biological Aspects 103
References 104

3 Nonaqueous Sol–Gel Routes to Nanocrystalline Metal Oxides 119
M. Niederberger and M. Antonietti

3.1 Overview 119
3.2 Introduction 119
3.3 Short Introduction to Aqueous and Nonaqueous Sol–Gel Chemistry 120
3.4 Nonaqueous Sol–Gel Routes to Metal Oxide Nanoparticles 121
3.4.1 Surfactant-controlled Synthesis of Metal Oxide Nanoparticles 121
3.5 Solvent-controlled Synthesis of Metal Oxide Nanoparticles 127
3.5.1 Introduction 127
3.5.2 Reaction of Metal Halides with Alcohols 127
3.5.3 Reaction of Metal Alkoxides with Alcohols 130
3.5.4 Reaction of Metal Alkoxides with Ketones and Aldehydes 131

3.5.5	Reaction of Metal Acetylacetonates with Various Organic Solvents	*132*
3.6	Selected Reaction Mechanisms	*133*
3.7	Summary and Outlook	*134*
	References	*135*

4 Growth of Nanocrystals in Solution *139*
R. Viswanatha and D. D. Sarma

4.1	Introduction	*139*
4.2	Theoretical Aspects	*140*
4.2.1	Theory of Nucleation	*140*
4.2.2	Mechanism of Growth	*141*
4.2.2.1	Diffusion Limited Growth: Lifshitz–Slyozov–Wagner (LSW) Theory and Post-LSW Theory	*143*
4.2.2.2	Reaction-limited Growth	*147*
4.2.2.3	Mixed Diffusion–Reaction Control	*148*
4.3	Experimental Investigations	*151*
4.3.1	Au Nanocrystals	*153*
4.3.2	ZnO Nanocrystals	*154*
4.3.3	Effect of Capping Agents on Growth Kinetics	*160*
4.3.3.1	Effect of Oleic Acid on the Growth of CdSe Nanocrystals	*161*
4.3.3.2	PVP as a Capping Agent in the Growth of ZnO Nanocrystals	*163*
4.3.3.3	Effect of Adsorption of Thiols on ZnO Growth Kinetics	*166*
4.4	Concluding Remarks	*167*
	References	*168*

5 Peptide Nanomaterials: Self-assembling Peptides as Building Blocks for Novel Materials *171*
M. Reches and E. Gazit

5.1	Overview	*171*
5.2	Introduction	*171*
5.3	Cyclic Peptide-based Nanostructures	*172*
5.4	Linear Peptide-based Nanostructures	*174*
5.5	Amyloid Fibrils as Bio-inspired Material: The Use of Natural Amyloid and Peptide Fragments	*177*
5.6	From Amyloid Structures to Peptide Nanostructures	*178*
5.7	Bioinspired Peptide-based Composite Nanomaterials	*180*
5.8	Prospects	*180*
	References	*181*

6 Surface Plasmon Resonances in Nanostructured Materials *185*
K. G. Thomas

6.1	Introduction to Surface Plasmons	*185*
6.1.1	Propagating Surface Plasmons	*186*

6.1.2	Localized Surface Plasmons 189
6.2	Tuning the Surface Plasmon Oscillations 190
6.2.1	Size of Nanoparticle 190
6.2.2	Shape of Nanoparticle 191
6.2.3	Dielectric Environment 194
6.3	Excitation of Localized Surface Plasmons 196
6.3.1	Multipole Resonances 197
6.3.2	Absorption vs. Scattering 200
6.4	Plasmon Coupling in Higher Order Nanostructures 204
6.4.1	Assembly of Nanospheres 204
6.4.2	Assembly of Nanorods 208
6.5	Summary and Outlook 215
	References 216

7	**Applications of Nanostructured Hybrid Materials for Supercapacitors** 219
	A. V. Murugan and K. Vijayamohanan
7.1	Overview 219
7.2	Introduction 219
7.3	Nanostructured Hybrid Materials 220
7.4	Electrochemical Energy Storage 222
7.5	Electrochemical Capacitors 223
7.5.1	Electrochemical Double Layer Capacitor vs. Conventional Capacitor 225
7.5.2	Origin of Enhanced Capacitance 226
7.6	Electrode Materials for Supercapacitors 229
7.6.1	Nanostructured Transition Metal Oxides 229
7.6.2	Nanostructured Conducting Polymers 230
7.6.3	Carbon Nanotubes and Related Carbonaceous Materials 231
7.7	Hybrid Nanostructured Materials 234
7.7.1	Conducting Polymer–Transition Metal Oxide Nanohybrids 235
7.7.2	Conducting Polymer–Carbon Nanotube Hybrids 237
7.7.3	Transition Metal Oxides–Carbon Nanotube Hybrids 238
7.8	Hybrid Nanostructured Materials as Electrolytes for Super Capacitors 241
7.8.1	Nanostructured Polymer Composite Electrolytes 242
7.8.2	Ionic Liquids as Supercapacitor Electrolytes 242
7.9	Possible Limitations of Hybrid Materials for Supercapacitors 243
7.10	Conclusions and Perspectives 244
	References 245

8	**Dendrimers and Their Use as Nanoscale Sensors** 249
	N. Jayaraman
8.1	Introduction 249
8.2	Synthetic Methods 250
8.3	Macromolecular Properties 262

8.3.1	Molecular Modeling and Intrinsic Viscosity Studies	*262*
8.3.2	Fluorescence Properties	*264*
8.3.3	*Endo*- and *Exo*-Receptor Properties	*265*
8.4	Chemical Sensors with Dendrimers	*267*
8.4.1	Vapor Sensing	*267*
8.4.2	Sensing Organic Amines and Acids	*270*
8.4.3	Vapoconductivity	*270*
8.4.4	Sensing CO and CO_2	*271*
8.4.5	Gas and Vapor Sensing in Solution	*272*
8.4.6	Chiral Sensing of Asymmetric Molecules	*275*
8.4.7	Fluorescence Labeled Dendrimers and Detection of Metal Cations	*277*
8.4.8	Anion Sensing	*279*
8.5	Dendrimer-based Biosensors	*281*
8.5.1	Acetylcholinesterase Biosensor	*281*
8.5.2	Dendrimers as Cell Capture Agents	*282*
8.5.3	Dendrimers as a Surface Plasmon Resonance Sensor Surface	*283*
8.5.4	Layer-by-Layer Assembly Using Dendrimers and Electrocatalysis	*283*
8.5.5	SAM–Dendrimer Conjugates for Biomolecular Sensing	*284*
8.5.6	Dendrimer-based Calorimetric Biosensors	*288*
8.5.7	Dendrimer-based Glucose Sensors	*289*
8.6	Conclusion and Outlook	*292*
	References	*292*

9 Molecular Approaches in Organic/Polymeric Field-effect Transistors *299*
K. S. Narayan and S. Dutta

9.1	Introduction	*299*
9.2	Device Operations and Electrical Characterization	*300*
9.3	Device Fabrication	*301*
9.3.1	Substrate Treatment Methods	*304*
9.3.2	Electrode Materials	*305*
9.4	Progress in Electrical Performance	*306*
9.5	Progress in p-Channel OFETs	*306*
9.6	Progress in n-Channel OFET	*309*
9.7	Progress in Ambipolar OFET	*310*
9.8	PhotoPFETs	*311*
9.9	Photoeffects in Semiconducting Polymer Dispersed Single Wall Carbon Nanotube Transistors	*313*
9.10	Recent Approaches in Assembling Devices	*314*
	References	*316*

10 Supramolecular Approaches to Molecular Machines *319*
M. C. Grossel

10.1	Introduction	*319*
10.2	Catenanes and Rotaxanes	*320*
10.2.1	Synthetic Routes to Catenanes and Rotaxanes	*321*

10.2.2	Aromatic π–π Association Routes to Catenanes and Rotaxanes	*322*
10.2.2.1	Preparation and Properties of [2]-Catenanes	*322*
10.2.2.2	Multiple Catenanes	*323*
10.2.2.3	Switchable Catenanes	*324*
10.2.2.4	Other Synthetic Routes to Paraquat-based Catenanes	*326*
10.2.2.5	Rotaxane Synthesis	*328*
10.2.2.6	Switchable Catenanes	*328*
10.2.2.7	Neutral Catenane Assembly	*329*
10.2.3	Ion Templating	*329*
10.2.3.1	Approaches to Redox-switchable Catenanes and Rotaxanes	*329*
10.2.3.2	Making More Complex Structures	*332*
10.2.3.3	Routes to [n]-Rotaxanes using Olefin Metathesis – Molecular Barcoding	*333*
10.2.3.4	Anion-templating	*335*
10.2.3.5	Other Approaches to Ion-templating	*337*
10.2.4	Hydrogen-bonded Assembly of Catenane, Rotaxanes, and Knots	*338*
10.2.4.1	Catenane and Knotane Synthesis	*338*
10.2.4.2	Routes to Functional Catenanes and Rotaxanes	*339*
10.2.4.3	Catenanes and Rotaxanes Derived from Dialkyl Ammonium Salts	*346*
10.2.5	Cyclodextrin-based Rotaxanes	*348*
10.2.5.1	Controlling Motion	*349*
10.3	Molecular Logic Gates	*349*
10.4	Conclusions	*352*
	References 352	
11	**Nanoscale Electronic Inhomogeneities in Complex Oxides**	***357***
	V. B. Shenoy, H. R. Krishnamurthy, and T. V. Ramakrishnan	
11.1	Introduction	*357*
11.2	Electronic Inhomogeneities – Experimental Evidence	*358*
11.3	Theoretical Approaches to Electronic Inhomogeneities	*364*
11.4	The ℓb Model for Manganites	*366*
11.5	The Extended ℓb Model and Effects of Long-range Coulomb Interactions	*370*
11.6	Conclusion	*381*
	References 382	

Index *385*

Preface

The subject of nanoscience and technology has had an extraordinary season of development and excitement in the last few years. Chemistry constitutes a major part of nanoscience research and without employing chemical techniques it is difficult, nay impossible, to synthesize or assemble most of the nanomaterials. Furthermore, many of the properties and phenomena associated with nanomaterials require chemical understanding, just as many of the applications of nanomaterials relate to chemistry. Because of the wide interest in the subject, we edited a book entitled Chemistry of Nanomaterials, which was published by Wiley-VCH in the year 2004. This book was extremely well received. In view of the increasing interest evinced all over the world in the chemistry of nanomaterials, we considered it appropriate to edit a book covering research on nanomaterials published in the last 2 to 3 years. This book is the result of such an effort. The book covers recent developments in nanocrystals, nanotubes and nanowires. The first two chapters dealing with nanocrystals, nanotubes and nanowires broadly cover all aspects of these three classes of nanomaterials. There are also chapters devoted to topics such as peptide nanomaterials, dendrimers, molecular electronics, molecular motors and supercapacitors. We believe that this book not only gives a status report of the subject, but also indicates future directions. The book should be a useful guide and reference work to all those involved in teaching and research in this area.

C. N. R. Rao
A. Müller
A. K. Cheetham

List of Contributors

M. Antonietti
Max Planck Institute of Colloids
and Interfaces
Colloid Chemistry
Research Campus Golm
14424 Potsdam
Germany

A. K. Cheetham
University of California
Materials Research Laboratory
Santa Barbara, CA 93106-5130
USA

S. Dutta
ISMN-CNR Bologna Division
Via P. Gobetti, 101
40129 Bologna
Italy

E. Gazit
Faculty of Life Sciences
Ramat Aviv
Tel Aviv Univeristy
Tel Aviv 69978
Israel

A. Govindaraj
Hon. Faculty Fellow
CSIR COE in Chemistry &
DST Unit on Nanoscience
JNCASR, Jakkur PO
Bangalore 560064
India

M. C. Grossel
School of Chemistry
University of Southampton
Highfield
Southampton
Hants
SO17 1BJ
UK

N. Jayaraman
Department of Organic Chemistry
Indian Institute of Science
Bangalore 560 012
India

H. R. Krishnamurthy
Centre for Condensed Matter Theory
Indian Institute of Science
Bangalore 560 012
India

A. Müller
University of Bielefeld
Faculty of Chemistry
P.O. Box 10 01 31
33501 Bielefeld
Germany

A. V. Murugan
Physical and Materials Chemistry
Division
National Chemical Laboratory
Pune 411 008
India

K.S. Narayan
Jawaharlal Nehru Centre for
Advanced Scientific Research
Bangalore 560 064
India

M. Niederberger
Max Planck Institute of Colloids
and Interfaces
Colloid Chemistry
Research Campus Golm
14424 Potsdam
Germany

P. O'Brien
School of Chemistry and School
of Materials
The University of Manchester
Oxford Road
Manchester
M139PL
UK

T.V. Ramakrishnan
Department of Physics
Banaras Hindu University
Varanasi 221 005
Uttar Pradesh
India

C.N.R. Rao
Jawaharlal Nehru Centre for
Advanced Scientific Research
CSIR Centre of Excellence in
Chemistry and Chemistry and
Physics of Materials Unit
Jakkur P.O.
Bangalore 560 064
India

M. Reches
Department of Molecular
Microbiology and Biotechnology
Tel Aviv University
Tel Aviv 69978
Israel

D.D. Sarma
Solid State and Structural Chemistry
Unit
Indian Institute of Science
Bangalore 560 012
India

V.B. Shenoy
Materials Research Centre and Centre
for Condensed Matter Theory
Indian Institute of Science
Bangalore 560 012
India

K.G. Thomas
Photosciences and Photonics
Chemical Sciences and Technology
Division
Regional Research Laboratory
Trivandrum 695 019
India

P.J. Thomas
School of Chemistry and School of
Materials
The University of Manchester
Oxford Road
Manchester
M139PL
UK

K. Vijayamohanan
Physical and Materials Chemistry
Division
National Chemical Laboratory
Pune 411 008
India

R. Viswanatha
Solid State and Structural Chemistry
Unit
Indian Institute of Science
Bangalore 560 012
India

S.R.C. Vivekchand
Jawaharlal Nehru Centre for
Advanced Scientific Research
Chemistry and Physics of Materials
Unit and DST Unit on Nanoscience
Jakkur P. O.
Bangalore 560 064
India

1
Recent Developments in the Synthesis, Properties and Assemblies of Nanocrystals

P.J. Thomas and P. O'Brien

1.1
Introduction

Nanocrystals of metals, oxides and semiconductors have been studied intensely in the last several years by different chemical and physical methods. In the past decade the realization that the electronic, optical, magnetic and chemical properties of nanocrystals depend on their size has motivated intense research in this area. This interest has resulted in better understanding of the phenomena of quantum confinement, mature synthetic schemes and fabrication of exploratory nanoelectronic devices. The past couple of years has seen heightened activity in this area, driven by advances such as the ability to synthesize nanocrystals of different shapes. In this chapter, we review the progress in the area in the above period with emphasis on growth of semiconductor nanocrystals. Illustrative examples of the advances are provided. Several reviews have appeared in this period, seeking to summarize past as well as current work [1–4].

1.2
Spherical Nanocrystals

1.2.1
Semiconductor Nanocrystals

There have been several successful schemes for the synthesis of monodisperse semiconductor nanocrystals, especially the sufides and selenides of cadmium. However, there is still plenty of interest in exploring new synthetic routes. Nanocrystals of lead, manganese, cadmium and zinc sulfides have been obtained by thermolysis of the corresponding metal-oleylamine complex in the presence of S dissolved in oleylamine. The metal oleylamine complexes were prepared by reacting the corresponding chlorides with oleylamine. Nanocrystals with elongated shapes such as bullets and hexagons were produced by varying the stoichiometry

Nanomaterials Chemistry. Edited by C.N.R. Rao, A. Müller, and A.K. Cheetham
Copyright © 2007 WILEY-VCH Verlag GmbH & Co. KGaA, Weinheim
ISBN: 978-3-527-31664-9

and concentration of the precursors [5]. Manganese sulfide nanocrystals in the form of spheres and other shapes such as wires and cubes have been obtained by thermolysis of the diethyldithiocarbamate in hexadecylamine [6]. Nanocrystals of cadmium, manganese, lead, copper and zinc sulphides have been obtained by thermal decomposition of hexadecylxanthates in hexadecylamine and other solvents. A highlight of this report is the use of relatively low temperatures (50–150 °C) and ambient conditions for synthesis of the nanocrystals [7]. Fluorescent CdSe nanocrystals have been prepared using the air stable single source precursor cadmium imino-bis(diisopropylphosphine selenide) [8]. Cadmium selenide nanocrystals have been prepared in an organic medium prepared without the use of phosphines or phosphine oxides using CdO in oleic acid and Se in octadecene [9]. Water soluble luminescent CdS nanocrystals have been prepared by refluxing a single source precursor [(2,2'-bipyridine)Cd(SCOPh)$_2$] in aqueous solution [10]. Nanocrystals of EuS exhibiting quantum confinement were synthesized for the very first time by irradiating a solution of the dithiocarbamate Na[Eu(S$_2$CNEt$_2$)$_4$]3.5H$_2$O in acetonitrile [11]. Following this initial report, a number of single source precursors have been used to synthesize EuS nanocrystals [12].

Peng and coworkers have followed the nucleation and growth of CdSe nanocrystals in real time by monitoring absorption spectra with millisecond resolution [13]. Extremely small CdSe nanocrystals with dimensions of about 1.5 nm, with magic nuclearity have been reproducibly synthesized starting with CdO [14]. Intense emission, due to defects, spread across the visible spectrum has been observed from these nanocrystals (see Fig. 1.1). These nanocrystals can therefore be used to obtain white emission. The nanocrystals are presumed to have the structure of highly stable CdSe clusters observed in mass spectrometric studies [15].

Chalcogenide semiconductor nanocrystals have been synthesized in microfabricated flow reactors [16, 17]. The reactors employ either a continuous stream of

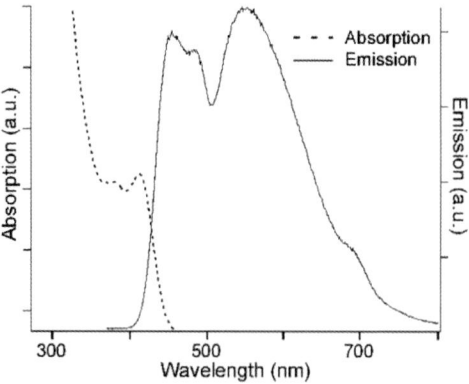

Fig. 1.1 Absorption and emission spectra of magic-sized CdSe nanocrystals (reproduced with permission from Ref. [14]).

liquid [16] or a segmented column containing gas and liquid droplets [17]. The use of small volumes and microscopic cells, provides new avenues for easy study of the nanocrystal growth process.

Gallium arsenide nanocrystals with diameters in the range 2.0–6.0 nm have been prepared in 4-ethylpyridine starting with $GaCl_3$ and $As(NMe_2)_3$ [18]. Highly luminescent InAs nanocrystals have been prepared in hexadecylamine by the use of the single source precursor $[^tBu_2AsInEt_2]_2$ [19].

Sardar and Rao [20] have prepared GaN nanoparticles of various sizes under solvothermal conditions, employing gallium cupferronate $(Ga(C_6H_5N_2O_2)_3)$ or chloride $(GaCl_3)$ as the gallium source and hexamethyldisilyzane as nitriding agent and toluene as solvent. This method is generally applicable for nitridation reactions. Nanocrystals of InP and GaP have been prepared using the thermolysis of the metal diorganophosphide $[M(P^tBu_2)_3]$ in hot 4-ethylpyridine [21], InP nanoparticles of different sizes (12–40 nm) have been prepared by a the solvothermal reaction involving $InCl_3$ and Na_3P [22]. Magnetic MnP nanocrystals with diameters of 6.7 nm have been produced by the treatment of $Mn_2(CO)_{10}$ with $P(SiMe_3)_3$ in trioctylphosphine oxide (TOPO)/myristic acid at elevated temperature [23]. Nanocrystals of iron and cobalt phosphides can also be synthesized by the use of the corresponding carbonyls.

There has been increasing interest in designing ligands that can effectively cap fluorescent semiconductor nanocrystals to make them water soluble, especially to bind to biological molecules or to facilitate easy incorporation into microscopic polymer beads. A family of ligands, which consist of a phosphine oxide binding site, long alkyl chains and a carbon–carbon double bond positioned such that the whole molecule can take part in a polymerisation reaction were used to cap CdSe nanocrystals (see Fig. 1.2). The nanocrystals, thus capped, can be incorporated into microscopic polymer beads by suspension polymerization reactions [24]. Oligomeric phosphines with methacrylate groups have been used to cap CdSe and CdSe–ZnS core–shell nanocrystals and to homogeneously incorporate them into polymer matrices [25]. A polymer ligand consisting of a chain of reactive esters has been successfully used to cap CdSe–ZnS core–shell nanocrystals [26], the pendant reactive groups can be substituted with molecules containing amino-functionalities.

Alivisatos and coworkers have carried out cation exchange reactions on nanocrystals of different shapes and sizes and find that complete and fully reversible

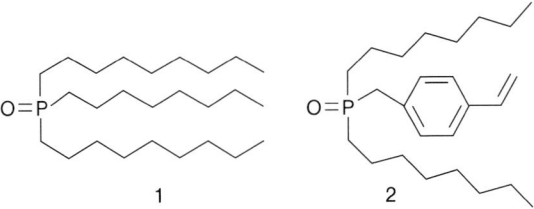

Fig. 1.2 Structure of polymerizable ligands used to cap CdSe nanocrystals.

cation exchange occurs in nanocrystals [27]. Thus, by treating a dispersion of CdSe nanocrystals in toluene with a small volume of methanolic solution of AgNO$_3$, Ag$_2$Se nanocrystals could be obtained in a few seconds. The rates of the cation exchange reactions on the nanoscale are much faster than in bulk cation exchange processes. This study has also identified a critical size above which the shapes of nanocrystals evolve toward the equilibrium shape with lowest energy during the exchange reaction.

1.2.2
Metal Nanocrystals

Metal nanocrystals have traditionally been prepared by reduction of metal salts. This method has been extremely successful in yielding noble and near-noble metals such as Au, Ag and Pt. However, more reactive metals such as Fe, Cu, Co are not readily obtainable. An organometallic route involving a combination of reduction and thermal decomposition of low or zero valent metal precursors is emerging as an attractive route for the synthesis of metal nanocrystals [28]. This route borrows from previous experience on thermal decomposition to obtain ferro fluids and the use of high boiling alcohols to synthesize metal particles (the polyol method).

Copper nanocrystals have been made by the thermolysis of [Cu(OCH(Me)CH$_2$NMe$_2$)$_2$] in trialkylphosphines or long chain amines [29]. Monodispersed gold nanocrystals with diameters of 9 nm have been prepared by reduction of HAuCl$_4$ in TOPO and octadecylamine [30] (see Fig. 1.3). Palladium nanocrystals in the size range 3.5–7.0 nm have been prepared by thermolysis of a complex with trioctylphosphine [31]. Mesityl complexes of copper, silver and gold have been used as precursors to nanocrystals capped with amines, or triphenyl-

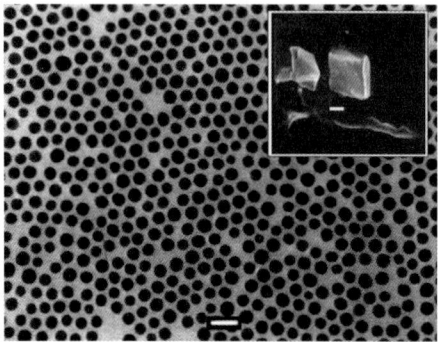

Fig. 1.3 Two-dimensional lattice of octadecylamine/trioctylphosphine oxide-capped gold nanocrystals, bar = 20 nm. Inset: SEM of cubic colloidal crystal prepared from octadecylamine/TOPO-capped gold nanocrystals (190 °C), bar = 80 μm. (reproduced with permission from Ref. [30]).

phosphine oxide [32]. Large nanocrystals of the semimetal bismuth with diameters of 100 nm have been prepared by reduction of bismuth 2-ethylhexanoate in the presence of oleic acid and trialkylphosphine [33]. Lead nanocrystals with diameters of 20 nm have been prepared by thermolysis of tetraethyllead in octanoic acid and trioctylphosphine [34]. Iridium nanocrystals have been prepared by thermolysis of the low-valent organometallic precursor (methylcyclopentadienyl)(1,5-cyclooctadiene)Ir in a mixture of amines [35].

Iron–platinum alloy nanocrystals have attracted a lot of attention, due to their magnetic properties [36]. Iron–platinum nanocrystals with 1:1 ratio of Fe:Pt, exist in two forms. An fcc (face centered cubic) form in which the Fe and Pt atoms are randomly distributed (A1 phase) or an fct (face centered tetragonal) form, in which Fe and Pt layers alternate along the $\langle 001 \rangle$ axis ($L1_0$ phase). The latter phase has the highest anisotropic constant among all known magnetic material. Synthesis of homogeneously alloyed Fe–Pt nanocrystals require that Fe and Pt are nucleated at the same time. This process was first accomplished by reducing platinum acetylacetonate with a long chain diol and decomposing $Fe(CO)_5$ in the presence of oleic acid and a long chain amine [37]. There have been several improvements to the original scheme [36]. The as-synthesized nanocrystals are present in the A1 phase and transform into the $L1_0$ phase upon annealing at 560 °C. The transition temperature can be varied by introducing other metal ions. Thus, $[Fe_{49}Pt_{51}]_{88}$ nanocrystals have been made by introduction of silver acetylacetonate in the reaction mixture [38, 39]. The A1 to $L1_0$ transition temperature is lowered to 400 °C by the introduction of Ag ions. Cobalt and copper ions introduced in Fe–Pt nanocrystals by the cobalt acetylacetonate or copper(II)bis(2,2,6,6-tetramethyl-3,5-heptanedionate) resulted in an increase in the A1 to $L1_0$ transition temperature [40, 41]. The successful synthesis of Fe–Pt nanocrystals using a combination of reduction and thermal decomposition to successfully generate homogeneous alloy nanocrystals has sparked a flurry of activity. Thus, CoPt, FePd, $CoPt_3$ have all been obtained [42, 43]. Manganese–platinum alloy nanocrystals have been obtained by using platinum acetylacetonate and $Mn_2(CO)_{10}$ using a combination of reduction and thermal decomposition brought about using 1,2-tetradecanediol in a dioctyl ether, oleic acid/amine medium [44]. Nickel–iron alloy nanocrystals have been obtained using iron pentacarbonyl and $Ni(C_8H_{12})_2$ [45]. Samarium–cobalt alloy nanocrystals have been prepared by thermolysis of cobalt carbonyl and samarium acetylacetonate in dioctyl ether with oleic acid [46]. Similarly, $SmCo_5$ nanocrystals have been prepared using a diol and oleic acid/amine [47].

As can be seen, a wide range of reactive metal and metal alloy nanocrystals have been obtained by thermolysis and/or reduction of organometallic precursors. The key advantage of this method seems be the ability to synthesize alloys of metals, whose reduction potentials are very different.

Alivisatos and coworkers have synthesized hollow nanocrystals by a process analogous to the Kirkendal effect observed in the bulk [48]. In bulk matter, pores are formed in alloying or oxidation reactions due to large differences in the solid-state diffusion rates of the constituents. By reacting Co nanocrystals with S, Se or oxygen, hollow Co nanocrystals have been obtained. By starting with core–shell nanocrystals of the form Pt–Co and carrying out the process of hole creation, egg-yolk-like nanostructures consisting of a Pt yolk-like core and a Co oxide shell have been obtained.

1.2.3
Nanocrystals of Metal Oxides

Nanocrystals of iron, cobalt, manganese, cobalt and nickel oxides have been synthesized by the thermolysis of the corresponding metal acetylacetonates in hexadecylamine [49]. The nanocrystals could be transferred into the aqueous phase using amine-modified poly(acrylic acid). Nickel oxide nanocrystals obtained by the above method were trigonal (see Fig. 1.4). Manganese oxide nanocrystals of diameter 7 nm have been synthesized by thermal decomposition of manganese acetate in the presence of oleic acid and trioctylamine at high temperature. The MnO nanocrystals so obtained were oxidized to Mn_3O_4 nanocrystals by the use of trimethylamine-N-oxide. FeO nanocrystals have also been using the same method [50]. Adopting a similar approach, a range of monodisperse oxide nano-

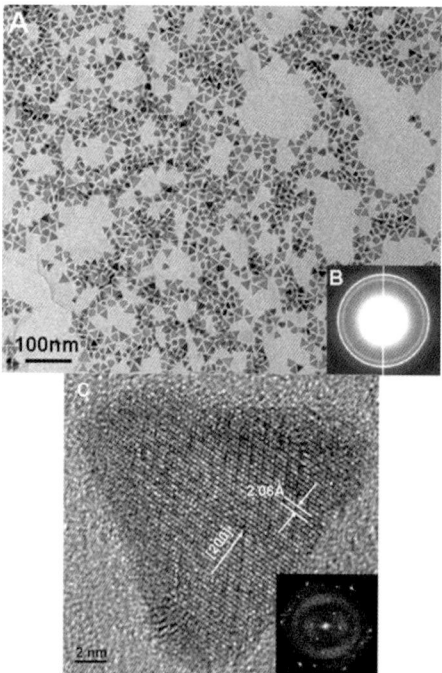

Fig. 1.4 A, TEM image of NiO nanocrystals. B, Electron diffraction pattern acquired from the NiO nanocrystals. C, HRTEM image of NiO nanocrystals, (insert) FFT of the HRTEM.

crystals such as iron oxide, CoO and MnO have been synthesized by decomposition of metal-oleate complexes in different solvents. The oleate complexes were prepared by reacting the corresponding chlorides with Na-oleate [51]. Tetragonal zirconia nanocrystals with diameters of 4 nm have been prepared by a sol–gel process using zirconium(IV) isopropoxide and zirconium(IV) chloride [52].

Biswas and Rao [53] have prepared metallic ReO_3 nanocrystals of different sizes by the solvothermal decomposition of rhenium(VII)oxide–dioxane complex $(Re_2O_7-(C_4H_8O_2)_x)$ in toluene. The diameter of the nanocrystals could be varied in the range 8.5–32.5 nm by varying the decomposition conditions. The metallic ReO_3 nanocrystals exhibit a plasmon band similar to Au nanocrystals.

Cobalt oxide nanoparticles with diameters in the 4.5–18 nm range have been prepared by the decomposition of cobalt cupferronate in decalin at 270 °C under solvothermal conditions. Magnetic measurements indicate the presence of ferromagnetic interaction in the small CoO nanoparticles [54]. Cubic and hexagonal CoO nanocrystals have been obtained starting from [Co(acetylacetonate)$_3$] [55]. Nanoparticles of MnO and NiO have been synthesized from cupferronate precursors under solvothermal conditions [56]. Octlyamine capped ZnO nanocrystals with band edge emission have been synthesized using single source precursors zinc cupferronate $(Zn(C_6H_5N_2O_2)_2)$, the Zn(II) salt of N-nitroso-N-phenylhydroxylamine) and a ketoacidoximate $(C_8H_{16}N_2O_8Zn$, diaquabis[2-(methoxyimino) propanoato]zinc(II)) [57].

Yi et al. have embedded magnetic nanocrystals and fluorescent quantum dots in a silica matrix using preformed nanocrystals and carrying out hydrolysis of silicate in a reverse microemulsion [58], building on previously published methods [59, 60]. The nanocomposites thus obtained are water soluble.

Thus, the main areas of research on spherical nanocrystals are directed at the synthesis of unusual materials such as EuS, simplifying synthetic schemes or facilitating the use of nanocrystals by appropriate functionalization.

1.3
Nanocrystals of Different Shapes

1.3.1
Anisotropic Growth of Semiconductor and Oxide Nanocrystals

Nanocrystals of oxides and semiconductors have been grown in different shapes. In a majority of these processes, anisotropic growth is achieved by maintaining a high concentration of the precursors, typically by introducing precursors by continuous or multiple injection. For example, spherical nanocrystals of Cd chalcogenides can be prepared using complexes of Cd and tetradecylphosphonic acid (prepared by reacting CdO with the phosphonic acid) and chalcogen dissolved in trioctylphosphine. By carrying out the same reaction, but maintaining a high concentration of precursors (by multiple injections), nanorods can be produced [61].

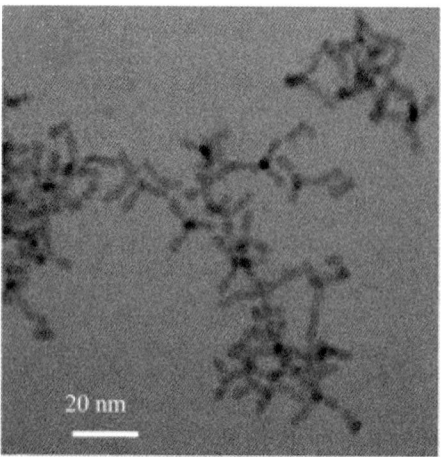

Fig. 1.5 TEM image of CdSe tetrapods (adapted with permission from Ref. [65]).

Cadmium selenide nanocrystals in the form of rods, arrows and branched forms like tetrapods were obtained by introducing the precursor in several steps (see Fig. 1.5) [62–65]. This scheme has been extended to synthesize CdS [66] and CdTe [67] nanocrystals of different shapes. A variety of precursors such as CdO, cadmium naphthenate and single source precursors such as cadmium ethylxanthate [68], cadmium diethyldithiocarbamate [69] have been used to synthesize rods and tetrapods of Cd chalcogenides. Heterostructured tetrapods consisting of CdS, CdSe and CdTe are prepared by alternating the precursors during the injections in the growth step. Thus, arms of CdSe have been grown on CdS nanorods [70]. Tetrapods and rods of MnS and PbS have been obtained by thermal decomposition of the corresponding diethyldithiocarbamate [73, 74]. ZnS nanorods and nanowires have been grown using zinc xanthates [71]. ZnSe nanorods and branched structures have been obtained by decomposing diethylzinc [72]. Narrow nanowires and nanorods of ZnS, CdS, CdSe, ZnSe have been obtained by microwave-assisted decomposition of metal xanthates or mixtures of metal acetates and selenourea [75].

Of particular interest in the synthesis of nanorods is the synthesis of cubic nanorods of sulfides and selenides. Nanorods of CdS have been obtained by reacting Cd-acetate with S in hexadecylamine at high temperatures. High-resolution transmission electron microscopy reveals that the obtained nanorods consist of cubic and hexagonal domains, with the cubic structure predominating [76] (see Fig. 1.6). Cubic CdS nanorods have been obtained solvothermally, starting with $CdCl_2$ and S [77]. Cubic CdSe nanrods have been synthesized by decomposition of the single source precursor $Li_2[Cd_{10}Se_4(SPh_{16})]$ (SPh-phenyl thiolate) [78]. Cubic ZnS nanorods have been obtained by solution-phase aging of spherical ZnS nanocrystals at 60 °C [79].

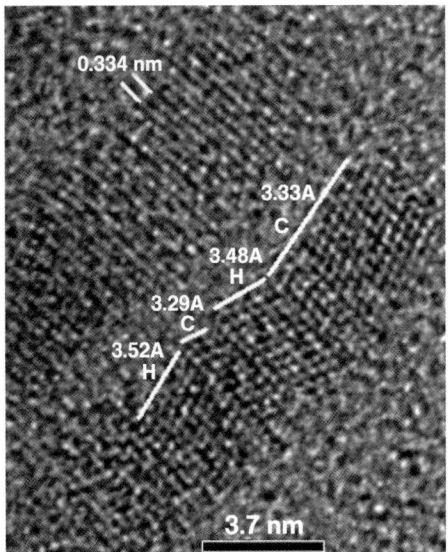

Fig. 1.6 HR-TEM image of a CdS nanorod showing lattice spacing of hexagonal(H) and cubic(C) phases.

A simple one-pot method has been used to prepare CoP nanowires by thermolyis of cobalt(II)acetylacetonate in alkylphosphonic acid in the presence of hexadecylamine and trioctylphosphine oxide [80]. The morphology of the resulting nanowires can be influenced by the ratio of the long chain amine and the phosphine oxide (see Fig. 1.7). Nanorods and nanowires of FeP have been prepared by a similar method [98].

Nanowires, nanotubes, nanowafers and other morphologies of a series of Sb, In and Bi chalcogenides have been obtained by reacting metal acetates with elemental chalcogens in amine [81–83] (see Fig. 1.8). Several of these metal chalcogen phases have been obtained in pure form for the very first time.

Little is understood about the factors determining the growth of anisotropic nanostructures, in particular tetrapods. The tetrapods consist of tetrahedral seeds with cubic structure and arms which are hexagonal. Despite the synthesis of a broad range of semiconductor matter, starting with a wide range of precursors, it is observed that the diameters of the arms of a tetrapod can only be experimentally varied in a narrow size range of 2.0–5.0 nm, leading us to suspect that a fundamental factor underpins the growth of tetrapods. Simple numerical calculations carried out by us suggest that the need to conserve the number of surface atoms plays a crucial role in determining the diameter of the arms of the tetrapods [84]. Branching of tetrahedral seeds is accompanied by a relative gain in the number of surface atoms when the dimensions of the seeds are in the range 2.0–5.0 nm (see Fig. 1.9). Seeds with higher or lower dimensions either suffer an increase in the number of surface atoms or experience marginal rises in the

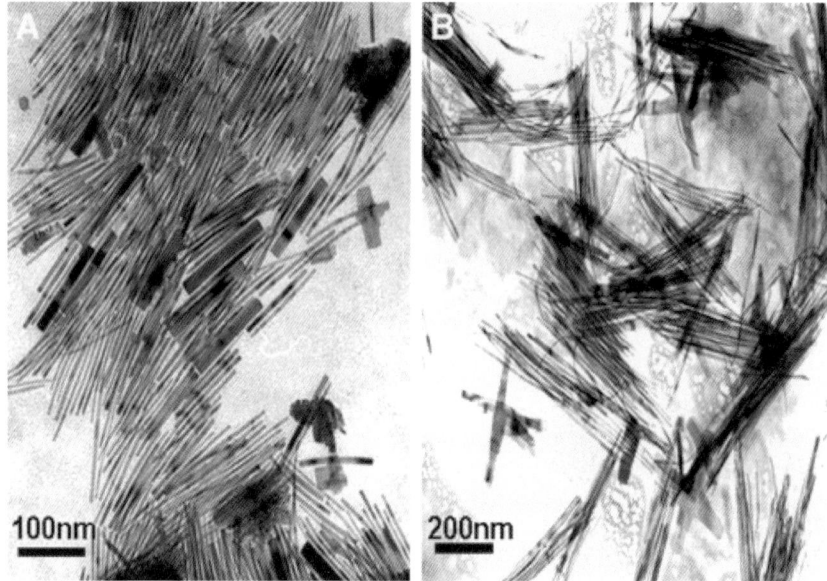

Fig. 1.7 TEM images of CoP nanorods of varying aspect ratios obtained using different ratios of hexadecylamine and triphenylphosphine oxide.

number of surface atoms following branching. The ability to conserve surface atoms lends extra stability to seeds of the right dimensions and plays a key role in determining the dimensions of the tetrapods.

The solution–liquid–solid process is employed for the synthesis of nanorods of more covalent semiconductors. In this method, a metal seed catalyzes the growth of nanorods. Under the growth conditions employed, the seeds form alloys with the nanowire material. The metal seed can either be generated *in situ* or be externally introduced [85]. Building on previous successes, Buhro and coworkers have synthesized GaAs nanowires by decomposition of the single-source precursor [tBu$_2$Gaμ-As(SiMe$_3$)$_2$]$_2$ [86]. Ga seeds are generated *in situ* during the decomposition of the precursor. Indium phosphide nanorods have been synthesized by the use of two single source precursors, one to generate the seeds and the other to sustain wire growth. Thus, InP nanorods with diameters controllable in the range 3–9 nm have been obtained using [tBu$_2$InP(SiMe$_3$)$_2$]$_2$ and [Cl$_2$InP(SiMe$_3$)$_2$]$_2$ [87]. Indium arsenide nanorods have been obtained by the use of Au nanocrystals as seeds and InCl$_3$ and As(trimethylsilane)$_3$ as precursors [88]. The diameters can be varied in the range 4–20 nm by varying the diameters of the seeds. Narrow CdSe nanowires have been obtained using low-melting Bi nanocrystals as seeds [89]. Narrow CdSe and PbSe nanowires have been synthesized using Au/Bi nanocrystals as seeds [90, 91].

Korgel and coworkers have devised a new way to synthesize semiconductor nanowires using supercritical fluids, high temperatures and pressures. By the

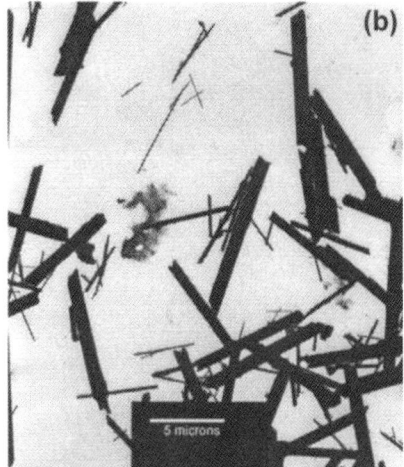
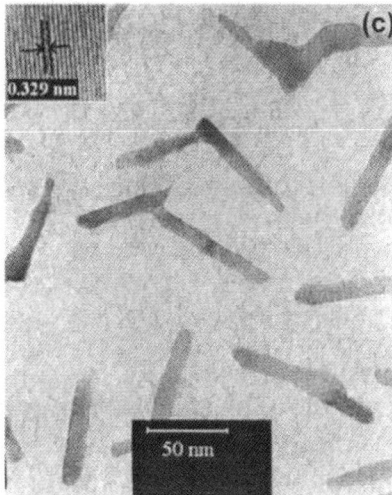

Fig. 1.8 (a) TEM image of InS nanorods (b) TEM image of Sb_2Se_3 nanorods and nanotubes prepared in octylamine (c) TEM image of In_2Te_3 nanorods.

use of this method, seeded growth of nanowires can be achieved in common solvents at high temperatures. Using this process Si [92], Ge [93], GaP [94], GaAs [95] nanowires seeded by either Au or Bi nanocrystals have been successfully obtained (see Fig. 1.10).

Nanorods of ZnO have been obtained by thermolysis of Zn-acetate in a mixture of amines [99]. Iron oxide nanocrystals in cubic, star and other shapes have been obtained by thermolysis of iron carbonyl and acetylacetonates [100]. A number of metal oxide nanocrystals in the form of rods have been obtained by acylhalide

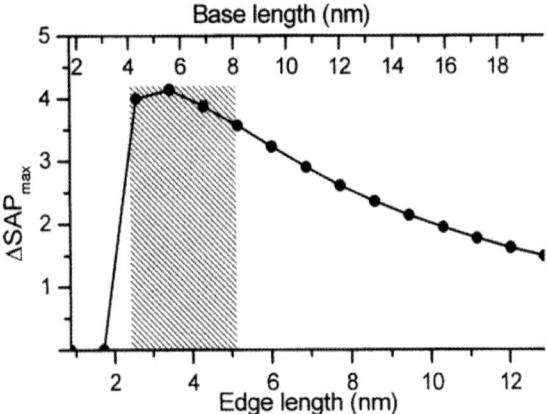

Fig. 1.9 Plot showing changes in the maximum difference in the surface atom percentage (ΔSAP_{max}) between the tetrahedron and the corresponding structures with four hexagonal branches grown from a CdSe seed. The edge length refers to the edge length of the hexagon; the base length pertains to the length of the tetrahedral face from which the branch grows. The size regime for experimentally obtained tetrapods is shaded.

elimination of metal halides in oleylamine [96]. Cubic $BaTiO_3$ nanowires have been synthesized by sol–gel reaction using $BaTi(O_2CCH_3)_6$ [97].

A method related to thermolysis has been used by Korgel and coworkers [101–103] to synthesize nanocrystals in different shapes. In this method, long chain

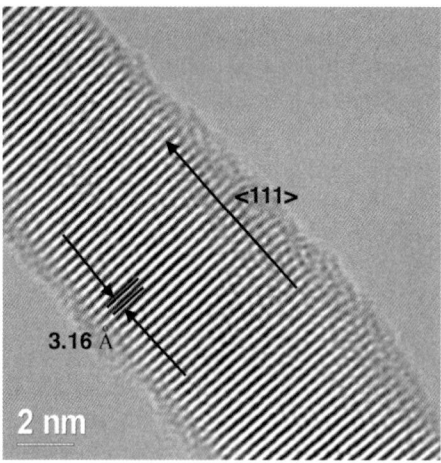

Fig. 1.10 High resolution TEM image of GaP nanowires (reproduced with permission from Ref. [94]).

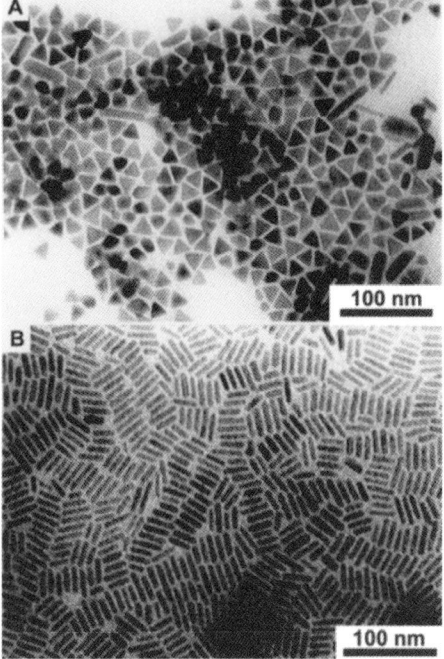

Fig. 1.11 TEM images showing nanoprisms and nanorods of NiS. The images correspond to different areas of the same sample (reproduced with permission from Ref. [103]).

alkanethiolates of metals such as Cu and Ni are mixed with fatty acids to obtain a waxy solid which is thermally decomposed. Thus, nanorods, nanodisks or nanoprisms of Cu_2S [101, 102] and NiS [103] have been prepared (see Fig. 1.11).

It is possible to rationally design templates for shape-controlled synthesis of nanocrystals using inverted micelles. By varying the relative concentrations of water:oil, cylindrical water pools of different dimensions can be obtained in micellar form. These micellar droplets can be used as templates to grow nanowires. Elongated Cu nanoparticles [104], nanowires of $BaCO_3$ [105] and $BaSO_4$ [106] have been obtained by the use of such templates (see Fig. 1.12). The particles present in the water pool can undergo further changes. Thus, triangular CdS [107] as well as prismatic $BaCrO_4$ [108] have been prepared using inverted micelles. This method of anisotropic growth has been recently reviewed [109].

Other ways to obtain anisotropic growth include controlled oxidation or reduction. Large tetrahedral Si nanocrystals have been obtained as exclusive products by a careful control of the reducing conditions [110]. Nanorods and nanodisks of ZnO have been grown by a controlled room-temperature oxidation of dicyclohexylzinc [111]. Zinc oxide nanocrystals with cone, hexagonal cone and rod shapes are obtained by the non-hydrolytic ester elimination sol–gel reaction [112]. In

Fig. 1.12 TEM micrograph of BaCO$_3$ nanowires obtained using inverted micelles (reproduced with permission from Ref. [103]).

this reaction, ZnO nanocrystals with various shapes were obtained by the reaction of zinc acetate with 1,12-dodecanediol in the presence of different surfactants.

1.3.2
Anisotropic Growth of Metal Nanocrystals

A number of synthetic schemes have been employed to obtain anisotropic metal nanocrystals [113–119]. A seed-mediated method has been adapted to produce nanorods [113, 114], nanowires [115–117] and other shapes [118, 119] of Au and Ag in aqueous media. The method involves two steps. In the first step, small citrate-capped Au or Ag nanocrystals are produced by borohydride reduction for use as seeds. In the second step, the particles are introduced into a solution containing the metal salt, CTAB (a structure-directing agent) and a mild reducing agent such as ascorbic acid. The use of a mild reducing agent is the key to achieving seed mediated growth. Under the reaction conditions, ascorbic acid is not sufficiently powerful to reduce the metal salt on its own. In the presence of seeds, a reduction mediated by the seed occurs, producing nanorods [113, 114] and nanowires [115–117]. A TEM image of Au nanorods produced by this method is shown in Fig. 1.13. The nanorod and nanowire structures are directed by the micellar structures adopted by CTAB. The chain length of the structure-directing agent plays an important role in determining the aspect ratio of the rod-shaped particles [116]. The presence of a small quantity of organic solvents leads to the formation of needle-shaped crystallites [117]. Addition of NaOH to the reaction mixture before reduction brings about dramatic changes in the product morphology. Hexagons, cubes and branched structures have been produced by using NaOH and varying the experimental parameters (see Fig. 1.14) [118]. Nanoplates of Ag have also been prepared by adopting a similar procedure [119]. Several different methods of synthesis of anisotropic noble metal nanostructures have been recently reviewed [120].

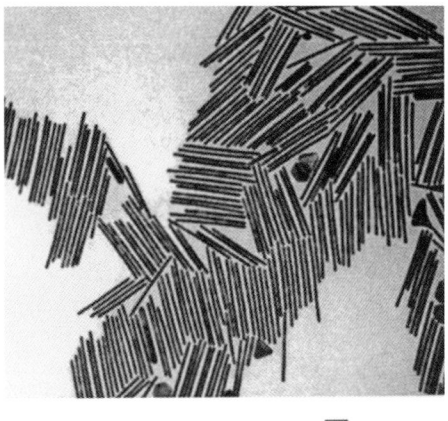

Fig. 1.13 TEM image of Au nanorods prepared by the seed-mediated growth method (reproduced with permission from Ref. [115]).

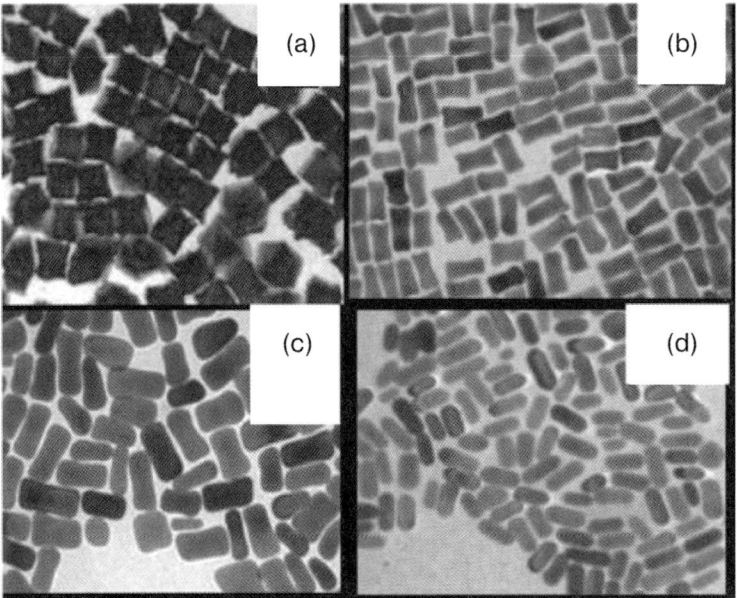

Fig. 1.14 TEM images showing cubic to rod-shaped gold particles produced with low concentrations of ascorbic acid in the presence of a small quantity of silver nitrate, by the seed-mediated growth method. The concentration of CTAB increases from 1.6×10^{-2} M(A), to 9.5×10^{-2} M (B,C,D). [Au^{3+}] decreases from (B) to (C), whereas the seed concentration increases from C to D. Scale bar is 100 nm (reproduced with permission from Ref. [118]).

By the use of NaOH, one can do away with both the surfactant and seeds and still obtain nanowires [121]. The mechanism operating in this reaction is not understood, but it is clearly not surfactant-directed. Polyvinylpyrrolidone-capped nanowires of Ag have been made by employing the polyol process with [122] and without seeds [123]. It is believed that PVP plays the role of the structure-directing agent. As with the seed-mediated method, other shapes have also been synthesized by this method. By tuning the experimental conditions, Sun and Xia obtained nanocubes of Au and Ag [124]. Triangular Au nanoparticles or nanoprisms have been grown using lemon grass extract [125]. Silver nanoparticles of a variety of morphologies are obtained by carrying out the reduction with silver binding peptides [126].

Mirkin and coworkers [127] have devised two different routes to nanoprisms of Ag. In the first method, Ag nanoprisms are produced by irradiating the mixture of the citrate and bis(p-sulfonatophenyl) phenylphosphine dihydrate dipotassium(BSPP)-capped Ag nanocrystals with a fluorescent lamp. By controlling the wavelength of irradiation, the nanoprisms can be induced to aggregate into large prisms in a controlled manner (see Fig. 1.15) [128]. In the second method, $AgNO_3$ is reduced with a mixture of borohydride and hydrogen peroxide [129]. The latter method has been extended to synthesize branched nanocrystals of Au (see Fig. 1.16) [130, 131]. An NAD(P)H-mediated reduction in the presence of ascorbic acid has been employed to synthesize dipods, tripods and tetrapods of Au [132]. Platinum nanocrystals in various branched forms, including the tetrapod, have been obtained by thermolysis of platinum 2,4-pentanedionate in organic solvents, in the presence of trace amounts of silver acetylacetonate [133]. Other methods such as the use of inverted micelles [134] can also be adapted to prepare nanoprisms or nanoplates of Ag. Silver salts reduced by DMF are also known to yield nanoprisms and plates [135].

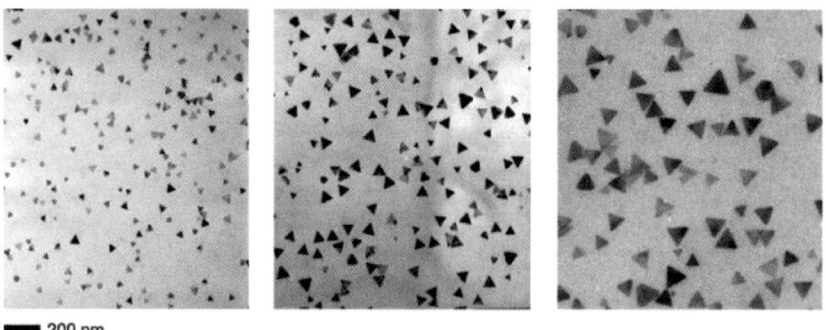

Fig. 1.15 Silver nanoprisms of different dimensions obtained by controlled irradiation of bis(p-sulfonatophenyl) phenylphosphine dihydrate dipotassium capped Ag nanoparticles (reproduced with permission from Ref. [128]).

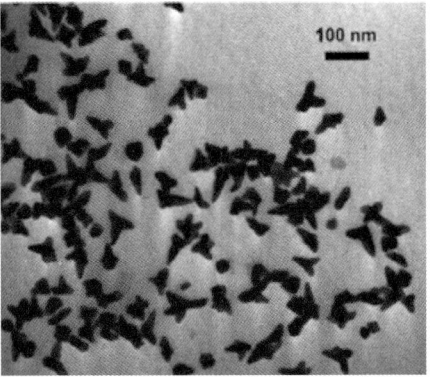 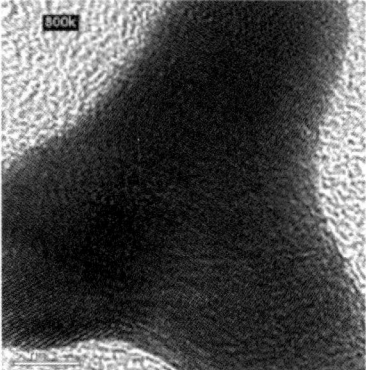

Fig. 1.16 Low and high magnification TEM images of branched Au nanocrystals (reproduced with permission from Ref. [130]).

An octylamine/water bilayer template has also been used to synthesize Ag nanoplates [136]. Gold nanoplates have been produced by using aspartate as the reducing agent [137]. Hydrogen reduction of olefinic Co compounds has yielded Co nanorods [138]. Photochemical reduction of In and Sn olefinic complexes in the presence of long chain amines yields In and In_3Sn nanowires [139]. Platinum nanocrystals in the form of cubes, octahedra and cuboctahedra have been prepared by the polyol method [140]. Branched structures and cubes of Rh have been obtained by reducing $RhCl_3$ in ethylene glycol, at different temperatures.

Clearly, there is tremendous interest in devising synthetic schemes to generate anisotropic structures in high, if not exclusive yield. As the guiding principles are yet to be fully understood, breakthroughs, with a few notable exceptions are on a trial and error basis.

1.4
Selective Growth on Nanocrystals

Banin and coworkers have been successful in selectively growing Au tips on either end of CdSe nanorods [142]. This was achieved by refluxing a solution of the rods with $AuCl_3$ dissolved in dodecylammonium bromide and dodecylamine. The optical spectra provide evidence for strong coupling between Au tips and CdSe rods. The resulting nanostructures, nanorods with Au tips, combine the easy ability to tune properties afforded by semiconductor nanostructures and the processability of Au nanocrystals. Others have been successful in growing PbSe tips on CdS or CdSe nanorods [143]. Dumbell and flower-shaped $Au-Fe_3O_4$ nanocrystals have been obtained by thermolysis of iron pentacarbonyl in the presence of Au nanocrystals [144]. FePt and CdS nanocrystals have been combined into a structure with similar shape in a one pot method [145].

1.5
Properties of Nanocrystals

1.5.1
Electronic and Optical Properties

The variation of bandgap with diameter in semiconductor nanocrystals is the most identifiable manifestation of quantum confinement effects. Probably as a consequence, it is also the most studied. According to the effective mass approximation, the onset of absorption is proportional to $1/R^2$ (where R is the radius of the quantum dot). Thus, we have

$$\Delta E_g = \frac{\hbar^2 \pi^2}{2R^2}\left[\frac{1}{m_e^*} + \frac{1}{m_h^*}\right] - \frac{1.786e^2}{\varepsilon R} - 0.248 E_{Ry}^* \tag{1.1}$$

where E_{Ry}^* is

$$\frac{e^4}{2\varepsilon^2 \hbar^2 (m_e^{*-1} + m_h^{*-1})} \tag{1.2}$$

Experimental variation of band gap by and large follows the trend. It is now well known that effective mass approximation overestimates the variation of band gap, especially at small diameters. A more accurate relationship for band gap variation has been obtained, based on tight binding methods [146, 147]. Accordingly, the band gap variation is reduced to an equation of the form

$$\Delta E_g = \frac{1}{aD^2 + bD + c} \tag{1.3}$$

where D is the diameter of the quantum dot. The a, b and c parameters are dependent on the system. Table 1.1 lists the a, b and c values for a few common semiconductors. The band gaps estimated in this way reproduce experimental results very accurately. Encouraged by the accuracy of the relations, the authors have developed a method for estimating the diameters and diameter-distribution from the UV spectra [148].

Recently, the discovery of the phenomena of multiple exciton generation in quantum dots of PbS and PbSe has generated considerable excitement [149–151]. It seems possible that lead chalcogenide quantum dots when irradiated with photons of four times the band gap energy generate up to three excitons possessing band gap energy. In other words, quantum yields of the order of 200–300% have been observed. The experimental results confirm the earlier theoretical prediction of Nozik [152, 153]. The experimental observation, though indirect, has tremendous implications for applications such as solar power generation.

Studies aimed at understanding the evolution of electronic structure from three-dimensionally confined quantum dots to one-dimensionally confined quan-

Table 1.1 Parameters for estimating the bandgap variation using Eq. (1.3).

Substance	E_g eV	a nm^{-2} eV^{-1}	b nm^{-1} eV^{-1}	c eV^{-1}
ZnS	3.57	0.2349	−0.0418	0.2562
ZnSe	2.822	0.0845	0.1534	0.2128
CdS	2.51	0.1278	0.1018	0.1821
CdSe	1.75	0.0397	0.1723	0.1111
GaP	2.272	0.1969	0.2631	0.0728
GaAs	1.424	0.0359	0.1569	0.1564
InP	1.344	0.0461	0.3153	0.0623
InAs	0.354	0.0374	0.2569	0.1009

tum wells have recently received a fillip following the synthesis of nanocrystals in different shapes. The ability to easily synthesize quantum wires with genuine confinement in two dimensions is particularly significant. It is clear that 2D confined nanorods have properties different from 3D confined spherical nanocrystals. For example, linearly polarized emission has been observed by fluorescence measurements on single rods [154]. Polarized lasing has also been observed [155].

Simple particle in a box models yield the following relationship for the variation of the band gap when the degree of quantum confinement changes from 1D to 3D.

$$\Delta E_g = \frac{\xi h^2}{8D^2} * \left[\frac{1}{m_e^*} + \frac{1}{m_h^*} \right] \quad (1.4)$$

Where, D is the diameter, m_e^* and m_h^* the electron and hole effective masses. The value of $\xi = 1.0$ for one-dimensionally confined wells, 1.17 for two-dimensionally confined wires and 2.0 for three-dimensionally confined dots [156–160]. In other words, the band gap increases with decreasing diameter is much more gradual as the degree of confinement increases from one to three dimensions. Despite the degree of approximation used in arriving at the relationship above, high level theoretical calculations using *ab initio* or less sophisticated methods yield similar results [161–164]. For example, *ab initio* calculations using density functional theory for a range of semiconductors indicate that the relationship in Eq. (1.4) is valid for III–V and II–IV semiconductors, with the exception of AlN [161].

In practice, the bandgap in quantum dots varies less steeply than $\frac{1}{D^2}$ as suggested in Eq. (1.4). One can fit the variation to the form indicated in Eq. (1.3) or simply to an equation of the form $\frac{1}{D^x}$, where $(2.0 \leq x \geq 1.0)$. Experimental studies on InP nanowires and nanorods have estimate that the bandgap varies as indicated in Eq. (1.4) [165]. Similar results have been obtained experimentally for CdSe nanowires [89]. The studies on InP and CdSe nanowires and nanorods have been a few aimed at tracking the change from two- to three-dimensional confinement as the length or diameter of the nanorods is changed. Experimental data suggest that a length of 30 nm was required for the third dimension of quantum confinement to fully vanish in CdSe rods with diameters in the range 3–6 nm. The optical properties of InAs nanorods with diameters of 4 nm, fractionated on the basis of length, have been studied to ascertain the effect of length on the band gap [88]. The bandgaps of InAs rods were found to decrease with increasing length/diameter ratio. The length dependent absorption spectra of cubic InAs nanowires are shown in Fig. 1.17. A clear red shift is discernible with increase in length. A much weaker dependence on length was seen in the case of CdSe nanowires with diameters in the range 1.5–3.0 nm and lengths in the range 11–60 nm. The band gaps were studied using optical spectroscopy as well as scanning tunneling spectroscopy. The observations on InAs and CdSe nanowires can be justified on the basis of consideration of the Bohr exciton diameter which is ≈ 5 nm for CdSe and ≈ 35 nm for InAs. Thus, CdSe rods fall in the medium to weak confinement regime, while InAs rods are in the strong confinement regime. As is quite apparent, more effort is needed to understand the effects of shape on quantum confinement.

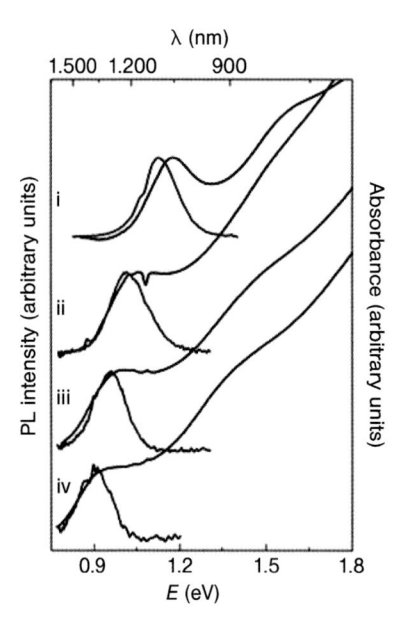

Fig. 1.17 The absorption and photoluminescence spectra in toluene solution of (i) InAs quantum dots, 4 nm in diameter; (ii) quantum rods 9.4W4.1 nm (mean length by diameter); (iii) quantum rods, 15.1W4.1 nm; (iv) quantum rods, 22.7W4.0 nm. (Reproduced with permission from Ref. [88].)

1.5.2
Magnetic Properties

The dominant feature of the magnetic properties of nanocrystals is superparamagnetism, induced by a reduction in size. Superparamagnetic behavior of ferromagnetic particles has set the limits on miniaturizing magnetic devices. Despite seeming detrimental to miniaturization efforts, the superparamagnetic nature has facilitated a number of applications, which shall be dealt with later. The interest in hard magnetic materials such as Fe–Pt has been sparked by their high anisotropy constant. It is expected that such nanocrystals will continue to be magnetic for sizes less than 10 nm. Other attempts to beat the superparamagnetic limit involve making use of exchange coupling interactions at the interface between ferromagnetic and antiferromagnetic systems [166]. Skumryev et al. have embedded Co in a CoO matrix and have succeeded in preparing ferromagnetic Co-CoO nanocrystals with diameters as small as 4 nm [167]. Many metal oxide nanopar-

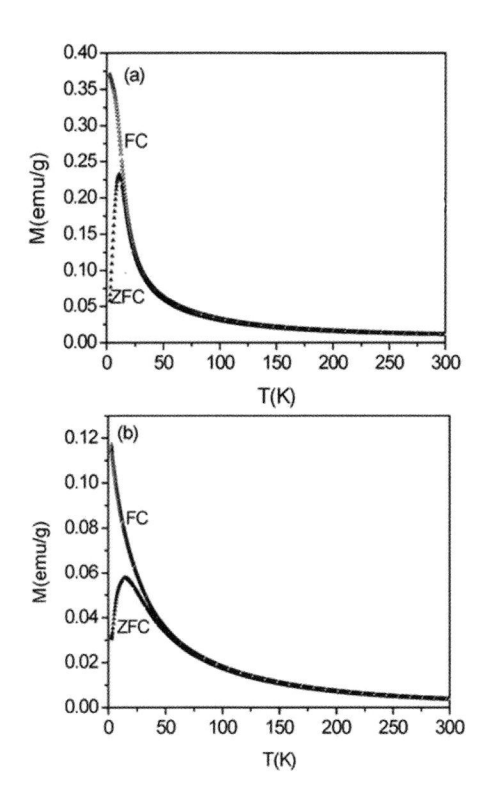

Fig. 1.18 The temperature dependence of dc magnetization of (a) 3 nm and (b) 7 nm NiO nanoparticles under zero-field-cooled and field-cooled conditions (H = 100 Oe). (Reproduced with permission from Ref. [56].)

ticles show evidence for the presence of ferromagnetic interactions at low temperatures. This is specially true of nanoparticles of antiferromagnetic oxides such as MnO, CoO and NiO [54, 56]. The magnetic interaction typically causes the zero-field cooled and field-cooled magnetization curves to diverge at low temperatures. It is also possible to observe magnetic hysteresis below a temperature of divergence (the blocking temperature). In Fig. 1.18, the magnetization behavior of NiO particles with diameters of 3 and 7 nm is shown [56]. The ferromagnetic interaction is thought to be due to spins acting under the influence of surface strain.

Dilute magnetic semiconductors, typically Mn-doped chalcogenides have evoked keen interest in recent times because of their potential uses in areas such as spin-dependent electronics. The ability to synthesize quantum confined magnetic semiconductor nanocrystals represents a key challenge. Efforts to produce nanocrystalline Mn doped II–VI magnetic quantum dots have yielded quantum dots with widely different magnetic characteristics, depending on the nature of preparation and post-synthesis treatment such as thermal annealing. For example, Strouse and coworkers have found that Mn-doped CdSe nanocrystals can be turned superparamagnetic by annealing the synthesized nanocrystals [168]. It has, till recently, been generally accepted that the difference in results arises from the inhomogeneities in doping caused by self-annealing. However, Eforos et al. have suggested that doping could be related to the binding energy of Mn ions to the growing faces of CdSe [169]. Thus, the magnitude of ferromagnetic exchange in Mn-doped semiconductor quantum dots is dependent on the clustering of ions, the size of the ion-cluster, the site occupation of the ion, the diameter of the quantum dot, and changes in the nature of the bonding in the host lattice in quantum-confined material.

1.6
Ordered Assemblies of Nanocrystals

1.6.1
One- and Low-dimensional Arrangements

One-dimensional organization of nanocrystals is brought about by the use of templates such as pores in alumina, steps on crystalline surfaces as well as polymer chains.

Single- and double-stranded DNA have been used to obtain linear assemblies of Au and other nanocrystals [170, 171]. These assemblies are mediated by the electrostatic interaction between the polyphosphate backbone of the DNA and the cationic charges resident on the surface of Au nanocrystals [170, 171]. Warner and Hutchison [171] have obtained closely-packed linear arrays, ribbons and branched chains of Au nanocrystals (see Fig. 1.19). ZnS nanocrystal aggregates have been organized into linear arrays by using a virus as template [172]. Similarly, Pt nanocrystals in the form of ribbons have been obtained using a cholesteric liquid crys-

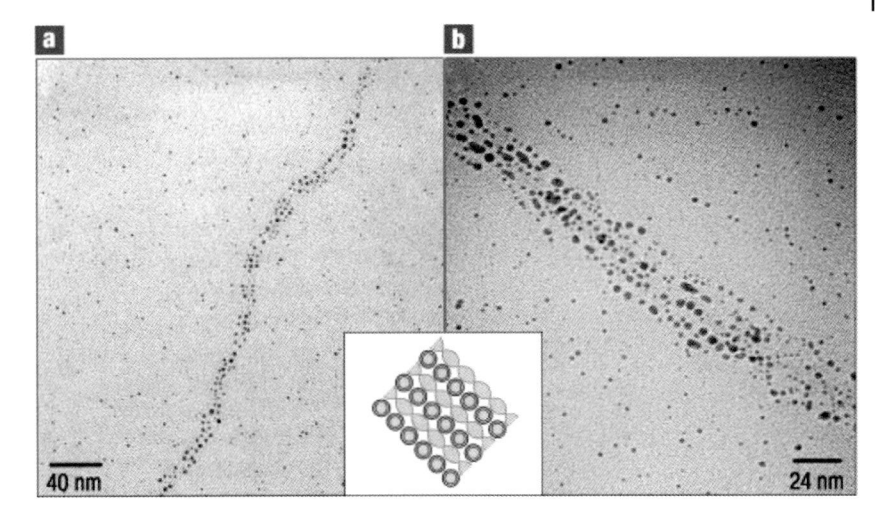

Fig. 1.19 Au nanocrystals organized into (a) ribbons and (b) thicker strands on the DNA backbone. The inset illustrates schematically, the relationship between the DNA chains and the Au nanocrystals (reproduced with permission from Ref. [171]).

talline template [173]. By reducing copper ions bound to a fiber-forming lipid, Kogiso et al. [174] have obtained a one-dimensional arrangement of Cu nanocrystals on the lipid backbone. It is suggested that the tobacco mosaic virus tubules can serve as templates for the growth of a 1D lattice of quantum dots [175].

Short gold nanorods have been linked to form chains by using a mercaptocarboxylic acid wherein the thiol end binds to the metal and the carboxyl ends form hydrogen bonds [176]. Gold nanocrystals have been assembled into one-dimensional chains by the controlled ligand exchange of citrate ions with 2-mercaptoethanol [177].

Oriented attachment of nanocrystals is thought to be responsible for the growth of one-dimensional and other complex nanostructures. Thus, nanotubes and nanowires of II–VI semiconductors have been synthesized using surfactants [178]. The nanorods or nanotubes of CdS and other materials produced in this manner actually consist of nanocrystals. The growth of TiO_2 nanowires from constituent nanocrystals has been investigated in detail [179].

Nanocrystals of metals, semiconductors and oxides could assemble into microscopic rings with walls consisting of several rows of nanocrystals [180–185]. There is evidence that such structures are brought about by Bénard–Marangoni instabilities caused by convection currents during evaporation of a thin film. The convection currents could lead to isolated as well as intersecting rings, depending on the concentration of the nanocrystals in the evaporating droplet. By adopting a complex process involving phase separation in binary fluids, instabilities and evaporation of microdroplets, rings of magnetic $CoPt_3$ nanocrystals have been prepared [186]. Small rings of Co nanocrystals with diameters in the range 50–100 nm have been obtained by capping the nanocrystals with a specially designed ligand, C-undecylcalix[4]resorcinarene [187]. Ring-like structures of semiconductors are found to occur by the self-assembly of nanocrystals in aqueous media [188]. This organization is considered to be due to the self-assembly of the ligands. By tailoring the surface of a block copolymer film and depositing the nanoparticles atop these films, nanoscopic rings of Au nanocrystals with diameters in the range 15–30 nm have been obtained [189]. Besides rings, a spontaneous formation of nanoparticle strip patterns have been observed on dewetting of a dilute film of polymer-coated nanocrystals on water [190]. DNA trimers and Au nanocrystal–DNA conjugates have been used to produce nanocrystal dendrites [191]. Self-assembly of triangular and hexagonal CdS nanocrystals into complex structures such as rods and arrows has been observed [192].

1.6.2
Two-dimensional Arrays

Two-dimensional lattices of nanocrystals have been obtained with metal, semiconductor and other nanocrystals of different sizes. These arrays have generally been prepared by simply evaporating dispersions of suitably functionalized, monodisperse nanocrystals on hydrophobic surfaces such as a carbon film. The cooperative assembly is brought about by a weak force that is the result of an attractive force between the nanocrystal cores in solution and a repulsive part arising from steric crowding of ligands as the nanocrystals are brought closer. To bring about large scale organization, a correct balance of forces is essential. Therefore the nanocrystals need to be monodisperse and capped with the right ligand molecules. Ligands such as long chain thiols, amines, phosphines or fatty acids have served as good candidates to bring about such an assembly. Nanocrystals that self-assemble into two-dimensional arrays often arrange themselves into superlattices consisting of a few layers of arrays. The self-assembly process of nanocrystals has been previously reviewed [194–196].

In the recent past, attempts have been made to extend this process to more unusual cases. Kimura and coworkers [197] have obtained two-dimensional arrays of Au nanocrystals from a hydrosol. This is the first report of a two-dimensional organization occurring in an aqueous medium. An interdigitated layer of octanethiol and ethanol is thought to be responsible for bringing about such an organization. Two-dimensional lattices of $CoPt_3$ of various sizes have been prepared by Weller and coworkers [198] using adamentane carboxylic acid as capping agent. An interesting aspect of this study is the use of a carboxylic acid without a long alkane chain. Ordered self-assembly of Au nanocrystals into arrays from water soluble micelles have been observed [199]. The method involves drying of water soluble Au nanocrystal micelles synthesized using a surfactant encapsulation technique conducted in an interfacially driven water-in-oil microemulsion process. Large Au nanocrystals with diameters of 15–90 nm have been organized

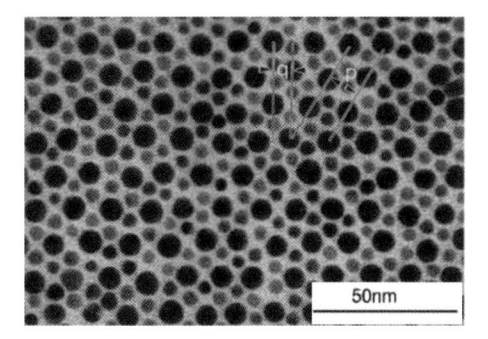

Fig. 1.20 An array of Au nanocrystals with the AB structure. The radius ratio of the nanocrystals is 0.58 (reproduced with permission from Ref. [203]).

into two-dimensional arrays using resorcinarene-based capping agents [200]. Highly ordered superlattices of iron nanocubes have also been prepared [201]. An array of cube-shaped Sn nanocrystals was made by exploiting the affinity of an acid to its conjugate base using hexadecylamine and the hexadecylamine–HCl adduct. The superlattice formation is supposed to occur immediately in solution rather than upon evaporation [202].

Ordered two-dimensional lattices containing thiolized spherical Au particles of two different sizes were first reported by Kiely et al. (see Fig. 1.20) [203]. The arrangement of the nanocrystals in these lattices correspond to the predicted atomic arrangement in metal alloys with similar radius ratios. Arrays corresponding to both AB and AB_2 prototypical lattices have been obtained. An AB_5 type of lattice has been observed when a solution containing $CoPt_3$ nanocrystals of diameters 4.5 nm and 2.6 nm was evaporated on a substrate [198]. Such an organization is also obtained starting with nanocrystals of different elements with appropriate radius ratios. Thus, two-dimensional arrays of AB and AB_2 type consisting of alternate rows of thiol-capped Au and Ag nanocrystals of different sizes have been made [204]. An AB type of superlattice has been obtained starting with oleic acid-capped Fe and dodecanethiol-capped Au nanocrystals [205].

Redl et al. [206] obtained three-dimensional binary lattices consisting of magnetic γ-Fe_2O_3 and semiconducting PbSe nanocrystals in ordered two-dimensional layers. The lattices obtained possessed prototypical structures such as AB_2, AB_5 and AB_{13} (see Fig. 1.21).

Bigioni et al. have used video microscopy to study the process of assembly of Au nanocrystals into extended two-dimensional arrays [207]. They suggest that the morphology of the drop-deposited nanocrystal films is controlled by evaporation kinetics and particle interactions with the liquid air interface. In the presence of an attractive particle–interface interaction, rapid early-stage evaporation dynamically produces a two-dimensional solution of nanocrystals at the liquid air interface, from which nanocrystal islands nucleate and grow.

{100}_{SL}

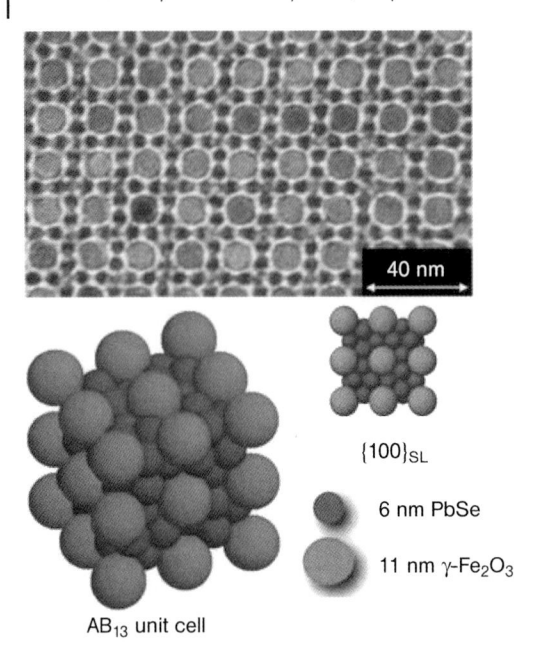

6 nm PbSe

11 nm γ-Fe$_2$O$_3$

AB$_{13}$ unit cell

Fig. 1.21 TEM showing a superlattice of PbSe and γ-Fe$_2$O$_3$ nanocrystals. The projection corresponds to ⟨100⟩ plane of the AB$_{13}$ structure. The AB$_{13}$ lattice and the structure of the ⟨100⟩ plane are illustrated below (reproduced with permission from Ref. [206]).

1.6.3
Three-dimensional Superlattices

Three-dimensional arrays of nanocrystals are obtained by the use of a layer-by-layer assembly technique. Multilayer deposition is achieved by the use of alternate layers of nanocrystals and linkers. The method of layer-by-layer deposition has been drawing a great deal of attention over the last few years, since it provides a convenient, low-cost means to prepare ultra-thin films of controlled thickness, suited for device applications [194, 208]. In a typical experiment, one end of a monolayer forming a bifunctional spacer, is tethered to a flat substrate such as gold, aluminum, indium tin oxide or glass, leaving the other end free to anchor nanocrystals. Subsequent layers can be introduced by dipping the substrate sequentially into the respective spacer molecule solution and the nanocrystal dispersion, with intermediate steps of washing and drying. The formation of a multilayer assembly is monitored by spectroscopic and microscopic methods. There are many advantages in the use of dithiols or similar crosslinkers that bind covalently to a metal substrate and at the other end to a nanocrystal. The layers consist of regularly arranged particles with high surface coverage and the interspers-

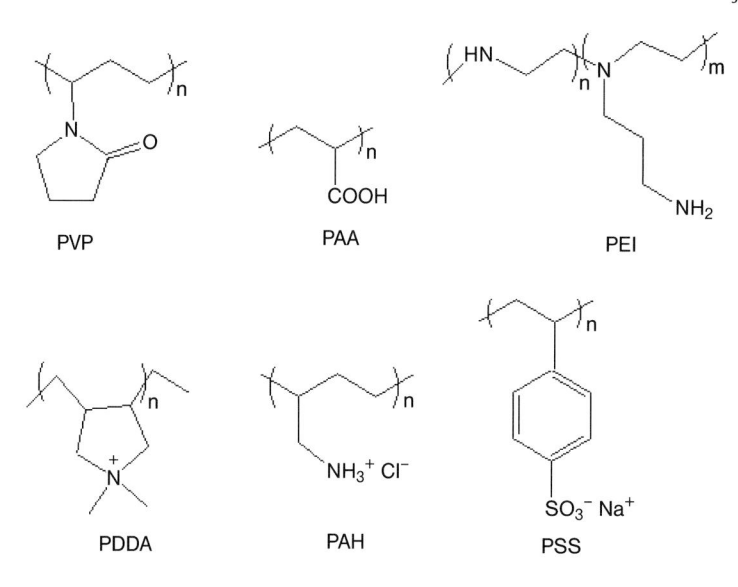

Fig. 1.22 The structures of commonly used polyelectrolytes. PEI-poly(ethyleneimine); PDDA-poly(diallyldimethylammonium chloride); PSS-poly(styrenesulfonate); PAH-poly(allylamine hydrochloride); PAA-poly(acrylic acid); PVP-poly(vinylpyrrolidone).

ing layers have a well-defined structure and thickness. Disadvantages include long deposition time per layer (~12 h), opacity of the substrate and the presence of defects such as pin holes in the dithiol layer that thwart charge transport studies. The use of dithiol linkers also restricts one to the use of organosols.

Polyelectrolytes such as poly(diallyldimethylammonium chloride) (PDDA) and polyethyleneimine(PEI) are used as alternative interspacers to build up multilayers of nanocrystals. The nanocrystals bind to the polyelectrolyte layer due to electrostatic interactions. A higher degree of disorder exists in each nanocrystalline layer. However, adsorption of nanocrystals can be carried out in only a few minutes. The structures of the commonly used polyelectrolytes are shown in Fig. 1.22. By the use of alternating layers of poly(styrene sulfonate)sodium salt(PSS) and poly(allylamine hydrochloride)(PAH), multilayers consisting of Au nanocrystals were obtained on PEI-coated Si or float glass substrates. A layer of cationic PSS is deposited on the PEI covered substrate. Anionic citrate-capped Au nanoparticles adhere to the cationic PAH layer by electrostatic interactions. Another layer of anionic PSS renders the substrate suitable for further deposition of PSS and Au layers. The deposition was followed, in this study by means of X-ray reflectivity and UV–visible spectroscopy measurements [209]. PAH and other polycationic polymers can be directly adsorbed on mica. On Si or glass substrates, PEI or aminosiloxanes are used as base layers to initiate the multilayer deposition process [210]. Fendler and coworkers [211] have used polycations to bring about the organization of semiconducting CdS, TiO$_2$ and PbS nanocrystals on quartz, Au and Teflon substrates.

Metal nanocrystals capped with an oxide layer such as silica or titania are useful ingredients of multilayered structures. The surface charge of the particles depends on the isoelectric point of the oxide shell and can be varied by choice of pH etc. In aqueous solutions, silica is negatively charged. Layer-by-layer assembly of Au nanocrystals coated with silica have been obtained with PDDA as the crosslinker. On the other hand, titania-coated nanocrystals can be organized using a polyanion such as PAA. Besides polyelectrolytes, small molecules with multiple charges can be used as crosslinkers. For example, Willner and coworkers [212–215] have carried out extensive studies on multilayers of Au and Ag nanocrystals prepared by crosslinking electroactive cyclophanes based on the 4,4'-bipyridinium cation.

In addition to electrostatic interactions, weak interactions such as hydrogen bonding and acid–base interactions are used to bring about multilayered organization. Au nanocrystals capped with 4-mercaptobenzoic acid have been deposited using PVP as crosslinker [216]. Murray and coworkers [217] have made use of acid–base interactions to build multilayer assemblies using Au nanocrystals capped with a mixed layer of hexanethiol and mercaptoundecanoic acid or 4-aminothiophenol. The carboxylic acid-capped particles were organized by the use of PAH interlayers while PSS was used to organize the amine-capped nanocrystals. UV–visible spectra showing layer-by-layer assembly of Au nanocrystals are shown in Fig. 1.23.

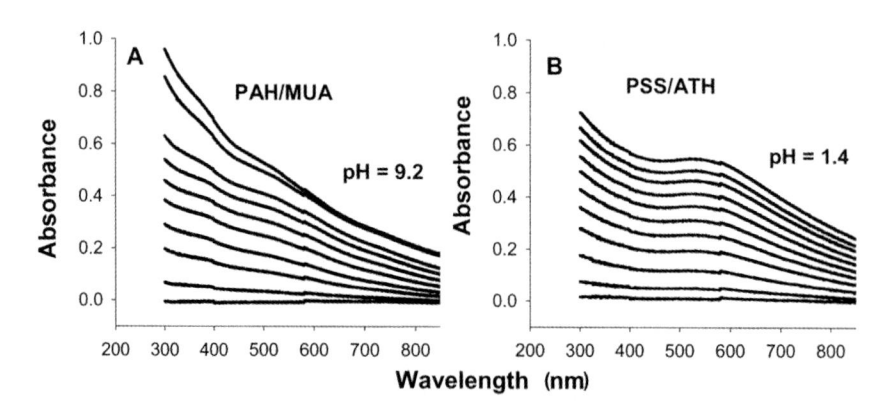

Fig. 1.23 UV–vis spectra showing layer-by-layer growth of polymer/Au nanocrystals on thiol-functionalized glass slides. (A) PAH/mercaptoundecanoicacid multilayer film formed by alternately exposing the slide to a solution of poly(allylamine hydrochloride) and another solution of Au nanocrystals with a mixed ligand shell consisting of hexanethiol and mercaptoundecanoic acid. (B) PSS/ATH multilayer film formed by alternately exposing the slide to a solution of poly(styrene sulfonic acid) and another solution of Au nanocrystals with a mixed ligand shell consisting of hexanethiol and 4-mercaptophenylamine (ATH) (reproduced with permission from Ref. [217]).

1.6.4
Colloidal Crystals

It has not been possible hitherto to prepare single crystals made up of nanocrystals. There is, however, a tendency of monodisperse nanocrystals to arrange into ordered three-dimensional arrays extending to a few microns. By tuning the crystallization conditions, crystallites of micrometer dimensions consisting of thousands of Au_{55} nanocrystals have been prepared [218]. Microcrystallites of CdSe [219] and Au [220] have been similarly obtained. Microcrystals of mercaptosuccinic acid-capped Au nanocrystals have been obtained from an aqueous medium [221], through an assembly mediated by hydrogen bonding interactions between the ligand molecules. Weller and coworkers [222] have devised a highly successful method of producing these microcrystals using what is called a three-layer technique of controlled oversaturation (see Fig. 1.24). In this method, a non-solvent such as methanol is allowed to diffuse slowly through a buffer solvent layer such as propanol to a toluene layer containing nanocrystals. By carrying out this process in long narrow tubes, microcrystals of nanocrystals are obtained after a few weeks. This technique has been used to prepare microcrystals of $CoPt_3$, Fe–Pt and CdSe nanocrystals (see Fig. 1.25) [186, 222, 223]. Careful experiments reveal that the arrangement of nanocrystals in such crystallites is polymorphic.

Kalsin et al. have used Au and Ag nanocrystals capped with oppositely charged ligands to grow microcrystallites [280]. This method of self-assembly termed electrostatic self-assembly is different from the organization brought about by uncharged ligands such as thiols in an organic medium. Surprisingly, the nanocrystals, despite having similar dimensions, choose to adopt a sphalerite (diamond) structure. The authors attribute this to screening of charges on nanocrystals.

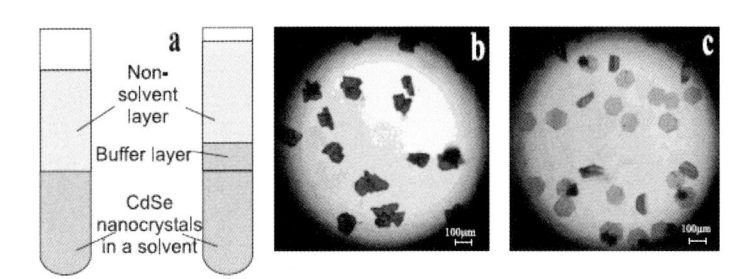

Fig. 1.24 (a) Schematic illustration of the three-layer technique of controlled oversaturation. (b) Photograph of CdSe microcrystals prepared by the two-layer technique (left test tube). (c) Photograph of CdSe microcrystals prepared by the three-layer technique (right test tube). The crystallites in (c) have better defined facets (reproduced with permission from Ref. [223]).

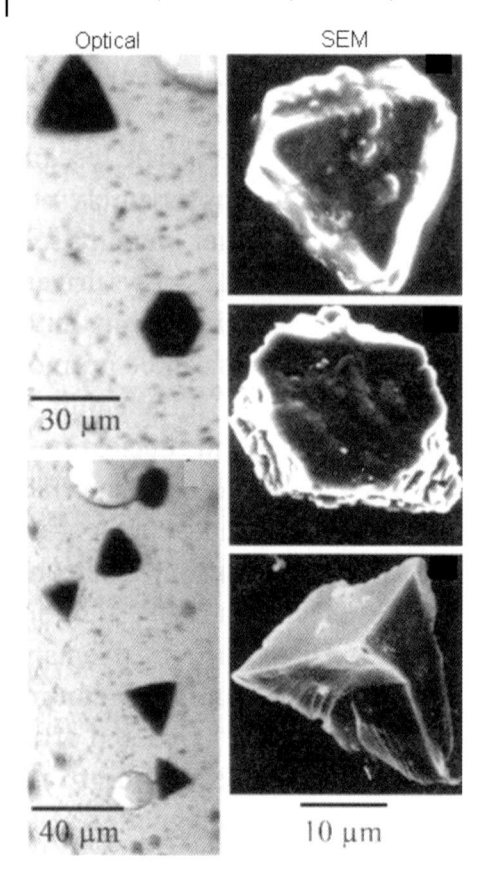

Fig. 1.25 Photographs and SEM images of microcrystals of Fe–Pt nanocrystals (reproduced with permission from Ref. [222]).

1.7
Applications

1.7.1
Optical and Electro-optical Devices

Devices based on composites of nanocrystals with polymers or polyelectrolytes are being fabricated to exploit electroluminescent and related properties of nanocrystalline media. These devices typically utilize thin films of composites obtained by layer-by-layer deposition using self-assembly, drop-casting or spin-casting methods.

Solar cells with efficiency comparable to the commercial cells have been made using poly(2-hexylthiophene)–CdSe nanorod multilayers [224]. Organic solar cells have been fabricated using porphyrins and fullerene units along with Au nanoparticles deposited on nanostructured SnO_2 electrodes [225]. Photocurrent generation and electroluminescence have been found in devices made up of CdS nanocrystals and polymers such as poly(3-hexylthiophene) and poly(2-methoxy-5-2(2-ethylhexoxy))-1,4-phenylene vinylene [226]. Electroluminescence has been observed in devices consisting of CdSe [227] and CdSe@ZnS [228] nanocrystals embedded in films of poly(vinylcarbazole) and an oxidiazole derivative. Light emitting diodes have been made with CdSe [230] and CdSe@CdS [231] nanocrystals. Full color emission has been achieved by using composites of CdSe@ZnS and CdS@ZnS in polylaurlymethacrylate [232]. The polymerization reaction was carried out in the presence of the nanocrystals to achieve optimal incorporation of the particles in the film. The correct configuration to yield full color emission has been studied [233]. In this study, CdTe multilayers were obtained on ITO glass using PDDA as the crosslinker. By varying the diameter of the nanocrystals in different layers, electroluminescence of different colors could be obtained. Efficient photodetectors based on poly[2-methoxy-5-(2-ethylhexyloxy)-1,4-phenylenevinylene] and PbSe nanocrystal composites have been fabricated [234]. The observed photocurrent gain is attributed to carrier multiplication in PbSe nanocrystals via multiple exciton generation and efficient charge transport through the polymer matrix.

Lasers have been fabricated by using composites containing titania and CdSe, CdSe@ZnS or CdS@ZnS core–shell nanocrystals with strong confinement [235, 236]. By varying the size of the CdSe nanocrystals from 1.7 to 2.7 nm or by using CdSe@ZnS nanocrystals, the wavelength of the stimulated emission is varied. The novelty is that lasers of different colors can be obtained using a single method of fabrication.

1.7.2
Nanocrystal-based Optical Detection and Related Devices

The optical response of metal nanocrystals to changes in the dielectric constant of the surrounding medium has been used to design probes that respond to a specific event such as oligonucleotide pairing with its complementary sequence, antibody–antigen binding or in general, protein–protein interactions [237, 238]. Gold nanocrystals of 13 nm diameter were made water soluble by derivatizing the surface with a thiol-terminated oligonucleotide sequence. A colorimetric response (color change from red to blue) was obtained when a complementary DNA strand was added. This color change can visibly indicate base–pair mismatches. The process is reversible, and the blue-colored nanocrystalline dispersion regains its original red color when the chains are sliced apart by heating. It is possible to observe a color change by spotting the nanocrystals on silica gels [239]. DNA fragments melt (lose their rigid conformation) over a much narrower temperature range when bound to metal particles rather than to conventional dyes [240]. Different color change patterns are obtained by using Au–Ag alloy or Ag@Au core–shell nanocrystals. This scheme of detection has received much attention. Small Au nanoparticles are shown to exhibit high photoluminescence upon irradiation with femtosecond pulses of 790 nm light, suggesting that the metal nanocrystals are potential alternatives to fluorophores for biological labeling and imaging [241].

Ag nanotriangles generated by nanosphere lithography have been used to detect biotin–streptavidin binding by immobilising biotin on the particles [242, 243]. Treatment with streptavidin solutions with concentrations in the picomolar range was sufficient to trigger the response. Nie and coworkers [244] have suggested a novel replacement for molecular beacons used to signal binding events through fluorescence quenching. In their scheme, either end of a oligonucleotide is functionalised – one end with a thiol group and the other end with a fluorophore. The thiol group binds to the Au nanocrystal by covalent linkage while the fluorophore loosely attaches itself to the surface of the nanocrystal; thereby losing its fluorescence. The oligonucleotide adapts a hairpin-like conformation. When a complementary DNA sequence is added, the fluorophore is detached from the surface regaining its fluorescence and thus acting as a beacon. Inhibition assays based on fluorescence resonance energy transfer (FRET) between streptavidin-conjugated quantum dots and biotin related Au nanoparticles have been described [245]. It has been suggested that larger nanoparticles could be very sensitively detected by measuring the light scattered by them [246–248]. The intensity of the scattered light by a single 80 nm Au particle has been compared to the total light emitted by 10^6 fluorescing molecules! Yguerabide and coworkers [246–248] have pioneered the use of large particles with diameters of ∼60 nm for carrying out tracer type biochemical assays. Third harmonic signals from large gold nanocrystals are also potentially useful for tracking individual biomolecules [249].

In addition to responding to the changes in the surrounding medium, absorption spectra of metal nanocrystals contain other important information. Dipolar coupling interactions in metal nanocrystal assemblies such as linear rows, lead to strong optical anisotropy. The transmitted light intensity thus depends on whether it is polarized parallel or perpendicular to the rows of nanocrystals. Such material could be useful as polarizing filters. Dirix et al. [250] have pioneered a simple method of preparing filters by stretching a polyethylene film impregnated with Ag nanocrystals. Another example is the planar assembly of CdSe nanocrystals emitting linearly polarized light in the plane of the assembly [251]. Atwater and coworkers [252] have suggested that a coupled plasmon mode could lead to coherent propagation of electromagnetic energy, i.e., a row of metal nanocrystals acting as a nanoscale waveguide. Using an array generated by means of e-beam lithography, they have provided a proof of concept demonstration. Nanoparticle assemblies connected by polymers can act as molecular springs and nanothermometers. Surface plasmon resonance and exciton–plasmon interaction are responsible for the nanothermometer function [253].

A touch sensor has been fabricated using alternating layers of Au (10 nm thick) and CdS (3 nm thick) nanocrystals, separated by polyelectrolytic layers, PSS and PAH [254]. By the use of metal and semiconductor nanocrystals, it is ensured that the change in current density through the film and the electroluminescent light intensity are linearly proportional to local stress. A stress image is obtained by pressing a substance such as a copper grid or coin on the device and focusing the resulting electroluminescent light directly on the charge-coupled device. It is claimed that the lateral and height resolution of texture of this device are comparable to the human finger at similar stress levels.

1.7.3
Nanocrystals as Fluorescent Tags

Quantum dots possess certain advantages over conventional fluorescent dyes. They are highly photostable, possess large Stoke shifts and emit in a narrower range (with emission peak widths about a third of the width seen in the case of molecular species). It is possible to simultaneously obtain intense emission at several different wavelengths by exciting at a single wavelength by the use of nanocrystals of the same material but with different sizes. Furthermore, emission from quantum dots is not easily susceptible to quenching.

Quantum dots capable of emitting light have been covalently attached to biomolecules by functionalizing their surface with carboxyl or related groups. For example, transferrin and IgG have been attached to CdSe@ZnS nanocrystals functionalised with mercaptoacetic acid [255]. We have been successful in attaching flourescent quantum dots to lymphocyte and macrophage cells using tutfisn and amine terminated poly(ethyleneglycol) ($H_2N-(CH_2CH_2O)n-NH_2$). It is also possible to attach quantum dots to biomolecules electrostatically. More sophisticated and specific attachment could be obtained by covalently linking a designer peptide incorporating a receptor sequence or moiety.

Several *in vivo* [256, 257] and *in vitro* [258–260] studies have been carried out using quantum dots as either specific or non-specific labels. It appears that quantum dots are likely to replace conventional dyes in the near future. The enhanced photostability of quantum dots also makes it possible to follow binding events in real time. Furthermore, the use of single-wavelength excitation to obtain emission at different wavelengths has applications in combinatorial searches.

1.7.4
Biomedical Applications of Oxide Nanoparticles

Magnetic oxide nanoparticles such as iron oxide are considered to be biocompatible as they possess no known toxicity [261]. Superparamagnetic oxide nanoparticles have therefore found several biomedical applications. By tailoring the ligand shell, the magnetic particles can be attached to a specific target molecule in a solution of different entities. The target can then be separated by the use of a magnetic field. Such aids are important in biochemical experiments where very low concentrations are generally employed. For example, by employing red blood cells labeled with iron oxide nanoparticles, the sensitivity of detection of malarial parasite is shown to be enhanced [262]. It has been proposed that hyperthermia, magnetic field induced heating of superparamagnetic particles, could be used to destroy diseased cells and thereby treat cancer. The particles are dispersed in the affected cells and an external AC magnetic field is applied to heat the particles to destroy the cells. Numerous studies have been carried out towards achieving this objective [263, 264]. A magnetic nanoprobe consisting of superparamagnetic nanoparticles coated with a specific molecule of interest has been used to study molecular interactions in live cells [265]. Magnetic nanoparticles also find applications as contrast enhancing agents in magnetic resonance imaging. This technology has been commercialized and a nanoparticle-based contrast agent called Feridex I.V is commercially available [261].

1.7.5
Nanoelectronics and Nanoscalar Electronic Devices

That single electron devices such as supersensitive electrometers and memory elements could be fabricated using nanocrystals has been one of the expectations of some practioners in the field. Proof of concept experiments to test chemically prepared nanocrystals in single electron devices have been carried out. For example, Murray and coworkers [266] have found that a single redox reaction taking place at the surface of Au nanocrystals induces an eight-fold increase in its capacitance. The concept of single electron transistors (SET), where the change in state is brought about by charging the transistor with the smallest quanta of charge, would be energy efficient transistors. Such transistors have been fabricated based on lithographically defined quantum dots. Single electron transistors have been fabricated with one or a few chemically prepared nanocrystals held between the electrodes [267, 268]. Some of them are operable at room temperature [268]. It is hoped that such transistors can be built by a self-assembly process in the future.

Schiffrin and coworkers [269] have obtained a nano-switch based on a layer of Au nanoparticles on a viologen moiety anchored to an Au substrate. The $I-V$ characteristics of the Au nanoparticles studied using *in situ* scanning tunneling spectroscopy(STS) revealed a dependence on the redox state of the viologen underneath the nanoparticle. By electrochemically altering the redox state, the conductivity of the circuit could be made high or low. A scanning tunneling spectroscopic study on two Pd nanocrystals linked with a conjugated thiol has been shown this combination to act as a switch. The $I-V$ spectra exhibit negative differential resistance, indicative of switching behavior [270].

A more tangible application of nanocrystals could be their use as vapor sensors. There have been a few attempts to obtain measurable electrical response for vapors of molecules that can adsorb to the surface of the nanocrystals. The molecules include thiols or simple solvent molecules such as toluene and ethanol. The sensor elements in these cases have been prepared by multilayer deposition [271] or spin coating techniques [271–274]. Murray and coworkers [275] have used Cu ions and thiolacid protected nanocrystals to build such a sensor element. These devices produce a reversible and rapid response to different kinds of vapors

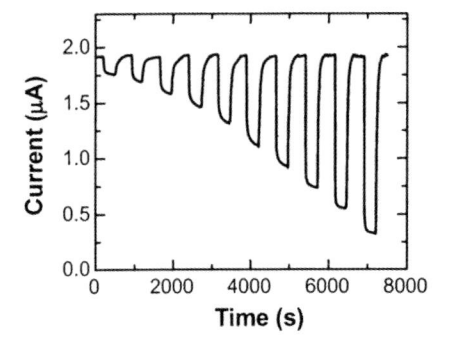

Fig. 1.26 Changes in the current over time when films were alternately exposed to nitrogen followed by ethanol vapor of increasing partial pressures. Ethanol vapor partial pressures were increased in fractions of 0.1 from 0.1–1.0 (reproduced with permission from Ref. [275]).

(see Fig. 1.26). In other cases, sensitivity of parts per million concentration has been achieved [276]. For example, nanoparticles of ZnO have been found to be excellent for sensing ethanol and hydrogen. Similarly, In_2O_3 nanoparticles are good sensors for NO_2.

Magnetic nanocrystals have been envisaged for use in recording. For example, ferromagnetic lattices made of Fe–Pt nanocrystals were found to sustain a high-density magnetization transition. An electrical transition induced by high electrical fields has been observed in a device consisting of 2-naphtolenethiol-capped Au nanoparticle/polystyrene composite sandwiched between two Al electrodes, an observation with potential application in memory devices [277]. Memory effects have also been observed in polyaniline nanofiber/Au nanoparticle composites [278] and in CdSe nanocrystal arrays [279].

1.8
Conclusions

The past couple of years have witnessed significant advances in the synthesis and assembly of metal, oxide and semiconductor nanocrystals. Burgeoning literature on the synthesis of nanocrystals of different sizes and shapes provides ample testimony. The advances in synthesis have overtaken our understanding of fundamental factors such as growth mechanisms and have generated a whole range of nanocrystalline matter in forms and shapes that had not even been envisaged previously. This period has also witnessed developments in our understanding of fundamental properties of nanocrystalline matter. Several devices relying on the unique properties of nanocrystals have been produced in the laboratory. It clearly is an exciting time to research nanocrystals.

References

1 C. Burda, X. Chen, R. Narayanan, M. A. El-Sayed: *Chem. Rev.* **105** 1025 (2005).
2 B. L. Cushing, V. L. Kolesnichenko, C. J. O'Connor: *Chem. Rev.* **104** 3893 (2004).
3 M.-C. Daniel, D. Astruc: *Chem. Rev.* **104** 293 (2004).
4 Y.-W. Jun, J.-S. Choi, J. Cheon: *Angew. Chem. Int. Ed.* **45** 3414 (2006).
5 J. Joo, H. B. Na, T. Yu, J. H. Yu, Y. W. Kim, F. Wu, J. Z. Zhang, T. Hyeon: *J. Am. Chem. Soc.* **125** 11100 (2003).
6 Y.-W. Jun, Y.-Y. Jung, J. Cheon: *J. Am. Chem. Soc.* **124** 615 (2002).
7 N. Pradhan, S. Efrima: *J. Am. Chem. Soc.* **125** 2050 (2003).
8 D. J. Crouch, P. O'Brien, M. A. Malik, P. J. Skabara, S. P. Wright: *Chem. Commun.* 1454 (2003).
9 J. Jasieniak, C. Bullen, J. van Embden, P. Mulvaney: *J. Phys. Chem. B* **109** 20665 (2005).
10 Z. H. Zhang, W. S. Chin, J. J. Vittal: *J. Phys. Chem. B* **108** 18569 (2004).
11 Y. Hasegawa, M. Afzaal, P. O'Brien, Y. Wada, S. Yanagida: *Chem. Commun.* 242 (2005).
12 T. Mirkovic, M. A. Hines, P. S. Nair, G. D. Scholes: *Chem. Mater.* **17** 3451 (2005).
13 L. Qu, W. W. Yu, X. Peng: *Nano Lett.* **4** 465 (2004).
14 M. J. Bowers, J. R. McBride, S. J. Rosenthal: *J. Am. Chem. Soc.* **127** 15378 (2005).
15 A. Kasuya, R. Sivamohan, Y. A. Barnakovi, I. M. Dmitruki, T. Nirasawa, V. R. Romanyuki, V. Kumar, S. V. Mamykin, K. Tohji, B. Jeyadevan, K. Shinoda, T. Kudo, O. Terasaki, Z. Liu, R. V. Belosludov, V. Sundararajan, Y. Kawazoe: *Nature Mater.* **3** 99 (2004).
16 E. M. Chan, R. A. Mathies, A. P. Alivisatos: *Nano Lett.* **3** 199 (2003).
17 B. K. H. Yen, A. Günther, M. A. Schmidt, K. F. Jensen, M. G. Bawendi: *Angew. Chem. Int. Ed.* **44** 5447 (2005).
18 M. A. Malik, P. O'Brien, S. Norager, J. Smith: *J. Mater. Chem.* **13** 2591 (2003).
19 M. A. Malik, P. O'Brien, M. Helliwell: *J. Mater. Chem.* **15** 1463 (2005).
20 K. Sardar, C. N. R. Rao: *Adv. Mater.* **16** 425 (2004).
21 M. Green, P. O'Brien: *J. Mater. Chem.* **12** 629 (2004).
22 K. Sardar, F. L. Deepak, A. Govindaraj, C. N. R. Rao: *Small* **1** 91 (2005).
23 S. C. Perera, G. Tsoi, L. E. Wenger, S. L. Brock: *J. Am. Chem. Soc.* **125** 13960 (2003).
24 P. O'Brien, S. S. Cummins, D. Darcy, A. Dearden, O. Masala, N. L. Pickett, S. Ryley, A. J. Sutherland: *Chem. Commun.* 2532 (2003).
25 S. Kim, M. G. Bawendi: *J. Am. Chem. Soc.* **125** 14652 (2003).
26 I. Potapova, R. Mruk, S. Prehl, R. Zentel, T. Basche, A. Mews: *J. Am. Chem. Soc.* **125** 320 (2003).
27 D. H. Son, S. M. Hughes, Y. Yin, A. P. Alivisatos: *Science* **306** 1009 (2004).
28 M. Green: *Chem. Commun.* 3002 (2005).
29 J. Hambrock, R. Becker, A. Birkner, J. Weiss, R. A. Fischer: *Chem. Commun.* 68 (2002).
30 M. Green and P. O'Brien: *Chem. Commun.* 183 (2000).
31 S.-W. Kim, J. Park, Y. Jang, Y. Chung, S. Hwang, T. Hyeon: *Nano Lett.* **3** 1289 (2003).
32 D. Bunge, T. J. Boyle, T. J. Headley: *Nano Lett.* **3** 901 (2003).
33 J. Fang, K. L. Stokes, W. L. Zhou, W. Wang, J. Lin: *Chem. Commun.* 1872 (2001).
34 I. S. Weitz, J. L. Sample, R. Ries, E. M. Spain, J. R. Heath: *J. Phys. Chem. B* **104** 4288 (2000).
35 C. A. Stowell, B. A. Korgel: *Nano Lett.* **5** 1203 (2005).
36 S. Sun: *Adv. Mater.* **18**, 393 (2006).
37 S. Sun, E. F. Fullerton, D. Weller, L. Folks, A. Maser: *Science* **287** 1989 (2000).
38 S. Kang, J. W. Harrell, D. E. Nikles: *Nano Lett.* **2** 1033 (2002).

39 S. S. Kang, D. E. Nikles, J. W. Harrell: *J. Appl. Phys.* **93** 7178 (2003).

40 M. Chen, D. E. Nikles: *Nano Lett.* **2** 211 (2002).

41 X. Sun, S. Kang, J. W. Harrell, D. E. Nikles, Z. R. Dai, J. Li, Z. L. Wang: *J. Appl. Phys.* **93** 7337 (2003).

42 M. Chen, D. E. Nikles: *J. Appl. Phys.* **91** 8477 (2002).

43 E. V. Shevchenko, D. V. Talapin, A. L. Rogach, A. Kornowski, M. Haase, H. Weller: *J. Am. Chem. Soc.* **124** 11480 (2002).

44 K. Ono, R. Okuda, Y. Ishii, S. Kamimura, M. Oshima: *J. Phys. Chem. B* **107** 1941 (2003).

45 F. Dumestre, S. Martinez, D. Zitoun, M.-C. Fromen, M.-J. Casanove, P. Lecante, M. Respaud, A. Serres, R. E. Benfield, C. Amiens, B. Chaudret: *Faraday Discuss.* **125** 265 (2003).

46 K. Ono, Y. Kakefuda, R. Okuda, Y. Ishii, S. Kamimura, A. Kitamura, M. Oshima: *J. Appl. Phys.* **91** 8480 (2002).

47 H. Gu, B. Xu, J. Rao, R. K. Zheng, X. X. Zhang, K. K. Fung and C. Y. C. Wong: *J. Appl. Phys.* **93** 7589 (2003).

48 Y. Yin, R. M. Rioux, C. K. Erdonmez, S. Hughes, G. A. Somorjai, A. P. Alivisatos: *Science* **304** 711 (2004).

49 Y. Li, M. Afzaal, P. O'Brien: *J. Mater. Chem.* **22** 2175 (2006).

50 M. Yin, S. O'Brien: *J. Am. Chem. Soc.* **125** 10180 (2003).

51 J. Park, K. An, Y. Hwang, J.-G. Park, H.-J. Noh, J.-Y. Kim, J.-H. Park, N.-M. Hwang, T. Hyeon: *Nature Mater.* **3** 891 (2004).

52 J. Joo, T. Yu, Y. W. Kim, H. M. Park, F. Wu, J. Z. Zhang, T. Hyeon: *J. Am. Chem. Soc.* **125** 6553 (2003).

53 K. Biswas, C. N. R. Rao: *J. Phys. Chem. B* **110** 842 (2006).

54 M. Ghosh, E. V. Sampathkumaran, C. N. R. Rao: *Chem. Mater.* **17** 2348 (2005).

55 W. S. Seo, J. H. Shim, S. J. Oh, E. K. Lee, N. H. Hur, J. T. Park: *J. Am. Chem. Soc.* **127** 6188 (2005).

56 M. Ghosh, K. Biswas, A. Sundaresan, C. N. R. Rao: *J. Mater. Chem.* **16** 106 (2006).

57 Y. S. Wang, P. J. Thomas, P. O'Brien: *J. Phys. Chem. B* **110** 4099 (2006).

58 D. K. Yi, S. T. Selvan, S. S. Lee, G. C. Papaefthymiou, D. Kundaliya, J. Y. Ying: *J. Am. Chem. Soc.* **127** 4990–4991 (2005).

59 R. C. Vestal, Z. J. Zhang, *Nano Lett.* **3** 1739 (2003).

60 H.-H. Yang, S.-Q. Zhang, X.-L. Chen, Z.-X. Zhuang, J.-G. Xu, X.-R. Wang, *Anal. Chem.* **76** 1315 (2004).

61 F. Shieh, A. E. Saunders, B. A. Korgel: *J. Phys. Chem. B* **109** 8538 (2005).

62 L. Qu, X. Peng: *J. Am. Chem. Soc.* **124** 2049 (2002).

63 L. Manna, E. C. Sher, A. P. Alivisatos: *J. Am. Chem. Soc.* **122** 12700 (2000).

64 X. Peng, L. Manna, W. Yang, J. Wickham, E. Scher, A. Kadavanich, A. P. Alivisatos: *Nature* **404** 59 (2000).

65 Q. Pang, L. Zhao, Y. Cai, D. P. Nguyen, N. Regnault, N. Wang, S. Yang, W. Ge, R. Ferreira, G. Bastard, J. Wang: *Chem. Mater.* **17** 5263 (2005).

66 S. D. Bunge, K. M. Krueger, R. H. Boyle, M. A. Rodriguez, T. J. Headley, V. L. Colvin: *J. Mater. Chem.* **13** 1705 (2003).

67 L. Manna, D. J. Milliron, A. Meisel, E. Scher, A. P. Alivisatos: *Nature Mater.* **2** 382 (2003).

68 Y. Li, X. Li, C. Yang, Y. Li: *J. Mater. Chem.* **13** 2641 (2003).

69 Y.-W. Jun, S.-M. Lee, N.-J. Kang, J. Cheon: *J. Am. Chem. Soc.* **123** 5150 (2001).

70 D. J. Milliron, S. M. Hughes, Y. Cui, L. Manna1, J. Li, L.-W. Wang, A. P. Alivisatos: *Nature* **430** 190 (2004).

71 N. Pradhan, S. Efrima: *J. Phys. Chem. B* **108** 11964 (2004).

72 P. D. Cozzoli, L. Manna, M. L. Curri, S. Kudera, C. Giannini, M. Striccoli, A. Agostiano: *Chem. Mater.* **17** 1296 (2005).

73 Y. Jun, Y. Jung, J. Cheon: *J. Am. Chem. Soc.* **124** 615 (2002).

74 S.-M. Lee, Y.-W. Jun, S.-N. Cho, J. Cheon: *J. Am. Chem. Soc.* **124** 11244 (2002).

75 A. B. Panda, G. Glaspell, M. S. El-Shall: *J. Am. Chem. Soc.* **128** 2790 (2006).

76 P. Christian, Paul O'Brien: *Chem. Commun.* 2817 (2005).
77 J. A. Ascencio, O. Santiago, Rendón, U. Pal: *Appl. Phys. A* **78** 5 (2004).
78 S. G. Thoma, A. Sanchez, P. P. Provencio, B. L. Abrams, J. P. Wilcoxon: *J. Am. Chem. Soc.* **127** 7611 (2005).
79 J. H. Yu, J. Joo, H. M. Park, S. Baik, Y. W. Kim, S. C. Kim, T. Hyeon: *J. Am. Chem. Soc.* **127** 5662 (2005).
80 Y. Li, M. A. Malik, P. O'Brien: *J. Am. Chem. Soc.* **127** 16020 (2005).
81 P. Christian, P. O'Brien: *J. Mater. Chem.* **15** 4949 (2005).
82 J. Tabenor, P. Christian, P. O'Brien: *J. Mater. Chem.* **21** 2082 (2006).
83 P. Christian, P. O'Brien: *J. Mater. Chem.* **15** 3021 (2005).
84 P. J. Thomas, P. O'Brien: *J. Am. Chem. Soc.* **128** 5614 (2006).
85 T. J. Trentler, K. M. Hickman, S. C. Goel, A. M. Viano, P. C. Gibbons, W. E. Buhro: *Science* **270** 1791 (1995).
86 H. Yu, W. E. Buhro: *Adv. Mater.* **15** 416 (2003).
87 S. P. Ahrenkiel, O. I. Micic, A. Miedaner, C. J. Curtis, J. M. Nedeljkovic, A. J. Nozik: *Nano Lett.* **3** 833 (2003).
88 S. Kan, T. Mokari, E. Rothenberg, U. Banin: *Nature Mater.* **2** 155 (2003).
89 H. Yu, J. Li, R. A. Loomis, P. C. Gibbons, L.-W. Wang, W. E. Buhro: *J. Am. Chem. Soc.* **125** 16168 (2003).
90 J. W. Grebinski, K. L. Hull, J. Zhang, T. H. Kosel, M. Kuno: *Chem. Mater.* **16** 5260 (2004).
91 K. L. Hull, J. W. Grebinski, T. H. Kosel, M. Kuno: *Chem. Mater.* **17** 4416 (2005).
92 J. D. Holmes, K. P. Johnston, R. C. Doty, B. A. Korgel: *Science* **287** 1471 (2000).
93 X. Lu, D. D. Fanfair, K. P. Johnston, B. A. Korgel: *J. Am. Chem. Soc.* **127** 15718 (2005).
94 F. M. Davidson, R. Wiacek, B. A. Korgel: *Chem. Mater.* **17** 230 (2005).
95 F. M. Davidson, A. D. Schriker, R. J. Wiacek, B. A. Korgel: *Adv. Mater.* **16** 646 (2004).
96 J.-W. Seo, Y.-W. Jun, S. J. Ko, J. Cheon: *J. Phys. Chem. B* **109** 5389 (2005).
97 J. J. Urban, W. S. Yun, Q. Gu, H. Park: *J. Am. Chem. Soc.* **124** 1186 (2002).
98 C. Qian, F. Kim, L. Ma, F. Tsui, P. Yang, J. Liu: *J. Am. Chem. Soc.* **126** 1195 (2004).
99 M. Yin, Y. Gu, I. L. Kuskovsky, T. Andelman, Y. Zhu, G. F. Neumark, S. O'Brien: *J. Am. Chem. Soc.* **126** 6206 (2004).
100 F. X. Redl, C. T. Black, G. C. Papaefthymiou, R. L. Sandstrom, M. Yin, H. Zeng, C. B. Murray, S. O'Brien: *J. Am. Chem. Soc.* **126** 14583 (2004).
101 T. H. Larsen, M. Sigman, A. Ghezelbash, R. C. Doty, B. A. Korgel: *J. Am. Chem. Soc.* **125** 5638 (2003).
102 M. B. Sigman, A. Ghezelbash, T. H. Hanrath, A. E. Saunders, F. Lee, B. A. Korgel: *J. Am. Chem. Soc.* **125** 16050 (2003).
103 A. Ghezelbash, M. B. Sigman, B. A. Korgel: *Nano Lett.* **4** 537 (2004).
104 J. Tanori, M.-P. Pileni: *Langmuir* **13** 639 (1997).
105 L. M. Qi, J. Ma, H. Chen, Z. Zhao: *J. Phys. Chem. B* **101** 340 (1997).
106 J. D. Hopwood, S. Mann: *Chem. Mater.* **9** 1819 (1997).
107 N. Pinna, K. Weiss, H. Sach-Kongehl, W. Vogel, J. Urban, M. P. Pileni: *Langmuir* **17** 7982 (2001).
108 M. Li, H. Schnablegger, S. Mann: *Nature* **402** 393 (1999).
109 I. Lisiecki: *J. Phys. Chem. B* **109** 12231 (2005).
110 R. K. Baldwin, K. P. Pettigrew, J. C. Garno, P. P. Power, G.-Y. Liu, S. M. Kauzlarich: *J. Am. Chem. Soc.* **124** 1150 (2002).
111 M. Monge, M. L. Kahn, A. Maisonnat, B. Chaudret: *Angew. Chem. Int. Ed.* **42** 5321 (2003).
112 J. Joo, S. G. Kwon, J. H. Yu, T. Hyeon: *Adv. Mater.* **17** 1873 (2005).
113 C. J. Murphy, N. R. Jana: *Adv. Mater.* **14** 80 (2002).
114 N. R. Jana, L. Gearheart, C. J. Murphy: *Chem. Commun.* 617 (2001).
115 N. R. Jana, L. Gearheart, C. J. Murphy: *J. Phys. Chem. B* **105** 4065 (2001).

116 J. Gao, C. M. Bender, C. J. Murphy: *Langmuir* **19** 9065 (2003).
117 N. R. Jana, L. Gearheart, C. J. Murphy: *Adv. Mater.* **13** 1389 (2001).
118 T. K. Sahu, C. J. Murphy: *J. Am. Chem. Soc.* **126** 8648 (2004).
119 S. Chen, D. L. Carroll: *J. Phys. Chem. B* **108** 5500 (2004).
120 C. J. Murphy, T. K. Sau, A. M. Gole, C. J. Orendorff, J. Gao, L. Gou, S. E. Hunyadi, T. Li: *J. Phys. Chem. B* **109** 13857 (2005).
121 K. K. Caswell, C. M. Bender, C. J. Murphy: *Nano Lett.* **3** 667 (2003).
122 Y. Sun, B. Gates, B. Mayers, Y. Xia: *Nano Lett.* **2** 165 (2002).
123 Y. Sun, B. Mayers, T. Herricks, Y. Xia: *Nano Lett.* **3** 955 (2003).
124 Y. Sun, Y. Xia: *Science* **298** 2176 (2002).
125 S. S. Shankar, A. Rai, B. Ankamwar, A. Singh, A. Ahmad, M. Sastry: *Nature Mater.* **3** 482 (2004).
126 R. K. Naik, S. J. Stringer, G. Agarwal, S. E. Jones, M. O. Stone: *Nature Mater.* **1** 169 (2002).
127 R. Jin, Y. Cao, C. A. Mirkin: *Science* **294** 1901 (2001).
128 R. Jin, Y. Cao, E. Hao, G. S. Mitraux, G. C. Schatz, C. A. Mirkin: *Nature* **425** 487 (2003).
129 G. S. Metraux, C. A. Mirkin: *Adv. Mater.* **17** 412 (2005).
130 E. Hao, R. C. Bailey, G. C. Schatz: *Nano Lett.* **4** 327 (2004).
131 C.-H. Kuo, M. H. Huang: *Langmuir* **21** 2012 (2005).
132 Y. Xiao, B. Shylahovsky, I. Popov, V. Pavlov, I. Willner: *Langmuir* **21** 5659 (2005).
133 X. Teng, H. Yang: *Nano Lett.* **5** 885 (2005).
134 M. Maillard, S. Giorgio, M.-P. Pileni: *Adv. Mater.* **14** 1084 (2002).
135 I. Pastoriza-Santos, L. L. Marzan: *Nano Lett.* **2** 903 (2002).
136 D. O. Yener, J. Sindel, C. A. Randell, J. H. Adair: *Langmuir* **18** 8692 (2002).
137 Y. Shao, Y. Jin, S. Dong: *Chem. Commun.* 1104 (2004).
138 F. Dumestre, B. Chaudret, C. Amiens, M.-C. Fromen, M.-J. Casanove, P. Renaud, P. Zurcher: *Angew. Chem. Int. Ed.* **41** 4286 (2002).
139 K. Soulantica, A. Maisonnat, F. Senocq, M.-C. Fromen, M.-J. Casanove, B. Chaudret: *Angew. Chem. Int. Ed.* **40** 2984 (2001).
140 H. Song, F. Kim, S. Connor: *J. Phys. Chem. B* **109** 188 (2005).
141 J. D. Hoefelmeyer, K. Niesz, G. A. Somarjai, T. D. Tilley: *Nano Lett.* **5** 435 (2005).
142 T. Mokari, E. Rothenberg, I. Popov, R. Costi, U. Banin: *Science* **304** 1787 (2004).
143 S. Kudera, L. Carbone, M. F. Casula, R. Cingolani, A. Falqui, E. Snoeck, W. J. Parak, L. Manna: *Nano Lett.* **5** 445–449 (2005).
144 H. Yu, M. Chen, P. M. Rice, S. X. Wang, R. L. White, S. Sun: *Nano Lett.* **5** 379 (2005).
145 H. Gu, R. Zheng, X. Zhang, B. Xu: *J. Am. Chem. Soc.* **126** 5664 (2004).
146 R. Viswanatha, S. Sapra, T. Saha-Dasgupta, D. D. Sarma: *Phys. Rev. B* **72** 045333 (2005).
147 S. Sapra, D. D. Sarma: *Phys. Rev. B* **69** 125304 (2004).
148 R. Viswanatha, D. D. Sarma: *Chem. Eur. J.* **12** 180 (2006).
149 R. D. Schaller, V. I. Klimov: *Phys. Rev. Lett.* **92** 186601 (2004).
150 R. J. Ellingson, M. C. Beard, J. C. Johnson, P. Yu, O. I. Micic, A. J. Nozik, A. Shabaev, A. L. Efros: *Nano Lett.* **5** 865 (2005).
151 P. Guyot-Sionnest: *Nature Mater.* **4** 653 (2005).
152 A. J. Nozik: *Annu. Rev. Phys. Chem.* **52** 193 (2001).
153 A. J. Nozik: *Physica E (Amsterdam)* **14** 115 (2002).
154 J. T. Hu, L.-S. Li, W. Yang, L. Manna, L.-W. Wang, A. P. Alivisatos: *Science* **292** 2060 (2001).
155 M. Kazes, D. Y. Lewis, Y. Ebenstein, T. Mokari, U. Banin: *Adv. Mater.* **14** 317 (2002).
156 R. Dingle: *Festkörperprobleme XV*, 21 (1975).
157 P. G. Harper, J. A. Hilder: *Phys. Status Solidi* **26** 69 (1968).
158 M. S. Gudiksen, J. Wang, C. M. Lieber: *J. Phys. Chem. B* **106** 4036 (2002).

159 K. K. Nanda, F. E. Kruis, H. Fissan: *Nano Lett.* **1** 605 (2001).
160 L. E. Brus: *J. Chem. Phys.* **80** 4403 (1984).
161 J. Li, L.-W. Wang: *Phys. Rev. B* **72** 125325 (2005).
162 A. D'Andrea, R. Del Sole: *Solid State Commun.* **74** 1121 (1990).
163 P. Lefebvre, P. Christol, H. Mathieu: *Phys. Rev. B* **48** 17308 (1993).
164 P. Lefebvre, P. Christol, H. Mathieu, S. Glutsch: *Phys. Rev. B* **52** 5756 (1995).
165 H. Yu, J. Li, R. A. Loomis, L.-W. Wang, W. E. Buhro: *Nature Mater.* **2** 517 (2003).
166 J. Nogués, J. Sort, V. Langlais, S. Doppiu, B. Dieny, J. S. Muñoz, S. Suriñach, M. D. Baró, S. Stoyanov, Y. Zhang: *Int. J. Nanotech.* **2** 23 (2005).
167 V. Skumryev, S. Stoyanov, Y. Zhang, G. Hadjipanayis, D. Givord, J. Nogues: *Nature* **423** 19 (2003).
168 D. Magana, S. C. Perera, A. G. Harter, N. S. Dalal, G. F. Strouse: *J. Am. Chem. Soc.* **128** 2931 (2006).
169 S. C. Erwin, L. Zu, M. I. Haftel, A. L. Efros, T. A. Kennedy, D. J. Norris: *Nature* **436** 91 (2005).
170 A. P. Alivisatos, K. P. Johnsson, X. Peng, T. E. Wilson, C. J. Loweth, M. P. Bruchez, P. G. Schultz: *Nature* **382** 609 (1996).
171 M. G. Warner, J. Hutchison: *Nature Mater.* **2** 272 (2003).
172 M. Mitov, C. Portet, C. Bourgerette, E. Snoeck, M. Verelst: *Nature Mater.* **1** 229 (2002).
173 S.-W. Lee, C. Mao, C. E. Flynn, A. M. Belcher: *Science* **296** 892 (2002).
174 M. Kogiso, K. Yoshida, K. Yase, T. Shimizu: *Chem. Commun.* 2492 (2002).
175 E. Dujardin, C. Peet, G. Stubbs, J. N. Culver, S. Mann: *Nano Lett.* **3** 413 (2003).
176 K. G. Thomas, S. Barazzouk, B. I. Ipe, S. T. S. Joseph, P. V. Kamat: *J. Phys. Chem. B* **108** 13066 (2004).
177 S. Lin, M. Li, E. Dujardin, C. Girard, S. Mann: *Adv. Mater.* **17** 2553 (2005).
178 C. N. R. Rao, A. Govindaraj, F. L. Deepak, N. A. Gunari, M. Nath: *Appl. Phys. Lett.* **78** 1853 (2001).
179 E. J. H. Lee, C. Ribeiro, E. Longo, E. R. Leite: *J. Phys. Chem. B* **109** 20842 (2005).
180 P. C. Ohara, J. R. Heath, M. W. Gelbart: *Angew. Chem. Int. Ed.* **36** 1077 (1997).
181 T. Vossmeyer, S.-W. Chung, W. M. Gelbart, J. R. Heath: *Adv. Mater.* **10** 351 (1998).
182 P. C. Ohara and W. M. Gelbart: *Langmuir* **14** 3418 (1998).
183 M. Maillard, L. Motte, A. T. Ngo, M. P. Pileni: *J. Phys. Chem. B* **104** 11871 (2000).
184 M. Maillard, L. Motte, M. P. Pileni: *Adv. Mater.* **16** 200 (2001).
185 C. Stowell, B. A. Korgel: *Nano Lett.* **1** 595 (2001).
186 L. V. Govor, G. H. Bauer, G. Reiter, E. Shevchenko, H. Weller, J. Parisi: *Langmuir* **19** 9573 (2003).
187 S. L. Tripp, S. V. Pusztay, A. E. Ribbe, A. Wei: *J. Am. Chem. Soc.* **124** 7914 (2002).
188 B. Liu, H. C. Zeng: *J. Am. Chem. Soc.* **127** 18262 (2005).
189 Z. Liu, R. Levicky: *Nanotecnology* **15** 1483 (2004).
190 J. Huang, F. Kim, A. R. Tao, S. Connor, P. Yang: *Nat. Mater.* **4** 896 (2005).
191 S. A. Claridge, S. L. Goh, J. M. J. Frechet, S. C. Williams, C. M. Michel, A. P. Alivisatos: *Chem. Mater.* **17** 1628 (2005).
192 J. H. Warner, R. D. Tilley: *Adv. Mater.* **17** 2997 (2005).
193 Y. Lin, A. Boker, J. He, K. Sill, H. Xiang, C. Abetz, X. Li, J. Wang, T. Emrick, S. Long, Q. Wang, A. Balazs, T. P. Russell: *Nature* **434** 55 (2005).
194 A. N. Shipway, E. Katz, I. Willner: *ChemPhysChem* **1** 18 (2000).
195 C. N. R. Rao, G. U. Kulkarni, P. J. Thomas, P. P. Edwards: *Chem. Soc. Rev.* **29** 27 (2000).
196 C. P. Collier, T. Vossmeyer, J. R. Heath: *Annu. Rev. Phys. Chem.* **49** 371 (1998).
197 S.-Y. Zhao, S. Wang, K. Kimura: *Langmuir* **20** 1977 (2004).
198 E. V. Shevchenko, D. V. Talapin, A. L. Rogach, A. Kornowski, M. Haase, H.

Weller: *J. Am. Chem. Soc.* **124** 11480 (2002).
199 H. Fan, E. Leve, J. Gabaldon, A. Wright, R. E. Haddad, C. J. Brinker: *Adv. Mater.* **17** 2587 (2005).
200 B. Kim, S. L. Tripp, A. Wei: *J. Am. Chem. Soc.* **123** 7955 (2001).
201 F. Dumestre, B. Chaudret, C. Amiens, P. Renaud, P. Fejes: *Science* **303** 821 (2004).
202 K. Soulantica, A. Maisonnat, M.-C. Fromen, M.-J. Casanove, B. Chaudret: *Angew. Chem. Int. Ed.* **42** 1945 (2003).
203 C. J. Kiely, J. Fink, M. Brust, D. Bethell, D. J. Schiffrin: *Nature* **396** 444 (1998).
204 C. J. Kiely, J. Fink, J. G. Zheng, M. Brust, D. Bethell, D. J. Schiffrin: *Adv. Mater.* **12** 640 (2000).
205 A. E. Saunders, B. A. Korgel: *ChemPhysChem* **6** 61 (2005).
206 F. X. Redl, K.-S. Cho, C. B. Murray, S. O'Brien: *Nature* **423** 968 (2003).
207 T. P. Bigioni, X.-M. Lin, T. T. Nguyen, E. I. Corwin, T. A. Witten, H. M. Jaeger: *Nature Mater.* **5** 265 (2006).
208 K. V. Sarathy, P. J. Thomas, G. U. Kulkarni, C. N. R. Rao: *J. Phys. Chem. B* **103** 399 (1999).
209 J. Schmitt, G. Decher, W. J. Dressick, S. L. Brandow, R. E. Geer, R. Shashidhar, J. M. Calvert: *Adv. Mater.* **9** 61 (1997).
210 J. Schmitt, P. Mächtle, D. Eck, H. Möhwald, C. A. Helm: *Langmuir* **15** 3256 (1999).
211 N. A. Kotov, I. Dékány, J. H. Fendler: *J. Phys. Chem.* **99** 13065 (1995).
212 R. Blonder, L. Sheeney, I. Willner: *Chem. Commun.* 1393 (1998).
213 M. Lahav, A. N. Shipway, I. Willner, M. B. Nielsen, J. F. Stoddart: *J. Electroanal. Chem.* **482** 217 (2000).
214 M. Lahav, A. N. Shipway, I. Willner: *J. Chem. Soc., Perkin Trans.* **2** 1925 (1999).
215 M. Lahav, R. Gabai, A. N. Shipway, I. Willner: *Chem. Commun.* 1937 (1999).
216 E. Hao, T. Lian: *Chem. Mater.* **12** 3392 (2000).
217 J. F. Hicks, Y. Seok-Shon, R. W. Murray: *Langmuir* **18** 2288 (2002).

218 G. Schmid, R. Pugin, T. Sawitowski, U. Simon, B. Marler: *Chem. Commun.* 1303 (1999).
219 C. B. Murray, C. R. Kagan, M. G. Bawendi: *Science* **270** 1335 (1995).
220 S. I. Stoeva, B. L. V. Prasad, S. Uma, P. K. Stoimenov, V. Zaikovski, C. M. Sorensen, K. J. Klabunde: *J. Phys. Chem. B* **107** 7441 (2003).
221 S. Wang, S. Sato, K. Kimura: *Langmuir* **15** 2445 (2003).
222 E. Shevchenko, D. Talapin, A. Kornowski, F. Wiekhorst, J. Kvtzler, M. Haase, A. Rogach, H. Weller: *Adv. Mater.* **14** 287 (2002).
223 D. V. Talapin, E. V. Shevchenko, A. Kornowski, N. Gaponik, M. Haase, A. L. Rogach, H. Weller: *Adv. Mater.* **13** 1868 (2001).
224 W. U. Huynh, J. J. Dittmer, A. P. Alivisatos: *Science* **295** 2425 (2002).
225 T. Hasobe, H. Imahori, P. V. Kamat, T. K. Ahn, S. K. Kim, D. Kim, A. Fujimoto, T. Hirakawa, S. Fukuzumi: *J. Am. Chem. Soc.* **127** 1216 (2005).
226 K. S. Narayan, A. G. Manoj, J. Nanda, D. D. Sarma: *Appl. Phys. Lett.* **74** 871 (1999).
227 B. O. Dabbousi, M. G. Bawendi, O. Onitsuka, M. F. Rubner: *Appl. Phys. Lett.* **66** 1316 (1995).
228 S. Chaudhary, M. Ozkan, W. C. W. Chan: *Appl. Phys. Lett.* **84** 2925 (2004).
229 M. S. Gudiksen, K. N. Maher, L. Ouyang, H. Park: *Nano Lett.* **5** 2257 (2005).
230 V. L. Colin, M. C. Schlamp, A. P. Alivisatos: *Nature* **370** 354 (1994).
231 M. C. Schlamp, X. Peng, A. P. Alivisatos: *J. Appl. Phys.* **82** 5837 (1997).
232 J. Lee, V. C. Sundar, J. R. Heine, M. G. Bawendi, K. F. Jensen: *Adv. Mater.* **12** 1102 (2000).
233 M. Gao, C. Lesser, S. Kirstein, H. Möhwald, A. L. Rogach, H. Weller: *J. Appl. Phys.* **87** 2297 (2000).
234 D. Qi, M. Fischbein, M. Drndic, S. Selmic: *Appl. Phys. Lett.* **86** 093103 (2005).
235 H.-J. Eisler, V. C. Sundar, M. G. Bawendi, M. Walsh, H. I. Smith, V. Klimov: *Appl. Phys. Lett.* **80** 4614 (2002).

236 Y. Chan, J. S. Steckel, P. T. Snee, J.-M. Caruge, J. M. Hodgkiss, D. G. Nocera, M. G. Bawendi: *Appl. Phys. Lett.* **86** 073102 (2005).
237 N. L. Rosi, C. A. Mirkin: *Chem. Rev.* **105** 1547 (2005).
238 C.-S. Tsai, T.-B. Yu, C.-T. Chen: *Chem. Commun.* 4273 (2005).
239 R. Elghanian, J. J. Storhoff, R. C. Mucic, R. L. Letsinger, C. A. Mirkin: *Science* **277** 1078 (1997).
240 R. C. Jin, G. Wu, Z. Li, C. A. Mirkin, G. C. Schatz: *J. Am. Chem. Soc.* **125** 1643 (2003).
241 R. A. Farrer, F. L. Butterfield, V. W. Chen, J. T. Fourkas: *Nano Lett.* **5** 1139 (2005).
242 A. J. Haes, R. P. van Duyne: *J. Am. Chem. Soc.* **124** 10596 (2002).
243 J. C. Riboh, A. J. Haes, A. D. McFarland, C. R. Yonzon, R. P. V. Duyne: *J. Phys. Chem. B* **107** 1772 (2003).
244 D. J. Maxwell, J. R. Taylor, S. Nie: *J. Am. Chem. Soc.* **124** 9606 (2002).
245 E. Oh, M.-Y. Hong, D. Lee, S.-H. Nam, H. C. Yoon, H.-S. Kim: *J. Am. Chem. Soc.* **127** 3270 (2005).
246 J. Yguerabide, E. E. Yguerabide: *Anal. Biochem.* **262** 137 (1998).
247 J. Yguerabide, E. E. Yguerabide: *Anal. Biochem.* **262** 157 (1998).
248 S. Schultz, D. R. Smith, J. J. Mock, D. A. Schultz: *Proc. Natl. Acad. Sci.* **97** 996 (2000).
249 M. Lippitz, M. A. van Dijk, M. Orrit: *Nano Lett.* **5** 799 (2005).
250 Y. Dirix, C. Bastiaansen, W. Caseri, P. Smith: *Adv. Mater.* **11** 223 (1999).
251 J. Y. Kim, H. Hiramatsu, F. E. Osterloh: *J. Am. Chem. Soc.* **127** 15556 (2005).
252 S. A. Maier, M. L. Brongersma, P. G. Kik, S. Meltzer, A. A. G. Requicha, H. A. Atwater: *Adv. Mater.* **13** 1501 (2001).
253 J. Lee, A. O. Govorov, N. A. Kotov: *Angew. Chem. Int. Ed.* **44** 7439 (2005).
254 V. Maheshwari, R. F. Saraf: *Science* **312** 1501 (2006).
255 W. C. W. Chan, S. Nie: *Science* **281** 2016 (1998).
256 S. Pathak, S.-K. Choi, N. Arnheim, M. E. Thompson: *J. Am. Chem. Soc.* **123** 4103 (2001).
257 H. Mattoussi, J. M. Mauro, E. R. Goldman, G. P. Anderson, V. C. Sundar, F. V. Mikulec, M. G. Bawendi: *J. Am. Chem. Soc.* **122** 12142 (2000).
258 Z. Chunyang, M. Hui, D. Yao, J. Lei, C. Dieyan, N. Shuming: *Analyst* **125** 1029 (2000).
259 J. O. Winter, T. Y. Liu, B. A. Korgel, C. E. Schmidt: *Adv. Mater.* **13** 1673 (2001).
260 M. Dahan, T. Laurence, F. Pinaud, D. S. Chemla, A. P. Alivisatos, M. Sauer, S. Weiss: *Opt. Lett.* **26** 825 (2001).
261 Q. A. Pankhurst, J. Connolly, S. K. Jones, J. Dobson: *J. Phys. D: Appl. Phys.* **36** R167 (2003).
262 F. Paul, D. Melville, S. Roath, D. Warhurst: *IEEE Trans. Magn. Mag.* **17** 2822 (1981).
263 A. Jordan, R. Scholz, P. Wust, H. Fähling, R. Felix: *J. Magn. Magn. Mater.* **201** 413 (1999).
264 A. Jordan, R. Scholz, K. Maier-Hauff, M. Johannsen, P. Wust, J. Nadobny, H. Schirra, H. Schmidt, S. Deger, S. Loening: *J. Magn. Magn. Mater.* **225** 118 (2001).
265 J. Won, M. Kim, Y.-W. Yi, Y. H. Kim, N. Jung, T. K. Kim: *Science* **309** 121 (2005).
266 S. J. Green, J. I. Stokes, M. J. Hostetler, J. Pietron, R. W. Murray: *J. Phys. Chem. B* **101** 2663 (1997).
267 D. L. Klein, R. Roth, A. K. L. Lim, A. P. Alivisatos, P. L. McEuen: *Nature* **389** 699 (1997).
268 S. H. M. Persson, L. Olofsson, L. Hedberg: *Appl. Phys. Lett.* **74** 2546 (1999).
269 D. L. Gittins, D. Bethell, D. J. Schiffrin, R. J. Nichols: *Nature* **408** 67 (2000).
270 V. V. Agrawal, R. Thomas, G. U. Kulkarni, C. N. R. Rao: *Pramana – J. Phys.* **65** 769 (2005).
271 N. Krasteva, I Besnard, B. Guse, R. E. Bauer, K. Mullen, A. Yasuda, T. Vossmeyer: *Nano Lett.* **2** 551 (2002).
272 S. M. Briglin, T. Gao, N. S. Lewis: *Langmuir* **20** 299 (2002).

273 H. Ahn, A. Chandekar, B. Kang, C. Sung, J. E. Whitten: *Chem. Mater.* **16** 3274 (2004).

274 J. W. Grate, D. A. Nelson, R. Skaggs: *Anal. Chem.* **75** 1864 (2003).

275 F. P. Zamborini, M. C. Leopold, J. F. Hicks, P. J. Kulesza, M. A. Malik, R. W. Murray: *J. Am. Chem. Soc.* **124** 8958 (2002).

276 H. Wohlhen, A. W. Snow: *Anal. Chem.* **70** 2856 (1998).

277 J. Ouyang, C.-W. Chu, D. Sieves, Y. Yang: *Appl. Phys. Lett.* **86** 123507 (2005).

278 R. J. Tseng, J. Huang, J. Ouyang, R. B. Kaner, Y. Yang: *Nano Lett.* **5** 1077 (2005).

279 M. D. Fischbein, M. Drndic: *Appl. Phys. Lett.* **86** 193106 (2005).

280 A. M. Kalsin, M. Fialkowski, M. Paszewski, S. K. Smoukov, K. J. M. Bishop, B. A. Grzybowski: *Science* **312** 420 (2006).

2
Nanotubes and Nanowires: Recent Developments

S. R. C. Vivekchand, A. Govindaraj, and C. N. R. Rao

2.1
Introduction

Nanocrystals, nanowires, nanotubes, nanowalls and nanofilms are different classes of nanomaterials. Interest in one-dimensional nanotubes and nanowires received a major boost with the discovery of carbon nanotubes in 1991 [1]. Since then there has been extensive work on carbon nanotubes and inorganic nanotubes. Nanowires of a large variety of inorganic materials have been synthesized and characterized in the last few years. The technological potential of nanotubes and nanowires has triggered research on these materials in a big way [2–4]. Various aspects of carbon nanotubes and inorganic nanotubes were covered in the two Wiley-VCH volumes on the Chemistry of Nanomaterials [2], and the recent monograph on nanotubes and nanowires [4] provides a survey of the one-dimensional materials up to 2004. In this chapter, we concentrate on the research findings of the last two years, covering synthesis, structure, properties and applications. In doing so, we have cited a large number of references which should help the reader to obtain greater details where necessary. In our effort to make the chapter complete by covering most of the recent literature it was necessary to be brief.

2.2
Carbon Nanotubes

2.2.1
Synthesis

Several methods of synthesis of carbon nanotubes have been reported in the literature. We shall examine the recent developments related to the synthesis of multi-walled, single-walled and other types of carbon nanotubes. Multi-walled carbon nanotubes (MWNTs) can be produced in high yields by the catalytic combus-

Nanomaterials Chemistry. Edited by C. N. R. Rao, A. Müller, and A. K. Cheetham
Copyright © 2007 WILEY-VCH Verlag GmbH & Co. KGaA, Weinheim
ISBN: 978-3-527-31664-9

tion of polypropylene in the presence of an organically-modified clay and a Ni catalyst [5]. In this method, polypropylene was mixed with an organic-modified clay, Ni catalyst supported on silica–alumina, and maleated polypropylene in a Brabender mixer and then the nanocomposite was placed in a crucible and heated with the flame of a gas lamp at 600 °C. Carbon nanotubes prepared by the catalytic decomposition of methane over Mo/Ni/MgO catalysts turn out to be fairly thin with 2–4 graphitic walls due the formation of small Ni–Mo catalyst particles (2–16 nm in diameter) [6]. Alignment of MWNTs can be achieved by making use of Fe_2O_3 nanoparticles and magnetic fields [7]. A controlled synthesis of aligned MWNTs can be achieved by a layer-by-layer assembly of multilayer catalyst precursor films with the desired number of layers and composition [8]. Poly(sodium-4-styrenesulfonate) (PSS) and iron oxide colloids have been used in this procedure to prepare catalyst precursor films wherein the catalysts retain their structural integrity even at high temperature and high catalyst particle densities. In Fig. 2.1, we show SEM images of aligned MWNT bundles obtained by this method.

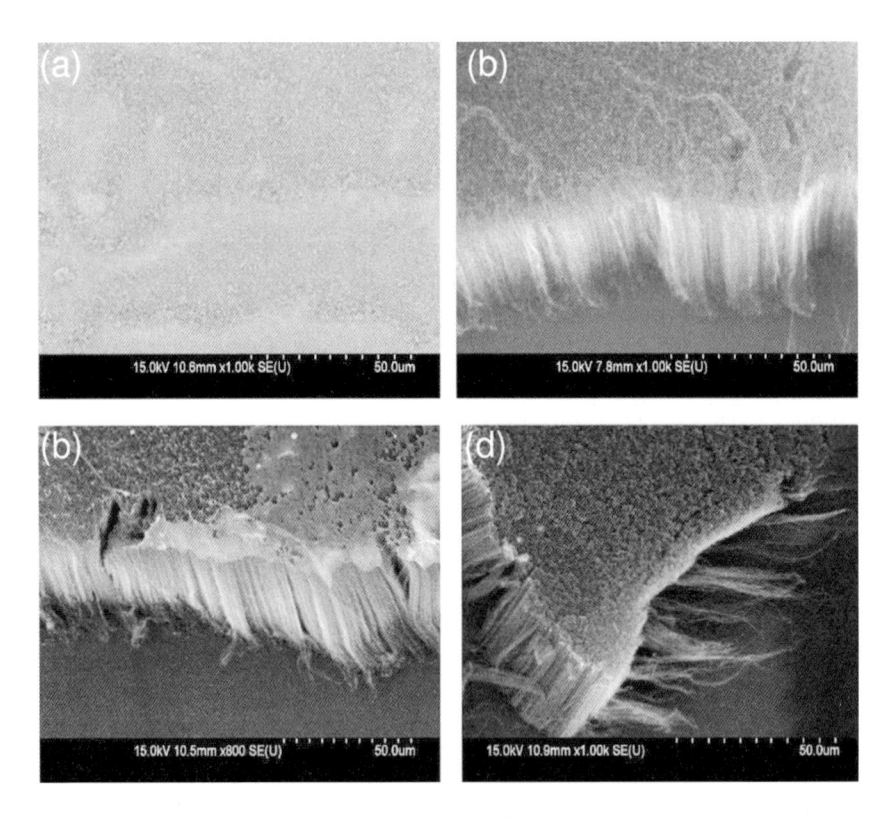

Fig. 2.1 SEM images of the carbon nanotubes grown from the (PSS/iron hydroxide colloid particles)$_n$ catalyst precursor films of different layer pairs (i.e., different *n*) after the calcination: a, b, c and d represent results for 1, 2, 3, and 5 layer pairs, respectively. (Reproduced from Ref. [8].)

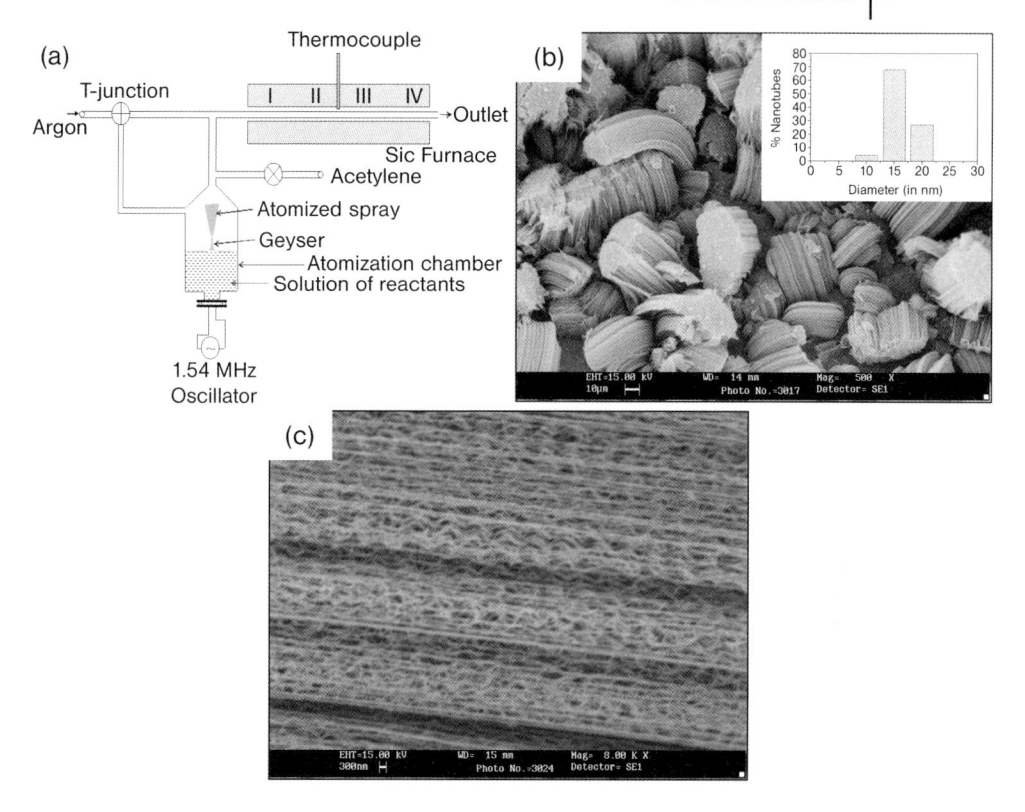

Fig. 2.2 (a) Schematic diagram of the nebulized spray pyrolysis experimental apparatus, (b) and (c) SEM images of MWNTs obtained by the nebulzied spray pyrolysis of a toluene solution of ferrocene (20 g L^{-1}) at 900 °C in the presence of acetylene (50 sccm). Inset in (b) shows the diameter distribution (Reproduced from Ref. [9(a)].)

MWNTs with fairly uniform diameters as well as aligned MWNT bundles have been obtained by nebulized spray pyrolysis using solutions of organometallics such as ferrocene in hydrocarbon solvents [9a]. Nebulized spray is a spray generated by an ultrasonic atomizer and a schematic diagram of the experimental set-up is shown in Fig. 2.2(a). SEM images of aligned MWNT bundles obtained by the pyrolysis of a nebulized spray of ferrocene–toluene–acetylene mixture are shown in Fig. 2.2(b) and (c). The advantage of using a nebulized spray is the ease of scaling into an industrial scale process, as the reactants can be fed continuously into the furnace. Successive layers of aligned MWNTs have been prepared by spray pyrolysis by the interruption of the reactant flow [9b]. Water-assisted selective etching of carbon atoms affords the growth of carbon nanotube stacks of up to 10 layers [10a]. In this method, etching takes place at the nanotube caps as well as at the interface between the nanotubes and metal catalyst particles. The overall growth process of carbon nanotube stacks is illustrated in Fig. 2.3. Multi-

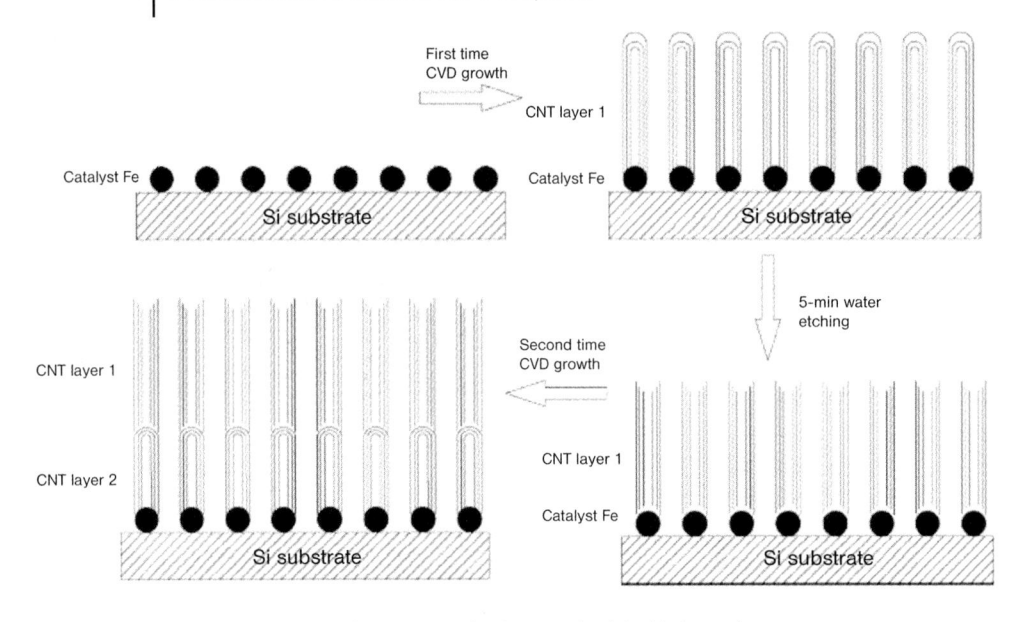

Fig. 2.3 Schematic of the process for the growth of double-layered CNT films with open ends. (Reproduced from Ref. [10(a)].)

ple layers of MWNTs have also been grown on partially oxidized Si substrates [10b].

Single-walled carbon nanotubes (SWNTs) have been grown using small catalytic iron nanoparticles synthesized within protein cages as catalysts [11]. Small diameter SWNTs with controlled density and ordered locations are obtained using a polyferrocenylsilane block copolymer as the precursor [12]. A low-temperature method for the growth of SWNTs by water plasma chemical vapor deposition (CVD) has been reported [13]. The water plasma lowers the growth temperature to 450 °C. A study of the kinetics of SWNT growth by water-assisted CVD has revealed the catalytic role of water [14]. SWNTs can be grown continuously from an ordered array of open-ended SWNTs [15]. In this method, nanometer-sized metal catalyst particles are attached to the open ends of SWNTs and subsequently activated to restart the nanotube growth.

A root-growth mechanism appears to operate in the synthesis of vertically aligned SWNTs by microwave plasma CVD [16]. SWNTs grow preferentially when CVD is carried out over SiO_2 spheres [17]. The diameter of the SWNTs and the proportions of the metallic and semiconducting nanotubes obtained by plasma assisted CVD appear to be related [18]. A higher percentage of semiconducting nanotubes is obtained when the parameters favor the smaller diameter nanotubes at relatively lower temperatures. Employing a CVD reactor, SWNTs coated with nanodiamond have been produced [19]. The tubular inner structures consists of bundles of SWNTs up to 15 µm in length, and the outer deposits con-

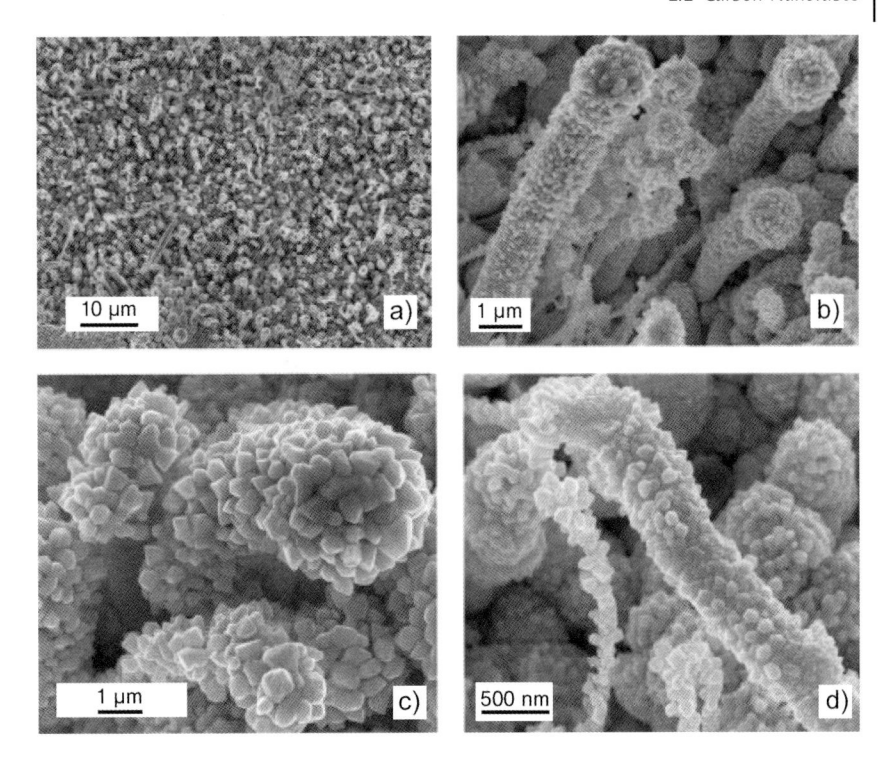

Fig. 2.4 SEM images showing: (a) the general view of nanodiamond coated SWNTs; (b–d) nanotube bundles covered by diamond nanocrystallites. (Reproduced from Ref. [19].)

sist of well-shaped diamond crystallites with diameters in the 20–100 nm range (Fig. 2.4). The aligned assembly of SWNTs on solid substrates is enhanced by an electric field [20]. Vertical growth of SWNTs of small diameter has been accomplished by heating small iron nanoparticles in the presence of an activated gas [21]. In Fig. 2.5, a schematic diagram of the experimental set-up and a SEM image of aligned SWNTs obtained by this method are shown. A substrate containing a thin layer of iron is quickly heated in the presence of an activated gas. The activated gas is generated by rapidly flowing gas mixtures of H_2 (400 sccm) and CH_4 (40 sccm) at pressures of 15–25 Torr, over a hot filament (temperature greater than 2000 °C) to create activated gas mixtures of hydrogen and carbon-containing species. The SWNTs have diameters in the 0.8–1.6 nm range. Real-time growth dynamics of SWNTs has been observed using ultra-high vacuum transmission electron microscopy (TEM) at 650 °C [22]. Three distinct growth regimes, incubation, growth and passivation have been observed. B-doped SWNTs have been obtained by the laser ablation of graphite targets loaded with boron as well as NiB [23].

SWNTs as well as double-walled carbon nanotubes (DWNTs) are formed

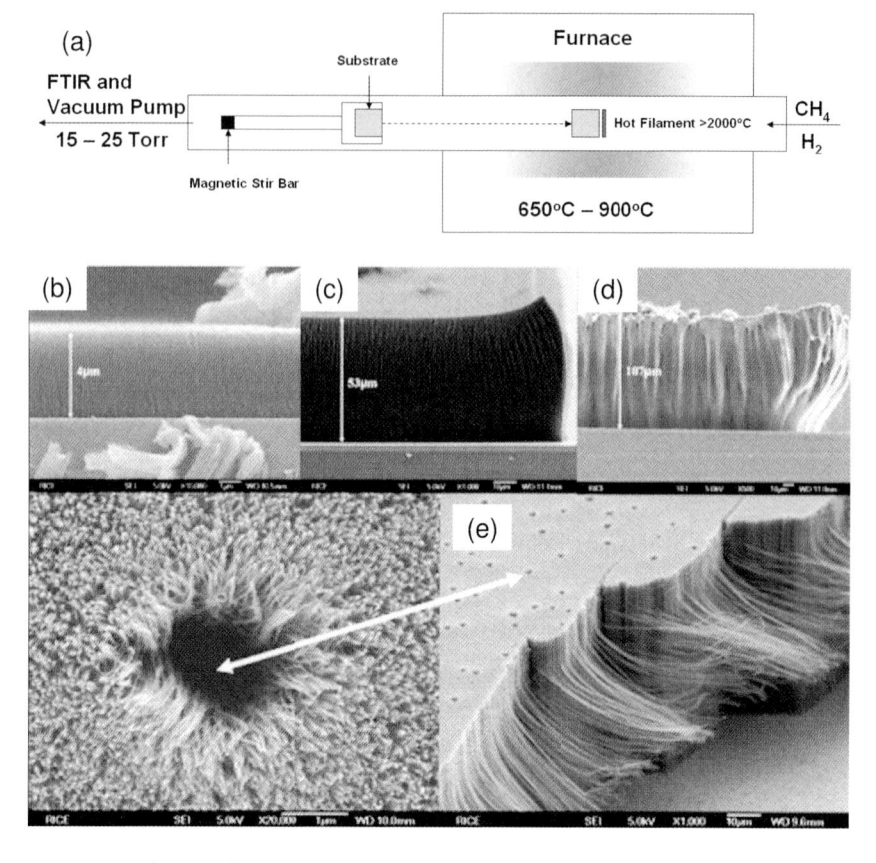

Fig. 2.5 (a) Schematic drawing of a 1 in hot filament CVD furnace system used for aligned SWNT growth. SEM images of carpet SWNTs grown with different time: (b) 1.3 min, scale bar 1 μm; (c) 20 min, scale bar 10 μm; (d) 40 min, scale bar 10 μm; (e) inside view of the sample grown with 40 min, scale bar 10 μm, inset image scale bar 1 μm. (Reproduced from Ref. [21].)

over Fe–Co alloy nanoparticles produced by the decomposition of methane over $Mg_{0.9}Fe_xCo_yO$ $(x + y = 0.1)$ solid solutions [24]. Aligned DWNT ropes (shown in Fig. 2.6) are obtained directly by sulfur-assisted floating catalytic decomposition of methane [25]. An aerosol based method has been employed for the synthesis of highly branched Y-junction carbon nanotubes [26].

2.2.2
Purification

Purification is an important problem faced in the use of SWNTs for various purposes. As-synthesized SWNTs prepared by processes such as arc-discharge,

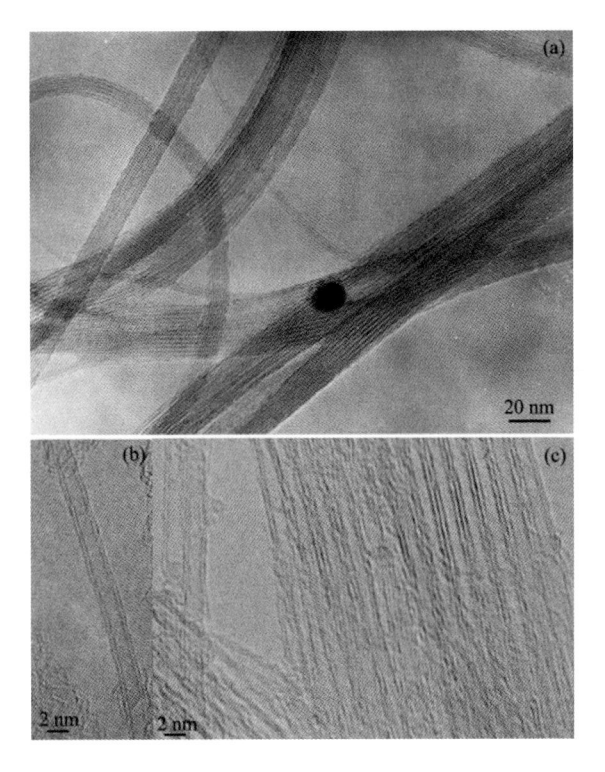

Fig. 2.6 (a) Low-magnification TEM image of the as-prepared DWNT ropes; (b) and (c) HREM images of an isolated DWNT and several DWNT bundles in a rope, respectively. (Reproduced from Ref. [25].)

laser-ablation, HiPco and pyrolysis of hydrocarbons or organometallic precursors, contain carbonaceous impurities, typically amorphous carbon and graphite nanoparticles, as well as particles of the transition metal catalyst [27]. Generally, dilute mineral acids are used to remove the catalyst metal nanoparticles, as concentrated acids tend to functionalize and even destroy the nanotubes. Amorphous carbon cannot be eliminated completely by air oxidation due to the presence of metal nanoparticles that catalyze the oxidation of SWNTs at relatively low temperatures. Heating acid-treated SWNTs in H_2 at elevated temperatures efficiently removes amorphous carbon. A typical procedure for the purification of SWNTs synthesized by the arc method is to carry out an initial air oxidation at 300 °C followed by acid washing, and then heating in H_2 at 700–1000 °C. Whereas, in air, oxidation converts amorphous carbon to CO_2, it is converted to CH_4 on hydrogen treatment. In Fig. 2.7, the Raman spectra of the SWNTs at various stages of purification are shown. The intensities of the G-band and radial breathing modes increase as the SWNTs are purified and the intensity of the D-band decreases as the amorphous carbon is eliminated. High-temperature CO_2 treatment is also used for the elimination of carbonaceous impurities. A controlled and scaleable

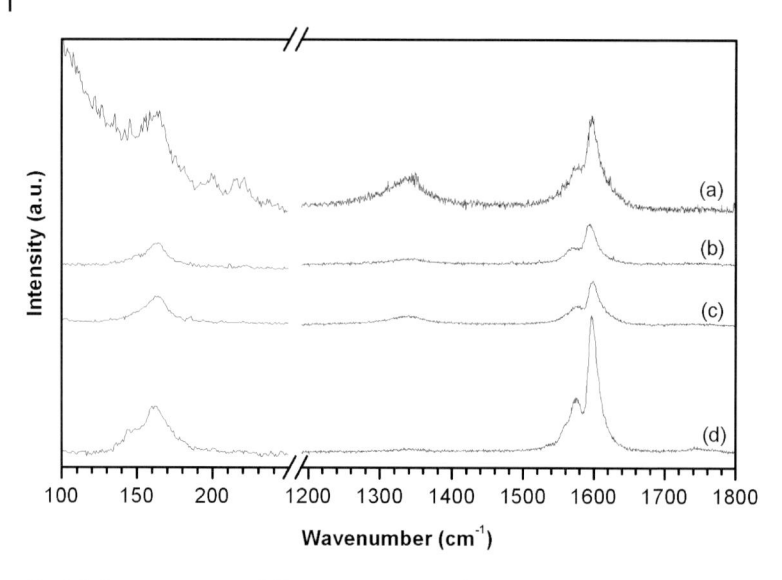

Fig. 2.7 Raman spectra at different stage in procedure B: (a) as-synthesized SWNTs, (b) acid-refluxed SWNTs, (c) H_2 treated SWNTs at 800 °C followed by acid washing and (d) final product after H_2 treatment at 1200 °C followed by acid washing. (Reproduced from Ref. [27].)

multi-step purification method to remove carbonaceous and metallic impurities from raw HiPco SWNTs has been reported [28]. The carbon-coated iron nanoparticles were exposed and oxidized by multi-step oxidation at increasing temperatures, with the oxidized iron nanoparticles being deactivated by using either $C_2H_2F_4$ or SF_6. Magnetic filtration has been employed for the removal of metal nanoparticle impurities in SWNTs [29]. Purification of nitric acid treated SWNTs by centrifugation at constant pH has been investigated [30]. The zeta potential seems to have an effect on the separation of carbonaceous impurities by centrifugation.

One of the problems in the purification of SWNTs is the evaluation of the purity of the sample after purification. Electron microscopy, thermogravimetric analysis (TGA), Raman spectroscopy and visible-NIR spectroscopy have been employed for the determination of the purity of SWNTs. The incapability of electron microscopy in the quantitative evaluation of purity of bulk SWNTs has been pointed out by Itkis et al. [31]. This is due to the small volume of the sample analyzed and the absence of algorithms to convert the images into numerical data. The use of solution-phase Raman spectroscopy and NIR spectroscopy for the quantitative evaluation of bulk purity is strongly recommended. A schematic illustration of electronic spectra of SWNTs is shown in Fig. 2.8. The main features in the spectra are due to the electronic transitions between the inter-van Hove singularities in both the metallic and the semiconducting SWNTs. The authors have

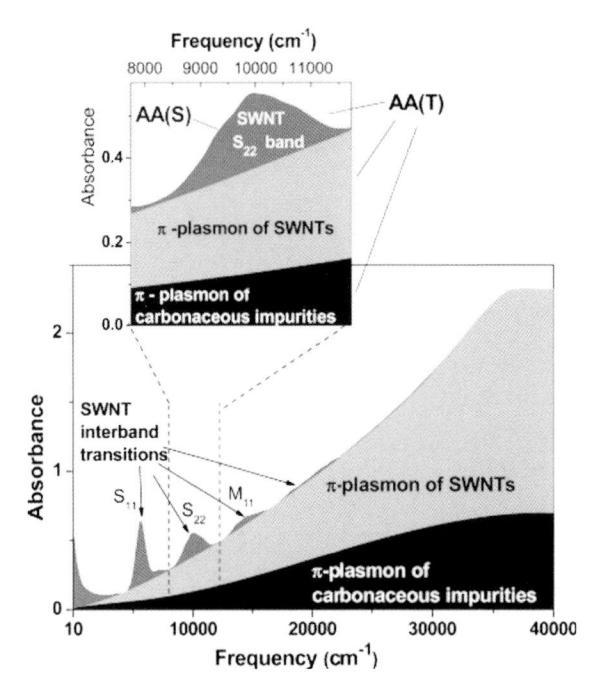

Fig. 2.8 Schematic illustration of the electronic spectrum of a typical SWNT sample produced by the electric arc method. The inset shows the region of the S_{22} interband transition utilized for NIR purity evaluation. In the diagram, AA(S) = area of the S_{22} spectral band after linear baseline correction and AA(T) = total area of the S_{22} band including SWNT and carbonaceous impurity contributions. The NIR relative purity is given by RP = (AA(S)/AA(T))/0.141 (Reproduced from Ref. [31].)

suggested a method to evaluate the relative purity of SWNTs based on the electronic absorption bands. Electronic absorption spectroscopy is especially useful in determining the purity [32a]. The way to treat the spectroscopic data has been discussed [32b].

Metallic SWNTs with small diameters are reported to be removed selectively by treatment with nitric and sulfuric acid mixtures [33]. Raman spectroscopy and electronic absorption spectroscopy have been employed to determine the relative concentrations of metallic and semiconducting SWNTs before and after separation using the acid mixtures (Fig. 2.9). Large-scale separation of metallic and semiconducting SWNTs has been achieved using a dispersion-centrifugation process using long chain alkyl amines [34]. In this process, HiPco nanotubes are dispersed in a solution of 1-octylamine in THF wherein the amine is strongly adsorbed on the metallic nanotubes. Centrifugation gives, selectively, an enrichment of the metallic nanotubes. Metallic and semiconducting SWNTs can be separated by selective functionalization using azomethine ylides [35]. Based on theoretical calculations, it is proposed that aromatic molecules such as naphthalene, anthracene and TCNQ interact strongly with metallic SWNTs [36]. Metallic

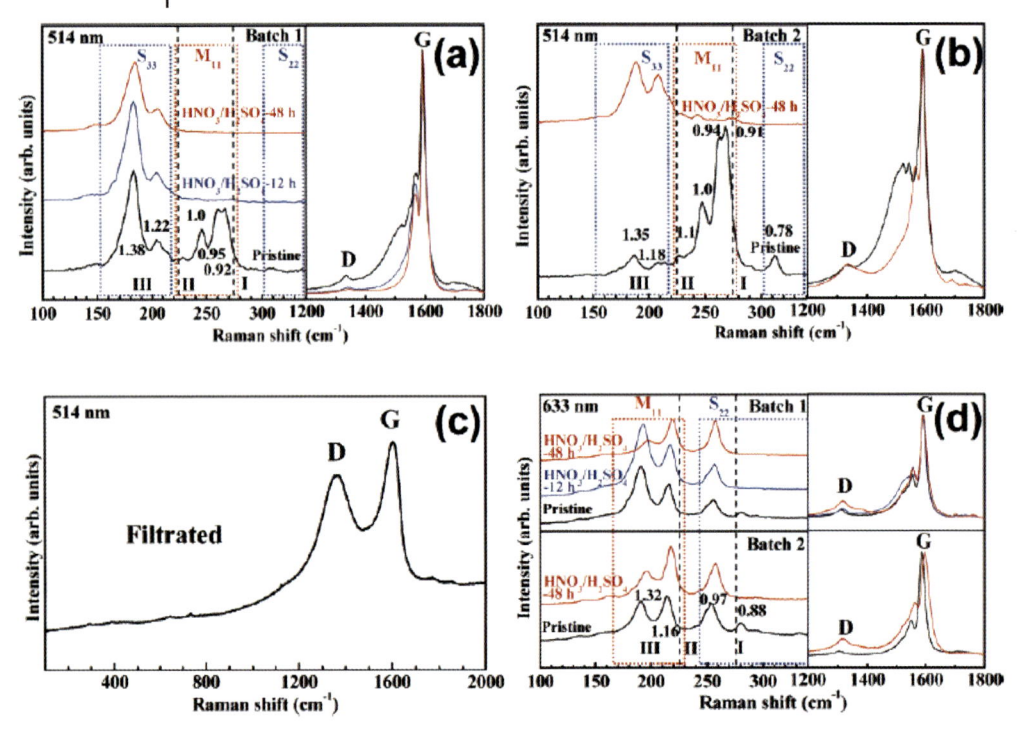

Fig. 2.9 Raman spectra for HiPco SWNTs. (a), (b) RBM and G-band Raman spectra from the leftover on the filter for batch 1 and 2 HiPco SWNTs at 514 nm excitation wavelength. The bottom (black), middle (blue), and top (red) lines are from the pristine, HNO_3/H_2SO_4 (1:9) treatment for 12 h, and HNO_3/H_2SO_4 (1:9) treatment for 48 h, respectively, followed by heat treatment at 900 °C in Ar atmosphere. (c) Raman spectrum of the filtrated sample after HNO_3/H_2SO_4 (1:9) treatment for 12 h at 514 nm excitation wavelength. (d) Corresponding RBM and G-band Raman spectra for batch 1 and 2 HiPco SWNTs at 633 nm excitation wavelength similar to (a). The RBMs were normalized with respect to the G-band. M_{ii} and S_{ii} correspond to metallic and semiconducting inter-band transitions. (Reproduced from Ref. [33].)

SWNTs are also preferentially destroyed by laser irradiation [37]. Optical absorption of metallic SWNTs is less affected by bundling effects than the semiconducting SWNTs, resulting in an enhanced absorption ratio of metallic and semiconducting SWNTs [38].

2.2.3
Functionalization and Solubilization

Functionalization of MWNTs using mineral acids such as H_2SO_4 and HNO_3 is well documented in the literature. Direct sidewall functionalization of MWNTs has been carried out using dilute HNO_3 under supercritical conditions [39]. Asymmetric end-functionalization of individual MWNTs, wherein each end is functionalized with a different chemical group, has been reported [40]. The procedure employed for the asymmetric end-functionalization of an aligned carbon nanotube film is shown in Fig. 2.10. Aligned MWNTs are first functionalized with 3′-azido-3′-deoxythymidine and then placed in a solution of perfluorooctyl iodide in tetrachloroethane with the unmodified side of the aligned MWNT film in contact with the tetrachloroethane solution for the end-attachment of the perfluorooctyl chains. Water soluble MWNTs with temperature-responsive shells have been synthesized by grafting thermosensitive poly(N-isoproplyacrylamide) to the sidewalls of the MWNT via reversible addition and a fragmentation chain-transfer polymerization process [41]. A liquid–liquid extraction process has been reported for the extraction of water-soluble SWNTs into the organic phase [42]. The extraction utilizes electrostatic interactions between tetraoctylammonium bromide and the sidewall functional groups on the nanotubes. SWNTs can be cut into short segments by controlled oxidation by piranha (HNO_3/H_2SO_4) solutions [43]. At high temperatures, piranha attacks the damage sites generating vacancies in the nanotube sidewalls, and consuming the oxidized vacancies to yield short, cut nanotubes. SWNTs can be readily dispersed in alcohol using sodium hydroxide [44]. Several block copolymers with different functional side groups have been employed for the preparation of aqueous dispersions of SWNTs and MWNTs [45]. Cross-linking of an amphiphilic α-peptide to SWNTs improves the solubility of the nanotubes in water [46].

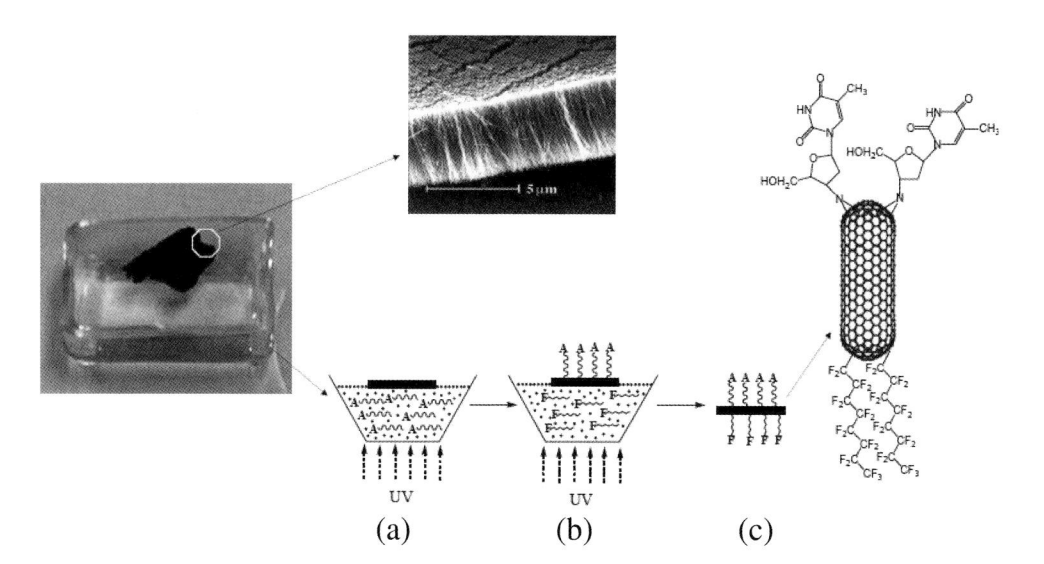

Fig. 2.10 A free-standing film of aligned MWNTs floating on the top surface of (a) an AZT solution in ethanol (2%) for UV irradiation at one side of the nanotube film for 1 h, and (b) a perfluorooctyl iodide solution in TCE (2%) for UV irradiation at the opposite side of the nanotube film for 1.5 h. (Reproduced from Ref. [40].)

Noncovalent functionalization of SWNTs can be carried out using the water-soluble porphyrin, meso-(tetrakis-4-sulfonatophenyl) porphine dihydrochloride [47]. Porphyrin–SWNT interaction is selective for the free base form, resulting in aqueous solutions that are stable for several weeks. SWNTs prepared by the HiPco process can be noncovalently modified with ionic pyrene and naphthalene derivatives to prepare water-soluble SWNT polyelectrolytes [48]. SWNTs can be exfoliated and functionalized by grinding them for minutes at room temperature with aryldiazonium salts in the presence of ionic liquids and K_2CO_3 (Fig. 2.11) [49]. Raman spectrum of the functionalized product shows a higher D-band, corresponding to functionalization of the SWNT (Fig. 2.11(c)) due to the increase in the number of sp^3-carbons that formed on the SWNTs after functionalization. Electrochemical functionalization of SWNTs at room temperature has been carried out [50]. An ionic liquid supported three-dimensional network SWNT electrode is employed in this method. Dispersion and noncovalent functionalization of SWNTs using self-assembling peptide amphiphiles has been demonstrated [51a]. Cyclic peptides are found to be useful in the solubilization and noncovalent functionalization of SWNTs in aqueous solutions [51b]. SWNTs with high solubility in water are obtained by a two-step process involving first their functionalization with phenyl groups, followed by sulfonation in oleum [52]. Aqueous solubilization of SWNTs is achieved by using polycyclic aromatic ammonium amphiphiles such as trimethyl-(2-oxo-2-phenylethyl)-ammonium bromide [53]. The interaction leading to the solubilization of SWNTs in the presence of *p*-terphenyl and anthracene has been studied using Raman and fluorescence spectroscopy [54]. SWNTs have also been solubilized using proteins [55].

SWNTs have been functionalized by diazonium salts produced *in situ* by the reaction of sodium nitrite with substituted aniline in a mixture of 96% sulfuric acid and ammonium persulfate [56]. Functionalization of SWNTs with sec-butyllithium and subsequent treatment with CO_2 leads to the formation of SWNTs functionalized with both alkyl and carboxyl groups [57]. Treatment of SWNTs with $SOCl_2$ affects the electrical and mechanical properties, thereby indicating a Fermi level shift in to the valence band [58]. SWNTs containing COOH groups can be chemically transformed to SWNTs containing amino groups [59]. Sidewall functionalization and subsequent solubilization of MWNTs has been achieved using microwave-assisted cycloaddition reaction of azomethine ylides [60]. Microwave nitrogen plasma has been employed for the functionalization of SWNTs [61]. MWNTs can be functionalized at pre-selected locations by initial irradiation with an ion beam followed by functionalization. This approach may be useful in the creation of hybrid devices [62]. Covalent modification of carbon nanotubes with imidazolium salts based ionic liquids has been reported [63]. Oxidation of SWNTs by ozone causes an irreversible increase in their electrical resistance [64]. Microwave-assisted functionalization of SWNTs in a mixture of nitric and sulfuric acids yields highly water-dispersible SWNTs [65].

DNA molecules can be used to prepare stabilized nanotube suspensions with liquid crystalline properties [66]. Carbon nanotubes have been functionalized with cleavable disulfide bonds for intracellular drug delivery of short-interfacing

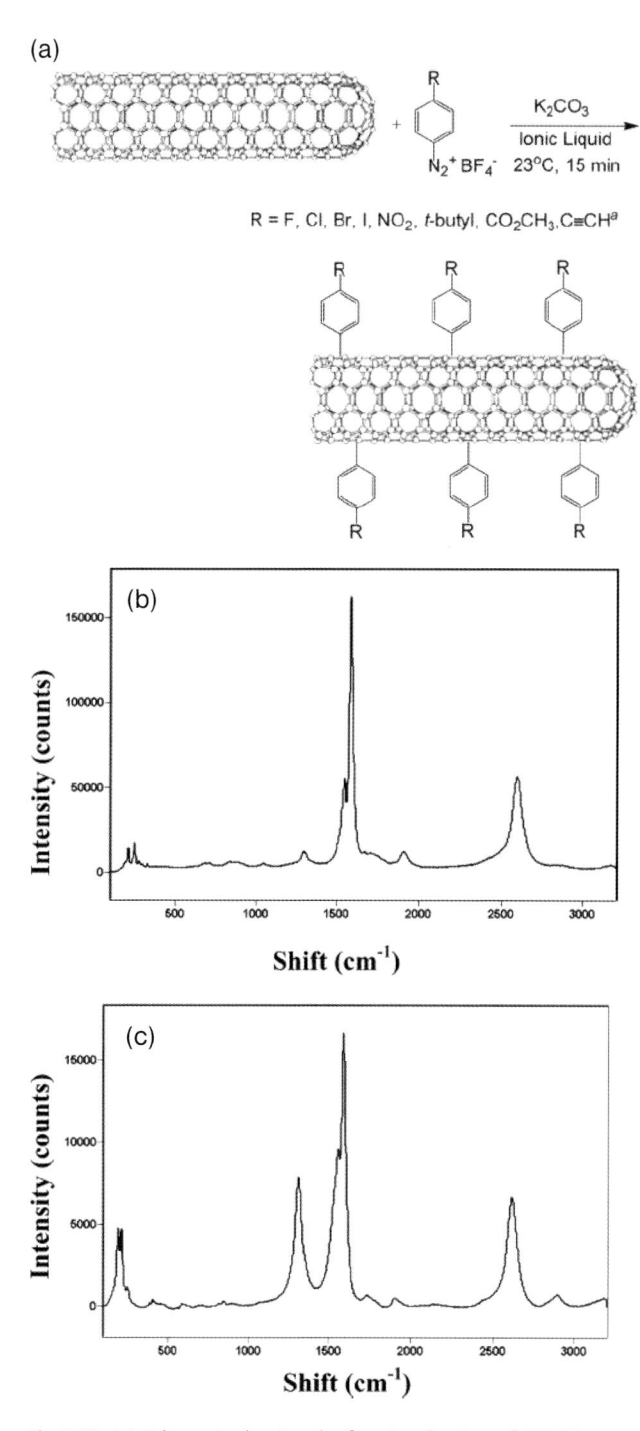

Fig. 2.11 (a) Schematic showing the functionalization of SWNTs in ionic liquid using a mortar and pestle and Raman spectra (633 nm, solid) of (b) purified SWNTs and (c) SWNTs functionalized with 4-chlorobenzenediazonium tetrafluoroborate in BMIPF$_6$. (Reproduced from Ref. [49].)

RNA (siRNA) and gene silencing [67]. A stable aqueous suspension of short SWNTs is first prepared by noncovalent adsorption of phospholipid molecules with poly(ethyleneglycol) (PL-PEG, MW of PEG = 2000) chains and terminal amine or maleimide groups. PL-PEG binds strongly to SWNTs via van der Waals and hydrophobic interactions between the two PL alkyl chains and the SWNT sidewalls, with the PEG chain extending into the aqueous phase to impart solubility in water. The siRNA and DNA molecules are coupled to the functionalized SWNTs through the incorporation of a disulfide bond. Blood-compatible carbon nanotubes for *in vivo* applications have been prepared through heparination [68]. DWNTs have been fluorinated to get $CF_{0.3}$ [69]. By controlling the crystallization conditions, polyethylene has been periodically crystallized over carbon nanotubes [70].

Various metal nanoparticles can be deposited on carbon nanotube surfaces. Thus, noble metal nanoparticles have been deposited onto the surface of SWNTs by electrodeposition [71]. Pt–carbon nanotube composites have been prepared in supercritical CO_2 using platinum(II) acetylacetonate as a precursor, for use as electrocatalysts [72]. Carbon nanotubes have been decorated with Ru nanoparticles using $RuCl_3$ in supercritical water [73]. SWNTs have been coated with fluorinated silica and the UV-visible-NIR spectrum shows a red shift in the first van Hove singularity transition, inferring that the SWNTs in SiO_2-coated SWNTs are in a more polarizable and inhomogeneous environment than that of surfactant solutions [74]. SWNTs have also been coated with cadmium chalcogenides [75] and Fe_3O_4 nanoparticles [76]. Photoactive donor–acceptor nanohybrids have been obtained by linking thioglycolic acid capped CdTe nanoparticles to SWNTs and MWNTs [77]. By reacting acid-treated carbon nanotubes with vapors of metal halides such as $TiCl_4$, $SiCl_4$, followed by reaction with water and calcination, chemically bonded ceramic oxide-coated carbon nanotubes can been prepared (Fig. 2.12) [78]. The thickness is controlled by the number of cycles of the reactions with the metal halide and H_2O.

Lattice defects and distortions in alkali metal iodides (MI, M = K, Cs) encapsulated in DWNTs have been examined [79]. Several lattice defects such as interstitials and vacancies, as well as crystal plane distortions like shearing and rotation, are found to be present. Nanopeapods of SWNTs encapsulating cesium metallofullerene (CsC_{60}) have been reported [80].

DNA-directed self-assembly of Au nanoparticles and SWNTs has been investigated [81]. Carbon nanotubes attached to single-stranded DNA are hybridized using complementary DNA-attached Au nanoparticles, leading to the formation of

Fig. 2.12 The ceramic coating process for (a) carbon nanotubes and (b) metal oxide nanowires; (c), (d) TEM images of the same sample after calcination at 350 °C; (e) TEM image of the nanotubes and nanowires after the removal of carbon nanotubes; and (f) TEM image of TiO_2-coated SWNTs after calcination at 350 °C. The arrow shows the oxide coating. (Reproduced from Ref. [78].)

2.2 Carbon Nanotubes | 59

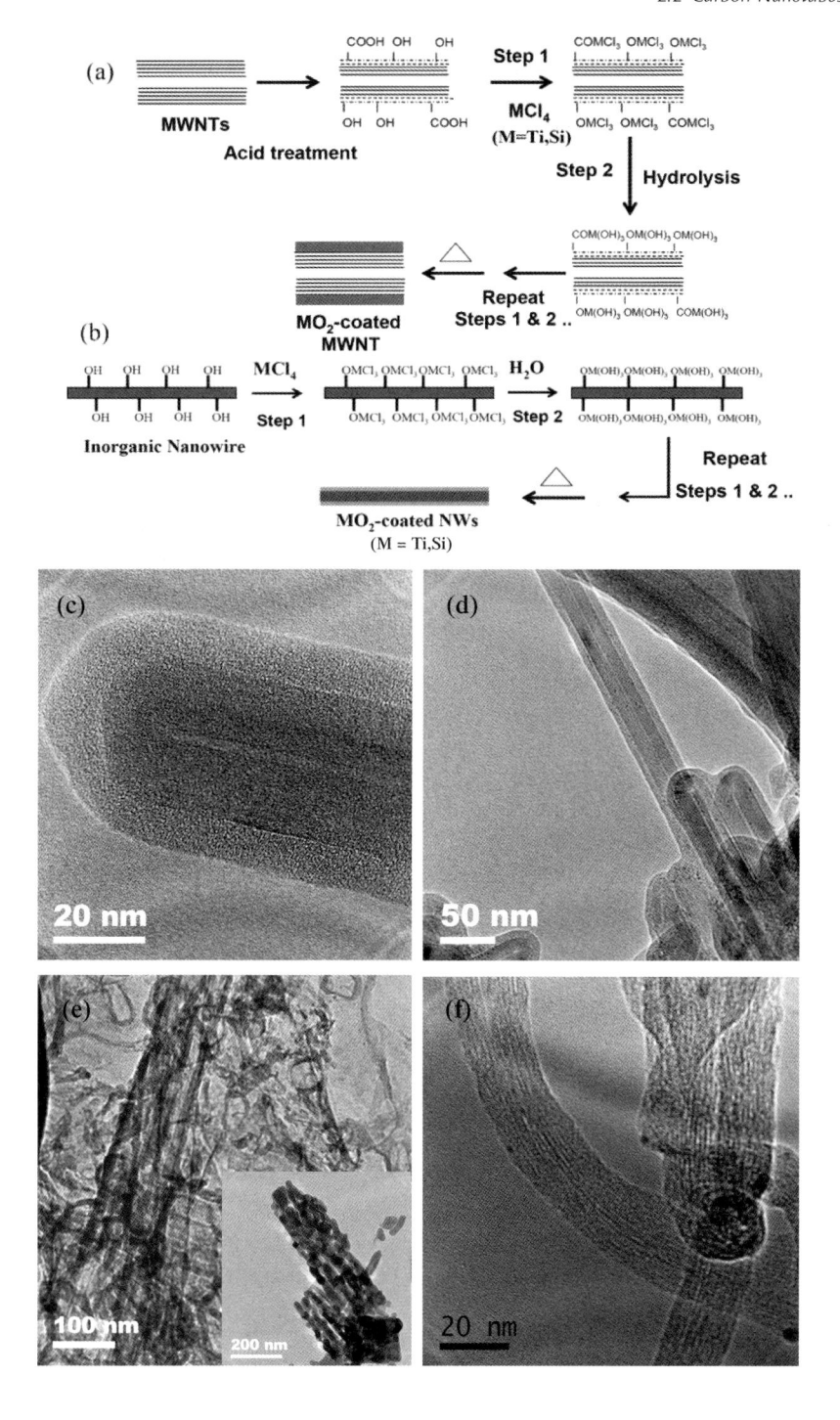

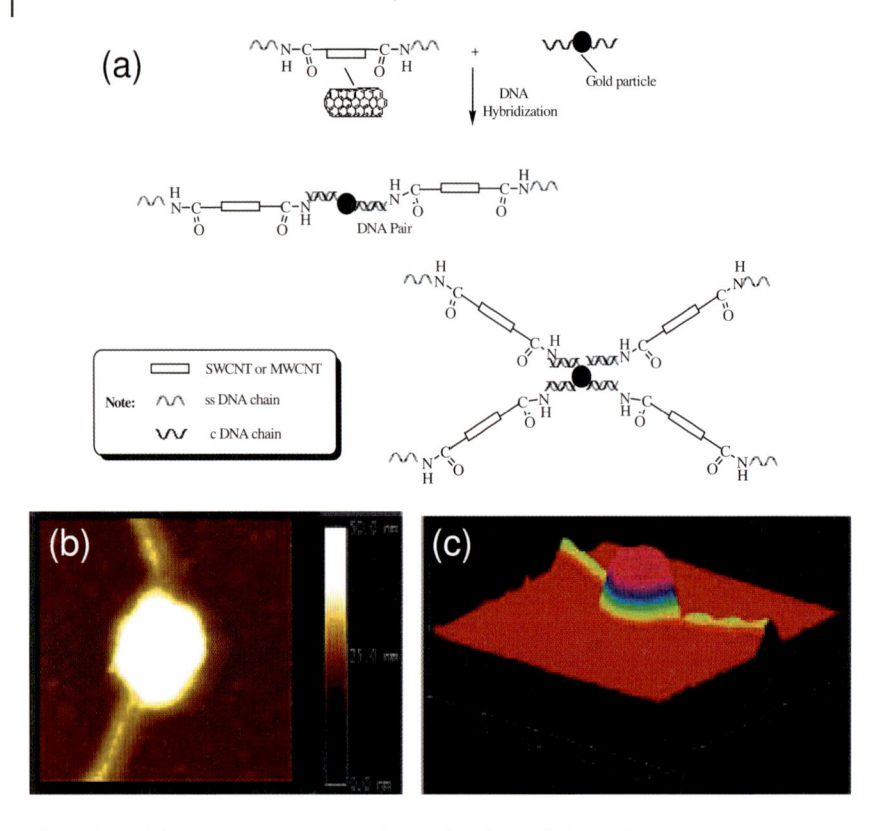

Fig. 2.13 (a) Schematic representation of procedure for DNA-directed self-assembling of multiple carbon nanotubes and nanoparticles. Typical AFM image of (b) the self-assembly of ssDNA–MWNTs and cDNA–Au nanoparticle (scanning area: 0.55 µm by 0.55 µm; vertical scale bar: 50 nm) and (c) the 3D surface plot of (a) with different color codes (scanning area: 0.85 µm by 0.85 µm). (Reproduced from Ref. [81].)

hybrid structures as seen in Fig. 2.13. Ordered arrays of nanotubes are obtained using liquid crystalline solutions [82]. The method involves the patterning of indanthrone disulfonate using dip-pen lithography and conversion of patterns into nanotubes by the thermal carbonization.

2.2.4
Properties and Applications

2.2.4.1 Optical, Electrical and Other Properties

Atomic-scale imaging with concurrent transport measurements has been carried out on individual MWNTs [83]. The study provides evidence for different breakdown sequences of the nanotube walls. It has been found that individual vertical

MWNTs have a large current-carrying capacity (7.27 mA) with a low resistance of 34.3 Ω, which is significantly higher than that of metallic SWNTs (∼25 μA) [84]. This behavior has been attributed to multichannel quasiballistic conducting behavior due to the participation of multiple walls in the electrical transport. Electronic transport spectroscopy on nanotubes in a magnetic field has been able to resolve the spin and orbital contributions to the magnetic moment [85].

Direct experimental evidence for exciton–phonon bound states in the photoluminescence excitation spectra of isolated SWNTs in aqueous suspensions has been obtained [86]. Femtosecond spectroscopy has yielded transient absorption spectra of SWNTs on optical excitation [87]. The excitonic origins of the spectrum and other features have been revealed by this study. Nanoscale optical imaging of excitons in SWNTs has been carried out by simultaneous near-field photoluminescence and Raman spectroscopy [88]. By combining electron diffraction and Rayleigh scattering spectroscopy the physical structure and optical transition energies of individual SWNTs have been determined [89]. The structural assignment of the SWNTs has been obtained by comparing the experimental electron diffraction and the simulated electron diffraction pattern as shown in Fig. 2.14(a)–(c). The Rayleigh spectra of the S_{33} and S_{44} transitions for (16,11) and (15,10) SWNTs are shown in Fig. 2.14(d) and (e). These two species have similar chiral angles (23.9° and 23.4°) but different diameters (1.83 and 1.71 nm). There is an upward shift of both the transition energies for the smaller diameter tube but little change in the ratio of the S_{44}/S_{33} transition energies. Intrinsic as well as extrinsic effects on the temperature dependence of photoluminescence of semiconducting SWNTs have been investigated [90]. A small energy shift of the emission line with temperature has been observed showing moderate chirality dependence. A clear chiral dependence due to strain on polymer wrapped nanotubes was observed when the effect of the temperature on the band gap was removed.

The visible luminescence of nanotubes and its dependence on functionalization have been examined [91]. A better dispersion and functionalization of the nanotubes results in more intense emissions. Optical properties of SWNTs can be manipulated without covalent functionalization, by wrapping them with fluorescently labeled polymer poly(vinylpyrrolidone) (PVP-1300) [92]. Individual SWNTs and DWNTs suspended in water using surfactants have been examined by using time-resolved photoluminescence spectroscopy [93]. This study has yielded information on the influence of tube chirality and diameter as well as the environment on the non-radioactive decay in smaller diameter tubes. Two-photon excitation spectroscopy has been used to measure exciton binding energies and band-gap energies in a range of individual species of semiconducting SWNTs [94]. These exciton binding energies are large and vary with the nanotube diameter whereas band-gap energies are significantly blue-shifted from values predicted by tight-binding calculations. Optical manipulation of SWNTs has been achieved using a linearly polarized infrared tweezers system [95]. Ultrafast relaxation of photoexcitations in semiconducting SWNTs has been investigated using polarized pump–probe photomodulation (with 10 fs time resolution) and

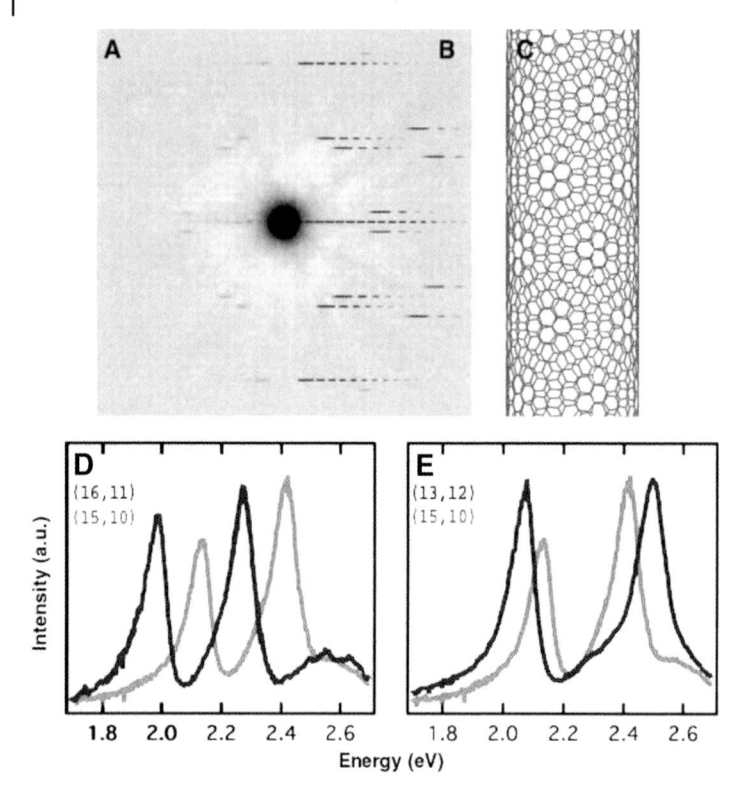

Fig. 2.14 Measurement of the carbon nanotube chiral indices. (A) Experimental TEM diffraction image from a semiconducting (16,11) SWNT and (B) the corresponding simulated diffraction pattern. (C) Model of the structure of the (16,11) SWNT. Measuring chirality and diameter effects in semiconducting SWNTs. (A) Comparison of the Rayleigh scattering spectra for the S_{33} and S_{44} transitions of the (16,11) and a (15,10) SWNTs. These SWNTs have chiral angles differing by only 0.5°, but diameters of 1.83 and 1.71 nm, respectively. We see an upward shift of both transition energies for the smaller diameter tube but little change in the ratio of the S_{44}/S_{33} transition energies. (B) Comparison of a (13,12) and a (15,10) SWNT. These SWNTs have nearly identical diameters, but chiral angles of 28.7° and 23.4°, respectively. In this case, the average transition energies for the S_{33} and S_{44} transitions in the two structures are almost identical. The ratios of the transition energies clearly differ and obey the family relation discussion in the text. (Reproduced from Ref. [89].)

cw polarized photoluminescence [96]. Near-infrared photoluminescence decay time measurements in different chirality SWNTs have provided excited-state carrier lifetimes [97]. Along with the results of resonant pump-and-probe spectroscopy it is shown that the carrier lifetime in the first excited state of semiconducting nanotubes exceeds 30 ps, which is one order of magnitude larger than the carrier dynamics observed in nanotube bundles. Near-edge X-ray absorption fine structure spectroscopy has been employed to study the degree of order and alignment in SWNTs [98].

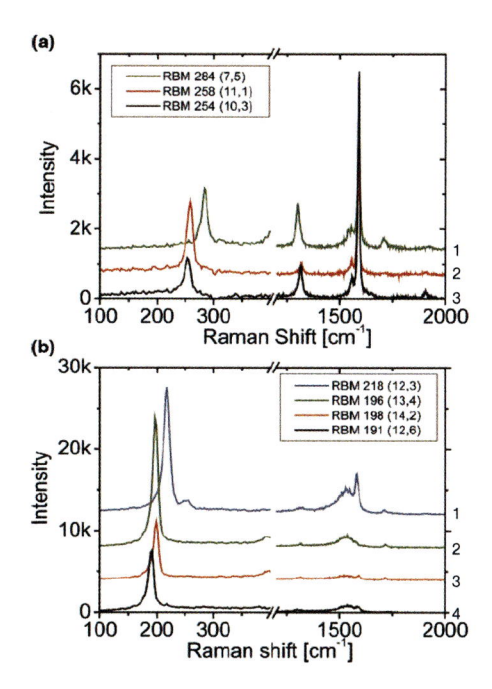

Fig. 2.15 (a) Raman spectra of isolated semiconducting HiPco SWNTs and (b) isolated metallic HiPco SWNTs and one small bundle (blue) spin-coated onto Si/SiO$_2$. (Reproduced from Ref. [99].)

Raman spectra of individual SWNTs obtained from various sources have been compared and the characteristic features of the G-modes of metallic and semiconducting tubes established (Fig. 2.15) [99]. The diameter dependence of tangential mode frequencies of free standing individual semiconducting SWNTs has been measured [100]. Raman spectroscopy of individual SWNT dispersion has been used to show that the Raman cross-sections of mod(n − m, 3) = 1 semiconducting SWNTs are disproportionately smaller than the nanotubes by an order of magnitude [101]. The temperature dependence of the Raman modes of individual SWNTs on SiO$_2$ substrates has been investigated [102]. Raman spectroscopy and imaging of ultra-long nanotubes consisting of both semiconducting and metallic tubes have been compared [103]. Resonance Raman spectra of semiconducting and metallic SWNTs in the 600–1100 cm^{-1} range have been investigated and band assignments have been made in these intermediate frequency modes [104]. A step-like dispersive behavior of these modes has been examined in some detail. Resonant Raman spectra of individual metallic and semiconducting SWNTs have been examined under uniaxial strain [105]. The D, G and G′ bands of the Raman modes have down shifted by upto 27, 15 and 40 cm^{-1} respectively. These relative strain induced shifts of D, G, and G′ bands vary significantly from nanotube to nanotube, implying that there is a strong chirality dependence of the relative shifts. The intermediate frequency modes of SWNTs and DWNTs have been investigated by Raman spectroscopy as well as *in situ* Raman spectroelectrochemistry [106]. Small-angle neutron scattering using high concentration method of SWNTs have been studied [107].

Atomic force microscopy (AFM) has been used to locally perturb and detect charge density in carbon nanotubes [108]. Such a measurement gives the local band gap of an intratube quantum-well structure. A large enhancement in the photoresponse of SWNT films is observed when the film is suspended in vacuum and this effect has been attributed to the heating of nanotubes networks [109]. Photoconductivity excitation spectra of individual semiconducting nanotubes incorporated as the channel of field-effect transistors have been measured [110]. The spectra show the presence of a weak sideband at ∼200 meV higher in energy than the main band arising from the second van Hove singularity. Electronic structure calculations indicate that the spectrum originates from the simultaneous excitation of an exciton (the main resonance) and a C–C bond stretching phonon (sideband).

Micrometer-long fully suspended SWNTs grown between metal contacts can be used as devices, which show well-defined characteristics over much wider energy ranges than nanotubes pinned on substrates (Fig. 2.16) [111]. Various low-temperature transport regimes in true-metallic, small- and large-bandgap semiconducting nanotubes are observed, including quantum states shell-filling, -splitting and -crossing in magnetic fields owing to the Aharonov–Bohm effect. The transport data show a correlation between the contact junction resistance and the various transport regimes in SWNT devices. Furthermore, electrical transport data can be used to probe the band structures of nanotubes, including nonlinear band dispersion.

A metal–semiconductor transition is induced in SWNTs by low-energy electron irradiation, due to the inhomogeneous electric fields arising from charging during electron irradiation [112]. Electronic properties of p-doped SWNTs have been studied by various methods including thermopower, optical reflectivity and Raman spectroscopy [113]. These results give the Fermi level downshifts due to doping. Electronic properties of fluorinated n-type CNTs prepared by CF_4 plasma fluorination and amino functionalizations have been studied [114]. The degree of amino functionalization is dependent on the degree of initial fluorination rather than the oxygen or carbon defects. Reaction at both ends of 1,2-diaminoethane was observed to increase with fluorine content. Back-gated SWNT devices have shown p-type semiconducting behavior of CF_4-functionalized SWNTs and n-type semiconducting behavior of amino-functionalized SWNTs.

Thermal properties of carbon nanotubes and their composite materials have attracted much attention [115]. The thermal conductivity of a single carbon nanotube has been measured [116]. Electrical characteristics of metallic nanotubes at high bias voltages between 300 and 800 K suggest the thermal conductivity to be 3500 W m^{-1} K^{-1} [117]. Thermal properties of individual MWNTs have also been investigated by using the phenomena of electrical breakdown in electrical transport measurements [118]. Thermal conductance and thermopower of an individual suspended SWNT have been measured [119].

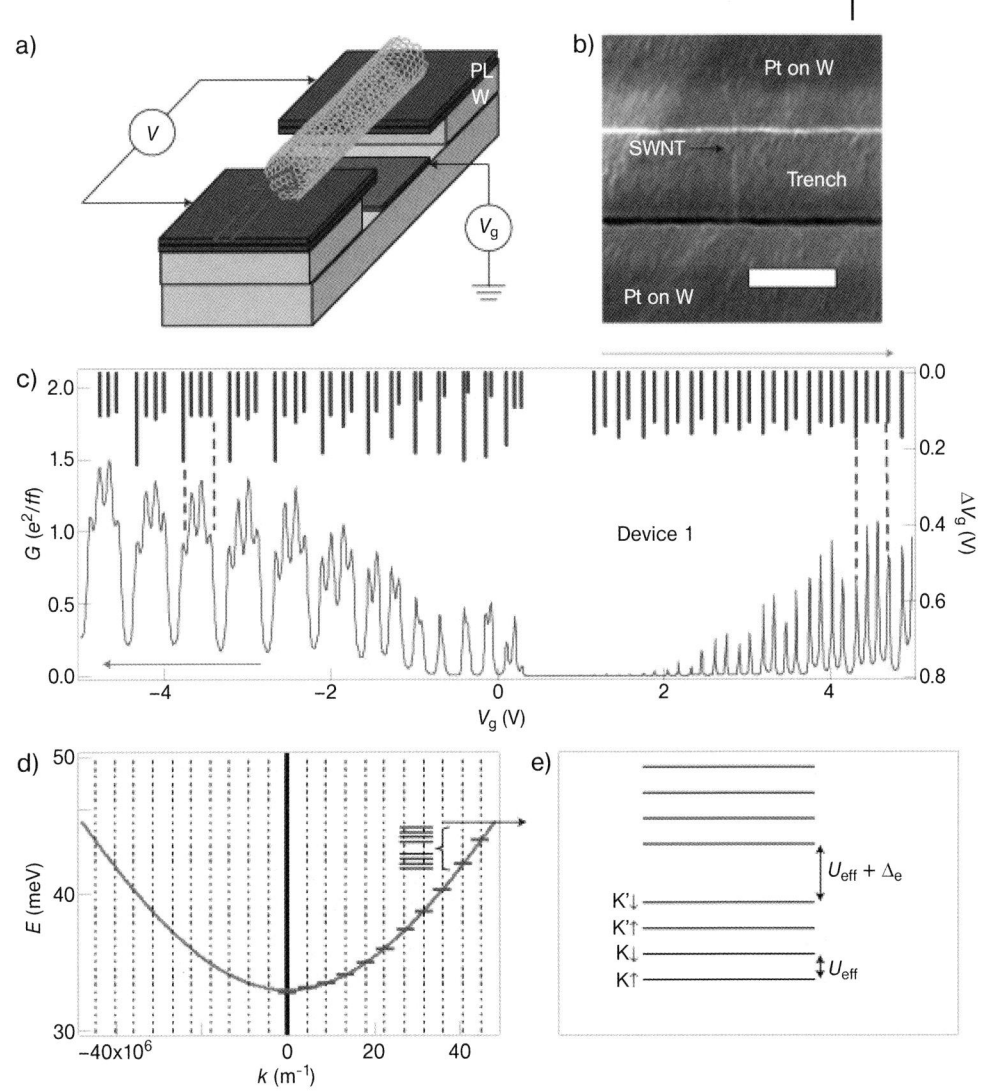

Fig. 2.16 (a) Schematic of the device with a local gate at the bottom of the trench. (b) A SEM image of the actual device, the scale bar is 0.5 μm. (c), G–V_g characteristics of a SWNT (Dev1) recorded at $T = 300$ mK under $V = 1$ mV and $B = 0$ T. The heights of the bars along the top axis correspond to peak spacings ΔV_g (right vertical axis) between conductance peaks along the V_g axis. A four-peak shell is highlighted (by dashed lines) for the p-channel (negative V_g side) and n-channel (positive V_g side). (d) Energy dispersion $E(k)$ for the valence band. Quantization of wavevectors along the length of the carbon nanotube ($k_n = n\pi/L$) is indicated by evenly spaced vertical lines. Each k_n gives rise to a shell (represented by the horizontal levels), each consisting of four states corresponding to the K and K' sub-bands and spin-up and spin-down. (e) Details of two of the shells in (c). Four electrons fill each shell with a charging energy of U_{eff}. To reach the next shell, in addition to U_{eff}, an energy difference between the quantized shells Δn needs to be paid. (Reproduced from Ref. [111].)

The torque created by an electric field gives rise to transient induced birefringence in SWNTs [120]. The pure orbital Kondo effect has been found in carbon nanotubes [121]. A magnetic field was used to tune the spin-polarized states into orbital degeneracy to find that the orbital quantum number is conserved during tunneling. An enhanced Kondo effect was also observed with a multiple splitting of the Kondo resonance at a finite field when the orbital and spin degeneracies were present simultaneously, as predicted by the so-called SU(4) symmetry.

2.2.4.2 Phase Transitions, Mechanical Properties, and Fluid Mechanics

Identification of the lyotropic phase in carbon nanotubes was reported earlier [122a]. An isotropic–nematic phase transition in dispersions of MWNTs has been observed recently [122b]. This phase transition is also observed in SWNTs dispersed in strong acid media [122c]. The critical concentration of the isotropic–nematic transition increases with acid strength. The hydrostatic pressure-induced structural transition that occurs in DWNTs has been studied by using constant-pressure molecular dynamics simulations [122d]. During the transition, the tube cross-section changes from circular to elliptical in shape, associated with the large reduction in the radial bulk modulus. Electrically controlled carbon nanotube switches have been fabricated using liquid crystal–nanotube dispersions [123].

SWNTs undergo superplastic deformation up to 280% of their original length at high temperature, as seen in Fig. 2.17 [124]. A new hot-drawing process for

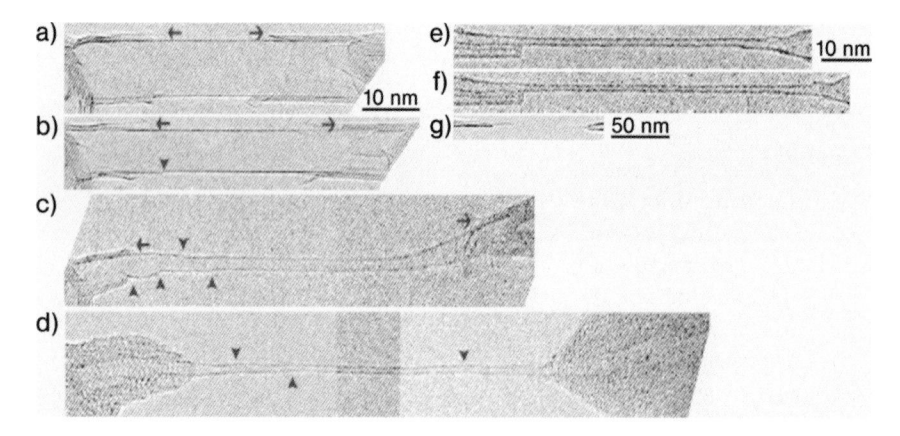

Fig. 2.17 *In situ* tensile elongation of individual single-walled carbon nanotubes viewed in a high-resolution transmission electron microscope. (a)–(d) Tensile elongation of a single-walled carbon nanotube (SWNT) under a constant bias of 2.3 V (images are all scaled to the same magnification). Arrowheads mark kinks; arrows indicate features at the ends of the nanotube that are almost unchanged during elongation. (e)–(g) Tensile elongation of a SWNT at room temperature without bias (images e and f are scaled to the same magnification). Initial length is 75 nm (e); length after elongation (f) and at the breaking point (g) is 84 nm; (g) low-magnification image of the SWNT breaking in the middle. (Reproduced from Ref. [124].)

treating wet-spun composite fibers made of single and multiwalled carbon nanotubes and poly(vinyl alcohol) (PVA) has been reported [125]. The inspiration for this process comes from textile technology and composite nanotube fibers prepared by this method exhibit a large strain-to-failure, and their toughness exceeds commercially available polyacramide fibers such as Kevlar or Twaron. Hot-drawn nanotube/PVA fibers hold great potential for a number of applications such as bulletproof vests, protective textiles, helmets, and so forth.

Mechanical properties of continuously spun fibers of carbon nanotubes from an aerogel, formed during synthesis by chemical vapor deposition, have been investigated [126]. The conditions were chosen to lie within the range satisfactory for continuous spinning, the catalyst concentration being varied within this range. Increasing proportions of SWNTs were found as the iron concentration was decreased; these conditions also produced fibers of the best strength and stiffness. The maximum tensile strength obtained was 1.46 GPa (equivalent to 0.70 N/tex, assuming a density of 2.1 g cm^{-2}). A carbon nanotube reinforced metal (Cu) matrix nanocomposite, fabricated by a novel fabrication process called "molecular-level mixing" showing extremely high strength, several times higher than the matrix, has been reported [127]. The spark plasma sintered CNTs/Cu nanocomposite powders show an extraordinary strengthening effect, higher than that of any other kind of reinforcement ever used for metal-matrix composites. Multifunctional conductive brushes made from carbon nanotube bristles grafted on fiber handles have been developed [128]. These brushes with nanotube bristles may be useful for several tasks such as cleaning nanoparticles from narrow spaces and devices, coating the inside of holes, selective chemical adsorption, movable electromechanical brush contacts and switches and selective removal of heavy metal ions.

The speed of fluid flow/transport through aligned carbon nanotube membranes approaches that through biological channels [129]. The extraordinarily fast flow makes the nanotube membrane a promising mimic of protein channels for transdermal drug delivery and selective chemical sensing. A method to fabricate integrated SWNT/microfluidic devices has been developed to study the mechanism of nanotube sensors [130]. This simple process could be used to prepare nanotube thin film transistors within the microfluidic channel, wherein various analytes can be introduced for sensing. This study has enabled us to understand the sensing mechanism of nanotube thin film field-effect transistors, where the sensing signal comes from the target molecules absorbed on or around the nanotubes. Superhydrophobicity on two-tier (micro and nanometer scales) rough surfaces fabricated by the controlled growth of aligned carbon nanotube arrays, coated with fluorocarbon, has been reported [131]. This study throws light on the wetting and hydrophobicity of water droplets on two-tier rough surfaces. Stable superhydrophobicity has been found in aligned carbon nanotubes coated with a ZnO thin film [132]. The wettability of the surface can be reversibly changed from superhydrophobicity to hydrophilicity by alternation of ultraviolet (UV) irradiation and dark storage. Electowetting studies of SWNTs with mercury show that open nanotubes wet reversibly and filling is also observed [133].

2.2.4.3 Energy Storage and Conversion

Nitrogen-containing carbon nanotubes with Pt nanoparticles dispersed on them exhibit high electrocatalytic activity with potential use in fuel cells [134]. SWNT-supported Pt nanoparticles exhibit selective electrocatalytic activity for O_2 reduction, suggesting possible use in H_2 and methanol fuel cells [135a]. A simple filtration method has been used to prepare superhydrophobic films of carbon nanotubes with catalytic Pt nanoparticles [135b]. The films exhibit enhanced electrocatalytic activity and improved mass transport within the film. Pt/Ru-carbon nanotube nanocomposites are effective in methanol fuel cells [135c]. Films of carbon nanotubes with ordered structures obtained from concentrated sulfuric acid can be used to fabricate electrodes for high power density supercapacitors [136]. Functionalized carbon nanotubes treated with pyrrole have been employed as electrode materials in supercapacitors [137]. A capacitance of 350 F g^{-1} with power and energy densities of 4.8 kW kg^{-1} and 3.3 kJ kg^{-1}, respectively, were obtained for these electrodes in 6 M KOH with a double layer capacity of 154 $\mu F\ cm^{-2}$. SWNTs modified with oxygenated carbon at their ends by acid treatment appear to exhibit favorable electrochemical properties [138]. Aligned MWNTs–RuO_2 nanocomposites have been investigated for use as supercapacitors [139]. A composite with psuedocapacitance properties and high electrical conductivity has been obtained by the pyrolysis of carbon nanotube/polyacrylonitrile blends [140]. Thin films formed by MWNTs with high packing density and alignment yield high power density supercapacitors [141]. The capacitance and conductance of SWNTs in the presence of different chemical vapors have been examined [142]. SWNT samples generally exhibit hydrogen adsorption capacities of less than 1 wt% at 25 °C for pressures up to 110 bar [143].

Splitting of water into hydrogen and oxygen occurs in the channels of SWNTs on irradiation with a flash light [144]. Stacked-cup carbon nanotubes have been used in photochemical solar cells [145]. SWNTs can be used as integrated building blocks for solar energy conversion [146]. In this study, SWNTs have been combined with porphyrin to prepared nanostructured devices. SWNT-CdSe nanocomposites may be useful for light harvesting and photoinduced charge transfer [147].

2.2.4.4 Chemical Sensors

An electrochemical sensing platform has been developed, based on the integration of redox mediators and carbon nanotubes in a polymer matrix, for the detection of β-nicotinamide adenine dinucleotide (NADH) [148]. DNA-decorated carbon nanotube transistors act as chemical sensors [149]. Single-strand DNA (ss-DNA) is also known to have high affinity for SWNTs due to a favorable π–π stacking interaction. The sensor consists of ss-DNA attached to a SWNT field effect transistor (FET), wherein the ss-DNA acts as a chemical recognition site and the SWNT FETs as an electronic read-out component. Responses of ss-DNA/SWNT FET differ in sign and magnitude for different gases as well as odors and can be tuned by choosing the base sequence of the ss-DNA. SWNTs functionalized with ss-DNA have been used for electrochemical detection and sensing of a low concentration of dopamine in the presence of excess ascorbic acid [150]. Water-soluble SWNTs can be used to study the reaction with H_2O_2 and suggest a possible use in H_2O_2 optical sensors [151]. CO_2 detection using SWNTs and a microelectromechanical system has been demonstrated [152]. Room temperature hydrogen sensing has been carried out using SWNT films [153].

2.2.5
Biochemical and Biomedical Aspects

Reversible biochemical switching of ion transport through aligned carbon nanotube membranes has been observed and found to mimic protein ion channels [154]. SWNT forests have been used for protein immunosensors [155]. Here, antibodies are attached to the carboxylate end of the nanotube forests. SWNT polyelectrolyte composite films provide a biochemical platform for plastic implants [156]. The thin films are mechanically compatible with tissues under constant flexural as well as shear stresses and biocompatible with neuronal cell cultures.

Osteoblast proliferation and bone growth based on chemically functionalized SWNT scaffolds has been investigated [157, 158]. SWNTs encapsulated in DNA oligonucleotide can be used as markers in live cells and are resistant to photobleaching for up to 3 months [159]. The DNA-encapsulated SWNT markers in cells have been analyzed using Raman and fluorescence spectroscopy. Functionalized carbon nanotubes have been used for targeted delivery of amphotericin B, an antibiotic effective in the treatment of chronic fungal infections [160]. An *in vivo* fluorescence detector for glucose based on an SWNT optical sensor has been developed [161]. Amperometric biosensors based on redox polymer–carbon nanotube–enzyme composites have been demonstrated [162].

Band gap fluorescence modulation of SWNTs can be used to detect DNA [163]. SWNTs can also be used for the optical detection of conformational polymorphism in DNA [164]. A red shift in the SWNT fluorescence is observed due to the change in the dielectric environment of the DNA-coated SWNTs on conformation change of the DNA (Fig. 2.18). An electrochemical sensor employing aligned MWNTs has been fabricated for the detection of the cholesterol level in blood [165]. An enzyme–carbon nanotube-based electrochemical sensor has been demonstrated for potential use in the detection of V-type nerve gases [166].

Cytotoxicity of SWNTs and MWNTs has been examined in comparison with C_{60} and quartz [167]. Significant cytotoxicity of SWNTs was observed in alveolar macrophage after 6 h exposure *in vitro*. The cytotoxicity increased by as much as $\sim 35\%$ when the dosage of SWNTs increased by 11.30 µg cm^{-2}. No significant toxicity was observed for C_{60} up to a dose of 226.00 µg cm^{-2}. The cytotoxicity apparently follows a sequence order on a mass basis: SWNTs > MWNT > quartz > C_{60}. SWNTs significantly impaired phagocytosis of AM at the low dose of 0.38 µg cm^{-2}, whereas MWNT and C_{60} induced injury only at the high dose of 3.06 µg cm^{-2}. The macrophages exposed to doses of SWNTs or MWNTs of 3.06 µg cm^{-2} showed characteristic features of necrosis and degener-

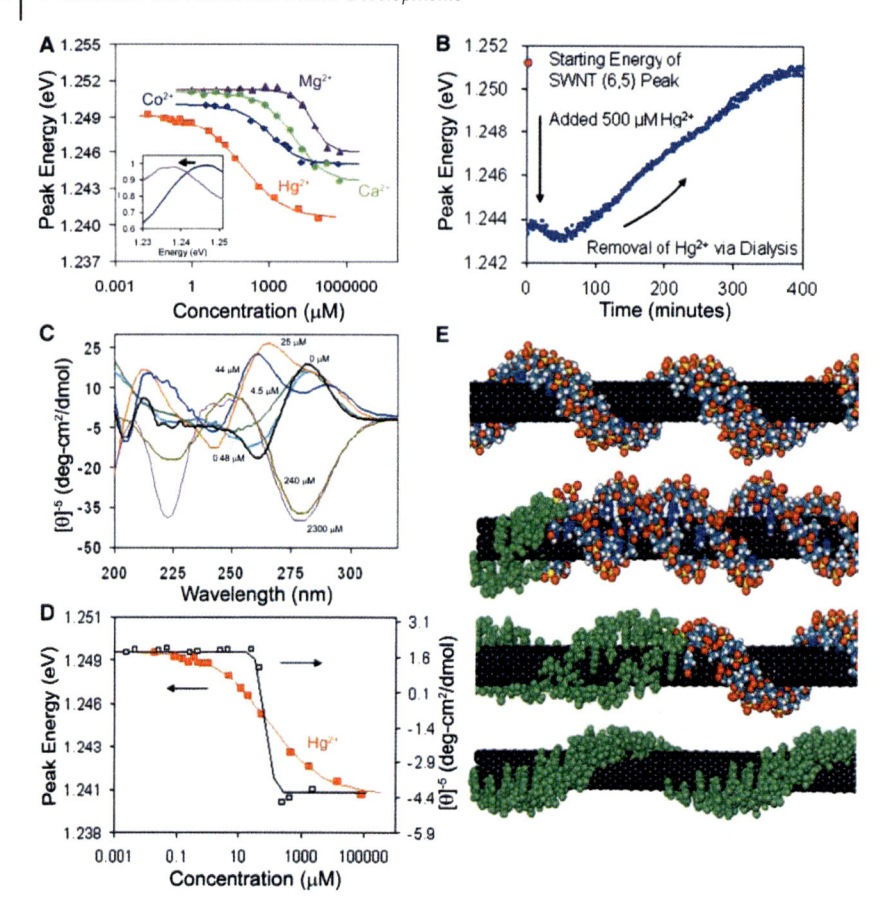

Fig. 2.18 (A) Concentration-dependent fluorescence response of the DNA-encapsulated (6,5) nanotube to divalent chloride counterions. The inset shows the (6,5) fluorescence band at starting (blue) and final (pink) concentrations of Hg^{2+}. (B) Fluorescence energy of DNA–SWNTs inside a dialysis membrane upon removal of Hg^{2+} during a period of 7 h by dialysis. (C) Circular dichroism spectra of unbound d(GT)15 DNA at various concentrations of Hg^{2+}. (D) DNA–SWNT emission energy plotted versus Hg^{2+} concentration (red curve) and the ellipticity of the 285 nm peak obtained via circular dichroism measurements upon addition of mercuric chloride to the same oligonucleotide (black curve). Arrows point to the axis used for the corresponding curve. (E) Illustration of DNA undergoing a conformational transition from the B form (top) to the Z form (bottom) on a carbon nanotube. (Reproduced from Ref. [164].)

ation. Carbon nanotubes modified with suitable biomimetic polymers have been found to be biocompatible while the uncoated carbon nanotubes lead to cell death [168].

Several proteins absorb on the sidewalls of acid-treated SWNTs leading to the formation of noncovalent protein–nanotube conjugates [169]. The proteins can be transported into various mammalian cells with the nanotube acting as the transporter. SWNT FETs can be used for biosensing with the use of artificial oligonucleotide (aptamers) [170]. The small size of the aptamers (1–2 nm) makes it possible for the aptamer–protein binding event to occur inside the electrical double layer in millimolar salt concentrations.

2.2.6
Nanocomposites

Polymer composites containing different amounts of MWNTs have been prepared using solution dispersion and melt-shear mixing [171]. The wrapping of nanotubes generally occurs when MWNTs are dissolved in organic solvents along with the polymers. MWNTs have been functionalized with alkylbenzoic acids and dispersed in ethylene glycol [172]. Subsequently, *in situ* polycondensation of ethylene glycol and terephthalic acid in the presence of functionalized MWNTs yields poly(ethylene terephthalate)–MWNT nanocomposites. Polyurea-functionalized MWNTs have been synthesized and characterized [173]. Poly(ethyleneimine) functionalized SWNTs have been synthesized [174]. They act as substrates for neuronal growth. Water-soluble graft copolymers of SWNTs have been prepared by covalent attachment [175]. Anisotropic films of SWNTs and amine polymers have been prepared and their electrical properties investigated [176]. Composite films of MWNTs with polyaniline have been grown electrochemically from aqueous acidic solutions and their electrochemical capacitance has been examined [177]. Aligned SWNT composite films are prepared by attaching Au nanoparticles to SWNTs followed by compression in a Langmuir–Blodgett trough [178]. Electrostatic matching has been used to coat carbon nanotubes with layers of oppositely charged polyelectrolytes [179]. The composite consists of a MWNT core coated with four functionalized layers that successively comprise protein-encapsulated iron oxide nanoparticles, the tetravalent biotin-binding protein, streptavidin, 24-base three-stranded biotin-terminated oligonucleotide duplexes, and oligonucleotidecoupled Au nanoparticles.

A novel hybrid material based on carbon nanotube–polyaniline–nickel hexacyanoferrate nanocomposite exhibiting good ion exchange capacity, stability and selectivity for Cs ions has been prepared [180]. Carbon nanotube–metal nitride composites with enhanced electrical properties have been fabricated [181]. Aligned polymer composites with controlled nanotube dispersion and alignment can be fabricated by a two-step process, involving the CVD growth of aligned MWNts and followed by *in situ* polymerization [182]. Wet spinning is used to obtain polymer-free carbon nanotube fibers using solutions consisting of nanotubes, surfactant and water [183]. SWNT–nylon composites fibers with improved mechanical properties have been prepared by a continuous spinning process [184]. Nanotube brushes consisting of N-doped MWNTs and polystyrene have been synthesized by covalent attachment using grafting techniques and subsequent polymerization [185].

The temperature dependence of the electrical conductivity of poly(3-octylthiophene)–SWNT composite films has been measured and the results described by a fluctuation induced tunneling model [186]. A multi-functional chemical vapor sensor has been fabricated using aligned MWNTs–polymer composites [187]. Near-infrared photovoltaic devices can be fabricated using SWNT–polymer composites as an active layer [188]. SWNTs have been incorporated in polythiophene films for potential use in photoconversion [189]. SWNTs with porphyrin polymer wrapping exhibit long-lived intracomplex charge separation for possible use as photoactive materials [190]. Optical properties of polyaniline–MWNT composite films prepared from solution have been investigated [191]. Photo-induced mechanical actuation has been observed in polymer–nanotube composites upon IR irradiation [192]. Transparent conducting SWNT films have been deposited on various substrates using a transfer printing technique [193]. The possible use of SWNT films as hole-conducting transparent electrodes in polymer–fullerene photovoltaic devices has been examined [194]. Electrical transport and chemical sensing properties of individual polypyrrole–SWNT nanocables have been investigated [195].

Nanomechanical properties of SiO_2-coated MWNT–poly(methyl methacrylate) (PMMA) composites have been studied by using the nanoindentation technique [196]. A significant increase in the hardness (by a factor of 2) and Young's modulus (by a factor of 3) is observed upon incorporation of 4 wt% of SiO_2-coated MWNTs to PMMA. A large improvement in damping without sacrificing the mechanical strength and stiffness of the polymer has been achieved by the incorporation of SWNTs [197]. Enhanced strain response at reduced electrical fields is observed on incorporation of SWNTs into electrostrictive poly(vinylidene fluoride–trifluoroethylene–chlororfluroethylene) [198]. Thermal conductivity and interface resistance of SWNT–epoxy composites have been studied [199]. The fabrication of aligned carbon nanotube–polymer composites with improved thermal conductivity has been described [200]. Nanotube composites are also fabricated in microelectromechanical systems. Mechanical tests reveal an increase in the composite modulus by a factor of 20 compared with pristine nanotubes [201]. The effects of sidewall functionalization on the dispersion and interfacial properties of fluorinated SWNT–polyethylene composites have been studied [202]. High strength actuators based on carbon nanotubes-reinforced polyaniline fibers have been prepared for potential use as artificial muscles [203].

Both MWNTs and SWNTs surpass nanoclays as effective flame-retardant additives in polymer nanocomposites if they form a jammed network structure in the polymer matrix [204]. Carbon nanotube–nylon composites can be used for the fabrication of a microcatheter. Their thrombogenicity and blood coagulation have been examined [205].

2.2.7
Transistors and Devices

Carbon nanotubes have been used as cathodes and their electron emission properties have been investigated by several workers [206]. Recently, a microwave diode that uses carbon nanotubes as a cold cathode electron source operating at high frequency and high current densities has been fabricated. Flexible 3-D arrays of carbon nanotube field emitters have been prepared by direct microwave irradiation on organic polymer substrates [207]. Electrical and field emission properties of chemically anchored SWNT patterns have been measured [208]. The patterned SWNT layers have high-density multilayer structures and excellent surface adhesion due to direct bonding to the substrates, resulting in high electrical conductivity.

Individual SWNTs can form ideal p–n junction diodes [209]. Under illumination, they show significant power conversion efficiencies. Such photovoltaic effects may be useful in electronic materials development. Vertically aligned carbon-nanotube arrays fabricated in thin-film anodic aluminium oxide (AAO) templates on silicon wafers exhibit Schottky behaviour at room temperature [210]. This structure can be a useful cornerstone in the fabrication of nanotransistors operating at room temperature.

Field effect transistors have been fabricated using high-density SWNT thin films and the effects of ionic surfactant adsorption on their surfaces examined [211]. SWNT device characteristics can be tuned to be sensitive to cationic or anionic surfactants by tailoring the surface properties of the SiO_2 substrates. The behavior of such devices is related to the surface charge densities around the SWNTs in aqueous solutions. Adsorption of ionic surfactants on the surface modulates the device characteristics such as the conductance. This effect could be useful in designing chemical and biological sensors. A floating-potential dielectrophoresis method has been used to achieve controlled alignment of individual semiconducting or metallic SWNTs between two electrical contacts with high repeatability [212]. This method is useful for fabricating single SWNT transistors and other devices. Field effect transistors have been fabricated at pre-selected locations by using chemically functionalized carbon nanotubes [213]. An integrated logic circuit has been assembled using single carbon nanotubes [214]. Quantum supercurrent carbon nanotubes FET fabricated with superconducting leads have been investigated [215].

There have been several studies on the CNT-based devices. Guided growth of large-scale, horizontally aligned arrays of SWNTs on single crystal quartz substrates and their use in thin-film transistors have been reported [216]. High performance n-type carbon nanotube FETs with chemically doped contacts have been fabricated [217]. These short channel (~80 nm) SWNT-FETs with potassium-doped source and drain regions and high-κ gate dielectrics act as n-MOSFET-like devices with high on-currents due to chemically suppressed Schottky barriers at the contacts, a sub-threshold swing of 70 mV decade^{-1}, negligible ambipolar conduction and high on/off ratios upto 10^6 at a bias voltage of 0.5 V. Polymer electrolyte gating of carbon nanotube network transistors showing high gate efficiencies, low operating voltages and absence of hysteresis have been reported [218]. Carbon nanotube FETs with sub-20 nm short channels and on/off current ratios of >10^6 have been demonstrated [219]. These nanotube transistors display on-currents in excess of 15 μm for a drain-source bias of 0.4 V. Self-aligned CNT transistors with charge transfer p-doping, which utilize one-electron oxidizing molecules, have been reported [220]. Transparent and flexible carbon nanotube transistors, where both the bottom gate and the conducting channel are CNT networks of different densities, with Parylene N as the gate insulator, have been described [221]. Device mobilities of 1 cm^2 V^{-1} s^{-1} and on/off ratios of 100 are obtained, with the latter influenced by the properties of the insulating layer. Transparent and flexible FETs using SWNT films exhibiting mobilities of 0.5 cm^2 V^{-1} s^{-1} with an on/off ratio of $\sim 10^4$ have been described [222]. The fabrication of a transparent flexible organic thin film transistor using printed SWNT electrodes and pentacene has also been reported [223]. Carbon nanotube FETs have been designed wherein the nanotubes are grown by a spatially selective guided growth process [224].

Improvement in the performance of the CNT-FETs has been achieved by chemical optimization/tuning of the nanotube/substrate and nanotube/electrode interfaces [225]. This method of selective replacement of individual SWNTs by a patterned aminosilane monolayer was used for the fabrication of self-assembled carbon nanotube transistors. The aminosilane monolayer reactivity can be used to improve carrier injection and the doping level of the SWNTs. These chemical treatments reduce the Schottky barrier height at the nanotube/metal interface down to that of an almost ohmic contact. Such self-assembled FETs open new prospects for gas sensors as demonstrated for the 20 ppb of triethylamine. The use of nanoscopic 3 D $\sigma-\pi$ self-assembled superlattices function as exceptionally good organic nanodielectrics for n- and p-channel low-voltage SWNT thin film transistors and complementary logic gates [226].

C_{60} peapods exhibit single-electron transistor properties [227]. The current behavior induced by the applied gate and source-drain voltages shows that the gate-dependent conductance is enhanced at negative gate voltages and is suppressed and oscillates at positive gate voltages. This behavior is ascribed to the modulation of the density of states by the insertion of C_{60} inside SWNT. SWNT-FETs are able to sense the changes in conformation when the molecules are switched [228]. The unique feature explored here is how to populate the surface of the CNTs with functional molecules that can be toggled back-and-forth between different molecular conformations.

Carbon nanotube–molecule–silicon junctions can be fabricated by covalently attaching individual SWNTs to Si surfaces via orthogonally functionalized oligo(phenylene ethynylene) (OPE) aryldiazonium salts without the use of the CVD growth process [229]. Novel electrical switching behaviour and logic in carbon nanotube Y-junctions has been reported recently [230]. The mutual interaction of the electron currents in the three branches of the Y-junctions is shown to be the basis for the new logic device, which works without the use of an external gate. These properties may be useful for novel transistor technologies. Three-terminal transistor-like operations of Y-junction CNT devices showing differential current amplification have been reported [231]. Y-juction SWNT-based FETs have been fabricated and the device exhibits an on/off ratio of 10^5 with a low off-state leakage current of $\sim 10^{-13}$ A [232]. Suspended SWNT quantum wires with two electrostatic gates per device exhibit little hysteresis related to environmental factors and act as cleaner Fabry–Perot interferometers or single-electron transistors [233].

Double-walled FETs show an ambipolar to unipolar transition by the adsorption of oxygen molecules [234]. The lowest unoccupied molecular state of the adsorbed oxygen molecules, which is around the midgap of the carbon nanotube, could suppress the electron channel formation and consequently induce unipolar transport behavior. Band engineering of carbon nanotube FETs can be carried out by exposing the center part or the contact of the nanotube devices to oxidizing or reducing gases [235]. A good control over the threshold voltage and sub-threshold swing have been achieved by the so-called selected area chemical gating. A printing process for high-resolution transfer of all the components for organic electronic devices on plastic substrates has been developed using carbon nanotube thin-film transistors [236]. Electrochemical actuators based on sheets of SWNTs have been known for some time and shown to generate stresses higher than that of natural muscle and excellent strains at low applied voltages. MWNT mats in which the nanotubes are randomly oriented within the plane of the film, have been shown to exhibit actuation [237]. Nanoscale torsional actuators consisting of metal mirrors bonded to oriented MWNTs have been fabricated [238].

An ambipolar random telegraph signal has been observed in ambipolar SWNT-FETs [239]. The ambipolar RTS can be used to extract the small band gap of the SWNT. The possibility of using CNTs as potential devices to improve neural signal transfer while supporting dendrite elongation and cell adhesion has been demonstrated [240]. These results strongly suggest that the growth of neuronal circuits on a CNT grid is accompanied by a significant increase in the network activity. The increase in the efficacy of the neural signal transmission may be related to the specific properties of CNTs. An individual CNT partially filled with liquid gallium can act as a miniaturized thermometer or temperature sensor and also as an electrical switch [241]. ZnO nanoparticle–MWNT nanohybrids exhibit ultrafast nonlinear optical switching [242].

2.3
Inorganic Nanotubes

2.3.1
Synthesis

The synthesis and characterization of nanotubes of inorganic materials including elements, oxides and chalcogenides have been reported extensively in the recent literature [243]. Several inorganic nanotubes have been synthesized during the last year by different strategies. Large-scale synthesis of Se nanotubes has been carried out in the presence of CTAB [244]. Thus, a low-temperature route for synthesizing highly oriented ZnO nanotubes/nanorod arrays has been reported [245]. In this work, a radio frequency magnetron-sputtering technique was used to prepare ZnO-film-coated substrates for subsequent growth of the oriented nanostructures. Controllable syntheses of SiO_2 nanotubes with dome-shaped interiors have been prepared by pyrolysis of silanes over Au catalysts [246]. High aspect-ratio, self-organized nanotubes of TiO_2 are obtained by anodization of titanium [247a]. These self-organized porous structures consist of pore arrays with a uniform pore diameter of ~100 nm and an average spacing of 150 nm. The pore mouths are open on the top of the layer while on the bottom of the structure the tubes are closed by the presence of an about 50 nm thick barrier of TiO_2. Electrochemical etching of titanium under potentiostatic conditions in fluorinated dimethyl sulfoxide and ethanol (1:1) under a range of anodizing conditions gives rise to ordered TiO_2 nanotube arrays [247b]. TiO_2-B nanotubes can be prepared by a hydrothermal method [248]. Lithium is readily intercalated into the TiO_2-B nanotubes up to a composition of $Li_{0.98}TiO_2$, compared with $Li_{0.91}TiO_2$ for the corresponding nanowires. Intercalation of alkali metals into titanate nanotubes has also been investigated [249]. Highly crystalline TiO_2 nanotubes have been synthesized by hydrogen peroxide treatment of low crystalline TiO_2 nanotubes prepared by hydrothermal methods [250]. TiO_2 nanotubes with rutile structure have been prepared by using carbon nanotubes as templates [251]. Anatase nanotubes can be nitrogen doped by ion-beam implantation [252]. RuO_2 nanotubes have been synthesized by the thermal decomposition of $Ru_3(CO)_{12}$ inside anodic alumina membranes [253].

Transition metal oxide nanotubes have been prepared in water using iced lipid nanotubes as the template [254]. Self-assembled cholesterol derivatives act as a template as well as a catalyst for the sol–gel polymerization of inorganic precursors to give rise to double-walled tubular structures of transition metal oxides [255]. Hydrothermal synthesis of single-crystalline α-Fe_2O_3 nanotubes has been accomplished [256]. Nanotubes of single crystalline Fe_3O_4 have been prepared by wet-etching of the MgO inner cores of MgO/Fe_3O_4 core–shell nanowires [257]. Cerium oxide nanotubes can be prepared by the controlled annealing of the as-formed $Ce(OH)_3$ nanotubes [258].

Boron nitride nanotubes can be grown directly on substrates at 873 K by a plasma-enhanced laser-deposition technique [259]. Nanotubes and onions of GaS and GaSe have been generated through laser and thermally induced exfoliation of the bulk powders [260]. In Fig. 2.19, electron microscope images of nanoscrolls and nanotubes of GaSe obtained by thermal exfoliation are shown. Single-wall nanotubes of $SbPS_{4-x}Se_x$ ($0 \leq x \leq 3$) with tunable bandgap have been synthesized [261]. GaP nanotubes with zinc blende structure have been obtained by the VLS growth [262]. Open-ended gold nanotube arrays have been obtained by the electrochemical deposition of Au onto an array of nickel nanorod templates, followed by selective removal of the templates [263]. Free-standing, electroconductive nanotubular sheets of indium tin oxide with different In/Sn ratios have been fabricated by using cellulose as the template [264].

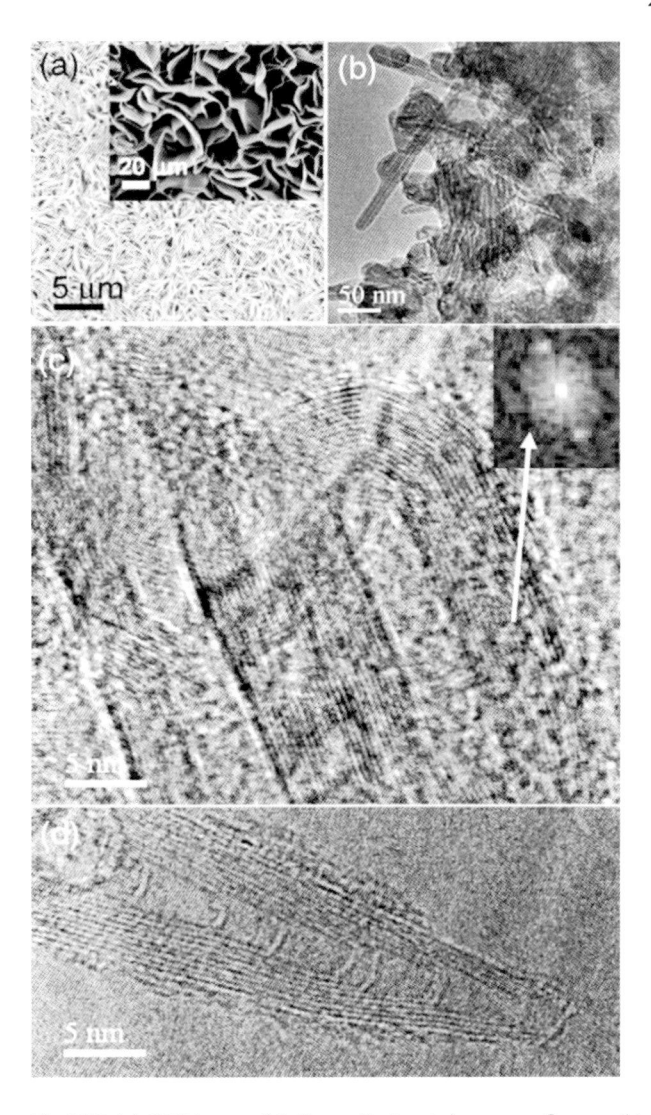

Fig. 2.19 (a) SEM image of GaSe scrolls. Inset shows nanoflowers. (b) TEM image of GaSe nanotubes obtained by thermal treatment. (c), (d) HREM images of GaSe nanotubes. (Reproduced from Ref. [260].)

2.3.2
Solubilization and Functionalization

Boron nitride nanotubes can be dissolved in organic solvents by wrapping them with a polymer [265]. BN nanotubes have been purified by polymer wrapping using a conjugated polymer [266]. Such a solution does not destroy the intrinsic properties due to the noncovalent functionalization of the BN nanotubes. Functionalization and solubilization of BN nanotubes have been carried out by making use of the interaction of amino groups with the nanotube surface [267]. Solubilization here is based on the interactions of the amino groups in oligomeric diamine-terminated poly(ethyleneglycol) with the BN nanotube surface. Covalent functionalizaion of BN nanotubes has been accomplished by a reaction between the COCl group of stearoylchloride and the amino groups on the BN nanotubes [268]. Chemical peeling and branching of BN nanotubes in dimethyl sulfoxide (DMSO) under solvothermal conditions has been observed [269]. The cycloaddition of DMSO to BN nanotubes has been suggested to weaken the B–N covalent bond and the subsequent peeling of the BN nanotubes. BN nanotubes have been functionalized with SnO_2 nanoparticles [270]. BN nanotubes have been fluorinated through the introduction of F atoms at the stage of the nanotube growth for possible applications in nanoelectronics [271]. Inorganic nanotubes have been integrated with microfluidic systems to create devices for single DNA molecule sensing [272]. A schematic and an electron microscope image of the device is shown in Fig. 2.20. Transient changes in ionic current are observed on DNA

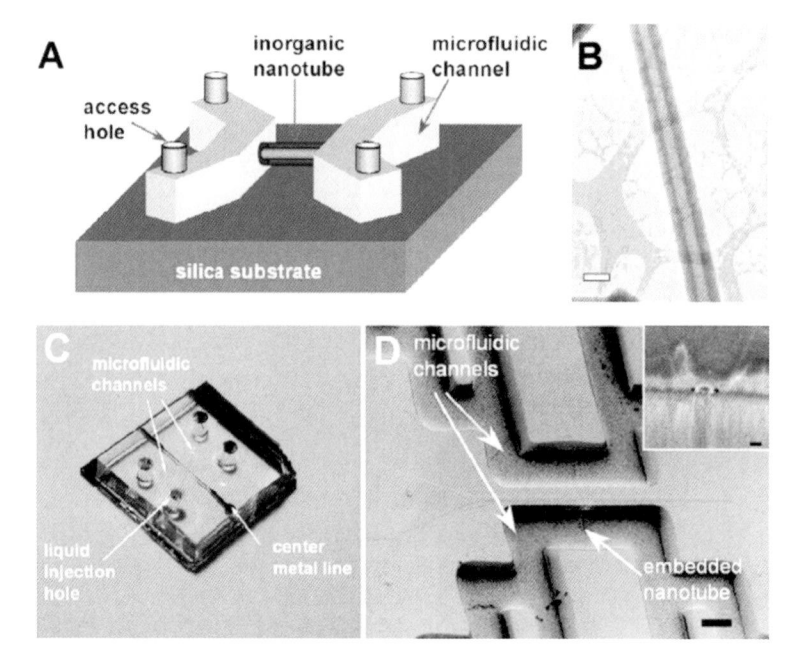

Fig. 2.20 Inorganic nanotube nanofluidic device: (A) Schematic of device structure features a single nanotube bridging two microfluidic channels to form a nanofluidic system. (B) Scanning electron micrograph of the nanofluidic device before cover bonding. Scale bar represents 10 μm. Inset shows cross-section view of the nanotube embedded between two silicon dioxide layers. Scale bar represents 100 nm. (C) A fully packaged nanofluidic device. (D) Scanning electron micrograph of the nanofluidic device before cover bonding. Scale bar represents 10 μm. Inset shows cross-section view of the nanotube embedded between two silicon dioxide layers. Scale bar represents 100 nm. (Reproduced from Ref. [272].)

translocation events and a transition from current decrease to current enhancement during translocation has been observed on changing the buffer concentration, suggesting interplay between electrostatic charge and geometric blockage effects. Hybrid nanotubes with concentric organic and inorganic layers are obtained by the self-assembly of glycolipids on silica nanotubes [273].

2.3.3
Properties and Applications

Electronic transport properties of Bi nanotubes in Al_2O_3 membranes have been investigated [274]. A metal to semiconductor transition occurs with decrease in the wall thickness, probably due to quantum confinement. BN nanotubes have been shown to chemisorb hydrogen [275]. Fluorescent silica nanotubes are suggested to be useful in gene delivery [276]. Protein biosensors based on biofunctionalized conical gold nanotubes have been fabricated [277]. Ferroelectric phase transitions in template-synthesized $BaTiO_3$ nanotubes and nanofibers have been examined [278]. Ferroelectric and piezoelectric properties of biferroic $BiFeO_3$ nanotube arrays have been studied [279].

TiO_2 nanotubes show very high sensitivity to H_2 gas [280]. Highly efficient α-Fe_2O_3 nanotube chemical sensors based on chemiluminescence have been fabricated to detect H_2S gas, using carbon nanotubes as templates [281]. These nanotubes have a high specific area and exhibit excellent sensitivity to reductive vapors and gases such as alcohol and hydrogen and superior electrochemical activity of 1415 mA h g^{-1} at 100 mA g^{-1} and 293 K [282]. $LiCoO_2$, $LiMn_2O_4$ and $LiNi_{0.8}Co_{0.2}O_2$ nanotubes, synthesized by the thermal decomposition of sol–gel precursors inside porous alumina templates have been examined as cathode materials for lithium ion batteries [283]. Inorganic nanotubes integrated into metal-oxide-solution field-effect transistors exhibit rapid field effect modulation of ionic conductance [284]. Halloysite nanotubes can be employed as hollow enzymatic nanoreactors [285]. TiO_2 nanotube arrays prepared by anodic oxidation of titanium thin films have been employed as photoanodes in dye-sensitized solar cells [286].

2.4
Inorganic Nanowires

2.4.1
Synthesis

Silicon nanowires (SiNWs) with diameters in the 5–20 nm range have been prepared, along with nanoparticles of $\sim$4 nm diameter, by arc-discharge in water [287]. SiNWs have also been synthesized in solution by using Au nanocrystals as seeds and silanes as precursors [288a]. In Fig. 2.21, electron microscope images of SiNWs obtained by the decomposition of diphenylsilane are shown. The HREM

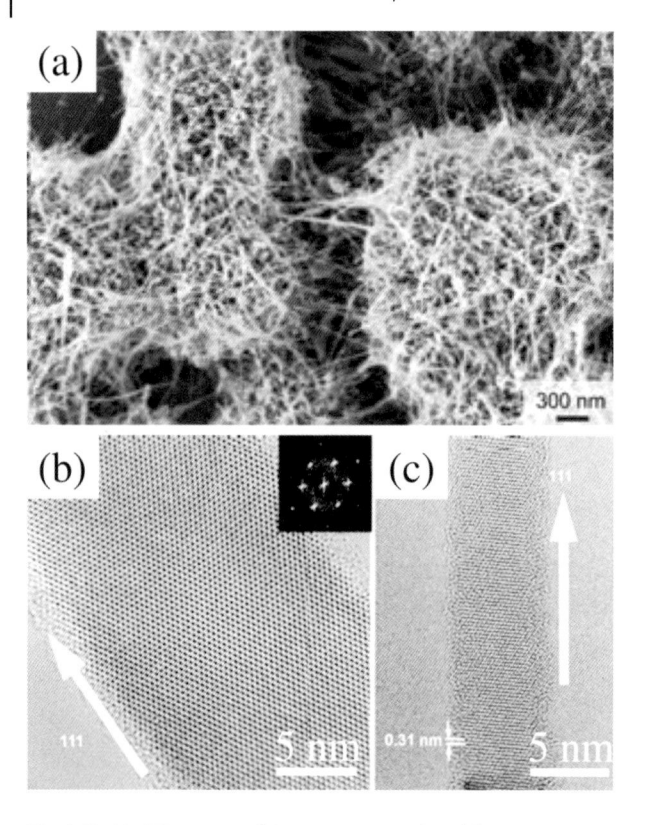

Fig. 2.21 (a) SEM image of Si nanowires produced from Au nanocrystals and diphenylsilane at 450 °C. (b), (c) HREM images of Si showing predominantly ⟨111⟩ orientation. (Reproduced from Ref. [288(a)].)

images reveal the presence of ⟨111⟩ oriented nanowires predominantly. A vapor–liquid–solid (VLS) type nanowire growth has been accomplished by increasing the pressure on the solvent and employing a reaction temperature above the Au/Si eutectic temperature (640 K). Vertically aligned SiNWs have been obtained by the chemical vapor deposition (CVD) of $SiCl_4$ on a gold colloid deposited Si (111) substrate [288b]. Gold colloids have been used for nanowire synthesis by the VLS growth mechanism. By manipulating the colloidal deposition of Au on the substrate, a controlled growth of aligned silicon nanowires was achieved. Solution–liquid–solid synthesis of germanium nanowires (GeNWs) gives high yields [289]. In this work, Bi nanocrystals were used as seeds for promoting nanowire growth in trioctylphosphine (TOP), by the decomposition of GeI_2 at ∼623 K. A solid-phase seeded growth with nickel nanocrystals has been employed to obtain GeNWs by the thermal decomposition of diphenylgermane in supercritical toluene [290]. A patterned growth of freestanding single-crystalline GeNWs with uniform distribution and vertical projection has been accomplished recently

[291]. Low temperature CVD has been employed to obtain high yields of GeNWs and nanowire arrays by using gold nanoseeds and patterned nanoseeds respectively [292]. GeNWs have been synthesized starting from the alkoxide, by using a solution procedure involving the injection of a germanium 2,6-dibutylphenoxide solution in oleylamine into a 1-octadecene solution heated to 300 °C under an argon atmosphere [293]. Experience in this laboratory has shown that several metals such as Ni, Ru, and Ir can be prepared from the respective actetate precursor by injecting them into a hydrocarbon such as decalin or, preferably, a long-chain amine at higher temperatures.

A seed-mediated surfactant method using a cationic surfactant has been developed to obtain pentagonal silver nanorods [294]. Linear Au–Ag nanoparticle chains are obtained by templated galvanostatic electrodeposition in the pores of anodic alumina membranes [295]. For particle-chain preparation, sacrificial Ni segments are included between the segments of noble metals (Au, Ag). During electrodeposition, the template pore diameter fixes the nanowire width, and the length of each metal segment is independently controlled by the amount of current passed before switching to the next plating solution for deposition of the subsequent segments. Nanowires are released by dissolution of the template, and subsequently coated with the SiO_2. Au nanorods obtained by the seed-mediated growth approach employ ~4 nm Au nanospheres as seeds, which react with the metal salt along with a weak reducing agent such as ascorbic acid in the presence of a directing surfactant [296]. The various reaction parameters can be used to control the shape. Addition of nitric acid significantly enhances the production of Au nanorods with high aspect ratios (~20) in seed-mediated synthesis [297]. A layer-by-layer deposition approach has been employed to produce polyelectrolyte-coated gold nanorods [298]. Au-nanoparticle-modified enzymes act as biocatalytic inks for growing Au or Ag nanowires on Si surfaces by using a patterning technique such as dip-pen nanolithography [299].

Single-crystalline Au nanorods can be shortened selectively by mild oxidation using 1 M HCl at 70 °C [300]. Aligned Au nanorods can be grown on a silicon substrate by employing a simple chemical amidation reaction on NH_2-functionalized Si(100) substrates [301]. The transformation of Au nanorods to Au nanochains by α,ω-alkenedithiols and the resultant interplasmon coupling have been investigated [302]. Above the critical concentration, a chain-up process proceeds through the interlocking of nanorods, initially to dimers and subsequently to oligomers, resulting in longitudinal interplasmon coupling. Depending on the surface chemical functionality of the coated gold nanorods, they can be selectively immobilized onto cationic or anionic surfaces. Micellar solutions of nonionic surfactants are employed to obtain nanowires and nanobelts of t-Se, which are single crystalline [303]. Se nanowires have been prepared at room temperature by using ascorbic acid as a reducing agent in the presence of β-cyclodextrin [304].

Single crystalline SiO_x nanowires with blue light emission can be prepared by a low temperature iron-assisted hydrothermal procedure [305]. A nanoribbon multicomponent precursor has been used to produce nanoparticle nanoribbons of ZnO [306]. The 1D porous structured nanoribbons are self-assembled by textured ZnO nanoparticles. Nanobelts of ZnO can be converted into superlattice-structured nanohelices by a rigid lattice rotation or twisting, as seen in Fig. 2.22 [307]. Well-aligned crystalline ZnO nanorods along with nanotubes can be grown from aqueous solutions on Si wafers, poly(ethylene terephthlate) and sapphire [308]. Atomic layer deposition was first used to grow a uniform ZnO film on the substrate of choice and to serve as a templating seed layer for the subsequent growth of nanorods and nanotubes. On this ZnO layer, highly oriented 2D ZnO nanorod arrays were obtained by solution growth using zinc nitrate and hexamethylenetetramine in aqueous solution. A seed-assisted chemical reaction at 368 K is found to yield uniform, straight, thin single-crystalline ZnO nanorods on a hectogram scale [309]. Controlled growth of well-aligned ZnO nanorod arrays has been accomplished by an aqueous ammonia solution method [310]. In this method, an aqueous ammonia solution of $Zn(NO_3)_2$ is allowed to react with a zinc-coated silicon substrate at a growth temperature 333–363 K. 3D interconnected networks of ZnO nanowires and nanorods have been synthesized by a high temperature solid–vapor deposition process [311]. Templated electrosynthesis of ZnO nanorods has been carried out, the procedure involving the electroreduction of either hydrogen peroxide or nitrate ions to alter the local pH within the pores of the membrane, and the subsequent precipitation of the metal oxide within the pores [312]. 1D ZnO nanostructures have been synthesized by oxygen assisted thermal evaporation of zinc on a quartz surface over a large area [313]. Ionic liquids such as 1-n-butyl-3-methylimidazolium tetrafluoroborate have been used to synthesize nanoneedles and nanorods of manganese dioxide (MnO_2) [314]. Variable-aspect-ratio, single-crystalline, 1D ZnO nanostructures (nanowires and nanotubes) can be prepared in alcohol/water solutions by reacting a Zn^{2+} precursor with an organic weak base, tetramethylammonium hydroxide [315].

Pattern and feature designed growth of ZnO nanowire arrays for vertical devices is accomplished by following a predesigned pattern and feature with controlled site, shape, distribution and orientation [316]. This technique relies on an integration of atomic force microscopy (AFM) nanomachining with catalytically activated vapor–liquid–solid (VLS) growth. IrO_2 nanorods can be grown by metal-organic chemical vapour deposition on a sapphire substrate consisting of patterned SiO_2 as the nongrowth surface [317]. By employing a hydrothermal route, uniform single-crystalline $KNbO_3$ nanowires have been obtained [318].

MgO nanowires and related nanostructures have been produced by carbothermal synthesis, starting with polycrystalline MgO or Mg with or without the use of metal catalysts [319]. This study has been carried out with different sources of carbon, all of them yielding interesting nanostructures such as nanosheets, nanobelts, nanotrees and aligned nanowires. Orthogonally branched single-crystalline MgO nanostructures have been obtained through a simple chemical vapor transport and condensation process in a flowing Ar/O_2 atmosphere [320].

Catalyst-assisted VLS growth of single-crystal Ga_2O_3 nanobelts has been accomplished by graphite-assisted thermal reduction of a mixture of Ga_2O_3 and SnO_2 powders under controlled conditions [321]. Zigzag and helical β-Ga_2O_3 1D nanostructures have been produced by the thermal evaporation of Ga_2O_3 in the pre-

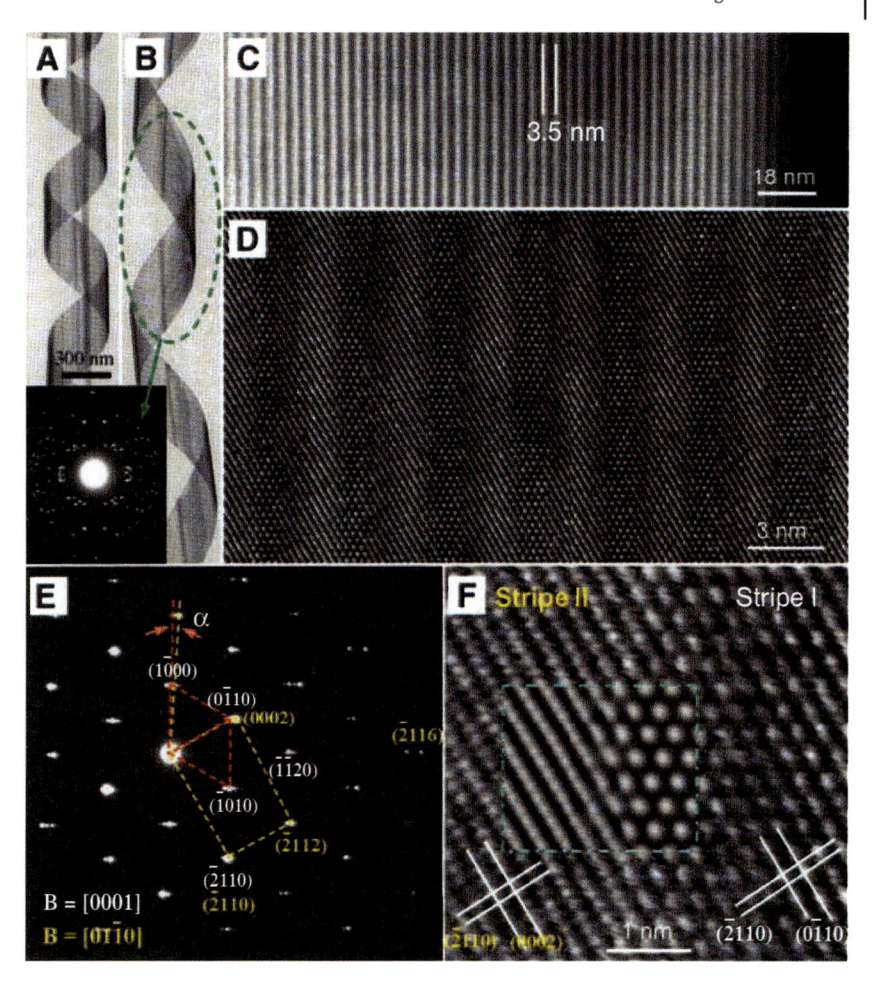

Fig. 2.22 Crystal structure of the nanohelix. (A) Typical low-magnification TEM image of a ZnO nanohelix, showing its structural uniformity. A straight nanobelt was enclosed inside the helix during the growth. (B) Low-magnification TEM image of a ZnO nanohelix with a larger pitch to diameter ratio. The selected area ED pattern (SAEDP, inset) is from a full turn of the helix. (C) Dark-field TEM image from a segment of a nanohelix, showing that the nanobelt that coils into a helix is composed of uniformly parallel, longitudinal, and alternatively distributed stripes at a periodicity of ~3.5 nm across its entire width. The edge at the right-hand side is the edge of the nanobelt. (D and E) High-magnification TEM image and the corresponding SAEDP of a ZnO nanohelix with the incident beam perpendicular to the surface of the nanobelt, respectively, showing the lattice structure of the two alternating stripes. The selected area ED pattern is composed of two sets of fundamental patterns, labeled and indexed in red for stripe I and yellow for stripe II. A careful examination of the image indicates that the true interface between the stripes is not edge-on with reference to the incident electron beam but at a relatively large angle. (F) Enlarged high-resolution TEM image showing the interface between the two adjacent stripes. The area within the dotted line is a simulated image using the dynamic electron diffraction theory. The interface proposed here is (-1-122) for stripe I, which is inclined with respect to the incident electron beam at an angle of 32°. (Reproduced from Ref. [307].)

sence of GaN [322]. TiO_2 nanorods can be obtained on a large scale by the non-hydrolytic sol–gel ester elimination reaction [323]. Here, the reaction is carried out between titanium (IV) isopropoxide and oleic acid at 543 K to generate 3.4 nm diameter crystalline TiO_2 nanorods. Single-crystalline and well facetted VO_2 nanowires with rectangular cross sections have been prepared by the vapor transport method, starting with bulk VO_2 powder [324]. Copious quantities of single-crystalline and optically transparent Sn-doped In_2O_3 (ITO) nanowires have been grown on gold-sputtered Si substrates by carbon-assisted synthesis, starting with a powdered mixture of the metal nitrates or with a citric acid gel formed by the metal nitrates [325]. Vertically aligned and branched ITO nanowire arrays which are single-crystalline have been grown on yttrium-stabilized zirconia substrates containing thin gold films of 10 nm thickness [326].

Bicrystalline nanowires of hematite (α-Fe_2O_3) have been synthesized by the oxidation of pure Fe [327]. Single-crystalline hexagonal α-Fe_2O_3 nanorods and nanobelts can be prepared by a simple iron–water reaction at 400 °C [328]. Networks of WO_{3-x} nanowires shown in Fig. 2.23 are produced by the thermal evaporation of W powder in the presence of oxygen [329]. The growth mechanism involves

Fig. 2.23 (a) Low-magnification and (b) high-magnification SEM images of tungsten oxide nanowires. (Reproduced from Ref. [329].)

ordered oxygen vacancies (100) and (001) planes which are parallel to the (010) growth direction. A general and highly effective one-pot synthetic protocol for producing 1D nanostructures of transition metal oxides such as $W_{18}O_{49}$, TiO_2, Mn_3O_4 and V_2O_5 through a thermally induced crystal growth process starting from mixtures of metal chlorides and surfactants, has been described [330]. A polymer-assisted hydrothermal synthesis of single crystalline tetragonal perovskite PZT ($PbZr_{0.52}Ti_{0.48}O_3$) nanowires has been carried out [331]. Nanowires of the type II superconductor $YBa_2Cu_4O_8$, have been synthesized by a biomimetic procedure [332]. The nanowires, produced by the calcination of the gelled reaction solutions containing the biopolymer chitosan and Y, Ba, Cu salt, have mean diameters of 50 ± 5 nm and lengths up to 1 µm. Other methods produce only an outgrowth of nanorods from the surface of large, irregularly shaped grains of the type II superconductor. Multi-micrometer-sized grains remain *in situ* and appear to comprise the bulk of the final material.

Oriented attachment of nanocrystals can be used to make one-dimensional as well as complex nanostructures. Thus, nanotubes and nanowires of II–VI semiconductors have been synthesized using surfactants [333]. The nanorods or nanotubes of CdS and other materials produced in this manner actually consist of nanocrystals. The synthesis of SnO_2 nanowires from nanoparticles has been investigated [334]. A variety of 1D nanostructures of CdS formed on Si substrates by a simple thermal evaporation route [335]. The shapes of the 1D CdS nanoforms were controlled by varying the experimental parameters such as temperature and position of the substrates. In Fig. 2.24, we show CdSe nanorods formed by redox-assisted asymmetric Ostwald ripening of CdSe dots to rods [336]. Nanorods of V_2O_5 prepared by a polyol process self-assemble into microspheres [337]. Cubic ZnS nanorods are obtained by the oriented attachment mechanism starting with diethylzinc, sulfur and an amine [338]. ZnS nanowires and nanoribbons with wurtzite structure can be prepared by the thermal evaporation of ZnS powder onto silicon substrates, sputter-coated with a thin (~25 Å) layer of Au film [339]. Thermal evaporation of a mixture of ZnSe and activated carbon powders in the presence of a tin-oxide based catalyst yields tetrapod-branched ZnSe nanorod architectures [340]. Nanorods of luminescent cubic CdS are obtained by injecting solutions of anhydrous cadmium acetate and sulfur in octylamine into hexadecylamine [341]. CdSe nanowires have been produced by the cation-exchange route [342]. By employing the cation-exchange reaction between Ag^{2+} and Cd^{2+}, Ag_2Se nanowires are transformed into single-crystal CdSe nanowires. A single-source molecular precursor has been used to obtain blue-emitting, cubic CdSe nanorods (~2.5 nm diameter and 12 nm length) at low temperatures [343]. Thin aligned nanorods and nanowires of ZnS, ZnSe, CdS and CdSe can be produced by using microwave-assisted methodology by starting from appropriate precursors [344]. An organometallic preparation of CdTe nanowires with high aspect ratios in the wurtzite structure has been described [345]. Thermal decomposition of copper-diethyldithiocarbamate (CuS_2CNEt_2) in a mixed binary surfactant solvent of dodecanethiol and oleic acid at 433 K gives rise to single-crystal line high aspect ratio ultrathin nanowires of hexagonal Cu_2S [346].

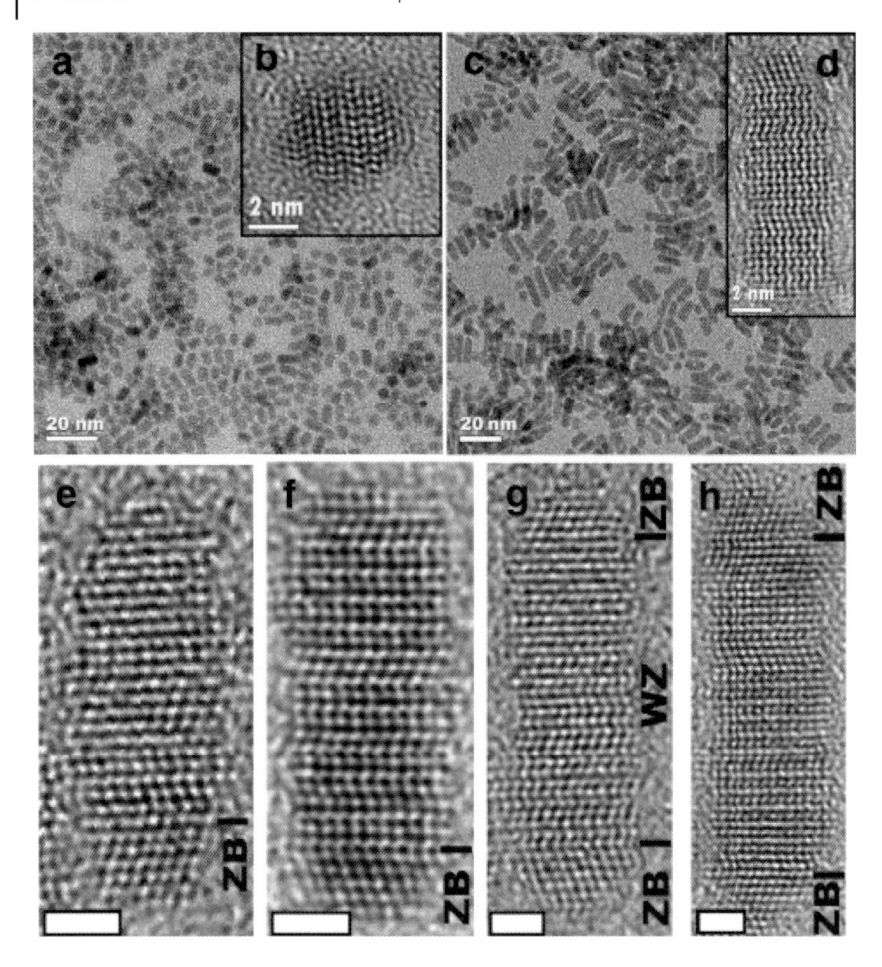

Fig. 2.24 Representative low- and high-magnification TEM images of CdSe NCs before (a), (b) and after (c)–(f) annealing at 135 °C in 3-amino-1-propanol/ H_2O (v/v) (9/1) with 0.1 M $CdCl_2$ for 48 h. Representative HREM images of rods (e)–(h) depict their zinc blende (ZB) tip(s) and stacking faults along the 002 axis. Scale bar = 2 nm. (Reproduced from Ref. [336].)

Atmospheric pressure CVD is employed to obtain arrays and networks of uniform PbS nanowires [347]. PbSe nanowires as well as complex 1D nanostructures shown in Fig. 2.25 can be obtained in solution through oriented attachment of PbSe nanocrystals [348]. Monodispersed PbTe nanorods of sub-10 nm diameter are obtained by sonoelectrochemical means by starting with a lead salt and TeO_2 along with nitrilotriacetic acid [349]. Using bismuth citrate and thiourea as the precursor material in DMF, well-segregated, crystalline Bi_2S_3 nanorods have been synthesized by a reflux process [350]. Single-crystalline Bi_2S_3 nanowires have also been obtained by using lysozyme which controls the morphology and directs the

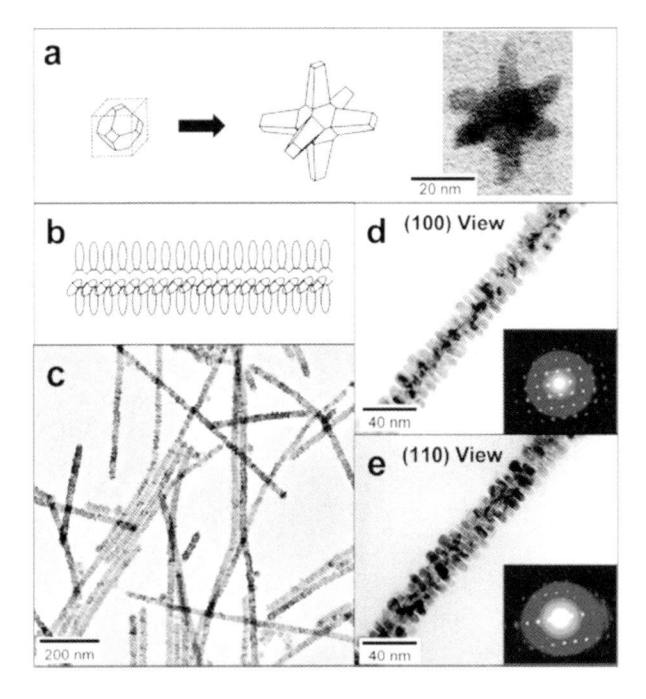

Fig. 2.25 (a) Star-shape PbSe nanocrystals and (b)–(e) radially branched nanowires. (d) TEM image of the (100) view of the branched nanowire and the corresponding selected area electron diffraction pattern. (e) TEM image of the (110) view of the branched nanowire and the corresponding selected area electron diffraction pattern. (Reproduced from Ref. [348].)

formation of the 1D inorganic material [351]. In this method, Bi(NO$_3$)$_3$ 5H$_2$O, thiourea and lysozyme are reacted together at 433 K under hydrothermal conditions. A solvent-less synthesis of orthorhombic Bi$_2$S$_3$ nanorods and nanowires with high aspect ratios (>100) has been accomplished by the thermal decomposition of bismuth alkylthiolate precursors in air around 500 K in the presence of a capping ligand species, octanoate [352]. Single-crystalline Bi$_2$Te$_3$ nanorods have been synthesized by a template free method at 100 °C by using a procedure involving addition of thioglycolic acid or L-cysteine to a bismuth chloride solution [353].

Single crystalline AlN, GaN and InN nanowires can be deposited on Si substrates covered with Au islands by using urea complexes formed with the trichlorides of Al, Ga and In as the precursors [354]. In Fig. 2.26, we show SEM and TEM images of AlN, GaN and InN nanowires obtained by this single-precursor route. Single crystalline GaN nanowires are also obtained by the thermal evaporation/decomposition of Ga$_2$O$_3$ powders with ammonia at 1423 K directly onto a Si substrate coated with an Au film [355]. Direction-dependent homoepitaxial growth of GaN nanowires, as shown in Fig. 2.27, has been achieved by controlling the

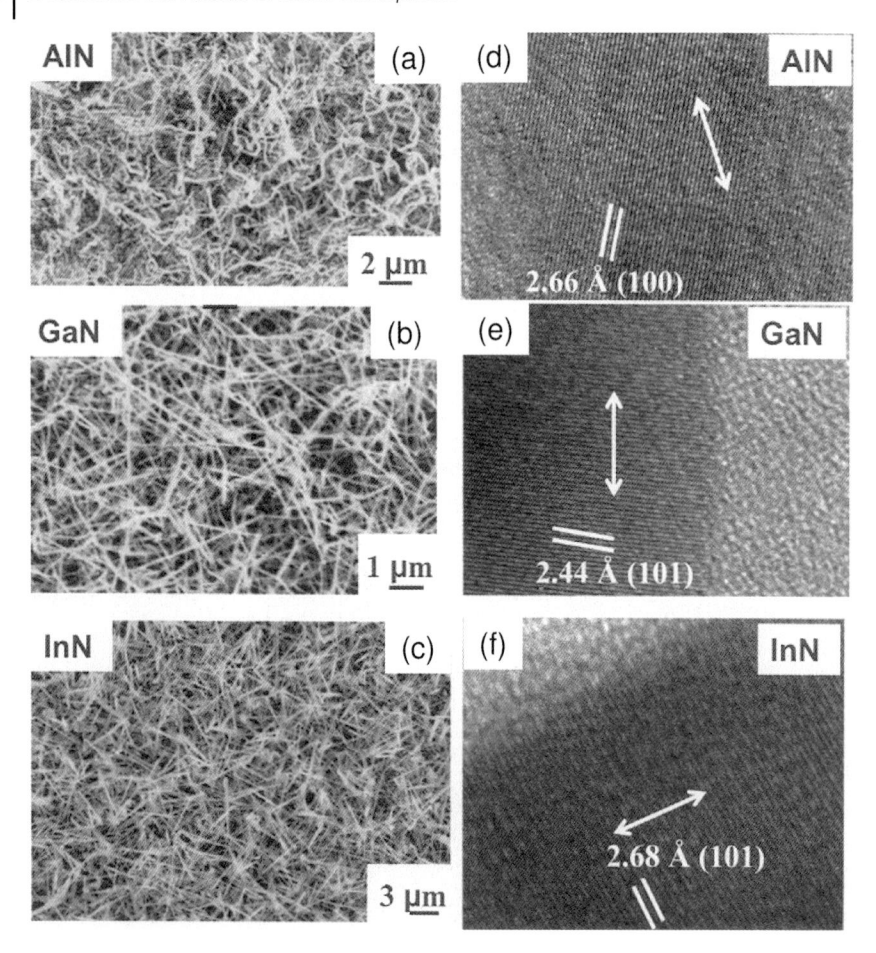

Fig. 2.26 SEM images of (a) AlN, (b) GaN, (c) InN; (d)–(f) HREM images of AlN, GaN and InN nanowires (double headed arrow indicates crystal long axis, and the spacing between two white lines gives the lattice spacing). (Reproduced from Ref. [354].)

Ga flux during direct nitridation in dissociated ammonia [356]. The nitridation of Ga droplets at a high flux leads to GaN nanowire growth in the c-direction (⟨1000⟩), while nitridation with a low Ga flux leads to growth in the a-direction (⟨10–10⟩). InN nanowires with uniform diameters have been obtained in large quantities by the reaction of In_2O_3 powders in ammonia [357]. A general method for the synthesis of Mn-doped nanowires of CdS, ZnS and GaN based on metal nanocluster-catalyzed chemical vapor deposition has been described [358]. Vertically aligned, catalyst-free InP nanowires have been grown on InP(III)B substrates by CVD of trimethylindium and phosphine at 623–723 K [359]. Homogeneous $InAs_{1-x}P_x$ nanowires as well as $InAs_{1-x}P_x$ heterostructure segments in

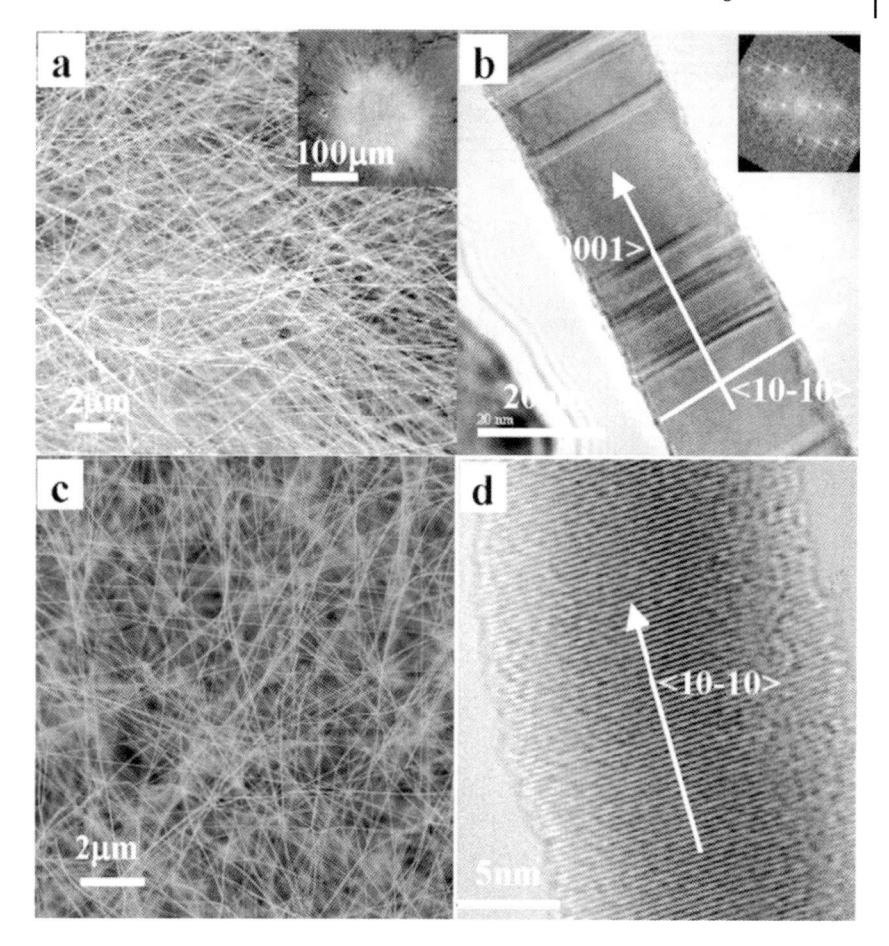

Fig. 2.27 (a) SEM images showing GaN nanowires with diameters less than 30 nm obtained from the direct reaction of Ga droplets and NH_3. The inset shows the spontaneous nucleation and growth of multiple nanowires directly from a larger Ga droplet. (b) High-resolution transmission electron microscopy (HREM) image of a GaN nanowire from the sample shown in (a) indicating that the growth direction is $\langle 0001 \rangle$. The inset is a fast Fourier transform of the HREM image, and the zone axis is $\langle 11\text{--}20 \rangle$. (c) SEM image showing GaN nanowires with diameters less than 30 nm resulting from the reactive-vapor transport of a controlled Ga flux in a NH_3 atmosphere. (d) TEM image of a GaN nanowire from the sample shown in (c) indicating that the growth direction is $\langle 10\text{--}10 \rangle$. (Reproduced from Ref. [356].)

InAs nanowires, with P concentration varying from 22% to 100%, have been grown and studied in detail as a function of reactant ratio, temperature and diameter of the nanowires [360].

Single-crystalline nanowires of LaB_6, CeB_6 and GdB_6 have been prepared and deposited on a Si substrate by the reaction of the rare-earth chlorides with BCl_3 in the presence of hydrogen [361]. Starting from BiI_3 and FeI_2, Fe_3B nanowires were synthesized on Pt and Pd (Pt/Pd) coated sapphire substrates by CVD at 800 °C [362]. The morphology of the Fe_3B nanowires can be controlled by manipulating the Pt/Pd film thickness and growth time, typical diameters are in the 5–50 nm range and lengths in the 2–30 µm range. Nanowires and nanoribbons of $NbSe_3$ have been obtained by the direct reaction of Nb and Se powders [363]. A one-pot metal-organic synthesis of single-crystalline CoP nanowires with uniform diameters has been reported [364]. The method involves the thermal decomposition of cobalt(II)acetylacetonate and tetradecylphosphonic acid in a mixture of TOPO and hexadecylamine. CoNi nanowires have been prepared by heterogeneous nucleation in liquid polyol [365].

2.4.2
Self Assembly and Functionalization

Surfactant-protected gold nanorods self-assemble into ordered structures in the presence of adipic acid [366]. Gold nanorods can be linked to each other in an end-to-end fashion by using cysteine as the molecular bridge [367]. End-to-end assembly of gold nanorods and nanospheres is also accomplished by oligonucleotide hybridization [368]. The rationale behind the selection of the mercaptoalkyloligonucleotide molecule is based on the fact that the thiol group binds to the ends of the nanorods, which assemble in an end-to-end fashion through hybridization with the target oligonucleotide. MWNTs can be effectively used as templates for aligning Au nanorods [369]. The longitudinal absorption band of the Au nanorods shifts to higher wavelengths, indicating a preference for preferential string-like alignment on the surface of the MWNTs. Alignment of gold nanorods in polymer composites and on polymer surfaces has been examined [370]. By employing the stretch-film method, it was found that, as the polymer was stretched in a direction, the nanorods became oriented with their long axis along the direction.

Silica nanowires can be assembled on silica aerogel substrates by employing a scanning tunneling microscope [371]. Ge nanowires prepared by CVD and functionalized with alkanethiols are found to be soluble in organic solvents and to readily assemble into close-packed Langmuir–Blodgett films as seen in Fig. 2.28 [372].

Crystalline WO_3 nanowires formed by the decomposition of tungsten isopropoxide in a solution of benzyl alcohol self-assemble into bundles with diameters in the 20–100 nm range and lengths in the 300–1000 nm range [373]. WO_3 nanostructures grow and assemble in the presence of deferoxamine mesylate under different reaction conditions [374]. Using a facile solution method, arrays of SnO_2 nanorods can be assembled on the surface of α-Fe_2O_3 nanotubes [375].

Silicon nanowires can be covalently modified with DNA oligonucleotides and such nanowires show biomolecular recognition properties [376]. Gold nanorods are stabilized and conjugated to antibodies for biological applications [377]. In_2O_3 nanowires have been selectively functionalized for biosensing

Fig. 2.28 Langmuir–Blodgett film of GeNWs. (a) SEM image of a GeNW film with dodecanethiol (C_{12}) functionalization. Inset: photo of a GeNW suspension in chloroform. (b) SEM image of a GeNW film with octyl (C_8) functionalization. (Reproduced from Ref. [372].)

applications by a simple and mild self-assembling process, making use of 4-(1,4-dihydroxybenzene)butyl phosphonic acid (HQ-PA) [378]. Cyclic voltammetry and fluorescence have been used to study the binding of DNA binding functionalized nanowire. For the formation of a linear oriented assembly of gold nanorods, use of antigens specifically binding to antibodies appears to be a feasible approach [379]. Anti-mouse IgG was immobilized on the {111} end faces of gold nanorods through a thioctic acid containing a terminal carboxyl group. The biofunctionalized nanorods are assembled using mouse IgG for biorecognition and binding. Nanowires and other nanostructures can be assembled utilizing highly engineered M13 bacteriophage as templates [380]. The phage clones with gold-binding motifs on the capsid and streptavidin-binding motifs at one end are used to assemble Au and CdSe nanocrystals into ordered one-dimensional arrays and more complex geometries.

2.4.3
Coaxial Nanowires and Coatings on Nanowires

A general procedure has been proposed for producing chemically bonded ceramic oxide coatings on carbon nanotubes and inorganic nanowires [78]. The ceramic oxide-coated structures are obtained by the reaction of reactive metal chlorides with acid-treated carbon nanotubes or metal oxide nanowires, followed by hydrolysis with water. On repeating the above process several times, followed by calcination, ceramic coatings of the desired thickness are obtained. Core–sheath CdS and polyaniline (PANI) coaxial nanocables with enhanced photoluminescence have been fabricated by an electrochemical method using a porous anodic alumina membrane as the template [381]. SiC nanowires can be coated with Ni and Pt nanoparticles ($\sim$3 nm) by plasma-enhanced CVD [382]. Single and double-shelled coaxial core–shell nanocables of GaP with SiO_x and carbon (GaP/SiO_x, GaP/C, $GaP/SiO_x/C$), with selective morphology and structure, have been synthesized by thermal CVD [383]. Silica-sheathed 3C-Fe_7S_8 has been prepared on silicon substrates with $FeCl_2$ and sulfur precursors at 873–1073 K [384].

Nanowires containing multiple GaP–GaAs junctions are grown by the use of metal – organic vapor phase epitaxy (MOVPE) on SiO_2 [385]. The VLS growth kinetics of GaP and GaAs in heterostructured GaP–GaS nanowires has been studied as a function of temperature and partial pressures of arsine and trimethylgallium. Silica-coated PbS nanowires have been deposited by CVD using $PbCl_2$ and S on silicon substrates at temperatures between 650 and 700 °C [386]. A novel silica-coating procedure has been devised for CTAB-stabilized gold nanorods and for the hydrophobation of the silica shell with octadecyltrimethoxysilane (OTMS) [387]. A combination of the polyelectrolyte layer-by-layer (LBL) technique and the hydrolysis and condensation of tetraethoxy orthosilicate (TEOS) in a 2-propanol–water mixture leads to homogeneous coatings with tight control on shell thickness. On the other hand, the strong binding of CTAB molecules to the gold surface makes surface hydrophobation difficult but the functionalization with OTMS, which contains a long hydrophobic hydrocarbon chain, allows the particles to be transferred into nonpolar organic solvents such as chloroform.

Fabrication of InP/InAs/InP core–multishell heterostructure nanowire arrays has been achieved by selective area metal-organic vapor phase epitaxy [388]. These core–multishell nanowires were designed to accommodate a strained InAs quantum well layer in a higher band gap InP nanowire (Fig. 2.29). Precise control over the nanowire growth direction and the heterojunction formation enabled the successful fabrication of the nanostructure in which all the three layers were epitaxially grown without the assistance of a catalyst.

2.4.4
Optical Properties

The dependence of the fluorescence intensity of Au nanorods on the aspect ratio has been examined in detail [389]. It appears that non-radiative processes domi-

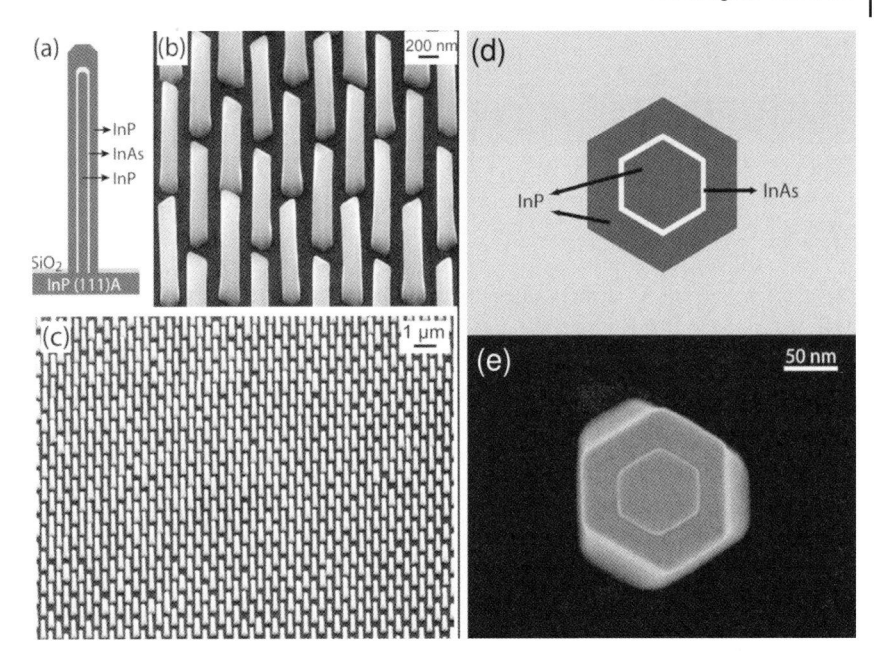

Fig. 2.29 (a) Schematic cross-sectional image of InP/InAs/InP core–multishell nanowire. (b) SEM image of periodically aligned InP/InAs/InP core–multishell nanowire array. (c) Low-angle inclined SEM image showing high dense ordered arrays of core–multishell nanowires. Schematic illustration and high resolution SEM cross-sectional image of a typical core–multishell nanowire observed after anisotropic dry etching and stain etching. (Reproduced from Ref. [388].)

nate the relaxation mechanism of the excited state. Absorption and scattering properties of gold nanoparticles of different size, shape, and composition have been calculated using Mie theory and the discrete dipole approximation method [390]. Absorption and scattering efficiencies and optical resonance wavelengths have been calculated for three commonly used classes of nanoparticles: gold nanospheres, silica–gold nanoshells, and gold nanorods. The calculated spectra clearly reflect the well-known dependence of the optical properties (viz. the resonance wavelength, the extinction cross section, and the ratio of scattering to absorption), on the nanoparticle dimensions. Gold nanorods show optical cross sections comparable to nanospheres and nanoshells, however, at much smaller effective size. To compare the effectiveness of nanoparticles of different sizes for real biomedical applications, size-normalized optical cross sections or per micron coefficients are calculated. Gold nanorods show per micron absorption and scattering coefficients that are an order of magnitude higher than those for nanoshells and nanospheres. Multiple higher-order plasmon resonances in colloidal cylindrical gold nanorods electrochemically deposited in anodic aluminum oxide templates (AAO) has been studied [391]. Homogeneous suspensions of nanorods with an average diameter of 85 nm and with varying lengths have been used. The AAO template provided a synthetic route that resulted in a homogeneous suspension of rods with the proper dimensions to observe these modes. The experimental optical spectra agree with discrete dipole approximation calculations (DDA) that have been modeled from the dimensions of the gold nanorods. As in the lithographically generated patterns, both the even and odd modes were detected up to the seventh order and were in good agreement with DDA.

ZnO nanowires prepared by a low-temperature aqueous pathway with low defect density can be used for room-temperature nanowire ultraviolet lasers [392]. The optimal synthesis conditions led to the low-temperature growth of ZnO nanowires that showed room-temperature ultraviolet lasing at a low threshold of pump fluence. Based on experimental results and optical waveguide theory, the control of the density of defects generated in aqueous solutions and the optimal microstructure of the grown nanowires to produce strong optical confinement are found to be necessary for realizing room-temperature ultraviolet lasing in ZnO nanowires.

Electroluminescence from ZnO nanowires in n-ZnO film/ZnO nanowire array/p-GaN film heterojunction light-emitting diodes has been observed [393]. ZnO nanowire-array-embedded n-ZnO/p-GaN heterojunction light-emitting diodes were fabricated by growing Mg-doped p-GaN films, ZnO nanowire arrays, and polycrystalline n-ZnO films consecutively. Electroluminescence emission having a wavelength of 386 nm was observed under forward bias in the heterojunction diodes and UV-violet light emerged from the ZnO nanowires. The heterojunction diode was thermally treated in hydrogen ambient to increase the electron injection rate from the n-ZnO films into the ZnO nanowires. A high concentration of electrons supplied from the n-ZnO films activated the radiative recombination in the ZnO nanowires, i.e., increased the light-emitting efficiency of the heterojunction diode.

A simple and effective approach for growing large-scale, high-density, and well-patterned conical boron nitride (BN) nanorods has been reported together with their cathodoluminescence (CL) properties [394]. CL spectra of these BN nanorods show two broad emission bands centered at 3.75 and 1.85 eV. Panchromatic CL images reveal clear patterned structures. Fabrication of a self-organized photosensitive gold nanoparticle chain encapsulated in a dielectric nanowire has been achieved by using a microreactor approach [395]. Such a hybrid nanowire shows pronounced surface plasmon resonance (SPR) absorption. More remarkably, a strong wavelength-dependent and reversible photoresponse has been demonstrated in a two-terminal device using an ensemble of gold nanopeapod silica nanowires under illumination, whereas no photoresponse was observed for the plain silica nanowires. These results show the potential of using gold nanopeapodded silica nanowires as wavelength-controlled optical nanoswitches. The microreactor approach can be applied to the preparation of a range of hybrid metal-dielectric 1D nanostructures that can be used as functional building blocks for nanoscale waveguiding devices, sensors and optoelectronics. These noble-metal nanoparticles embedded in dielectric matrices are considered to have practical applications in ultrafast all-optical switching devices owing to their enhanced third-order nonlinear susceptibility, especially near the SPR frequency.

Quantum efficiency and other aspects of ZnO nanowire nanolasers have been investigated [396]. The nanowires were prepared on sapphire and Si using pulsed laser ablation and the differential external quantum efficiency was as high as 60%. Aligned CdS nanowires are shown to exhibit optical waveguide behavior on continuous-wave laser excitation [397]. The mechanism of lasing action in single

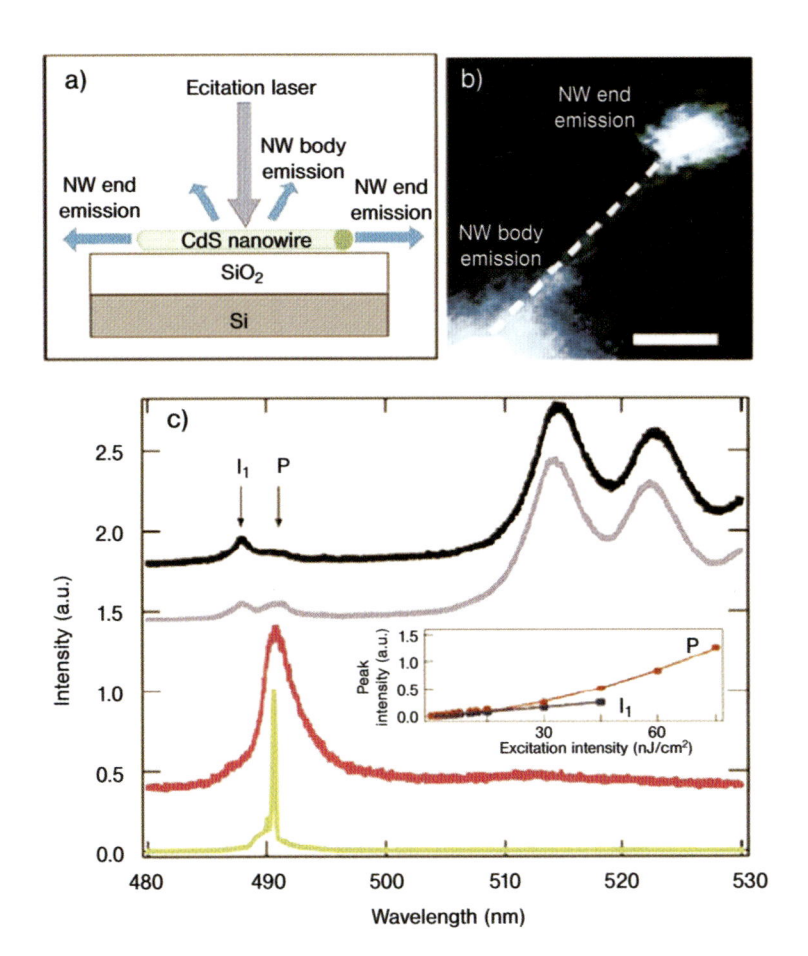

Fig. 2.30 (A) Schematic of single NW optical experiments. (B) PL image showing luminescence from the excitation area (lower left) and one end (upper right) of a CdS NW. The NW was excited with a focused beam (~5 μm in diameter) with a power of 10 nJ cm^{-2}; scale bar, 5 μm. (C) PL spectra of CdS NW end emission recorded at 4.2 K with excitation powers of 0.6, 1.5, 30, and 240 nJ cm^{-2} for the black, blue, red, and green curves, respectively. Inset shows peak intensity of I_1 (black squares) and P (red circles) bands vs. incident laser power. Solid lines are fits to experimental data with power exponents of 0.95 for I_1 and 1.8 for P. (Reproduced from Ref. [398].)

CdS nanowire cavities has been elucidated by temperature-dependent and time-resolved photoluminescence measurements (Fig. 2.30) [398]. Temperature-dependent photoluminescence studies show rich spectral features and reveal that an exciton–exciton interaction is critical to lasing up to 75 K, while an exciton–phonon process dominates at higher temperatures. Electric-field modulation of the visible and ultraviolet nanoscale lasers composed of single CdS and GaN nanowires has been achieved using integrated, microfabricated electrodes [399]. Optically pumped room-temperature lasing in GaN nanowires with low lasing thresholds has been reported [400]. Nanoscale light-emitting diodes with colors ranging from ultraviolet to near-infrared have been prepared using a solution-based approach in which electron-doped semiconductors are assembled with hole-doped silicon nanowires in a crossed nanowire architecture [401].

GaN nanowires have been shown to act as ring resonator lasers [402]. Recent advances in nanomanipulation have made it possible to modify the shape of GaN structures from a linear to a pseudo-ring conformation. Changes to the optical boundary conditions of the lasing cavity affect the structure's photoluminescence, photon confinement, and lasing as a function of ring diameter. For a given cavity, ring-mode redshifting is observed to increase with decreasing ring diameter. Significant shifts are observed during optical pumping of a ring resonator nanolaser compared to its linear counterpart. The shift appears to result from conformational changes of the cavity rather than effects such as band-gap renormalization, allowing the mode spacing and position to be tuned with the same nanowire gain medium. Photoluminescence of CdS_xSe_{1-x} nanobelts can be tuned, affording emission varying from 500 to 700 nm [403]. In these nanowires, the band-gap is engineered by controlling the composition [404]. Single-crystal ZnO nanowires can be used as ultraviolet photodetectors [405].

It has been demonstrated that light force, irrespective of the polarization of the light, can be used to run a simple nanorotor, as revealed in Fig. 2.31 [406]. While the gradient force of a single beam optical trap is used to hold an asymmetric nanorod, the utilization of the scattering force generates a torque on the nanorod, making it rotate about the optic axis. The inherent textural irregularities or morphological asymmetries of the nanorods give rise to the torque under the radiation pressure. Even a small surface irregularity with non-zero chirality is sufficient to produce enough torque for moderate rotational speed. Different sized rotors can be used to set the speed of rotation over a wide range with fine tuning possible through the variation of the laser power. Optical trapping and integration of semiconductor nanowire assemblies in water has been achieved [407]. It is shown that an infrared single-beam optical trap can be used to individually trap, transfer and assemble high-aspect-ratio semiconductor nanowires into arbitrary structures in a fluid environment. Nanowires with diameters as small as 20 nm and aspect ratios of more than 100 can be trapped and transported in three dimensions, enabling the construction of nanowire architectures that may function as active photonic devices. Nanowire structures can be assembled in physiological environments, offering new forms of chemical, mechanical and optical stimulation of living cells.

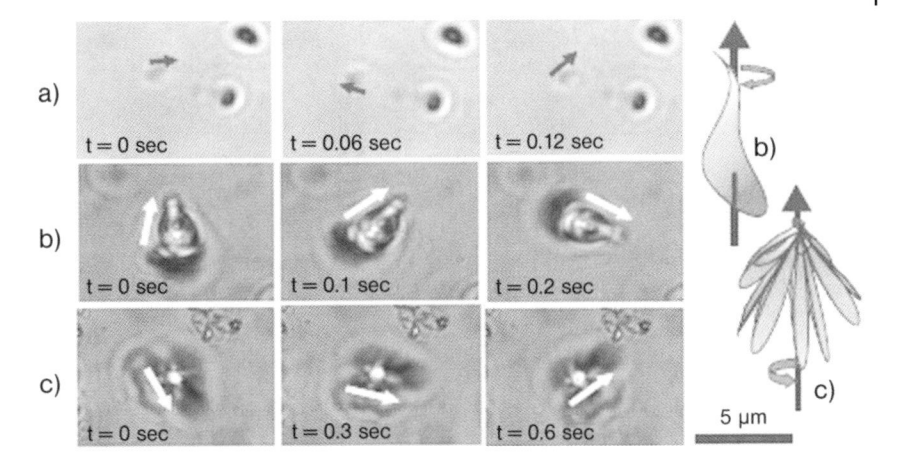

Fig. 2.31 Time sequences of different sized and shaped rotors are shown here. In each time frame the orientation of the rotor is indicated by an arrow. Panels (A), (B) and (C) represent rotations of three Al$_2$O$_3$ rotors. The rotor in panel (A) is a typical nanorod whereas the rotor in panel (B) is a bigger asymmetric nanorod, and in panel (C), the rotor is a nanorod bundle. The predicted structures of the rotors in panels (B) and (C) are depicted in the rightmost column. A size bar is shown at the bottom right-hand corner. Magnification factors of all the images are the same. The images shown in panel (A) are diffraction-limited images and hence they do not convey the real size of the rotor. (Reproduced from Ref. [406].)

2.4.5
Electrical and Magnetic Properties

Size-dependent transport and thermoelectric properties of individual polycrystalline Bi nanowires have been reported [408]. The combination of nanofabrication methods and device architecture has allowed four-point electric, thermoelectric, magnetic-field and electric-field-effect measurements on individual Bi nanowires. No clear semimetal-to-semiconductor transition or enhancement in thermoelectric power has been observed, probably due to the polycrystalline nature of Bi nanowires. Individual SiNWs exhibit coulomb blockade features, with coherent charge transport through discrete single particle quantum levels extending across the whole device [409]. The application of the superlattice nanowire pattern transfer (SNAP) technique to the fabrication of arrays of aligned silicon nanowires has been reported [410]. By the selection of appropriate silicon-on-insulator substrates, careful reactive-ion etching, and spin-on glass doping so fabricated Si nanowire arrays (10–20 nm width and 40–50 nm pitch) have resistivity values comparable to the bulk.

Reproducible interconnects of dielectrophoretic nanowires assembled from gold nanorods that vary in their conductance by ±10% have been fabricated by using cleanroom-based lithographic procedures [411]. The current–voltage profiles of these interconnect exhibited barriers to charge transport at temperatures less than ~225 K. Furthermore, their conductance increased exponentially with temperature with activation energy comparable to the nanorod charging energy. These results indicate that the Coulomb blockade associated with individual nanorods in interconnects are the primary conductance-limiting feature.

Preparation and electrical properties of uniform and well-crystallized n-type semiconductor β-Ga_2O_3 nanowires have been reported [412]. The ultrafine nanowires are prepared by reacting metal Ga with water vapor based on the VLS mechanism. The contact properties of individual Ga_2O_3 nanowires with Pt or Au/Ti electrodes have been studied and show that Pt can form Schottky barrier junctions while Au/Ti is advantageous for fabricating ohmic contacts with individual Ga_2O_3 nanowires. In ambient air, the conductivity of the Ga_2O_3 nanowires is about 1 $(\Omega\ m)^{-1}$, while with adsorption of NH_3 (or NO_2) molecules, the conductivity can increase (or decrease) dramatically at room temperature. Electrical properties of single GaN nanowires have been characterized [413]. Gate-dependent one-dimensional transport in a single-crystal In_2O_3 nanowire field-effect transistor has been studied at low temperatures [414]. Field-effect transistors of ZnO nanowires have been fabricated (Fig. 2.32) and studied in vacuum and in a variety of ambient gases [415]. Field-effect transistors based on the self-assembly of colloidal ZnO nanorods have also been examined [416]. ZnO nanowire field-effect transistors composed of individual ZnO nanowires can be made using a self-assembled superlattice as the gate insulator, a device of possible use in flexible display and logic technologies [417]. Top-gated field-effect transistors of core–shell structured GaP nanowires have been fabricated [418]. Storage of electrons occurs in semiconductor nanowires (InAs/InP) epitaxially grown from Au nanoparticles [419]. Electrical properties of inorganic nanowire–polymer composites such as ZnO, RuO_2 and Ag with polyaniline as well as with polypyrrole (PPY) have been measured [420]. Thermoelectric properties of electrodeposited bismuth telluride nanowires have been reported [421].

Field-emission properties of P-doped GaN nanowires synthesized via a simple thermal evaporation process have been examined [422]. Large-area nanowires of organic charge-transfer complexes such as Ag-TCNQ and Cu-TCNQ show field-emission properties [423]. Field-emission and cathodoluminescence properties of well-aligned AlN nanorods with multi-tipped (hairy) surfaces grown by a vapor–solid (VS) process have been studied [424]. Field-emission (FE) of electrons from $\langle 001 \rangle$ oriented single-crystalline LaB_6 grown by chemical vapor deposition (CVD) have shown excellent FE characteristics [425]. The enhancement of secondary electron emission (SEE) yields from Group III nitride/ZnO coaxial nanorod heterostructures (ultrafine coaxial GaN/ZnO and AlN/ZnO) has been reported [426]. The significant enhancement in SEE yields is due to the inherited nanostructure obtained from the ZnO nanoneedle template.

Superconductivity is suppressed in Zn nanowires [427]. Photovoltaic applications of aligned silicon nanowire arrays have been explored [428]. Dye-sensitized solar cells incorporating ZnO nanowires have been assembled [429, 430]. Dye-sensitized solar cells using TiO_2 single-crystalline nanorod electrodes provide efficient photocurrent generation in a quasi-solid-state, with a conversion efficiency

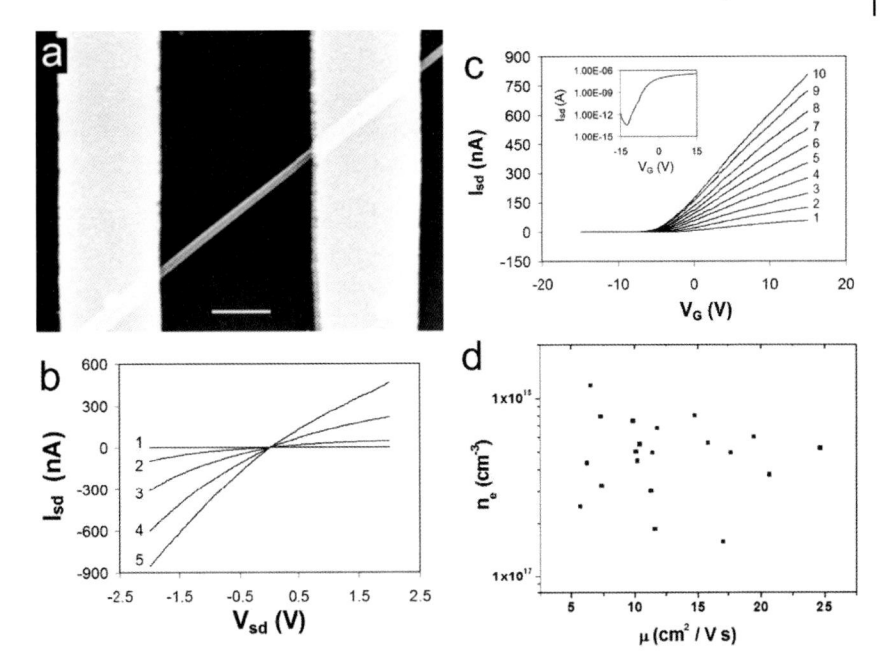

Fig. 2.32 (a) SEM micrograph of a 101 nm diameter ZnO nanowire device (b) Current (I_{sd}) vs. voltage (V_{sd}) curves recorded at different gate voltages for the device shown in (a). Curves 1–5 correspond to gate voltages of −10, −5, 0, +5, and +10 V, respectively. (c) Current (I_{sd}) vs. gate voltage (V_G) of the same device measured at bias voltages from 0.1 to 1.0 V. Curves 1–10 correspond to bias voltages of 0.1, 0.2, 0.3, 0.4, 0.5, 0.6, 0.7, 0.8, 0.9, and 1.0 V, respectively. The inset plots the current (I_{sd}) vs. gate voltage (V_G) measured at a bias voltage of 0.5 V_{sd} on a logarithmic scale. (d) Carrier concentration (n_e) vs. mobility (μ) for all transistor devices fabricated. (Reproduced from Ref. [415].)

of 6.2% under 100 mW cm^{-2} [431]. Self-powered synthetic nanorotors have been prepared from bar-coded gold–nickel nanorods having the gold end anchored to the surface of a silicon wafer. Constant velocity circular movements are observed when hydrogen peroxide is catalytically decomposed to oxygen at the unattached nickel end of the nanorod [432]. A high-throughput procedure is reported for lithographically processing 1D nanowires [433]. This procedure has been called on-wire lithography.

Mn-doped GaP nanowires appear to exhibit ferromagnetism with a Curie temperature higher than room temperature [434]. The magnetic tuning of the electrochemical reactivity through controlled surface orientation of catalytic Ni nanowires has been reported [435]. It is demonstrated how one can modulate the electrocatalytic activity by orienting catalytic nanowires at different angles. The dynamics of field-driven domain-wall propagation in ferromagnetic nanowires ($Ni_{80}Fe_{20}$) have been investigated [436]. Magnetoresistive properties of $La_{0.67}Sr_{0.33}MnO_3$ nanowires have been reported [437].

2.4.6
Some Chemical Aspects and Sensor Applications

Intercalation of lithium ions in TiO_2–B nanowires has been carried out without any structural degradation or loss of nanowire morphology [438]. Arrayed gold nanowires provide a useful platform for the electrochemical detection of DNA [439]. Remote-controlled autonomous movement of stripped metallic nanorods (catalytic nanomotors) has been reported [440]. ZnO nanorods integrated with a microchannel show a sensitivity (change in conductance) to the pH of the medium, suggesting that they can be used in sensor applications [441]. ZnO nanorods, nanowires and nanotubes prepared by different procedures, including electrochemical deposition in alumina membranes, have been investigated for hydrogen and ethanol sensing characteristics [442]. The sensing characteristics of these nanostructures were also investigated after impregnating them with 1% Pt. The nanowires exhibit excellent hydrogen sensing characteristics at relatively low temperatures ($\sim$150 °C), especially when impregnated with Pt, as seen in Fig. 2.33(a). One type of nanorod shows a dependence of the sensitivity on the hydro-

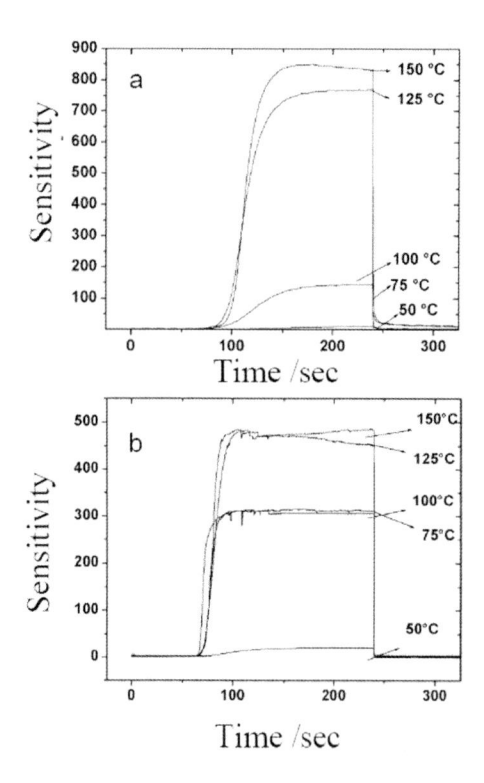

Fig. 2.33 Sensing characteristics of Pt-impregnated ZnO nanowires for (a) 1000 ppm of H_2 and (b) 1000 ppm of ethanol. (Reproduced from Ref. [442].)

gen concentration, along with short recovery and response times. The nanorods and nanowires impregnated with Pt show high sensitivity for 1000 ppm of ethanol at or below 150 °C (Fig. 2.33(b)), with short recovery and response times. Both nanorods and nanowires exhibit good sensing characteristics for 1000 ppm of ethanol at or below 423 K, whereas Pt impregnated nanorods and nanowires show better sensitivity for both hydrogen and ethanol. Field-effect transistors fabricated with ZnO nanowires have been used for sensing NO_2 and ammonia at room temperature [443]. $LiMo_3Se_3$ nanowire films show variation of conductivity in the presence of various chemical vapors, suggesting possible use as chemical sensors [444].

In_2O_3 nanowires along with carbon nanotubes can be usefully employed for the complementary detection of prostate-specific antigen biomolecules [445]. SnO_2 nanobelts have been integrated with microsystems for nerve agent detection [446]. Using silicon nanowire field-effect devices, multiplexed electrical detection of cancer markers has been achieved [447]. Single Pd nanowires, fabricated by electrodeposition in electrolyte channels patterned with electron-beam lithography, have been investigated for hydrogen sensing [448]. The fabrication technique can produce nanowires with controlled dimensions, positions, alignment and chemical compositions, and enables the use of a wide range of materials for creating arrays of single nanowire sensors.

Linear ethanol sensing properties of SnO_2 nanorods synthesized through a hydrothermal route have been reported [449]. The SnO_2 nanorod sensor with a diameter down to 3 nm, when exposed to 300 ppm ethanol vapor in air shows high sensitivity up to 83.8. Ultrasensitive and highly selective gas sensors using 3D tungsten oxide nanowire networks have been reported [450]. Utilizing the 3D hierarchical structure of the networks as well as the high-surface area of the material, high sensitivity has been obtained towards NO_2, revealing the capability of the material to detect concentration as low as 50 ppb. Distinctive selectivity at different working temperatures is observed for various gases. By using Au nanorod solutions in sodium borohydride, a quick and direct method for determining mercury in tap water samples at the parts per trillion level has been reported [451]. The outstanding selectivity and sensitivity results from the amalgamation process which occurs between mercury and gold nanorods.

2.4.7
Mechanical Properties

Gold nanowires show a Young's modulus which is independent of diameter whereas the yield strength is largest for the smallest diameter nanowires [452]. The elastic modulus of (0001) ZnO nanowires grown on a sapphire surface has been measured (29 ± 8 GPa) using atomic force microscopy [453]. ZnS nanobelts exhibit 79% increase in hardness and 52% decrease in elastic modulus compared to bulk ZnS [454]. A large increase in the elastic strength ($\sim 90\%$) and tensile strength ($\sim 70\%$) has been observed on incorporation of inorganic nanowires of SiC and Al_2O_3 in poly(vinyl alcohol) [455]. In Fig. 2.34(a), we show the variation

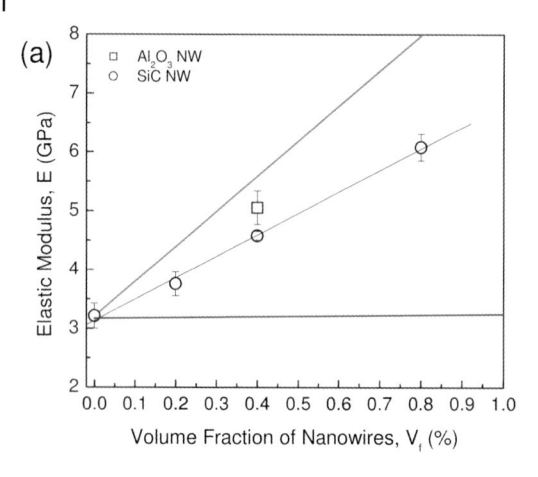

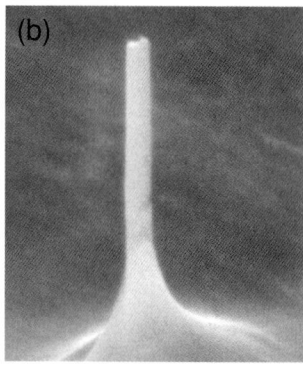

Fig. 2.34 (a) Variation of elastic modulus, E (measured with the DMA technique), as a function of the nanowires volume fraction, V_f. The upper- and lower-bound predictions (made using iso-strain and iso-stress models, respectively) are also plotted. (b) Field emission SEM images obtained from the fractured PVA–SiC nanowires (0.8 vol%) composite showing pull-out of the nanowires as well as stretching of the matrix along with the nanowire. (Reproduced from Ref. [455].)

of elastic modulus as a function of nanowire loading. A linear increase in the elastic modulus has been observed in the case of SiC nanowires. A significant increase in the strength of the composite with the addition of nanowires is also observed. This is due to the significant pull-out of the nanowires and the corresponding stretching of the matrix due to the complete wetting of the NW surface by the polymer as seen in Fig. 2.34(b). The increase in tensile strength is found to saturate at higher vol% of NW addition due to the reduced propensity for shear-band induced plastic deformation.

2.4.8
Transistors and Devices

The operation of single-electron tunneling (SET) transistors with gate-induced tunable electrostatic barriers using silicon nanowire metal-oxide-semiconductor field-effect transistor (MOSFET) structures has been achieved [456]. By using the flexible control of the tunable barriers, the systematic evolution from a single charge island to double islands was observed with gate capacitance values of the order of 10 aF with a variation smaller than 1 aF. The direct vertical integration of Si nanowire arrays into surrounding gate field effect transistors without the need for postgrowth nanowire assembly processes has been reported [457]. The device fabrication allows Si nanowire channel diameters to be readily reduced to the 5 nm regime. These first-generation vertically integrated nanowire field effect transistors exhibit electronic properties that are comparable to other horizontal nanowire field effect transistors. A generic process for fabricating a vertical surround-gate field-effect transistor based on epitaxially grown silicon nanowires is reported [458]. Piezoelectric nanogenerators based on ZnO nanowire arrays have been shown to convert nanoscale mechanical energy into electrical energy with efficiency of the power generation of 17 to 30% [459]. The coupling of piezoelectric and semiconducting properties of zinc oxide creates a strain field and charge separation across the nanowire as a result of its bending. The rectifying characteristic of the Schottky barrier formed between the metal tip and the nanowire leads to electrical current generation.

A ZnO nanowire photodetector with a fast photoresponse time was fabricated by a simple method of growing ZnO nanowires by bridging the gap of two patterned zinc electrodes [460]. The nanowire growth is self-catalytic, involving the direct heating of patterned Zn electrodes at 700 °C in an O_2/Ar gas flow for 3 h. The fabricated photodetector demonstrated fast response of less than 0.4 ms to UV illumination in air, which could be attributed to the adsorption, desorption, and diffusion of water molecules in the air onto the nanowire significantly influencing the photoresponse. Individual β-Ga_2O_3 nanowires as solar-blind photodetectors have been investigated [461]. The conductance of the nanowire increases by about three orders of magnitude with less than 254 nm ultraviolet illumination, the upper limits of the response and recovery time being 0.22 and 0.09 s, respectively.

A method based on template synthesis has been used for the construction of an array of coplanar fuel cells wherein each cell is 200 nm in diameter [462]. An array of nano fuel cells is produced by utilizing two arrays of porous Pt electrodes in between a polymer electrolyte membrane or an electrolyte support matrix sandwich. Electrodeposition of Pt–Cu nanowires inside a porous AAO membrane and the subsequent treatment with fuming HNO_3 acid gives rise to an array of porous platinum electrodes. This method of producing an array of coplanar fuel cells allows the series connection of fuel cells outside the array and eliminates the need for fuel and air manifolds, thereby reducing the overall system complexity. Initial prototypes utilizing an aqueous solution of $NaBH_4$ as a fuel have produced power densities of around 1 mW cm^{-2}.

2.4.9
Biological Aspects

Cancer cell imaging and photothermal therapy in the near-infrared region using Au nanorods have been reported [463]. Due to strong electric fields at the surface, the absorption and scattering of electromagnetic radiation by noble metal nanoparticles are strongly enhanced. These unique properties provide the potential for designing novel optically active reagents for simultaneous molecular imaging and photothermal cancer therapy. It is desirable to use agents that are active in the near-infrared region of the spectrum to minimize light extinction by intrinsic chromophores in the native tissue. Gold nanorods with suitable aspect ratios can absorb and scatter strongly in the 650–900 nm region. Nanorods are synthesized and conjugated to the anti-epidermal growth factor receptor (anti-EGFR) monoclonal antibodies and incubated in cell cultures with a nonmalignant epithelial cell line (HaCat) and two malignant oral epithelial cell lines (HOC 313 clone 8 and HSC 3). The anti-EGFR antibody-conjugated nanorods bind specifically to the surface of the malignant-type cells with a much higher affinity due to the over-expressed EGFR on the cytoplasmic membrane of the malignant cells. As a result of the strongly scattered red light from the gold nanorods in a dark field, the malignant cells are clearly visualized (with an ordinary laboratory microscope) and distinguished from the nonmalignant cells. It is found that, after exposure to continuous red laser at 800 nm, malignant cells require about half the laser energy to be photothermally destroyed than the nonmalignant cells. Thus, both efficient cancer cell diagnostics and selective photothermal therapy are realized at the same time.

DNA–Au nanorod conjugates for use in the remote control of localized gene expression by near-infrared irradiation have been reported [464]. Gold nanorods are attached to the gene of enhanced green fluorescence protein (EGFP) for the remote control of gene expression in living cells. UV–visible spectroscopy, electrophoresis, and transmission electron microscopy are used to study the optical and structural properties of the EGFP DNA and gold nanorod (EGFP–GNR) conjugates before and after femtosecond NIR laser irradiation. Upon NIR irradiation, the gold nanorods of EGFP–GNR conjugates underwent shape transformation that resulted in the release of EGFP DNA. When EGFP–GNR conjugates are delivered to cultured HeLa cells, induced GFP expression is specifically observed in cells that are locally exposed to NIR irradiation. These results demonstrate the feasibility of using gold nanorods and NIR irradiation as a means of remote control of gene expression in specific cells. This approach has potential applications in biological and medical studies.

References

1 S. Iijima, Nature, **1991**, *356*, 54–58.
2 *The Chemistry of Nanomaterials*, C. N. R. Rao, A. Muller, A. K. Cheetham (Eds.), Wiley-VCH, Weinheim. Vol. 1 & 2, 2004.
3 *Carbon Nanotubes: Basic concepts and Fundamentals*, S. Reich, C. Thomsen, C. Maultsch, Wiley-VCH, Weinheim, 2004.
4 *Nanotubes and Nanowires*, C. N. R. Rao, A. Govindaraj, Royal Society of Chemistry, Cambridge, 2005.
5 T. Tang, X. Chen, X. Meng, H. Chen, Y. Ding, Angew. Chem. Int. Ed., **2005**, *44*, 1517–1520.
6 L.-P. Zhou, K. Ohta, K. Kuroda, N. Lei, K. Matsuishi, L. Gao, T. Matsumoto, J. Nakamura, J. Phys. Chem. B, **2005**, *109*, 4439–4447.
7 M. A. C. Duarte, M. Grzelczak, V. S. Maceira, M. Giersig, L. M. L. Marzan, M. Farle, K. Sierazdki, R. Diaz, J. Phys. Chem. B, **2005**, *109*, 19060–19063.
8 J. Liu, X. Li, A. Schrand, T. Ohashi, L. Dai, Chem. Mater., **2005**, *17*, 6599–6604.
9 (a) S. R. C. Vivekchand, L. M. Cele, F. L. Deepak, A. R. Raju, A. Govindaraj, Chem. Phys. Lett., **2004**, *386*, 313–318; (b) C. P. Deck, K. S. Vecchio, J. Phys. Chem. B, **2005**, *109*, 12353–12357.
10 (a) L. Zhu, Y. Xiu, D. W. Hess, C.-P. Wong, Nano Lett., **2005**, *5*,

2641–2645; (b) X. Li, A. Cao, Y. J. Jung, R. Vajtai, P. M. Ajayan, *Nano Lett.*, **2005**, *5*, 1997–2000.
11 R. M. Kramer, L. A. Sowards, M. J. Pender, M. O. Stone, R. J. Naik, *Langmuir*, **2005**, *21*, 8466–8470.
12 J. Q. Lu, T. E. Kopley, N. Moll, D. Roitman, D. Chamberlin, Q. Fu, J. Liu, T. P. Russell, D. A. Rider, I. Manners, M. A. Winnik, *Chem. Mater.*, **2005**, *17*, 2227–2231.
13 Y.-S. Min, E. J. Bae, B. S. Oh, D. Khang, W. Park, *J. Am. Chem. Soc.*, **2005**, *127*, 12498–12499.
14 D. N. Futaba, K. Hata, T. Yamada, K. Mizuno, M. Yumura, S. Iijima, *Phys. Rev. Lett.*, **2005**, *95*, 056104.
15 Y. Wang, M. J. Kim, H. Shan, C. Kittrell, H. Fan, L. M. Ericson, W.-F. Hwang, S. Arepalli, R. H. Hauge, R. E. Smalley, *Nano Lett.*, **2005**, *5*, 997–1002.
16 T. Iwasaki, G. Zhong, T. Aikawa, T. Yoshida, H. Kawarada, *J. Phys. Chem. B*, **2005**, *109*, 19556–19559.
17 W. Zhou, Y. Zhang, X. Li, S. Yuan, Z. Jin, J. Xu, Y. Li, *J. Phys. Chem. B*, **2005**, *109*, 6963–6967.
18 Y. Li, S. Peng, D. Mann, J. Cao, R. Tu, K. J. Cho, H. Dai, *J. Phys. Chem. B*, **2005**, *109*, 6968–6971.
19 M. L. Terranova, S. Orlanducci, A. Fiori, E. Tamburri, V. Sessa, M. Rossi, A. S. Barnard, *Chem. Mater.*, **2005**, *17*, 3214–3220.
20 Z. Chen, Y. Yang, Z. Wu, G. Luo, L. Xie, Z. Liu, S. Ma, W. Guo, *J. Phys. Chem. B*, **2005**, *109*, 5473–5477.
21 Y.-Q. Xu, E. Flor, M. J. Kim, B. Hamadani, H. Schmidt, R. E. Smalley, R. H. Hauge, *J. Am. Chem. Soc.*, **2006**, *128*, 6560–6561.
22 M. Lin, J. P. Y. Tan, C. Boothroyd, K. P. Loh, E. S. Tok, Y.-L. Foo, *Nano Lett.*, **2006**, *6*, 449–452.
23 J. F. Blackburn, Y. Yan, C. Engtrakul, P. A. Parilla, K. Jones, T. Gennett, A. C. Dillon, M. J. Haben, *Chem. Mater.*, **2006**, *18*, 2558–2566.
24 P. Coquay, A. Peiney, E. D. Grave, E. Flauhaut, R. E. Vandenberghe, C. Laurent, *J. Phys. Chem. B*, **2005**, *109*, 17813–17824.
25 W. Ren, H.-M. Cheng, *J. Phys. Chem. B*, **2005**, *109*, 7169–7173.
26 O. T. Heyning, P. Bernier, M. Glerup, *Chem. Phys. Lett.*, **2005**, *409*, 43–47.
27 S. R. C. Vivekchand, R. Jayakanth, A. Govindaraj, C. N. R. Rao, *Small*, **2005**, *1*, 920–923.
28 Y.-Q. Xu, H. Peng, R. H. Hauge, R. E. Smalley, *Nano Lett.*, **2005**, *5*, 163–168.
29 Y. Kim, D. E. Luzzi, *J. Phys. Chem. B*, **2005**, *109*, 16636–16643.
30 H. Hu, A. Yu, E. Kim, B. Zhao, M. E. Itkis, E. Bekyarova, R. C. Haddon, *J. Phys. Chem. B*, **2005**, *109*, 11520–11524.
31 M. E. Itkis, D. Perea, R. Jung, S. Nijogi, R. C. Haddon, *J. Am. Chem Soc.*, **2005**, *127*, 3439–3448.
32 (a) M. E. Itkis, D. E. Perea, S. Niyogi, S. M. Rickard, M. A. Hamon, H. Hu, B. Zhao, R. C. Haddon, *Nano Lett.*, **2003**, *3*, 309–314; (b) B. J. Landi, H. J. Ruf, C. M. Evans, C. D. Cress, R. P. Raffaelle, *J. Phys. Chem. B*, **2005**, *109*, 9952–9965.
33 C.-M. Yang, J. S. Park, K. H. An, S. C. Lim, K. Seo, B. Kim, K. A. Park, S. Han, C. Y. Park, Y. H. Lee, *J. Phys. Chem. B*, **2005**, *109*, 19242–19248.
34 Y. Maeda, S. Kimura, M. Kanda, Y. Hirashima, T. Hasegawa, T. Wakahara, Y. Lian, T. Nakahodo, T. Tsuchiya, T. Akasaka, J. Lu, X. Zhang, Z. Gao, Y. Yu, S. Nagase, S. Kazaoui, N. Minani, T. Shimizu, H. Tokumoto, R. Saito, *J. Am. Chem. Soc.*, **2005**, *127*, 10287–10290.
35 C. M. Moyon, N. Izard, E. Doris, C. Mioskowski, *J. Am. Chem. Soc.*, **2006**, *128*, 6552–6553.
36 J. Lu, S. Nagase, X. Zhang, D. Wang, M. Ni, Y. Maeda, T. Wakahara, T. Nakahodo, T. Tsuchiya, T. Akasaka, Z. Gao, D. Yu, H. Ye, W. N. Mei, Y. Zhou, *J. Am. Chem. Soc.*, **2006**, *128*, 5114–5118.
37 H. Huang, R. Maruyama, K. Noda, H. Kajiura, K. Kadono, *J. Phys. Chem. B*, **2006**, *110*, 7316–7320.
38 H. Huang, H. Kajiura, R. Maruyama, K. Kadono, K. Noda, *J. Phys. Chem. B*, **2006**, *110*, 4686–4690.

39 K. C. Park, T. Hayashi, H. Tomiyasu, M. Endo, M. S. Dresselhaus, *J. Mater. Chem.*, **2005**, *15*, 407–411.
40 K. M. Lee, L. Li, L. Dai, *J. Am. Chem. Soc.*, **2005**, *127*, 4122–4123.
41 C.-Y. Hong, Y.-Z. You, C.-Y. Pan, *Chem. Mater.*, **2005**, *17*, 2247–2254.
42 K. J. Ziegler, D. J. Schmidt, U. Rauwald, K. N. Shah, E. L. Flor, R. H. Hauge, R. E. Smalley, *Nano Lett.*, **2005**, *5*, 2355–2359.
43 K. J. Ziegler, Z. Gu, H. Peng, E. L. Flor, R. H. Hauge, R. E. Smalley, *J. Am. Chem. Soc.*, **2005**, *127*, 1541–1547.
44 Q. Li, I. A. Kinloch, A. H. Windle, *Chem. Commun.*, **2005**, 3283–3285.
45 V. A. Sinani, M. K. Gheith, A. A. Yaroslavov, A. A. Raknyanskaya, K. Sun, A. A. Mamedov, J. P. Wicksted, N. A. Kotov, *J. Am. Chem. Soc.*, **2005**, *127*, 3463–3472.
46 H. Xie, A. O. Acevedo, V. Zorbas, R. H. Baughman, R. K. Draper, I. H. Musselman, A. B. Dalton, G. R. Dieckmann, *J. Mater. Chem.*, **2005**, *15*, 1734–1741.
47 J. Chen, C. P. Collier, *J. Phys. Chem. B*, **2005**, *109*, 7605–7609.
48 H. Paloniemi, T. Aaritalo, T. Laiho, H. Like, N. Kocharova, K. Haapakka, F. Terzi, R. Seeber, J. Lukkari, *J. Phys. Chem. B*, **2005**, *109*, 8634–8642.
49 B. K. Price, J. L. Hudson, J. M. Tour, *J. Am. Chem. Soc.*, **2005**, *127*, 14867–14870.
50 Y. Zhang, Y. Shen, J. Li, L. Niu, S. Dong, A. Ivaska, *Langmuir*, **2005**, *21*, 4797–4800.
51 (a) M. S. Arnold, M. O. Guler, M. C. Hersam, S. I. Stupp, *Langmuir*, **2005**, *21*, 4705–4709; (b) A. O. Acevedo, H. Xie, V. Zorbas, W. M. Sampson, A. B. Dalton, R. H. Baughman, R. K. Drapper, I. H. Musselman, R. Dieckmann, *J. Am. Chem. Soc.*, **2005**, *127*, 9512–9517.
52 F. Liang, J. M. Beach, P. K. Rai, W. Guo, R. H. Hauge, M. Pasquali, R. E. Smalley, W. E. Billups, *Chem. Mater.*, **2006**, *18*, 1520–1524.
53 Y. Tomonari, H. Murakami, N. Nakashima, *Chem. Eur. J.*, **2006**, *12*, 4027–4034.
54 T. G. Hedderman, S. M. Keogh, G. Chambers, H. J. Byrne, *J. Phys. Chem. B*, **2006**, *110*, 3895–3901.
55 S. S. Karajanagi, H. Yang, P. Asuri, E. Sellitto, J. S. Dordick, R. S. Kane, *Langmuir*, **2006**, *22*, 1392–1295.
56 J. J. Stephenson, J. L. Hudson, S. Azad, J. M. Tour, *Chem. Mater.*, **2006**, *18*, 374–377.
57 S. Chen, W. Shen, G. Wu, D. Chen, M. Jiang, *Chem. Phys. Lett.*, **2005**, *402*, 312–317.
58 U. D. Weglikowska, V. Skakalova, R. Graupner, S. H. Jhang, B. H. Kim, H. J. Lee, L. Ley, Y. W. Park, S. Berber, D. Tomanek, S. Roth, *J. Am. Chem. Soc.*, **2005**, *127*, 5125–5131.
59 A. Gromov, S. Dittmer, J. Svensson, O. A. Nerushev, S. A. P. Garcia, L. L. Jimenez, R. Rychwalski, E. E. B. Campbell, *J. Mater. Chem.*, **2005**, *15*, 3334–3339.
60 J. Li, H. Grennberg, *Chem. Eur. J.*, **2006**, *12*, 3869–3875.
61 B. Khare, P. Wilhite, B. Tran, E. Teixeira, K. Fresquez, D. N. Mvondo, C. Bauschlicher Jr. and M. Meyyappan, *J. Phys. Chem. B*, **2005**, *109*, 23466–23472.
62 M. S. Raghuveer, A. Kumar, M. J. Frederick, G. P. Louie, P. G. Ganesan, G. Ramanath, *Adv. Mater.*, **2006**, *18*, 547–552.
63 M. J. Park, J. K. Lee, B. S. Lee, Y.-W. Lee, I. S. Choi, S.-G. Lee, *Chem. Mater.*, **2006**, *18*, 1546–1551.
64 J. M. Simmons, B. M. Nichols, S. E. Baker, M. S. Marcus, O. M. Castellini, C.-S. Lee, R. J. Hammers, M. A. Eriksson, *J. Phys. Chem. B*, **2006**, *110*, 7113–7118.
65 Y. Wang, Z. Iqbal, S. Mitra, *J. Am. Chem. Soc.*, **2006**, *128*, 95–99.
66 S. Badaire, C. Zakri, M. Maugey, A. Derre, J. N. Barisci, G. Wallace, P. Poulin, *Adv. Mater.*, **2005**, *17*, 1673–1676.
67 N. W. S. Kam, Z. Liu, H. Dai, *J. Am. Chem. Soc.*, **2005**, *127*, 12492–12493.
68 S. Murugesan, T.-J. Park, H. Yang, S. Mousa, R. J. Linhardt, *Langmuir*, **2006**, *22*, 3461–3463.
69 H. Maramatsu, Y. A. Kim, T. Hayashi, M. Endo, A. Yonemoto, H. Arikai,

F. Okino, H. Touhara, *Chem. Commun.*, **2005**, 2002–2004.

70 C. Y. Li, L. Li, W. Cai, S. L. Kodjie, K. K. Tenneti, *Adv. Mater.*, **2005**, *17*, 1198–1202.

71 B. M. Quinn, C. Dekker, S. G. Lemay, *J. Am. Chem. Soc.*, **2005**, *127*, 6146–6147.

72 Y. Lin, X. Cui, C. Yen, C. M. Wai, *J. Phys. Chem. B*, **2005**, *109*, 14410–14415.

73 Z. Sun, Z. Liu, B. Han, Y. Wang, J. Du, Z. Xie, G. Han, *Adv. Mater.*, **2005**, *17*, 928–932.

74 E. A. Whitsitt, V. C. Moore, R. E. Smalley, A. Barron, *J. Mater. Chem.*, **2005**, *15*, 4678–4687.

75 R. Loscutova, A. R. Barron, *J. Mater. Chem.*, **2005**, *15*, 4346–4353.

76 F. Stoffelbach, A. Aqil, C. Jerome, R. Jerome, C. Detrembleur, *Chem. Comm.*, **2005**, 4532–4533.

77 D. M. Guldi, G. M. A. Rahman, V. Sgobba, N. A. Kotov, D. Bonifazi, M. Prato, *J. Am. Chem. Soc.*, **2006**, *128*, 2315–2323.

78 A. Gomathi, S. R. C. Vivekchand, A. Govindaraj, C. N. R. Rao, *Adv. Mater.*, **2005**, *17*, 2757–2561.

79 P. M. F. J. Costa, S. Friedrichs, J. Sloan, M. L. H. Green, *Chem. Mater.*, **2005**, *17*, 3122–3129.

80 B.-Y. Sun, Y. Sato, K. Suenaga, T. Okazaki, N. Kishi, T. Sugai, S. Bhandow, S. Iijima, H. Shinohara, *J. Am. Chem. Soc.*, **2005**, *127*, 17972–17973.

81 S. Li, P. He, J. Dong, Z. Guo, L. Dai, *J. Am. Chem. Soc.*, **2005**, *127*, 14–15.

82 M. E. Sousa, S. G. Cloutier, K. Q. Jian, B. S. Weissman, R. H. Hurt, G. P. Crawford, *Appl. Phys. Lett.*, **2005**, *87*, 173115.

83 J. Y. Huang, S. Chen, S. H. Jo, Z. Wang, D. X. Han, G. Chen, M. S. Dresselhaus, Z. F. Ren, *Phys. Rev. Lett.*, **2005**, *94*, 236802.

84 H. J. Li, W. G. Lu, J. J. Li, X. D. Bai, C. Z. Gu, *Phys. Rev. Lett.*, **2005**, *95*, 086601.

85 P. J. Herrero, J. Kong, H. S. J. Van der Zant, C. Dekker, L. P. Kouwenhoven, S. De Franceschi, *Phys. Rev. Lett.*, **2005**, *94*, 156802.

86 F. Plentz, H. B. Ribeiro, A. Jorio, M. S. Strano, M. A. Pimenta, *Phys. Rev. Lett.*, **2005**, *95*, 247401.

87 Y.-Z. Ma, L. Valkunas, S. L. Dexheimer, S. M. Bachilo, G. R. Fleming, *Phys. Rev. Lett.*, **2005**, *94*, 157402.

88 A. Hartschuh, H. Qian, A. J. Meixner, N. Anderson, L. Novotny, *Nano Lett.*, **2005**, *5*, 2310–2313.

89 M. Y. Sfeir, T. Beetz, F. Wang, L. Huang, X. M. Huang, M. Huang, J. Hone, S. O'Brien, J. A. Misewich, T. F. Heinz, L. Wu, Y. Zhu, L. E. Brus, *Science*, **2006**, *312*, 554–556.

90 D. Karaiskaj, C. Engtrakul, T. McDonald, M. J. Heben, A. Mascarenhas, *Phys. Rev. Lett.*, **2006**, 106805.

91 Y. Lin, B. Zhou, R. B. Martin, K. B. Henbest, B. A. Harruff, J. E. Riggs, Z.-X. Guo, L. F. Allard, Y.-P. Sun, *J. Phys. Chem. B*, **2005**, *109*, 14779–14782.

92 V. V. Didenko, V. C. Moore, D. S. Baskin, R. E. Smalley, *Nano Lett.*, **2005**, *5*, 1563–1567.

93 T. Hertel, A. Hagen, V. Talalaev, K. Arnold, F. Hennrich, M. Kappes, S. Rosenthal, J. McBride, H. Ulbricht, E. Flahaut, *Nano Lett.*, **2005**, *5*, 511–514.

94 G. Dukovic, F. Wang, D. Song, M. Y. Sfeir, T. F. Heinz, L. E. Brus, *Nano Lett.*, **2005**, *5*, 2314–2318.

95 J. Zhang, H. I. Kim, C. H. Oh, X. Sun, H. Lee, *Appl. Phys. Lett.*, **2006**, *88*, 053123.

96 C.-X. Sheng, Z. V. Vardeny, A. B. Dalton, R. H. Baughman, *Phys. Rev. B*, **2005**, *71*, 125427.

97 S. Reich, M. Dworzak, A. Hoffmann, C. Thomsen, M. S. Strano, *Phys. Rev. B*, **2005**, *71*, 033402.

98 S. Banerjee, T. H. Benny, S. Sambasivan, D. A. Fischer, J. A. Misewich, S. S. Wong, *J. Phys. Chem. B*, **2005**, *109*, 8489–8495.

99 F. Hennrich, R. Krupke, S. Lebedkin, K. Arnold, R. Fischer, D. E. Resasco, M. M. Kapps, *J. Phys. Chem. B*, **2005**, *109*, 10567–10573.

100 M. Paillet, S. Langlois, J.-L. Sauvajol, L. Marty, A. Iaia, C. Naud, V. Bouchait,

A. M. Bonnot, *J. Phys. Chem. B*, **2006**, *110*, 164–169.

101 S. K. Doorn, D. A. Heller, P. W. Barone, M. L. Usrey, M. S. Strano, *Appl. Phys. A*, **2004**, *78*, 1147–1155.

102 Z. Zhou, X. Dou, L. Ci, L. Song, D. Liu, Y. Gao, J. Wang, L. Liu, W. Zhou, S. Xie, D. Wan, *J. Phys. Chem. B*, **2006**, *110*, 1206–1209.

103 S. K. Doorn, L. Zheng, M. J. O'Connell, Y. Zhu, S. Huang, J. Liu, *J. Phys. Chem. B*, **2005**, *109*, 3751–3758.

104 C. Fantini, A. Jorio, M. Souza, R. Saito, Ge. G. Samsonidze, M. S. Dresselhaus, M. A. Pimenta, *Phys. Rev. B*, **2005**, *72*, 085446.

105 S. B. Cronin, A. K. Swan, M. S. Unlu, B. B. Goldberg, M. S. Dresselhaus, M. Tinkham, *Phys. Rev. B*, **2005**, *72*, 035425.

106 M. Kalbac, L. Kavan, M. Zukalova, L. Dunsch, *Chem. Eur. J.*, **2006**, *12*, 4451–4457.

107 B. J. Bauer, E. K. Hobbie, M. L. Becker, *Macromolecules*, **2006**, *39*, 2637–2642.

108 J. Heo, M. Bockrath, *Nano Lett.*, **2005**, *5*, 853–857.

109 M. E. Itkis, F. Borondics, A. Yu, R. C. Haddon, *Science*, **2006**, *312*, 413–416.

110 X. Qiu, M. Freitag, V. Perebeinos, P. Avouris, *Nano Lett.*, **2005**, *5*, 749–752.

111 J. Cao, Q. Wang, H. Dai, *Nat. Mater.*, **2005**, *4*, 745–749.

112 A. Vijayaraghavan, K. Kanzaki, S. Suzuki, Y. Kobayashi, H. Inokawa, Y. Ono, S. Kar, P. M. Ajayan, *Nano Lett.*, **2005**, *5*, 1575–1579.

113 W. Zhou, J. Vavro, N. M. Nemes, J. E. Fischer, F. Borondics, K. Kamaras, D. B. Tanner, *Phys. Rev. B*, **2005**, *71*, 205423.

114 N. O. V. Plank, G. A. Forrest, R. Cheung, A. J. Alexander, *J. Phys. Chem. B*, **2005**, *109*, 22096–22101.

115 J. Hone, M. C. Llaguno, M. J. Biercuk, A. T. Johnson, B. Batlogg, Z. Benes, J. E. Fischer, *Appl. Phys. A*, **2002**, 339–343.

116 M. Fujii, X. Zhang, H. Xie, H. Ago, K. Takahashi, T. Ikuta, H. Abe, T. Shimizu, *Phys. Rev. Lett.*, **2005**, *95*, 065502.

117 H.-Y. Chiu, V. V. Deshpande, H. W. Ch. Postma, C. N. Lau, C. Miko, L. Forro, M. Bockrath, *Phys. Rev. Lett.*, **2005**, *95*, 226101.

118 E. Pop, D. Mann, Q. Wang, K. Goodson, H. Dai, *Nano Lett.*, **2006**, *6*, 96–100.

119 C. Yu, L. Shi, Z. Yao, D. Li, A. Majumdar, *Nano Lett.*, **2005**, *5*, 1842–1846.

120 K. J. Donovan, K. Scott, *Phys. Rev. B*, **2005**, *72*, 195432.

121 P. J. Herrero, J. Kong, H. S. J. van der Zant, C. Dekker, L. P. Kouwenhoven, S. De Franceschi, *Nature*, **2005**, *434*, 484–488.

122 (a) W. Song, I. A. Kinloch, A. H. Windle, *Science*, **2003**, *302*, 1363; (b) W. Song, A. H. Windle, *Macromolecules*, **2005**, *38*, 6181–6188; (c) P. K. Rai, R. A. Pinnick, A. N. G. P. Vasquez, V. A. Devis, H. K. Schmidt, R. H. Hauge, R. E. Smalley, M. Pasquali, *J. Am. Chem. Soc.*, **2006**, *128*, 591–595; (d) X. Ye, D. Y. Sun, X. G. Gong, *Phys. Rev. B*, **2005**, *72*, 035454.

123 I. Dierking, G. Scalia, P. Morales, *J. Appl. Phys.*, **2005**, *97*, 044309.

124 J. Y. Huang, S. Chen, Z. Q. Wang, K. Kempa, Y. W. Wang, S. H. Jo, G. Chen, M. S. Dresselhaus, Z. F. Ren, *Nature*, **2006**, *439*, 281.

125 P. Miaudet, S. Badaire, M. Maugey, A. Derre, V. Pichot, P. Launois, P. Poulin, C. Zakri, *Nano Lett.*, **2005**, *5*, 2212–2215.

126 M. Motta, Y.-L. Li, L. Kinloch, A. Windle, *Nano Lett.*, **2005**, *5*, 1529–1533.

127 S. I. Cha, K. T. Kim, S. N. Arshad, C. B. Mo, S. H. Hong, *Adv. Mater.*, **2005**, *17*, 1377–1381.

128 A. Cao, V. P. Veedu, X. Li, Z. Yao, M. N. G. Nejhad, P. M. Ajayan, *Nat. Mater.*, **2005**, *4*, 540–545.

129 M. Majumder, N. Chopra, R. Andrews, B. J. Hinds, *Nature*, **2005**, *438*, 44.

130 Q. Fu, J. Liu, *J. Phys. Chem. B*, **2005**, *109*, 13406–13408.

131 L. Zhu, Y. Xiu, J. Xu, P. A. Tamirisa, D. W. Hess, C.-P. Wong, *Langmuir*, **2005**, *21*, 11208–11212.

132 L. Huang, S. P. Lau, H. Y. Yang, E. S. P. Leong, S. F. Yu, S. Prawer, *J. Phys. Chem. B*, **2005**, *109*, 7746–7748.

133 J. Y. Chen, A. Kutuna, C. P. Collier, K. P. Giapis, *Science*, **2006**, *310*, 1480–1483.

134 C.-L. Sun, L.-C. Chen, M.-C. Su, L.-S. Hong, O. Chyan, C.-Y. Hsu, K.-H. Chen, T.-F. Chang, L. Chang, *Chem. Mater.*, **2005**, *17*, 3749–3753.

135 (a) A. Kongkanand, S. Kuwabata, G. Girishkumar, P. Kamat, *Langmuir*, **2006**, *22*, 2392–2396; (b) W. Li, X. Wang, Z. Chen, M. Waje, Y. Yan, *Langmuir*, **2005**, *21*, 9386–9389; (c) Y. Lin, X. Cui, C. H. Yen, C. M. Wai, *Langmuir*, **2005**, *21*, 11474–11479.

136 C. Du, J. Yah, N. Pan, *J. Mater. Chem.*, **2005**, *15*, 548–550.

137 C. Zhou, S. Kumar, C. D. Doyle, J. M. Tour, *Chem. Mater.*, **2005**, *17*, 1997–2002.

138 A. Chou, T. Bocking, N. K. Singh, J. J. Gooding, *Chem. Commun.*, **2005**, 842–844.

139 J.-S. Ye, H. F. Cui, X. Liu, T. M. Lim, W.-D. Zhang, F.-S. Sheu, *Small*, **2005**, *1*, 560–565.

140 F. Beguin, K. Szostak, G. Lota, E. Frackowiak, *Adv. Mater.*, **2005**, *17*, 2380–2384.

141 C. Du, J. Yeh, N. Pan, *Nanotechnology*, **2005**, *16*, 350–353.

142 E. S. Snow, F. K. Perkins, *Nano Lett.*, **2005**, *5*, 2414–2417.

143 M. K. Haas, J. M. Zielinski, G. Danstin, C. G. Coe, G. P. Pez, A. C. Copper, *J. Mater. Res.*, **2005**, *20*, 3214–3223.

144 D.-Z. Guo, G.-M. Zhang, Z.-X. Zhang, Z.-Q. Xue, Z.-N. Gu, *J. Phys. Chem. B*, **2006**, *110*, 1571–1575.

145 T. Hasobe, S. Fukuzumi, P. V. Kamat, *Angew. Chem. Int. Ed.*, **2006**, *45*, 755–759.

146 D. M. Guldi, G. M. A. Rahman, M. Prato, N. Jux, S. Qin, W. Ford, *Angew. Chem. Int. Ed.*, **2005**, *44*, 2015–2018.

147 I. Robel, B. A. Bunker, P. V. Kamat, *Adv. Mater.*, **2005**, *17*, 2458–2463.

148 M. Zhang, W. Gorski, *Anal. Chem.*, **2005**, *77*, 3960–3965.

149 C. Staii, A. T. Johnson Jr., M. Chen, A. Gelperin, *Nano Lett.*, **2005**, *5*, 1774–1778.

150 C. Hu, Y. Zhang, G. Bao, Y. Zhang, M. Liu, Z. L. Wang, *J. Phys. Chem. B*, **2005**, *109*, 20072–20076.

151 C. Song, P. E. Pehrsson, W. Zhao, *J. Phys. Chem. B*, **2005**, 109, 21634–21639.

152 A. Zribi, A. Knobloch, R. Rao, *Appl. Phys. Lett.*, **2005**, 86, 203112.

153 J. S. Oakley, H.-T. Wang, B. S. Kang, Z. Wu, F. Ren, A. G. Rinzler, S. J. Pearton, *Nanotechnology*, **2005**, *16*, 2218–2221.

154 P. Nednoor, N. Chopra, V. Gavalas, L. G. Bachas, B. J. Hinds, *Chem. Mater.*, **2005**, *17*, 3595–3599.

155 X. Yu, S. N. Kim, F. Papadimitrakopoulos, J. F. Rusling, *Mol. BioSyst.*, **2005**, *1*, 70–78.

156 M. K. Gheith, V. A. Sinani, J. P. Wicksted, R. L. Matts, N. A. Kotov, *Adv. Mater.*, **2005**, *17*, 2663–2670.

157 B. Zhao, H. Hu, S. K. Mandal, R. C. Haddon, *Chem. Mater.*, **2005**, *17*, 3235–3241.

158 L. P. Zenello, B. Zhao, H. Hu, R. C. Haddon, *Nano Lett.*, **2006**, *6*, 562–567.

159 D. A. Heller, S. Baik, T. E. Eurell, M. S. Strano, *Adv. Mater.*, **2005**, *17*, 2793–2799.

160 W. Wu, S. Weikowski, G. Pastorin, M. Benincasa, C. Klumpp, J.-P. Briand, R. Gennaro, M. Prato, A. Bianco, *Angew. Chem. Int. Ed.*, **2005**, *44*, 6358–6362.

161 P. W. Barone, R. S. Parker, M. S. Strano, *Anal. Chem.*, **2005**, *77*, 7556–7562.

162 P. P. Joshi, S. A. Merchant, Y. Wang, D. W. Schmidtke, *Anal. Chem.*, **2005**, *77*, 3183–3188.

163 E. S. Jeng, A. E. Moll, A. C. Roy, J. B. Gastala, M. S. Strano, *Nano Lett.*, **2006**, *6*, 371–375.

164 D. A. Heller, E. S. Jeng, T.-K. Yeung, B. M. Martinez, A. E. Moll, J. B. Gastala, M. S. Strano, *Science*, **2006**, *311*, 508–511.

165 S. Roy, H. Vedala, W. Choi, *Nanotechnology*, **2006**, *17*, S14–S18.

166 K. A. Joshi, M. Prouza, M. Kum, J. Wang, J. Tang, R. Haddon, W.

Chen, A. Mulchandani, *Anal. Chem.*, **2006**, *78*, 331–336.

167 G. Jia, H. Wang, L. Yan, X. Wang, R. Pei, T. Yan, Y. Zhao, X. Guo, *Environ. Sci. Technol.*, **2005**, *39*, 1378–1383.

168 X. Chen, U. C. Tam, J. L. Czlapinski, G. S. Lee, D. Rabuka, A. Zettl, C. R. Bertozzi, *J. Am. Chem. Soc.*, **2006**, *128*, 6292–6293.

169 N. W. S. Kam, H. Dai, *J. Am. Chem. Soc.*, **2005**, *127*, 6021–6026.

170 H.-M. So, K. Won, Y. H. Kim, B.-K. Kim, B. H. Ryu, P. S. Na, H. Kim, J.-O. Lee, *J. Am. Chem. Soc.*, **2005**, *127*, 11906–11907.

171 D. Baskaran, J. W. Mays, M. S. Bratcher, *Chem. Mater.*, **2005**, *17*, 3389–3397.

172 H.-J. Lee, S.-J. Oh, J.-Y. Choi, J. W. Kim, J. Han, L.-S. Tan, J.-B. Baek, *Chem. Mater.*, **2005**, *17*, 5057–5064.

173 C. Gao, Y. Z. Jin, H. Kong, R. L. D. Whitby, S. F. A. Acquah, G. Y. Chen, H. Qian, A. Hartschuh, S. R. P. Silva, S. Henley, P. Fearon, H. W. Kroto, D. R. M. Walton, *J. Phys. Chem. B*, **2005**, *109*, 11925–11932.

174 H. Hu, Y. Ni, S. K. Mandal, V. Montana, B. Zhao, R. C. Haddon, V. Parpura, *J. Phys. Chem. B*, **2005**, *109*, 4285–4289.

175 B. Zhao, A. Yu, D. Perea, R. C. Haddon, *J. Am. Chem. Soc.*, **2005**, *127*, 8197–8203.

176 N. I. Kovtyukhova, T. E. Mallouk, *J. Phys. Chem. B*, **2005**, *109*, 2540–2545.

177 M. Wu, G. A. Snook, V. Gupta, M. Shaffer, D. J. Fray, G. Z. Chen, *J. Mater. Chem.*, **2005**, *15*, 2297–2303.

178 J. Cui, C. P. Daghlian, U. J. Gibson, *J. Phys. Chem. B*, **2005**, *109*, 11456–11460.

179 M. Li, E. Dujardin, S. Mann, *Chem. Comm.*, **2005**, 4952–4954.

180 Y. Lin, X. Cui, *J. Mater. Chem.*, **2006**, *16*, 585–592.

181 L. Jiang, L. Gao, *J. Mater. Chem.*, **2005**, *15*, 260–266.

182 N. R. Ravikkar, L. S. Schadler, A. Vijayaraghavan, Y. Zhao, B. Wei, P. M. Ajayan, *Chem. Mater.*, **2005**, *17*, 974–983.

183 M. E. Kozlov, R. C. Capps, W. S. Sampson, V. H. Ebron, J. P. Ferrais, R. H. Baughman, *Adv. Mater.*, **2005**, *17*, 614–617.

184 J. Gao, M. E. Itkis, A. Yu, E. Bekyarova, B. Zhao, R. C. Haddon, *J. Am. Chem. Soc.*, **2005**, *127*, 3847–3854.

185 M. Dehonor, K. M. Varlot, A. G. Montiel, C. Gautheir, J. Y. Cavaille, H. Terrones, M. Terrones, *Chem. Comm.*, **2005**, 5349–5351.

186 E. Kymakis, G. A. J. Amaratunga, *J. Appl. Phys.*, **2006**, *99*, 084302.

187 C. Wei, L. Dai, A. Roy, T. B. Tolle, *J. Am. Chem. Soc.*, **2006**, *128*, 1412–1413.

188 S. Kazaoui, N. Minami, B. Nalini, Y. Kim, K. Hara, *J. Appl. Phys.*, **2005**, *98*, 084314.

189 G. M. A. Rahman, D. M. Guldi, R. Cagnoli, A. Mucci, L. Schenetti, L. Vaccari, M. Prato, *J. Am. Chem. Soc.*, **2005**, *127*, 10051–10057.

190 D. M. Guldi, H. Taieb, G. M. A. Rahman, N. Tagmatarchis, M. Prato, *Adv. Mater.*, **2005**, *17*, 871–875.

191 M. i. h. Panhuis, R. Sainz, P. C. Innis, L. A. P. K. Maguire, A. M. Benito, M. T. Martinez, S. E. Moulton, G. G. Wallace, W. K. Maser, *J. Phys. Chem. B*, **2005**, *109*, 22725–22729.

192 S. V. Ahir, E. M. Terentjev, *Nat. Mater.*, **2005**, *4*, 491–495.

193 Y. Zhou, L. Hu, G. Gruner, *Appl. Phys. Lett.*, **2006**, *88*, 123109.

194 A. D. Pasquier, H. E. Unalan, A. Kanwal, S. Miller, M. Chhowalla, *Appl. Phys. Lett.*, **2005**, *87*, 203511.

195 X. Liu, J. Ly, S. Han, D. Zhang, A. Requicha, M. E. Thompson, C. Zhou, *Adv. Mater.*, **2005**, *17*, 2727–2732.

196 M. Olek, K. Kempa, S. Jurga, M. Giersig, *Langmuir*, **2005**, *21*, 3146–3152.

197 J. Suhr, N. Koratkar, P. Keblinski, P. M. Ajayan, *Nat. Mater.*, **2005**, *4*, 134–137.

198 S. Zhang, N. Zhang, C. Huang, K. Ren, Q. Zhang, *Adv. Mater.*, **2005**, *17*, 1897–1901.

199 M. B. Bryning, D. E. Milkie, M. F. Islam, J. M. Kikkawa, A. G. Yodh, *Appl. Phys. Lett.*, **2005**, *87*, 161909.

200 H. Huang, C. Liu, Y. Wu, S. Fan, *Adv. Mater.*, **2005**, *17*, 1652–1656.
201 W. Fang, H.-Y. Chu, W.-K. Hsu, T.-W. Cheng, N.-H. Tai, *Adv. Mater.*, **2005**, *17*, 2987–2992.
202 M. L. Shofner, V. N. Khabashesku, E. V. Barrera, *Chem. Mater.*, **2006**, *18*, 906–913.
203 G. M. Spinks, V. Mottaghitalab, M. B. Samani, P. G. Whitten, G. G. Wallace, *Adv. Mater.*, **2006**, *18*, 637–640.
204 T. Kashiwagi, F. Du, J. F. Douglas, K. I. Winey, R. H. Harris Jr, J. R. Shields, *Nat. Mater.*, **2005**, *4*, 928–933.
205 M. Endo, S. Koyama, Y. Matsuda, T. Hayashi, Y.-A. Kim, *Nano Lett.*, **2005**, 101–105.
206 K. B. K. Teo, E. Minoux, L. Hudanski, F. Peauger, J.-P. Schnell, L. Gangloff, P. Legagneux, D. Dieumegard, G. A. J. Amaratunga, W. I. Milne, *Nature*, **2005**, *437*, 968.
207 B.-J. Yoon, E. H. Hong, S. E. Jee, D.-M. Yoon, D.-S. Shim, G.-Y. Son, Y. J. Lee, K.-H. Lee, H. S. Kim, C. G. Park, *J. Am. Chem. Soc.*, **2005**, *127*, 8234–8235.
208 M.-S. Jung, Y. K. Ko, D.-H. Jung, D. H. Choi, H.-T. Jung, J. N. Heo, B. H. Sohn, Y. W. Jin, J. Kim, *Appl. Phys. Lett.*, **2005**, *87*, 013114.
209 J. U. Lee, *Appl. Phys. Lett.*, **2005**, *87*, 073101.
210 S.-H. Jung, S.-H. Jeong, S.-U. Kim, S.-K. Hwang, P.-S. Lee, K.-H. Lee, J.-H. Ko, E. Bae, D. Kang, W. Park, H. Oh, J.-J. Kim, H. Kim, C.-G. Park, *Small*, **2005**, *1*, 553–559.
211 Q. Fu, J. Liu, *Langmuir*, **2005**, *21*, 1162–165.
212 L. Dong, V. Chirayos, J. Bush, J. Jiao, V. M. Dubin, R. V. Chebian, Y. Ono, J. F. Conley Jr, B. D. Ulrich, *J. Phys. Chem. B*, **2005**, *109*, 13148–13153.
213 C. Klinke, J. B. Hannon, A. Afzali, P. Avouris, *Nano Lett.*, **2006**, *6*, 906–910.
214 Z. Chen, J. Appenzeller, Y.-M. Lin, J. S. Oakley, A. G. Rinzler, J. Tang, S. J. Wind, P. M. Solomon, P. Avouris, *Science*, **2006**, *311*, 1735.
215 P. J. Herrero, J. A. v. Dam, L. P. Kouwenhoven, *Nature*, **2006**, *439*, 953–956.
216 C. Kocabas, S.-H. Hur, A. Gaur, M. A. Meitl, M. Shim, J. A. Rogers, *Small*, **2005**, *1*, 1110–1116.
217 A. Javey, R. Tu, D. B. Farmer, J. Guo, R. G. Gordon, H. Dai, *Nano Lett.*, **2005**, *5*, 345–348.
218 T. Ozel, A. Gaur, J. A. Rogers, M. Shim, *Nano Lett.*, **2005**, *5*, 905–911.
219 R. V. Seidel, A. P. Graham, J. Kretz, B. Rajasekharan, G. S. Duesberg, M. Liebau, E. Unger, F. Kreupl, W. Hoenlein, *Nano Lett.*, **2005**, *5*, 147–150.
220 J. Chen, C. Klinke, A. Afzali, P. Avouris, *Appl. Phys. Lett.*, **2005**, *86*, 123108.
221 E. Artukovic, M. Kaempgen, D. S. Hecht, S. Roth, G. Gruner, *Nano Lett.*, **2005**, *5*, 757–760.
222 T. Takenobu, T. Takahashi, T. Kanbara, K. Tsukagoshi, Y. Aoyagi, Y. Iwasa, *Appl. Phys. Lett.*, **2006**, *88*, 033511.
223 Q. Cao, Z.-T. Zhu, M. G. Lemaitre, M.-G. Xia, M. Shim, J. A. Rogers, *Appl. Phys. Lett.*, **2006**, *88*, 113511.
224 C. Kocabas, M. Shim, J. A. Rogers, *J. Am. Chem. Soc.*, **2006**, *128*, 4540–4541.
225 S. Auvray, V. Derycke, M. Golfman, A. Filoramo, O. Jost, J.-P. Bourgoin, *Nano Lett.*, **2005**, *5*, 451–455.
226 S.-H. Hur, M.-H. Yoon, A. Gaur, M. Shim, A. Facchetti, T. J. Marks, J. A. Rogers, *J. Am. Chem. Soc.*, **2005**, *127*, 13808–13809.
227 H. Y. Yu, D. S. Lee, S. H. Lee, S. S. Kim, S. W. Lee, Y. W. Park, U. D. Weglikowskaand, S. Roth, *Appl. Phys. Lett.*, **2005**, *87*, 163118.
228 X. Guo, L. Huang, S. O'Brein, P. Kim, C. Nuckolls, *J. Am. Chem. Soc.*, **2005**, *127*, 15045–15047.
229 A. K. Flatt, B. Chen, J. M. Tour, *J. Am. Chem. Soc.*, **2005**, *127*, 8918–8919.
230 P. R. Bandaru, C. Daraio, S. Jin, A. M. Rao, *Nat. Mater.*, **2005**, *4*, 663–666.
231 B. R. Perkins, D. P. Wang, D. Soltman, A. J. Yin, J. M. Xu,

Zaslavsky, *Appl. Phys. Lett.*, **2005**, *87*, 123504.

232 D.-H. Kim, J. Huang, B. K. Rao, W. Choi, *J. Appl. Phys.*, **2006**, *99*, 056106.

233 J. Cao, Q. Wang, D. Wang, H. Dai, *Small*, **2005**, *1*, 138–141.

234 D. Kang, N. Park, J. Hyun, E. Bae, J. Ko, J. Kim, W. Park, *Appl. Phys. Lett.*, **2005**, *86*, 093105.

235 X. Liu, Z. Luo, S. Han, T. Tang, D. Zhang, C. Zhou, *Appl. Phys. Lett.*, **2005**, *86*, 243501.

236 D. R. Hines, S. Mezhenny, M. Breban, E. D. Williams, V. W. Ballarotto, G. Esen, A. Southard, M. S. Fuhrer, *Appl. Phys. Lett.*, **2005**, *86*, 163101.

237 M. Hughes, G. M. Spinks, *Adv. Mater.*, **2005**, *17*, 443–446.

238 T. D. Yuzvinsky, A. M. Fennimore, A. Kis, A. Zettl, *Nanotechnology*, **2006**, *17*, 434–438.

239 F. Liu, M. Bao, K. L. Wang, X. Liu, C. Li, C. Zhou, *Nano Lett.*, **2005**, *5*, 1333–1336.

240 V. Lovat, D. Pantarotto, L. Lagostena, B. Cacciari, M. Grandolfo, M. Righi, G. Spalluto, M. Prato, L. Ballerini, *Nano Lett.*, **2005**, *5*, 1107–1110.

241 P. S. Dorozhkin, S. V. Tovstonog, D. Golberg, J. Zhan, Y. Ishikawa, M. Shiozawa, H. Nakanishi, K. Nakata, Y. Bando, *Small*, **2005**, *1*, 1088–1093.

242 Y. Zhu, H. I. Elim, Y.-L. Foo, T. Yu, Y. Liu, W. Ji, J.-Y. Lee, Z. Shen, A. T. S. Wee, J. T. L. Thong, C. H. Sow, *Adv. Mater.*, **2006**, *18*, 587–592.

243 C. N. R. Rao, M. Nath, *Dalton Trans.*, **2003**, 1–24.

244 S.-Y. Zhang, Y. Li, X. Ma, H.-Y. Chen, *J. Phys. Chem. B*, **2006**, *110*, 9041–9047.

245 H. Yu, Z. Zhang, M. Han, X. Hao, F. Zhu, *J. Am. Chem. Soc.*, **2005**, *127*, 2378–2379.

246 C. Li, Z. Liu, C. Gu, X. Xu, Y. Yang, *Adv. Mater.*, **2006**, *18*, 228–234.

247 (a) J. M. Macak, H. Tsuchiya, P. Schumuki, *Angew. Chem. Int. Ed.*, **2005**, *44*, 2100–2102; (b) C. Ruan, M. Paulose, O. K. Varghese, G. K. Mor, C. A. Grimes, *J. Phys. Chem B*, **2005**, *109*, 15754–15759.

248 G. Armstrong, A. R. Armstrong, J. Canales, P. G. Bruce, *Chem. Comm.*, **2005**, 2454–2456.

249 R. Ma, T. Sasaki, Y. Bando, *Chem. Comm.*, **2005**, 948–950.

250 M. A. Khan, H.-T. Jung, O.-B. Yang, *J. Phys. Chem. B*, **2006**, *110*, 626–6630.

251 D. Eder, I. A. Kinloch, A. H. Windle, *Chem. Comm.*, **2006**, 1448–1450.

252 A. Ghicov, J. M. Macak, H. Tsuchiya, J. Kunze, V. Haeublein, L. Frey, P. Schmuki, *Nano Lett.*, **2006**, *6*, 1080–1082.

253 H. Tan, E. Ye, W. Y. Fan, *Adv. Mater.*, **2006**, *18*, 619–623.

254 Q. Ji, T. Shimizu, *Chem. Comm.*, **2005**, 4411–4413.

255 J. H. Jung, T. Shimizu, S. Shinkai, *J. Mater. Chem.*, **2005**, *15*, 3979–3986.

256 C.-J. Jia, L.-D. Sun, Z.-G. Yan, L.-P. You, F. Luo, X.-D. Han, Y.-C. Pang, Z. Zhang, C. H. Yan, *Angew. Chem. Int. Ed.*, **2005**, *44*, 4328–4333.

257 Z. Liu, D. Zhang, S. Han, C. Li, B. Lei, W. Lu, J. Fang, C. Zhou, *J. Am. Chem. Soc.*, **2005**, *127*, 6–7.

258 C. Tang, Y. Bando, B. Liu, D. Goldberg, *Adv. Mater.*, **2005**, *17*, 3005–3009.

259 J. Wang, V. K. Kayastha, Y. K. Yap, Z. Fan, J. G. Lu, Z. Pan, I. N. Ivanov, A. A. Puretzky, D. B. Geohegan, *Nano Lett.*, **2005**, *5*, 2528–2532.

260 U. K. Gautam, S. R. C. Vivekchand, A. Govindaraj, G. U. Kulkarni, N. R. Selvi, C. N. R. Rao, *J. Am. Chem. Soc.*, **2005**, *127*, 3658–3659.

261 C. D. Malliakas, M. G. Kantzidis, *J. Am. Chem. Soc.*, **2006**, *128*, 6538–6539.

262 Q. Wu, Z. Hu, C. Liu, X. Wang, Y. Chen, Y. Lu, *J. Phys. Chem. B*, **2005**, *109*, 19719–19722.

263 M. S. Sander, H. Gao, *J. Am. Chem. Soc.*, **2005**, *127*, 12158–12159.

264 Y. Aoki, J. Huang, T. Kunitake, *J. Mater. Chem.*, **2006**, *16*, 292–297.

265 C. Zhi, Y. Bando, C. Tang, R. Xie, T. Sekiguchi, D. Goldberg, *J. Am. Chem. Soc.*, **2005**, *127*, 15996–15997.

266 C. Zhi, Y. Bando, C. Tang, S. Honda, K. Sato, H. Kuwahara, D. Goldberg,

J. Phys. Chem. B, **2006**, *110*, 1525–1528.

267 S.-Y. Xie, W. Wang, K. A. S. Fernando, X. Wang, Y. Lin, Y.-P. Sun, *Chem. Comm.*, **2005**, 3670–3672.

268 C. Zhi, Y. Bando, C. Tang, S. Honda, K. Sato, H. Kuwahara, D. Goldberg, *Angew. Chem. Int. Ed.*, **2005**, *44*, 7932–7935.

269 Q. Huang, Y. Bando, C. Zhi, D. Goldberg, K. Kurashima, F. Xu, L. Gao, *Angew. Chem. Int. Ed.*, **2006**, *45*, 2044–2047.

270 C. Zhi, Y. Bando, C. Tang, D. Goldberg, *J. Phys. Chem. B*, **2006**, *110*, 8548–8550.

271 C. Tang, Y. Bando, Y. Huang, S. Yue, C. Gu, F. Xu, D. Goldberg, *J. Am. Chem. Soc.* **2005**, *127*, 6552–6553.

272 R. Fan, R. Karnik, M. Yue, D. Li, A. Mujumdar, P. Yang, *Nano Lett.*, **2005**, *5*, 1633–1637.

273 Q. Ji, S. Kamiya, J.-H. Jung, T. Shimizu, *J. Mater. Chem.*, **2005**, *15*, 743–748.

274 L. Li, Y. W. Yang, X. H. Huang, G. H. Li, R. Ang, L. D. Zhang, *Appl. Phys. Lett.*, **2006**, *88*, 103119.

275 X. Chen, X. P. Gao, H. Zhang, Z. Zhou, W. K. Hu, G. L. Pan, H. Y. Zhu, T. Y. Yan, D. Y. Song, *J. Phys. Chem. B*, **2005**, *109*, 11525–11529.

276 C.-C. Chen, Y.-C. Liu, C.-H. Wu, C.-C. Yeh, M.-T. Su, Y.-C. Wu, *Adv. Mater.*, **2005**, *17*, 404–407.

277 Z. Siwy, L. Trofin, P. Kohli, L. A. Baker, C. Trautmann, C. R. Martin, *J. Am. Chem. Soc.*, **2005**, *127*, 5000–5001.

278 B. A. H. Sancez, K.-S. Chang, M. T. Scancella, J. L. Burris, S. Kohli, E. R. Fischer, P. K. Dorhout, *Chem. Mater.*, **2005**, *17*, 5909–5919.

279 X. Y. Zhang, C. W. Lai, X. Zhao, D. Y. Wang, Y. Y. Dai, *Appl. Phys. Lett.*, **2005**, *87*, 143102.

280 M. Paulose, O. K. Varghese, G. K. Mor, C. A. Grimes, K. G. Ong, *Nanotechnology*, **2006**, *17*, 398–402.

281 Z. Sun, H. Yaun, Z. Liu, B. Han, X. Zhang, *Adv. Mater.*, **2005**, *17*, 2993–2997.

282 J. Chen, L. Xu, W. Li, X. Gou, *Adv. Mater.*, **2005**, *17*, 582–586.

283 X. Li, F. Cheng, B. Guo, J. Chen, *J. Phys. Chem. B*, **2005**, *109*, 14017–12024.

284 R. Fan, M. Yue, R. Karnik, A. Mujumdar, P. Yang, *Phys. Rev. Lett.*, **2005**, *95*, 086607.

285 D. G. Shchukin, G. B. Sukhurukov, R. R. Price, Y. M. Lvov, *Small*, **2005**, *1*, 510–513.

286 G. K. Mor, K. Shankar, M. Paulose, O. K. Varghese, C. A. Grimes, *Nano Lett.*, **2006**, *6*, 215–218.

287 S.-M. Liu, M. Kobayashi, S. Sato, K. Kimura, *Chem. Comm.*, **2005**, 4690–4692.

288 (a) D. C. Lee, T. Hanrath, B. A. Korgel, *Angew. Chem. Int. Ed.*, **2005**, *44*, 3573–3577; (b) A. I. Hochbaum, R. Fan, R. He, P. Yang, *Nano Lett.*, **2005**, *5*, 457–460.

289 X. Lu, D. D. Fanfair, K. P. Johnston, B. A. Korgel, *J. Am. Chem. Soc.*, **2005**, *127*, 15718–15719.

290 H.-Y. Tuan, D. C. Lee, T. Hanrath, B. A. Korgel, *Chem. Mater.*, **2005**, *17*, 5705–5711.

291 P. Nguyen, H. T. Ng, M. Meyyappan, *Adv. Mater.*, **2005**, *17*, 549–553.

292 D. Wang, R. Tu, L. Zhang, H. Dai, *Angew. Chem. Int. Ed.*, **2005**, *44*, 2925–2929.

293 H. Gerung, T. J. Boyle, L. J. Tribby, S. D. Bunge, C. J. Brinker, S. M. Han, *J. Am. Chem. Soc.*, **2006**, *128*, 5244–5250.

294 C. Ni, P. A. Hassan, E. W. Kaler, *Langmuir*, **2005**, *21*, 3334–3337.

295 J. A. Sioss, C. D. Keating, *Nano Lett.*, **2005**, *5*, 1779–1783.

296 L. Gou, C. J. Murphy, *Chem. Mater.*, **2005**, *17*, 3668–3672.

297 H.-Y. Wu, H.-C. Chu, T.-J. Kuo, C.-L. Kuo, M. L. H. Huang, *Chem. Mater.*, **2005**, *17*, 6447–6451.

298 A. Gole, C. J. Murphy, *Chem. Mater.*, **2005**, *17*, 1325–1330.

299 B. Basnar, Y. Weizmann, Z. Cheglakov, I. Willner, *Adv. Mater.*, **2006**, *18*, 713–718.

300 C.-K. Tsung, X. Kou, Q. Shi, J. Zhang, M. H. Yeung, J. Wang, G. D. Stucky,

J. Am. Chem. Soc., **2006**, *128*, 5352–5353.

301 A. J. Mieszawska, G. W. Slawinski, F. P. Zamborini, *J. Am. Chem. Soc.*, **2006**, *128*, 5622–5623.

302 S. T. S. Joseph, B. I. Ipe, P. Pramod, K. G. Thomas, *J. Phys. Chem. B*, **2006**, *110*, 150–157.

303 Y. Ma, L. Qi, W. Shen, J. Ma, *Langmuir*, **2005**, *21*, 6161–6164.

304 Q. Li, V. W.-W. Yam, *Chem. Commun.*, **2006**, 1006–1008.

305 P. Chen, S. Xie, N. Ren, Y. Zhang, A. Dong, Y. Chen, Y. Tang, *J. Am. Chem. Soc.*, **2006**, *128*, 1470–1471.

306 Z. Gui, J. Liu, Z. Wang, L. Song, Y. Hu, W. Fan, D. Chen, *J. Phys. Chem. B*, **2005**, *109*, 1113–1117.

307 P. X. Gao, Y. Ding, W. Mai, W. L. Hughes, C. Lao, Z. L. Wang, *Science*, **2005**, *309*, 1700–1704.

308 Q. Li, V. Kumar, Y. Li, H. Zhang, T. J. Marks, R. P. H. Chang, *Chem. Mater.*, **2005**, *17*, 1001–1006.

309 H. Zhang, D. Yang, X. Ma, N. Du, J. Wu, D. Gue, *J. Phys. Chem. B*, **2006**, *110*, 827–830.

310 Y. Tak, K. Yong, *J. Phys. Chem. B*, **2005**, *109*, 19263–19269.

311 P. X. Gao, C. S. Lao, W. L. Hughes, Z. L. Wang, *Chem. Phys. Lett.*, **2005**, *408*, 174–178.

312 M. Lai, D. J. Riley, *Chem. Mater.*, **2006**, *18*, 2233–2237.

313 S. Kar, B. N. Pal, S. Chaudhuri, D. Chakravorty, *J. Phys. Chem. B*, **2006**, *110*, 4605–4611.

314 L.-X. Yang, Y.-J. Zhu, W.-W. Wang, H. Tong, M.-L. Ruan, *J. Phys. Chem. B*, **2006**, *110*, 6609–6614.

315 B. Cheng, W. Shi, J. M. R. Tanner, L. Zhang, E. T. Samulski, *Inorg. Chem.*, **2006**, *45*, 1208–1214.

316 J. H. He, J. H. Hsu, C. W. Wang, H. N. Lin, L. J. Chen, Z. L. Wang, *J. Phys. Chem. B*, **2006**, *110*, 50–53.

317 G. Wang, D.-S. Tsai, Y.-S. Huang, A. Korotcov, W.-C. Yeh, D. Susanti, *J. Mater. Chem.*, **2006**, *16*, 780–786.

318 A. Magrez, E. Vasco, J. W. Seo, C. Dieker, N. Setter, L. Forro, *J. Phys. Chem. B*, **2006**, *110*, 58–61.

319 K. P. Kalyanikutty, F. L. Deepak, C. Edem, A. Govindaraj, C. N. R. Rao, *Mater. Res. Bull.*, **2005**, *40*, 831.

320 Y. Hao, G. Meng, C. Ye, X. Zhang, L. Zhang, *J. Phys. Chem B*, **2005**, *109*, 11204–11208.

321 J. Zhang, F. Jiang, Y. Yang, J. Li, *J. Phys. Chem. B*, **2005**, *109*, 13143–13147.

322 J. Zhan, Y. Bando, J. Hu, F. Xu, D. Goldberg, *Small*, **2005**, *1*, 883–888.

323 J. Joo, S. G. Kwon, T. Yu, M. Cho, J. Lee, J. Yoon, T. Hyeon, *J. Phys. Chem. B*, **2005**, *109*, 15297–15302.

324 B. S. Guiton, Q. Gu, A. L. Prieto, M. S. Gudiksen, H. Park, *J. Am. Chem. Soc.*, **2005**, *127*, 498–499.

325 K. P. Kalyanikutty, G. Gundiah, C. Edem, A. Govindaraj, C. N. R. Rao, *Chem. Phys. Lett.*, **2005**, *408*, 389–394.

326 Q. Wan, M. Wei, D. Zhi, J. L. MacManus-Driscoll and M. G. Blamire, *Adv. Mater.*, **2006**, *18*, 234–238.

327 R. Wang, Y. Chen, Y. Fu, H. Zhang, C. Kisielowski, *J. Phys. Chem. B*, **2005**, *109*, 12245–12249.

328 Y. M. Zhao, Y.-H. Li, R. Z. Ma, M. J. Roe, D. G. McCartney, Y. Q. Zhu, *Small*, **2006**, *2*, 422–427.

329 J. Zhou, Y. Ding, S. Z. Deng, L. Gong, N. S. Xu, Z. L. Wang, *Adv. Mater.*, **2005**, *17*, 2107–2110.

330 J.-W. Seo, Y.-W. Jun, S. J. Ko, J. Cheon, *J. Phys. Chem. B*, **2005**, *109*, 5389–5391.

331 G. Xu, Z. Ren, P. Du, W. Weng, G. Shen, G. Han, *Adv. Mater.*, **2005**, *17*, 907–910.

332 S. R. Hall, *Adv. Mater.*, **2006**, *18*, 487–490.

333 C. N. R. Rao, A. Govindaraj, F. L. Deepak, N. A. Gunari, M. Nath, *Appl. Phys. Lett.*, **2001**, *78*, 1853–1855.

334 E. J. H. Lee, C. Ribeiro, E. Longo, E. R. Leite, *J. Phys. Chem. B*, **2005**, *109*, 20842–20846.

335 S. Kar and S. Chaudhuri, *J. Phys. Chem. B*, **2006**, *110*, 4542–4547.

336 R. Li, Z. Luo and F. Papadimitrakopoulos, *J. Am. Chem. Soc.*, **2006**, *128*, 6280–6281.

337 A.-M. Cao, J.-S. Hu, H.-P. Liang, L.-J. Wan, *Angew. Chem. Int. Ed.*, **2005**, *44*, 4391–4395.

338 J. H. Yu, J. Joo, H. M. Park, S.-II. Baik, Y. W. Kim, S. C. Kim, T. Hyeon, *J. Am. Chem. Soc.*, **2005**, *127*, 5662–5670.

339 S. Kar, S. Chaudhuri, *J. Phys. Chem. B*, **2005**, *109*, 3298–3302.

340 J. Hu, Y. Bando, D. Goldberg, *Small*, **2005**, *1*, 95–99.

341 P. Christian, P. O'Brien, *Chem. Comm.*, **2005**, 2817–2819.

342 Y. Jeong, Y. Xia, Y. Yin, *Chem. Phys. Lett.*, **2005**, *416*, 246–250.

343 S. G. Thoma, A. Sanchez, P. Provencio, B. L. Abrams, J. P. Wilcoxon, *J. Am. Chem. Soc.*, **2005**, *127*, 7611–7614.

344 A. B. Panda, G. Glaspell, M. S. El-Shall, *J. Am. Chem. Soc.*, **2006**, *128*, 2790–2791.

345 S. Kumar, M. Ade, T. Nann, *Chem. Eur. J.*, **2005**, *11*, 2220.

346 Z. Liu, D. Xu, J. Liang, J. Shen, S. Zhang, Y. Qian, *J. Phys. Chem. B*, **2005**, *109*, 10699–10704.

347 J.-P. Ge, J. Wang, H.-X. Zhang, X. Wang, Q. Peng, Y. Li, *Chem. Eur. J.*, **2005**, *11*, 1889–1894.

348 K.-S. Cho, D. V. Talapin, W. Gaschler, C. B. Murray, *J. Am. Chem. Soc.*, **2005**, *127*, 7140–7147.

349 X. Giu, Y. Lou, A. C. S. Samia, A. Devadoss, J. D. Burgess, S. Dayal, C. Burda, *Angew. Chem. Int. Ed.*, **2005**, *44*, 5855–5857.

350 R. Chen, M. H. So, C.-M. Che, H. Sun, *J. Mater. Chem.*, **2005**, *15*, 4540–4545.

351 F. Gao, Q. Lu, S. Komarneni, *Chem. Comm.*, **2005**, 531–533.

352 M. B. Sigman, B. A. Korgel, *Chem. Mater.*, **2005**, *17*, 1655–1660.

353 A. Purkayastha, F. Lupo, S. Kim, T. Borca-Tasciuc, G. Ramanath, *Adv. Mater.*, **2006**, *18*, 496–500.

354 K. Sardar, M. Dan, B. Schwenzer, C. N. R. Rao, *J. Mater. Chem.*, **2005**, *15*, 2175–2177.

355 B. Liu, Y. Bando, C. Tang, F. Xu, J. Hu, D. Goldberg, *J. Phys. Chem. B*, **2005**, *109*, 17082–17085.

356 H. Li, A. H. Chin, M. K. Sunkara, *Adv. Mater.*, **2006**, *18*, 216–220.

357 S. Luo, W. Zhou, Z. Zhang, L. Liu, X. Dou, J. Wang, X. Zhao, D. Liu, Y. Gao, L. Song, Y. Xiang, J. Zhou, S. Xie, *Small*, **2005**, *1*, 1004–1009.

358 P. V. Radonanvic, C. J. Barrelet, S. Gradecak, F. Qian, C. M. Lieber, *Nano Lett.*, **2005**, *5*, 1407–1411.

359 C. J. Novotny, P. K. L. Yu, *Appl. Phys. Lett.*, **2005**, *87*, 203111.

360 A. I. Persson, M. T. Björk, S. Jeppesen, J. B. Wagner, L. R. Wallenberg, L. Samuelson, *Nano Lett.*, **2006**, *6*, 403–407.

361 H. Zhang, Q. Zhang, G. Zhao, J. Tang, O. Zhou, L.-C. Qin, *J. Am. Chem. Soc.*, **2005**, *127*, 13120–13121.

362 Y. Li, E. Tevaarwerk, R. P. H. Chang, *Chem. Mater.*, **2006**, *18*, 2552–2557.

363 Y. S. Hor, Z. L. Xiao, U. Welp, Y. Ito, J. F. Mitchell, R. E. Cook, W. K. Kwok, G. W. Crabtree, *Nano Lett.*, **2005**, *5*, 397–401.

364 Y. Li, M. A. Malik, P. O'Brien, *J. Am. Chem. Soc.*, **2005**, *127*, 16020–16021.

365 D. Ung, G. Viau, C. Ricolleau, F. Warmont, P. Gredin, F. Fievet, *Adv. Mater.*, **2005**, *17*, 338–344.

366 C. J. Ordendoff, P. L. Hankins, C. J. Murphy, *Langmuir*, **2005**, *21*, 2022–2026.

367 X. Hu, W. Cheng, T. Wang, E. Wang, S. Dong, *Nanotechnology*, **2005**, *16*, 2164–2169.

368 B. Pan, L. Ao, F. Gao, H. Tian, R. He, D. Cui, *Nanotechnology*, **2005**, *16*, 1776–1780.

369 M. A. C. Duarte, J. P. Juste, A. S. Iglesias, M. Giersig, L. M. L. Marzan, *Angew. Chem. Int. Ed.*, **2005**, *44*, 4375–4378.

370 C. J. Murphy, C. J. Orendorff, *Adv. Mater.*, **2005**, *17*, 2173–2177.

371 L. Tong, J. Lou, R. R. Gattass, S. He, X. Chen, L. Liu, E. Mazur, *Nano Lett.*, **2005**, *5*, 259–265.

372 D. Wang, Y.-L. Chang, Z. Liu, H. Dai, *J. Am. Chem. Soc.*, **2005**, *127*, 11871–11875.

373 J. Polleux, A. Gurlo, N. Barsan, U. Weimar, M. Antonietti, M. Niederberger, *Angew. Chem. Int. Ed.*, **2005**, *44*, 261–265.

374 D.-F. Zhang, L.-D. Sun, C.-J. Jia, Z.-G. Yan, L.-P. You, C.-H. Yan, *J. Am. Chem. Soc.*, **2005**, *127*, 13492–13493.

375 J. Polleux, N. Pinna, M. Antonietti, M. Niederberger, *J. Am. Chem. Soc.*, **2005**, *127*, 15595–15601.

376 J. A. Streifer, H. Kim, B. M. Nichols, R. J. Hammers, *Nanotechnology*, **2005**, *16*, 1868–1873.

377 H. Liao, J. H. Hafner, *Chem. Mater.*, **2005**, *17*, 4636–4641.

378 M. Curreli, C. Li, Y. Sun, B. Lei, M. A. Gundersen, M. E. Thompson, C. Zhou, *J. Am. Chem. Soc.*, **2005**, *127*, 6922–6923.

379 J.-Y. Chang, H. Wu, H. Chen, Y.-C. Ling, W. Tan, *Chem. Comm.*, **2005**, 1092–1094.

380 Y. Huang, C.-Y. Chiang, S. K. Lee, Y. Gao, E. L. Hu, J. D. Yoreo, A. M. Belcher, *Nano Lett.*, **2005**, *5*, 1429–1434.

381 Y. Xi, J. Zhou, H. Guo, C. Cai, Z. Lin, *Chem. Phys. Lett.*, **2005**, *412*, 60–64.

382 A. D. LaLonde, M. G. Norton, D. N. McIlroy, D. Zhang, R. Padmanabhan, A. Alkhateeb, H. Man, N. Lane, Z. Holman, *J. Mater. Res.*, **2005**, *20*, 549–553.

383 S. Y. Bae, H. W. Seo, H. C. Choi, D. S. Han, J. Park, *J. Phys. Chem. B*, **2005**, *109*, 8496–8502.

384 H.-X. Zhang, J.-P. Ge, J. Wang, Z. Wang, D.-P. Yu, Y.-D. Li, *J. Phys. Chem. B*, **2005**, *109*, 11585–11591.

385 M. A. Verheijen, G. Immink, T. de Smet, M. T. Borgström, E. P. A. M. Bakkers, *J. Am. Chem. Soc.*, **2006**, *128*, 1353–1359.

386 M. Afzaal, P. O'Brien, *J. Mater. Chem.*, **2006**, *16*, 1113–1115.

387 I. Pastoriza-Santos, J. Pérez-Juste, L. M. Liz-Marzán, *Chem. Mater.*, **2006**, *18*, 2465–2467.

388 P. Mohan, J. Motohisa, T. Fukui, *Appl. Phys. Lett.*, **2006**, *88*, 133105.

389 S. Eustis, M. El-Sayed, *J. Phys. Chem. B*, **2005**, *109*, 16350–16356.

390 P. K. Jain, K. S. Lee, I. H. El-Sayed, M. A. El-Sayed, *J. Phys. Chem. B*, **2006**, *110*, 7238–7248.

391 E. K. Payne, K. L. Shuford, S. Park, G. C. Schatz, C. A. Mirkin, *J. Phys. Chem. B*, **2006**, *110*, 2150–2154.

392 S. Hirano, N. Tekeuchi, S. Shimada, K. Masuya, K. Ibe, H. Tsunakawa, M. Kuwabara, *J. Appl. Phys.*, **2005**, *98*, 094305.

393 M.-C. Jeong, B.-Y. Oh, M.-H. Ham, J.-M. Myoung, *Appl. Phys. Lett.*, **2006**, *88*, 202105.

394 H. Z. Zhang, M. R. Phillips, J. D. F. Gerald, J. Yu, Y. Chen, *Appl. Phys. Lett.*, **2006**, *88*, 093117.

395 M.-S. Hu, H.-L. Chen, C.-H. Shen, L.-S. Hong, B.-R. Huang, K.-H. Chen, L.-C. Chen, *Nat. Mater.*, **2006**, *5*, 102–106.

396 Y. Zhang, R. E. Russo, S. S. Mao, *Appl. Phys. Lett.*, **2005**, *87*, 043106.

397 A. Pan, R. Liu, Q. Yang, Y. Zhu, G. Yang, B. Zou, K. Chen, *J. Phys. Chem. B*, **2005**, *109*, 24268–24272.

398 R. Agarwal, C. J. Barrelet, C. M. Lieber, *Nano Lett.*, **2005**, *5*, 917–920.

399 A. B. Greytak, C. J. Barrelet, Y. Li, C. M. Lieber, *Appl. Phys. Lett.*, **2005**, *87*, 151103.

400 S. Gradecak, F. Qian, Y. Li, H.-G. Park, C. M. Lieber, *Appl. Phys. Lett.*, **2005**, *87*, 173111.

401 Y. Huang, X. Duan, C. M. Lieber, *Small*, **2005**, *1*, 142–147.

402 P. J. Pauzauskie, D. J. Sirbuly and P. Yang, *Phys. Rev. Lett.*, **2006**, *96*, 143903.

403 A. Pan, H. Yang, R. Liu, R. Yu, B. Zou, Z. Wang, *J. Am. Chem. Soc.*, **2005**, *127*, 15692–15693.

404 Y. Liang, L. Zhai, X. Zhao, D. Xu, *J. Phys. Chem. B*, **2005**, *109*, 7120–7123.

405 C.-L. Hsu, S.-J. Chang, Y.-R. Lin, P.-C. Li, T.-S. Lin, S.-Y. Tsai, T.-H. Lu, I.-C. Chen, *Chem. Phys. Lett.*, **2005**, *416*, 7578.

406 M. Khan, A. K. Sood, F. L. Deepak, C. N. R. Rao, *Nanotechnology*, **2006**, *17*, S287–S290.

407 P. J. Pauzauskie, A. Radenovic, E. Trepagnier, H. Shroff, P. Yang, J. Liphardt, *Nat. Mater.*, **2006**, *5*, 97–101.

408 A. Boukai, K. Xu, J. R. Heath, *Adv. Mater.*, **2006**, *18*, 864–869.

409 Z. Zhong, Y. Fang, W. Lu, C. M. Lieber, *Nano Lett.*, **2005**, *5*, 1143–1146.

410 R. A. Beckman, E. J. Halperin, N. A. Melosh, Y. Luo, J. E. Green, J. R. Heath, *J. Appl. Phys.*, **2004**, *96*, 5921–5923.
411 B. Ozturk, C. Blackledge, B. N. Flanders, *Appl. Phys. Lett.*, **2006**, *88*, 073108.
412 Y. Huang, S. Yue, Z. Wang, Q. Wang, C. Shi, Z. Xu, X. D. Bai, C. Tang, C. Gu, *J. Phys. Chem. B*, **2006**, *110*, 796–800.
413 E. Stern, G. Cheng, E. Cimpoiasu, R. Klie, S. Guthrie, J. Klemic, I. Kretzschmar, E. Steinlauf, D. T. Evans, E. Broomfield, J. Hyland, R. Koudelka, T. Boone, M. Young, A. Sanders, R. Munden, T. Lee, D. Routenberg, M. A. Reed, *Nanotechnology*, **2005**, *16*, 2941–2953.
414 F. Liu, M. Bao, K. L. Wang, C. Li, B. Lei, C. Zhou, *Appl. Phys. Lett.*, **2005**, *86*, 213101.
415 J. Goldberger, D. J. Sirbuly, M. Law, P. Yang, *J. Phys. Chem. B*, **2005**, *109*, 9–14.
416 B. Sun, H. Sirringhaus, *Nano Lett.*, **2005**, *5*, 2408–2413.
417 S. Ju, K. Lee, D. B. Janes, M.-H. Yoon, A. Facchetti, T. J. Marks, *Nano Lett.*, **2005**, *5*, 2281–2286.
418 B.-K. Kim, J.-J. Kim, J.-O. Lee, K.-J. Kong, H. J. Seo, C. J. Lee, *Phys. Rev. B*, **2005**, *71*, 153313.
419 C. Thelander, H. A. Nilsson, L. E. Jensen, L. Samuelson, *Nano Lett.*, **2005**, *5*, 635–638.
420 S. R. C. Vivekchand, K. C. Kam, G. Gundiah, A. Govindaraj, A. K. Cheetham, C. N. R. Rao, *J. Mater. Chem.*, **2005**, *15*, 4922–4927.
421 J. Zhou, C. Jin, J. H. Seol, X. Li, L. Shi, *Appl. Phys. Lett.*, **2005**, *87*, 133109.
422 B. D. Liu, Y. Bando, C. C. Tang, F. F. Xu, D. Goldberg, *J. Phys. Chem. B*, **2005**, *109*, 21521–21524.
423 H. Liu, Q. Zhao, Y. Li, Y. Liu, F. Lu, J. Zhuang, S. Wang, L. Jiang, D. Zhu, D. Yu, L. Chi, *J. Am. Chem. Soc.*, **2005**, *127*, 1120–1121.
424 J. H. He, R. Yang, Y. L. Chueh, L. J. Chou, L. J. Chen. Z. L. Wang, *Adv. Mater.*, **2006**, *18*, 650–654.
425 H. Zhang, J. Tang, Q. Zhang, G. Zhao, G. Yang, J. Zhang, O. Zhou, L.-C. Qin, *Adv. Mater.*, **2006**, *18*, 87–91.
426 S. P. Lau, L. Huang, S. F. Yu, H. Yang, J. K. Yoo, S. J. An, G.-C. Yi, *Small*, **2006**, *2*, 736–740.
427 M. Tian, N. Kumar, S. Xu, J. Wang, J. S. Kurtz, M. H. W. Chan, *Phys. Rev. Lett.*, **2005**, *95*, 076802.
428 K. Peng, Y. Xu, Y. Wu, Y. Yan, S.-T. Lee, J. Zhu, *Small*, **2005**, *1*, 1062–1067.
429 M. Law, L. E. Green, J. C. Johnson, R. Saykally, P. Yang, *Nat. Mater.*, **2005**, *4*, 455–459.
430 J. B. Baxter, E. S. Aydil, *Appl. Phys. Lett.*, **2005**, *86*, 053114.
431 M. Y. Song, Y. R. Ahn, S. M. Jo, D. Y. Kim, J.-P. Ahn, *Appl. Phys. Lett.*, **2005**, *87*, 113113.
432 S. F. Bidoz, A. C. Arsenault, I. Manners, G. A. Ozin, *Chem. Comm.*, **2005**, 441–443.
433 L. Qian, S. Park, L. Huang, C. A. Mirkin, *Science*, **2005**, *309*, 113–115.
434 D. S. Han, S. Y. Bae, H. W. Seo, Y. J. Kang, J. Park, G. Lee, J.-P. Ahn, S. Kim, J. Chang, *J. Phys. Chem. B*, **2005**, *109*, 9311–9316.
435 J. Wang, M. Scampicchio, R. Laocharoensuk, F. Valentini, O. G. Garcia, J. Burdick, *J. Am. Chem. Soc.*, **2006**, *128*, 4562–4563.
436 G. S. D. Beach, C. Nistor, C. Knutson, M. Tsoi, J. L. Erskine, *Nat. Mater.*, **2005**, *4*, 741–744.
437 C. Li, B. Lei, Z. Luo, S. Han, Z. Liu, D. Zhang, C. Zhou, *Adv. Mater.*, **2005**, *17*, 1548–1553.
438 A. R. Armstrong, G. Armstrong, J. Canales, R. Garcia, P. G. Bruce, *Adv. Mater.*, **2005**, *17*, 862–865.
439 M. A. L. Devlin, C. L. Asher, B. J. Taft, R. Gasparac, M. A. Roberts, S. O. Kelly, *Nano Lett.*, **2005**, *5*, 1051–1055.
440 T. R. Kline, W. F. Paxton, T. E. Mallouk, A. Sen, *Angew. Chem. Int. Ed.*, **2005**, *44*, 744–746.
441 B. S. Kang, F. Ren, Y. W. Heo, L. C. Tien, D. P. Norton, S. J. Pearton, *Appl. Phys. Lett.*, **2005**, *86*, 112105.

442 C. S. Rout, S. H. Krishna, S. R. C. Vivekchand, A. Govindaraj, C. N. R. Rao, *Chem. Phys. Lett.*, **2006**, *418*, 586–590.

443 Z. Fan, J. G. Lu, *Appl. Phys. Lett.*, **2005**, *86*, 123510.

444 X. Qi, F. E. Osterloh, *J. Am. Chem. Soc.*, **2005**, *127*, 7666–7667.

445 C. Li, M. Curreli, H. Lin, B. Lei, F. N. Ishikawa, R. Datar, R. J. Cote, M. E. Thompson, C. Zhou, *J. Am. Chem. Soc.*, **2005**, *127*, 12484–12485.

446 C. Yu, Q. Hao, S. Saha, L. Shi, X. Kong, Z. L. Wang, *Appl. Phys. Lett.*, **2005**, *86*, 063101.

447 G. Zheng, F. Patolsky, Y. Cui, W. U. Wang, C. M. Lieber, *Nat. Biotech.*, **2005**, *23*, 1294–1301.

448 Y. Im, C. Lee, R. P. Vasquez, M. A. Bangar, N. V. Myung, E. J. Menke, R. M. Penner, M. Yun, *Small*, **2006**, *3*, 356–358.

449 Y. J. Chen, L. Nie, X. Y. Xue, Y. G. Wang, T. H. Wang, *Appl. Phys. Lett.*, **2006**, *88*, 083105.

450 A. Ponzoni, E. Comini, G. Sberveglieri, J. Zhou, S. Z. Deng, N. S. Xu, Y. Dong, Z. L. Wang, *Appl. Phys. Lett.*, **2006**, *88*, 203101.

451 M. Rex, F. E. Hernandez, A. D. Campiglia, *Anal. Chem.*, **2006**, *78*, 445–451.

452 B. Wu, A. Heidelberg, B. J. Boland, *Nat. Mater.*, **2005**, *4*, 525–529.

453 J. Song, X. Wang, E. Riedo, Z. L. Wang, *Nano Lett.*, **2005**, *5*, 1954–1958.

454 X. Li, X. Wang, Q. Xiong, P. C. Eklund, *Nano Lett.*, **2005**, *5*, 1982–1986.

455 S. R. C. Vivekchand, U. Ramamurthy, C. N. R. Rao, *Nanotechnology*, **2006**, *17*, S344–S350.

456 A. Fujiwara, H. Inokawa, K. Yamazaki, H. Namatsu, Y. Takahashi, N. M. Zimmerman, S. B. Martin, *Appl. Phys. Lett.*, **2006**, *88*, 053121.

457 J. Goldberger, A. I. Hochbaum, R. Fan, P. Yang, *Nano Lett.*, **2006**, *6*, 973–977.

458 V. Schmidt, H. Riel, S. Sanz, S. Karg, W. Riess, U. Gosele, *Small*, **2006**, *2*, 85–88.

459 Z. L. Wang, J. Song, *Science*, **2006**, *312*, 242–246.

460 J. B. K. Law, J. T. L. Thong, *Appl. Phys. Lett.*, **2006**, *88*, 133114.

461 P. Feng, J. Y. Zhang, Q. H. Li, T. H. Wang, *Appl. Phys. Lett.*, **2006**, *88*, 153107.

462 K. W. Lux, K. J. Rodriguez, *Nano. Lett.*, **2006**, *6*, 288–295.

463 X. Huang, I. H. El-Sayed, W. Qian, M. A. El-Sayed, *J. Am. Chem. Soc.*, **2006**, *128*, 2115–2120.

464 C.-C. Chen, Y.-P. Lin, C.-W. Wang, H.-C. Tzeng, C.-H. Wu, Y.-C. Chen, C.-P. Chen, L.-C. Chen, Y.-C. Wu, *J. Am. Chem. Soc.*, **2006**, *128*, 3709–3715.

3
Nonaqueous Sol–Gel Routes to Nanocrystalline Metal Oxides

M. Niederberger and M. Antonietti

3.1
Overview

Although metal oxides constitute one of the most important classes of functional materials, their synthesis on the nanoscale under mild reaction conditions and with control over particle size, shape, and crystallinity remains a major task of nanochemistry. In the last few years, a valuable alternative to the well-known aqueous sol–gel processes has been developed in the form of nonaqueous solution routes, which offer several advantages such as high crystallinity at low temperatures, robust synthesis parameters and the ability to control the crystal growth as well as the crystal morphology to a striking extent.

Nonaqueous processes can roughly be divided into two methodologies, namely surfactant- and solvent-controlled preparation routes. In the first case, the synthesis is typically performed in molten surfactants or in surfactants dissolved in high boiling solvents. In the second approach, the synthesis temperatures are lower and the major role is played by a common organic solvent, which often provides the oxygen for the metal oxide, controls the crystal growth, influences the particle shape and, in some cases, also determines the assembly behavior. Both synthesis strategies offer some peculiar features with advantages and limitations. In this chapter we will review and compare these two procedures, however with some focus on solvent-controlled processes, highlighting the versatility and the power of these protocols by means of selected examples from recent literature.

3.2
Introduction

Among all the functional materials, metal oxides play an outstanding role in many fields of technology including catalysis [1], sensing [2], energy storage [3] and conversion [4], or electroceramics [5], to name but a few. In order to obtain metal oxides in the form of nanoparticles with well-defined shape, size, crystallinity and also surface functionality, the traditional solid state routes are hardly suited, and novel innovative strategies have to be developed. Soft-chemistry routes presumably represent the most attractive alternatives, because they allow good control from the molecular precursor to the final product at low processing temperatures, result in the formation of nanomaterials with high purity and compositional homogeneity, and give access to kinetically controlled materials [6]. In this regard it is obvious, that the concept of sol–gel chemistry, which was particularly successful in the preparation of bulk metal oxides [7], has been adapted for metal oxide synthesis on the nanoscale. However, in spite of great efforts, the number of oxidic nanoparticles obtained by aqueous sol–gel chemistry is still marginal compared to the variety of compounds obtained by solid state routes, mainly due to the fact that the as-synthesized products often lack crystallinity. Nonaqueous solution routes are able to overcome this problem. Switching from aqueous sol–gel chemistry and its high reactivity of water to nonhydrolytic processes drastically decreases the reaction rates and leads to controlled crystallization. In contrast to aqueous systems, where the smallest changes in the experimental conditions result in alteration of the products, nonaqueous procedures are very robust within the same system. As a consequence, most of these processes are highly reproducible, easy to scale up to gram quantities and applicable to a broad family of metal oxides. On the other hand, the morphology of the final product depends strongly on the precursor and solvent used, i.e., metal oxides with the same composition and crystal structure but obtained from different precursors and/or solvents, are often characterized by different particle sizes and shapes. This observation highlights the crucial role of the organic side of the process, but also provides a precious tool to tailor the particle morphology. Furthermore, and most importantly, the chemistry of the oxygen–carbon bond is well-known from organic chemistry and therefore, nonaqueous routes open the possibility to adapt reaction principles from organic chemistry to the synthesis of inorganic nanomaterials. In comparison to the complex aqueous chemistry, the synthesis of metal oxide nanoparticles in organic solvents offers the possibility to better understand and to control the reaction pathways on a molecular level – crucial steps towards a rational synthesis design for inorganic nanoparticles.

3.3
Short Introduction to Aqueous and Nonaqueous Sol–Gel Chemistry

The sol–gel process can shortly be defined as the conversion of a precursor solution into an inorganic solid by chemical reactions. In most cases, the precursor or starting compound is either an inorganic metal salt (acetate, chloride, nitrate, sulfate, …) or a metal organic species such as a metal alkoxide. In aqueous sol–gel chemistry metal alkoxides constitute the most widely used class of precursors and their chemistry is well established [8, 9]. Upon hydrolysis, the metal alkoxide is transformed into a sol (dispersion of colloidal particles in a liquid), which reacts further via condensation to a gel, an interconnected, rigid and porous inorganic network enclosing a continuous liquid phase. Further information about aqueous sol–gel processes is outside the scope of this chapter and can be found elsewhere [7, 10–13].

Although aqueous sol–gel processes have been investigated for decades, the simultaneous occurrence of hydrolysis and condensation reactions leads to a wide variety of different species [14], which cannot be identified and consequently, aqueous sol–gel chemistry is not yet fully controllable. For most transition metal oxide precursors, the fast hydrolysis and condensation rates result in loss of morphological and also structural control over the final oxide material, often resulting in the formation of amorphous products. Furthermore, the different reactivities of metal alkoxides make it difficult to control the composition and the homogeneity of complex multicomponent oxides by the aqueous sol–gel process.

One possibility to decrease the reaction rates is to perform the synthesis procedures in organic solvents without the presence of water, which is usually referred to as nonhydrolytic or nonaqueous sol–gel chemistry. The oxygen is provided by donors such as ethers, alcohols or alkoxides rather than water, which leads to a completely different chemistry based on the reactivity of the oxygen–carbon bond [12]. Although nonaqueous processes have become an integral part of modern nanoparticle synthesis in the last few years, their investigation started as early as 1928, when Dearing and Reid presented their work on alkyl orthosilicates involving the reaction of silicon tetrachloride with alcohols [15]. However, it was only at the beginning of the 1990s, that these concepts attracted broader attention, mainly in the fields of metal oxide gels [16, 17] and nanoparticulate powders for catalytic applications [18]. The syntheses of zincite [19], zirconia [20], and titania nanocrystals [21–24] were further steps on the way to making nonaqueous reaction pathways an integral and rapidly growing part of nanochemistry.

This chapter is entitled "Nonaqueous sol–gel routes to nanocrystalline metal oxides", although some of the processes presented are, strictly speaking, not based on sol–gel chemistry but rather on precipitation. Nevertheless, we denote all these processes as "sol–gel", because on the one hand they involve the chemical transformation of a molecular precursor into the final oxidic compound by chemical reactions, and on the other hand we want to highlight the analogy to aqueous sol–gel procedures.

3.4
Nonaqueous Sol–Gel Routes to Metal Oxide Nanoparticles

3.4.1
Surfactant-controlled Synthesis of Metal Oxide Nanoparticles

Surfactant-controlled synthesis routes involve the transformation of the precursor species into the oxidic compound in the presence of stabilizing ligands in a typical temperature range of 250–350 °C. In the most cases, the process is based on the hot injection method, where the reagents are injected into a hot surfactant solution, which was particularly successfully applied in the synthesis of semiconductor nanocrystals [25]. The surfactants, generally consisting of a coordinating head group and a long alkyl chain, have to fulfill several tasks. The coating of the nanoparticles prevents agglomeration due to steric repulsion during synthesis and offers good colloidal stability of the final products in organic solvents. Dynamic adsorption and desorption of surfactant molecules onto particle surfaces during particle growth, sometimes combined with selectivity towards specific crystal faces, enables control over particle size, size distribution and morphology [26, 27]. Furthermore, the surface capping agents in the final material can be exchanged for other ones, offering control over the surface properties. On the other hand, the surface functionalization with surfactants bears the problem that it influences the toxicity of the nanoparticles [28].

The major parameters determining the characteristics of the final oxidic products are reaction time, temperature, concentration of reagents, nature and concentration of surfactants or surfactant mixtures. By a careful adjustment of these conditions, surfactant-controlled synthesis approaches result in the formation of metal oxide nanoparticles with outstanding monodispersity and astonishing particle morphologies.

The family of metal oxide nanoparticles synthesized via surfactant-controlled processes is growing rapidly [26, 29–31] and only selected examples, mainly from recent literature, are presented here to give an impression of what can be achieved by the method. There is no doubt that one of the main advantages lies in the monodispersity of the final products, nicely illustrated in the case of magnetic nanocrystals. The synthesis of monodisperse magnetic nanoparticles has always been of particular interest, because these materials possess a wide variety of technological applications [29]. In the last few years, several papers have reported the nonhydrolytic preparation of maghemite, magnetite and ferrite nanoparticles with highly uniform particle sizes and shapes. The decomposition of an iron oleate complex, prepared by reacting $Fe(CO)_5$ and oleic acid, led to the formation of metallic iron nanoparticles, which are subsequently transformed into γ-Fe_2O_3 by the mild oxidant trimethylamine oxide [32]. Figure 3.1(a) displays a TEM overview image of the highly monodisperse 11 nm nanocrystallites, arranged in a two-dimensional hexagonal assembly. In an analogous approach, the decomposition of an iron-oleate complex, prepared from iron chloride and sodium oleate, at 320 °C in a high boiling solvent gave access to large quantities of magnetite (Fe_3O_4) nanocrystals [33]. Spinel cobalt ferrite nanocrystals were obtained by a seed-mediated growth process [34]. In a first step, spherical $CoFe_2O_4$ nanocrystals with a diameter of 5 nm were prepared starting from cobalt and iron acetylacetonate in phenyl ether, 1,2-hexadecanediol, oleic acid and oleylamine (Fig. 3.1(b)). These primary nanocrystals were used as seeds in additional precursor solutions to grow larger particles, either spherical nanoparticles of 8 nm (Fig. 3.1(c)) or cubic ones of 9 nm (Fig. 3.1(d)) and 11 nm (Fig. 3.1(e)).

The nanoparticles closely packed on the TEM grid in Fig. 3.1 already point to the possibility to self-organize highly monodisperse nanoparticles into two-dimensional supercrystals by simply placing a drop of a colloidal solution on a

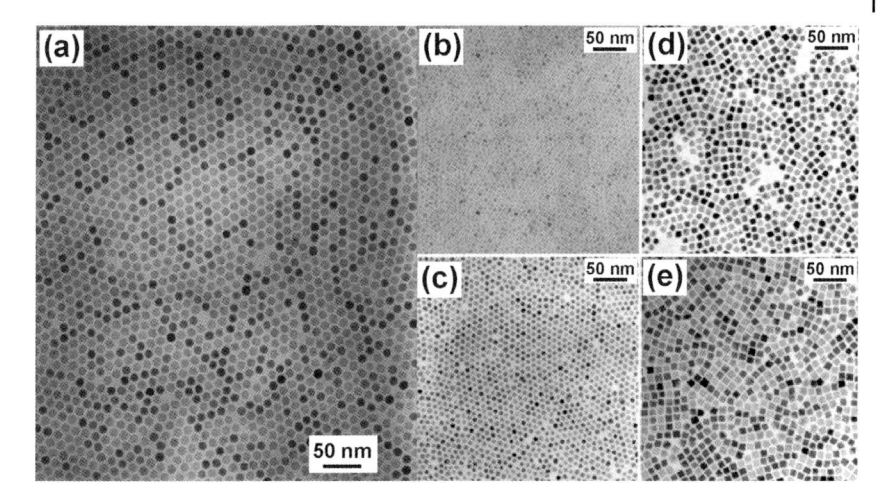

Fig. 3.1 (a) TEM overview image of 11 nm γ-Fe_2O_3 nanocrystals. The highly uniform particle size leads to the formation of a 2D supercrystal on the TEM grid. Image taken from Ref. [32] with permission of the American Chemical Society. TEM overview images of $CoFe_2O_4$ nanocrystals with spherical morphology of (b) 5 nm and (c) 8 nm, and of cubic shape, (d) 9 nm and (e) 11 nm in size. Images taken from Ref. [34] with permission of the American Chemical Society.

suitable support and slowly evaporating the solvent. If nanoparticle dispersions with bimodal size distributions are used, the composition of the obtained superstructures follows that of intermetallic alloys [35]. However, not only have nanoparticle assemblies consisting of particles with different sizes been reported, but also assemblies comprising particles with distinctly different properties. The deposition of 5 nm gold nanoparticles and 13.4 nm γ-Fe_2O_3 nanocrystals yielded precisely ordered, large single domains of AB superlattices, isostructural with NaCl (Fig. 3.2(a)) [36]. The NaCl superlattices are usually represented by (111) projections (Fig. 3.2(a) and (e)), however smaller domains of (100) projections were also observed (Fig. 3.2(a) upper inset) and (c). Figure 3.2(g) shows the (001) projections of an AlB_2-type superlattice assembled from 13.4 nm γ-Fe_2O_3 and 6.2 nm $CoPt_3$ nanocrystals, together with a 3D sketch of the AlB_2 unit cell (Fig. 3.2(h)) and its (001) plane (Fig. 3.2(i)). The combination of semiconducting, metallic and magnetic nanoparticles as building blocks provided great structural diversity in binary nanoparticle superlattices with precisely controlled chemical composition and tight placement of the components [37].

In addition to uniform particle sizes, the use of surfactants or surfactant mixtures also enables the preparation of metal oxide nanoparticles with complex and extraordinary crystal morphologies. Figure 3.3 presents selected examples in the systems ZnO, MnO, $W_{18}O_{49}$, TiO_2, CeO_2 and Co_3O_4. The thermal treatment of zinc acetate in dioctyl ether, trioctylphosphine oxide and oleic acid at 200 °C, followed by the addition of 1,12-dodecanediol resulted in the formation of

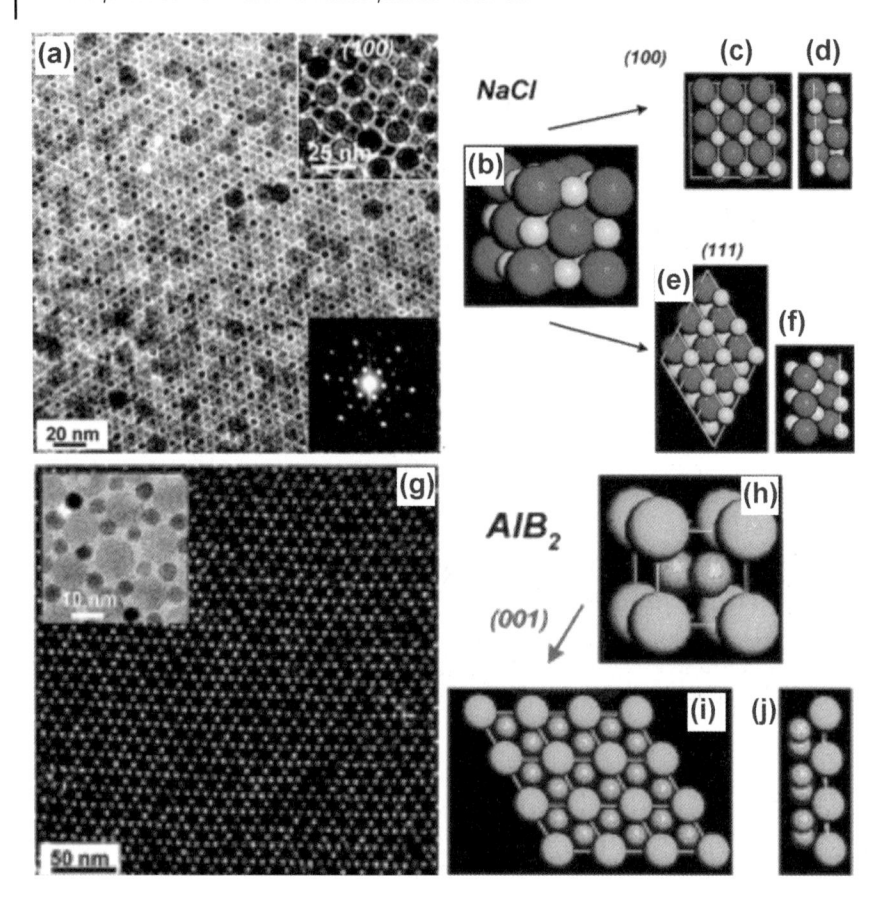

Fig. 3.2 (a) TEM micrographs of AB superlattices of 13.4 nm γ-Fe$_2$O$_3$ and 5 nm Au nanocrystals, isostructural with NaCl, (111) projection. Upper inset: (100) projection. Lower inset: Electron diffraction pattern. (b) 3D sketch of NaCl unit cell, (c) and (e) depictions of the (100) and (111) planes, (d) and (f) depictions of minimum number of layers in the (100) and (111) projections necessary for the formation of the experimentally observed patterns. (g) TEM image of the (001) plane of 13.4 nm γ-Fe$_2$O$_3$ and 6.2 nm CoPt$_3$ nanocrystals, isostructural with AlB$_2$. Inset: Higher magnification. (h) 3D sketch of AlB$_2$ unit cell, (i) (001) plane, (j) minimum number of layers in the (001) plane, leading to the observed patterns. All images taken from Ref. [35] with permission of the American Chemical Society.

ZnO nanocrystals [38]. The TEM image in Fig. 3.3(a) shows that the nanocrystals exhibit cone-like particle morphology with an average size of about 70 nm (base) × 170 nm (height). The dissolution of WCl$_4$ in oleic acid and oleylamine yielded, after heat treatment at 350 °C, tungsten oxide (W$_{18}$O$_{49}$) nanorods, 4.5 nm in width and 28 nm in length (Fig. 3.3(b)) [39]. MnO multipods were synthesized from a manganese oleate complex in oleic acid and oleylamine at 320 °C (Fig. 3.3(c)) [40]. The authors proposed a growth model based on the oriented at-

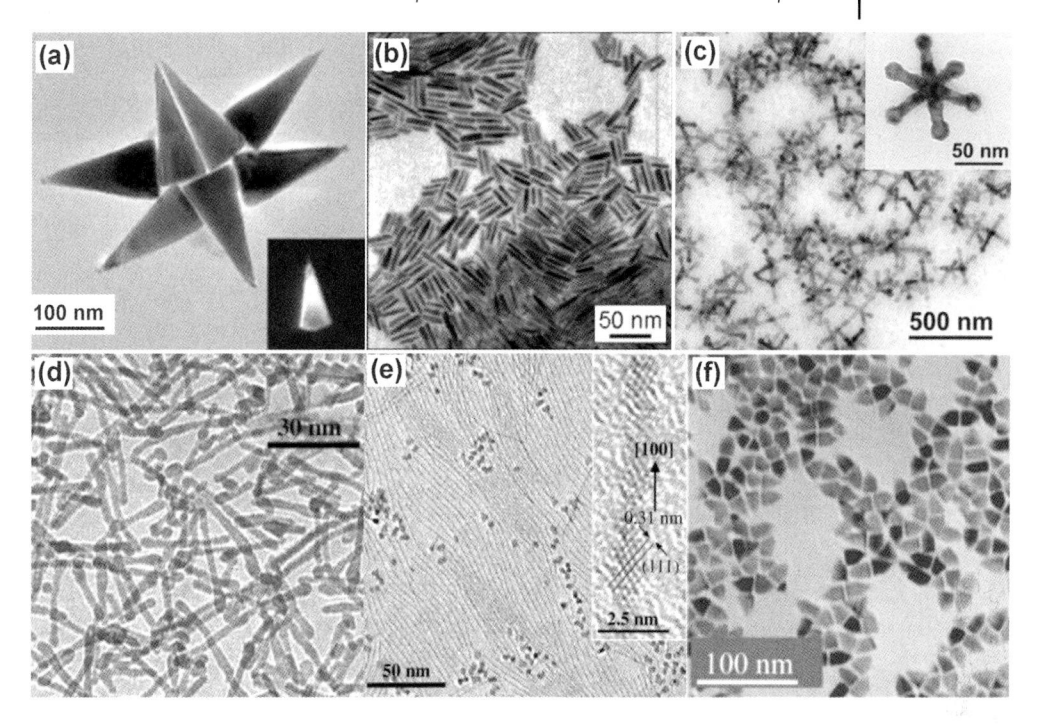

Fig. 3.3 (a) TEM image of cone-like ZnO nanocrystals (inset: dark field TEM image of one crystal). Image taken from Ref. [38] with permission of Wiley-VCH. (b) TEM image of tungsten oxide nanorods. Image taken from Ref. [39] with permission of the American Chemical Society. (c) TEM image of MnO multipods (inset: hexapod). Images taken from Ref. [40] with permission of the American Chemical Society. (d) TEM image of TiO_2 nanorods. Image taken from Ref. [42] with permission of Wiley-VCH. (e) TEM image of ceria nanowires (inset: HRTEM image of one nanowire). Image taken from Ref. [43] with permission of Wiley-VCH. (f) TEM image of Co_3O_4 nanocrystals. Image taken from Ref. [44] with permission of the American Chemical Society.

tachment process, involving the spontaneous self-organization of preformed nanoparticles in a crystallographically controlled way, so that the final structure behaves like a single crystal [41]. In the case of titania nanorods, the synthesis was based on the solvothermal reaction of a mixture of titanium butoxide, linoleic acid, triethylamine, and cyclohexane [42]. A TEM image shows titania nanorods with uniform diameters of 3.3 nm, length up to 25 nm and often with a spherical nanoparticle attached to one side of the rod (Fig. 3.3(d)). The reaction of cerium nitrate in oleic acid, oleylamine and diphenyl ether at 320 °C resulted in the formation of ceria nanowires with diameter of 1.2 nm and 71 nm in length (Fig. 3.3(e)) [43]. The diameter of these nanowires is comparable to that of single-walled carbon nanotubes and is among the thinnest ever reported for nanowires. The TEM image in Fig. 3.3(e) also shows the presence of some spherical nanoparticles, which are often connected to the nanowires. The last example includes the preparation of cobalt oxide nanoparticles with a bullet-like morphology

Table 3.1 Selected examples of metal oxide nanoparticles synthesized via nonaqueous and surfactant-controlled processes.

Metal Oxide	Precursors	Solvents and surfactants	Temperature	Product morphology	Reference
Cu_2O	copper acetate	OLA, TOA	270 °C	spherical	47
γ-Fe_2O_3	$M(Cup)_3$	OA	300 °C	spherical	48
γ-Fe_2O_3	$Fe(CO)_5$	octyl ether, OLA or LA	300 °C	spherical	32
Fe_3O_4	$Fe(acac)_3$	2-pyrrolidone	reflux	spherical	49
Fe_3O_4, MFe_2O_4 (M = Fe, Co, Mn)	$Fe(acac)_3$ $M(acac)_2$	polyalcohols, OLA, OA	305 °C	spherical	50
HfO_2	$Hf(OiPr)_4$ and $HfCl_4$	TOPO	360 °C	spherical and nanorods	51
In_2O_3	$In(acac)_3$	OA	250 °C	spherical	52
MnO, Mn_3O_4	$Mn(acac)_2$	OA	180 °C	spherical	53
SnO_2, ZnO	tin or zinc 2-ethylhexanoate	diphenyl ether and various amines	230–250 °C	spherical	54
TiO_2	$TiCl_4$ and $Ti(OiPr)_4$	dioctyl ether TOPO, LA	300 °C	bullet-shaped to nanorods	55
TiO_2	$Ti(COT)_2$	DMSO, TBP or TBPO or TOPO	120 °C	spherical	56
TiO_2	$Ti(OiPr)_4$	1-octadecene, OLA and OA	260 °C	nanorods	57
$W_{18}O_{49}$	$W(CO)_6$	OA	270 °C	nanorods	58
ZnO	diethyl zinc	decane, octylamine, TOPO	200 °C	spherical	59
ZnO	zinc acetate	various alkylamines, tert-butylphosphonic acid	220–300 °C	spherical	60
ZrO_2	$Zr(OiPr)_4$ and $ZrCl_4$	TOPO	340 °C	spherical	61

Abbreviations: TOPO: Trioctylphosphine oxide; LA: lauric acid; COT: cyclooctatetraene; DMSO: dimethyl sulfoxide; TBP: tributylphosphine; TBPO: tributylphosphine oxide; OLA: oleic acid; OA: oleylamine; acac: acetylacetonate; Cup: N-nitrosophenylhydroxylamine; TOA: trioctylamine.

(Fig. 3.3(f)), obtained from the decomposition of cobalt stearate in octadecene at 320 °C [44].

The use of surface-capping agents makes it possible to modify the nanoparticle surface by ligand exchange reactions. The feasibility of this idea was proven for iron oxide nanoparticles and titania nanorods. In the case of iron oxide, the surface monolayer consisting of alkyl amines could be exchanged by harder Lewis base ligands like alcohols [45], whereas the oleic acid coating of the titania nanorods was exchanged by capping with an alkylphosphonic acid [46]. This procedure increased the concentration as well as the stability of the obtained titania dispersions. Generally surfactant exchange reactions allow the introduction of a wide range of chemical functionalities on the surface of the nanoparticles.

The family of metal oxide nanoparticles synthesized via nonaqueous and surfactant-controlled processes has increased rapidly in the last few years, so that it is impossible to give a complete overview in this short chapter. However, some additional and representative examples, which have not been discussed here, are summarized in Table 3.1, together with precursors, reaction conditions and product characteristics. However, we have to point out that even this rather exhaustive table is far from being complete.

3.5 Solvent-controlled Synthesis of Metal Oxide Nanoparticles

3.5.1 Introduction

An elegant alternative to surfactants is the use of common organic solvents, which act as reactant as well as control agent for particle growth, enabling the synthesis of high-purity nanomaterials in a surfactant-free medium. We will discuss the following reaction systems in more detail: (i) metal halides with alcohols, (ii) metal alkoxides with alcohols, (iii) metal alkoxides with ketones and aldehydes, (iv) metal acetylacetonates with various organic solvents such as alcohols, amines and nitriles.

3.5.2 Reaction of Metal Halides with Alcohols

Probably the most widely explored approach to the synthesis of metal oxide nanoparticles in a nonaqueous medium in the absence of any surfactants is based on the use of metal halides and alcohols. In particular, the reaction between $TiCl_4$ with various alcohols is well documented [62–64], however other metal chlorides like $FeCl_3$ [65], $VOCl_3$ [66], $CoCl_2$ [67], $SnCl_4$ [68], $HfCl_4$ [69], $NbCl_5$ [69], $TaCl_5$ [69], and WCl_6 [66, 70], also readily react with alcohols to give the corresponding metal oxides. Among the various alcohols, benzyl alcohol plays an outstanding role, because it enables the synthesis of several metal oxide nanoparticles with

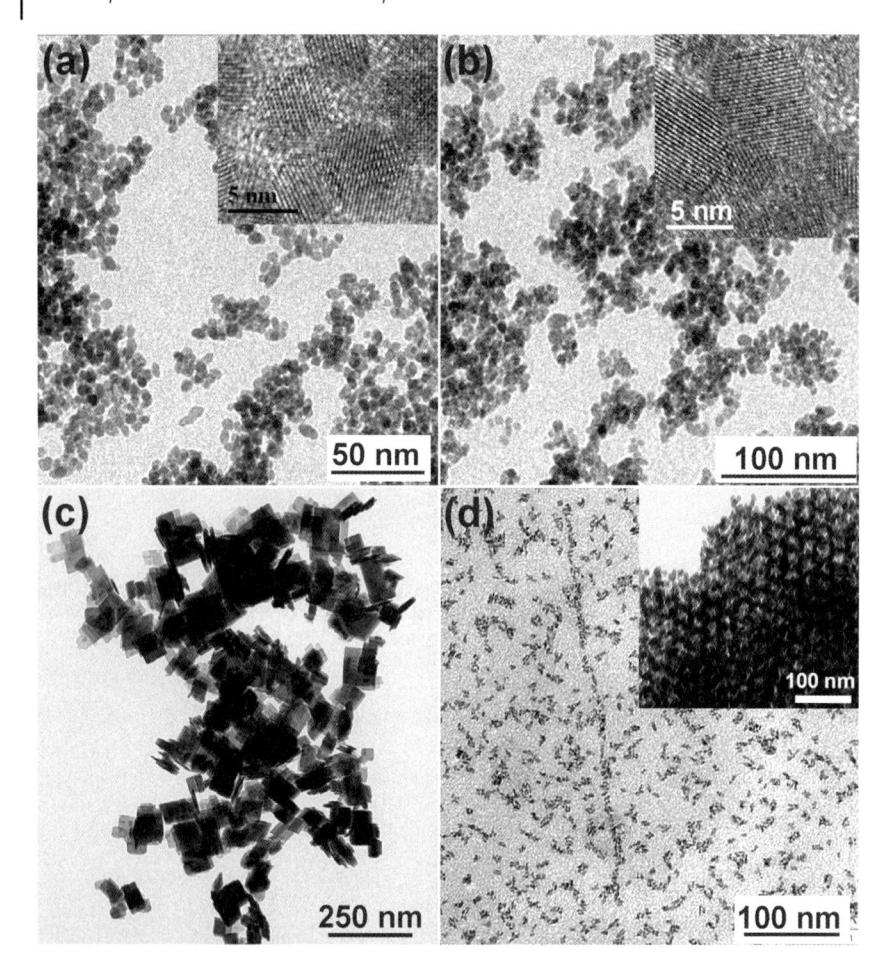

Fig. 3.4 TEM images of various metal oxide nanoparticles prepared from the corresponding metal chlorides in benzyl alcohol. (a) HfO_2, inset: HRTEM image; (b) Ta_2O_5, inset: HRTEM image; (c) $WO_3 \cdot H_2O$; (d) SnO_2, inset: SnO_2 nanoparticles assembled into mesoporous materials after calcination.

good control not only over particle size [66, 71], but also over surface properties [72] and assembly behavior [73, 74]. Figure 3.4 displays several examples of metal oxide nanoparticles obtained from the corresponding metal chlorides in benzyl alcohol. In the case of hafnium (Fig. 3.4(a)) and tantalum oxide (Fig. 3.4(b)), the reaction was performed under solvothermal conditions at 220 °C, yielding spherical and highly crystalline products with uniform size and shape in the range of a few nanometers [69]. The reaction of WCl_6 with benzyl alcohol at 100 °C resulted in the formation of tungstite platelets with side lengths between 30 and 100 nm (Fig. 3.4(c)) [66, 75], whereas $SnCl_4$ reacted to give crystalline, cassiterite

SnO$_2$ nanoparticles of 2–5 nm diameter (Fig. 3.4(d)) [68]. The tin oxide nanoparticles are well dispersible in organic solvents like tetrahydrofuran without any additional stabilizer and can be assembled in the presence of block copolymers as templates into ordered mesoporous materials via evaporation-induced self-assembly (Fig. 3.4(d), inset) [68].

The low reaction temperature of the metal halide–benzyl alcohol system offers several possibilities either to functionalize the surface of the nanoparticles or to influence the particle morphology by the addition of organic molecules. The addition of enediol ligands such as dopamine or 4-*tert*-butylcatechol to the reaction mixture provides a simple route to surface-functionalized nanoparticles (Fig. 3.5(a)), which are redispersible either in water (Fig. 3.5(b)) or in organic solvents (Fig. 3.5(c)) [72]. If the same reaction is performed in the presence of 2-amino-2-(hydroxymethyl)-1,3-propanediol (HOCH$_2$)$_3$CNH$_2$ (Trizma), the as-synthesized

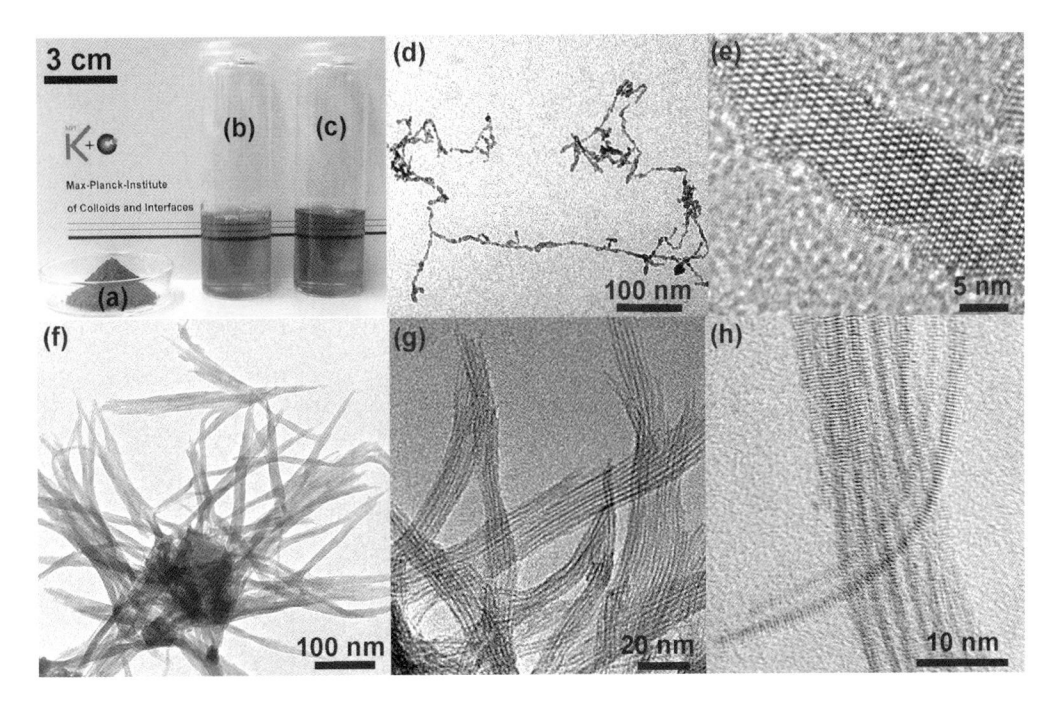

Fig. 3.5 Photograph of enediol-functionalized titania nanoparticles: (a) dopamine-functionalized TiO$_2$ in the form of the isolated powder, (b) dopamine-functionalized TiO$_2$ dispersed in water, (c) 4-*tert*-butylcatechol-functionalized TiO$_2$ dispersed in tetrahydrofuran. Image taken from Ref. [72] with permission of the American Chemical Society. (d) TEM image of titania nanoparticle assemblies in water, composed of Trizma-functionalized TiO$_2$, (e) HRTEM image of a part of such a nanostructure. Images taken from Ref. [73] with permission of Wiley-VCH. TEM images of tungsten oxide nanostructures synthesized at 175 °C in the presence of deferoxamine mesylate, (f) overview, (g) higher magnification and (h) HRTEM image. Images taken from Ref. [75] with permission of the American Chemical Society.

titania nanocrystals assemble into pearl-necklace structures upon redispersion in water (Fig. 3.5(d)) [73, 74]. Interestingly, these nanowire-like arrangements are composed of a continuous string of precisely ordered nanoparticles. HRTEM investigations give evidence that the nanoparticles assemble along the [001] direction via oriented attachment exhibiting monocrystal-like lattice fringes (Fig. 3.5(e)). The addition of the siderophore deferoxamine mesylate to the WCl_6–benzyl alcohol mixture changes the particle shape completely from the quasi-2D to a highly anisotropic 1D morphology (Fig. 3.5(f)–(h)) [75]. The fibers displayed in Fig. 3.5(f) consist of assembled nanowires (Fig. 3.5(g)), which are single-crystalline and have a uniform diameter of 1.3 nm (Fig. 3.5(h)).

3.5.3
Reaction of Metal Alkoxides with Alcohols

Metal halides as metal oxide precursors have the disadvantage that halide residues lead to impurities in the final oxidic materials, which is detrimental for applications in catalysis and gas sensing. The problem can be avoided by the use of halide-free routes involving the solvothermal reaction of metal alkoxides with alcohols, including various glycols. This approach is quite successful in the case of binary metal oxides such as ZrO_2 [20, 76], TiO_2 [77, 78], V_2O_3 [79], Nb_2O_5 [79], Ta_2O_5 [80], HfO_2 [80], SnO_2 [81], In_2O_3 (Fig. 3.6(a)) [81], CeO_2 (Fig. 3.6(b)) [76], however, it can be extended to more complex oxides like perovskites and ilmenites [82]. The synthesis of $BaTiO_3$, one of the technologically most important perovskites, involves the dissolution of metallic barium in benzyl alcohol, addition of titanium isopropoxide and subsequent solvothermal treatment at 200 °C for 48 h [83, 84]. This procedure leads to barium titanate nanoparticles with rather uniform shapes and diameters of 5–8 nm (Fig. 3.6(c)).

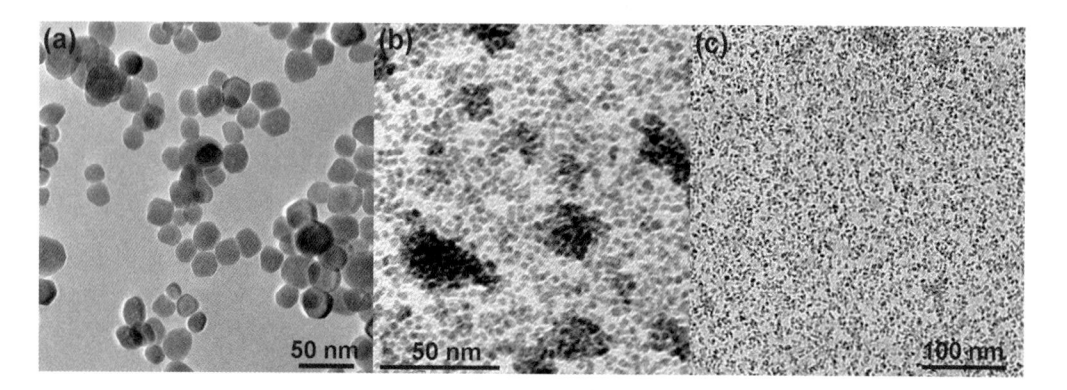

Fig. 3.6 TEM overview images of (a) In_2O_3, (b) CeO_2, and (c) $BaTiO_3$ nanocrystals obtained from the reaction of the metal alkoxides with benzyl alcohol.

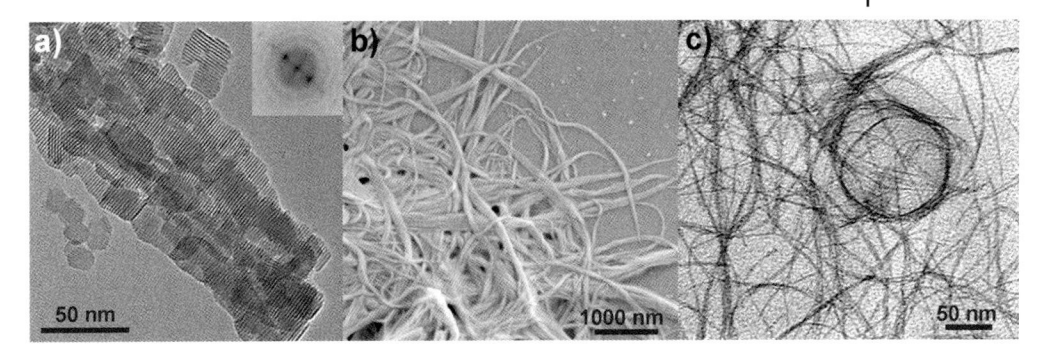

Fig. 3.7 (a) TEM image of lamellar yttrium oxide–benzoate nanohybrid. Image taken from Ref. [85] with permission of Wiley-VCH. (b) SEM image of tungsten oxide nanowire bundles. (c) TEM image of separated individual tungsten oxide nanowires. Image taken from Ref. [86] with permission of Wiley-VCH.

In the most cases, the reaction between metal alkoxides and benzyl alcohol leads to the formation of spherical nanoparticulate products. However, sometimes more complex morphologies are formed. The reaction between yttrium isopropoxide and benzyl alcohol yields a lamellar nanohybrid consisting of crystalline, 0.6 nm thick yttrium oxide layers with intercalated benzoate molecules [85]. Obviously, the yttrium oxide species is able to catalyze the oxidation of benzyl alcohol to benzoic acid, thus limiting the growth in thickness of the yttrium oxide lamellae (Fig. 3.7(a)). Doping with europium leads to strong red luminescence [85]. Tungsten isopropoxide dissolved in benzyl alcohol forms nanowire bundles without any additional structure-directing template (Fig. 3.7(b)) [86]. Infrared spectroscopy investigations point to the presence of benzaldehyde in between the nanowires, stemming from the oxidation of benzyl alcohol by tungsten oxide. Formamide is able to split up the bundles into individual nanowires. Due to the small diameter of about 1 nm the crystalline nanowires are highly flexible without breaking upon bending (Fig. 3.7(c)).

3.5.4
Reaction of Metal Alkoxides with Ketones and Aldehydes

The major limitation of the "benzyl alcohol route" lies in the reducing properties of alcohols. Metal oxide precursors comprising reduction-sensitive metals like copper or lead transform into the respective metals rather than into the oxides. The answer to this problem is the use of non-reductive solvents like ketones or aldehydes. The applicability of this approach has been proven for ZnO, TiO_2, In_2O_3, $BaTiO_3$, $PbTiO_3$, $Pb(Zr,Ti)O_3$, $PbZrO_3$ and $BaSnO_3$.

The addition of acetone to benzene solutions of zinc dialkoxides resulted in the formation of transparent gels, which transformed into 4 nm zincite nanocrystals [19]. The use of common ketones and aldehydes as oxygen-supplying agents made it possible to tailor the particle size of anatase nanocrystals in the range 7–20 nm, as sterically unfavorable alicyclic and aromatic ketones and aldehydes resulted in smaller particles than did aliphatic compounds [87]. Nanocrystalline $BaTiO_3$ was obtained by transformation of the bimetallic alkoxide complex $[(BaTiO)_4(iPrO)_{16}]$ $3iPrOH$ in acetone under stirring at room temperature for 21 days [88]. $PbTiO_3$, $Pb(Zr,Ti)O_3$ and $PbZrO_3$ were obtained from lead acetylacetonate and titanium isopropoxide in 2-butanone as amorphous powders with the correct metal ratios, which could be transformed into the nanocrystalline components by annealing at relatively moderate temperatures between 400 and 600 °C [89]. Another interesting example is the preparation of $BaSnO_3$ in ketones [82]. Although in this case none of the components is sensitive towards reduction, the synthesis was not successful in benzyl alcohol.

3.5.5
Reaction of Metal Acetylacetonates with Various Organic Solvents

Metal acetylacetonates perfectly complement the family of metal oxide precursors discussed before, because they are easy to handle due to their air- and moisture-stability and they are often commercially available at low cost. Furthermore, they react with various organic solvents to give the oxidic compounds, as will be shown in the following examples.

Iron acetylacetonate forms magnetite nanocrystals after solvothermal treatment in benzyl alcohol [90]. If benzyl amine is used as solvent, various metal acetylacetonates can be transformed into the respective metal oxides, for example γ-Ga_2O_3, zincite ZnO and cubic In_2O_3 [91]. Indium oxide is a particularly illustrative case to highlight the role of precursors and solvents in determining the morphological characteristics of the final oxidic compounds. Figure 3.8(a) shows a TEM overview of In_2O_3 nanoparticles obtained from indium acetylacetonate in benzylamine

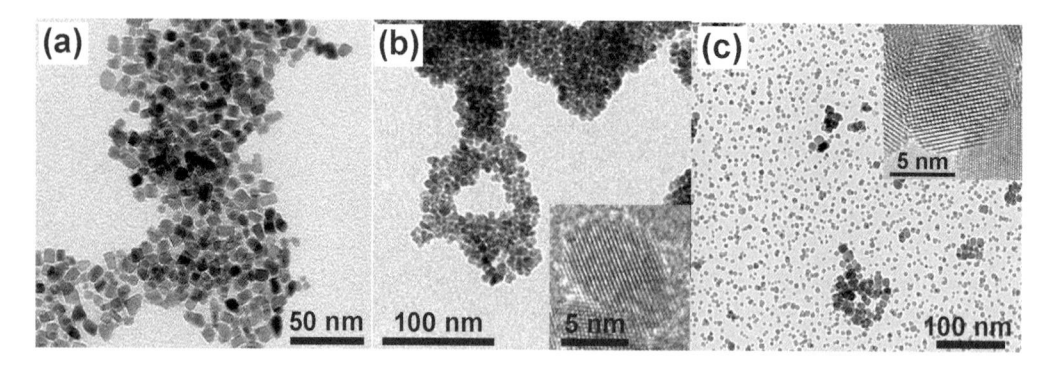

Fig. 3.8 TEM overview images of (a) In_2O_3 nanoparticles obtained from indium acetylacetonate and benzylamine, (b) In_2O_3 synthesized from indium acetylacetonate and acetonitrile, inset: HRTEM of one particle, (c) SnO_2-doped In_2O_3 (10 wt% SnO_2), inset: HRTEM of one particle.

[76]. The size of the particles is in the range 5–10 nm with pronounced crystal facets. The reaction of indium acetylacetonate with acetonitrile results in much smaller particles of about 3–5 nm and with nearly spherical morphology (Fig. 3.8(b)). To increase the electrical conductivity of indium oxide, the nanoparticles can be doped with SnO_2. This can be achieved by mixing indium acetylacetonate and tin *tert*-butoxide in benzyl alcohol [92]. After solvothermal treatment, indium tin oxide nanocrystals are obtained in the size range 5–10 nm (Fig. 3.8(c)). The solution route allows the variation of the tin oxide content in the range 2–30 wt%, with a maximum electrical conductivity obtained at a doping level of 15 wt% [92].

3.6
Selected Reaction Mechanisms

Keeping in mind that the majority of these reaction systems do not make use of water, the question arises where the oxygen for the oxide formation comes from. The most frequently found reaction pathways are displayed in Scheme 3.1.

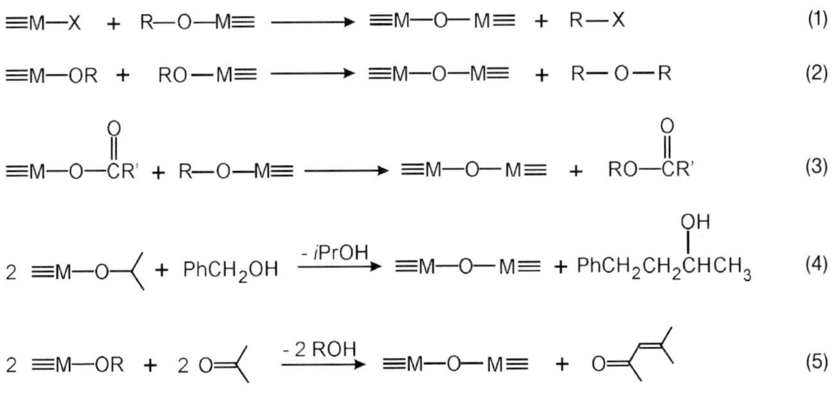

$$\equiv M-X \ + \ R-O-M\equiv \ \longrightarrow \ \equiv M-O-M\equiv \ + \ R-X \quad (1)$$

$$\equiv M-OR \ + \ RO-M\equiv \ \longrightarrow \ \equiv M-O-M\equiv \ + \ R-O-R \quad (2)$$

$$\equiv M-O-\underset{\underset{O}{\|}}{C}R' \ + \ R-O-M\equiv \ \longrightarrow \ \equiv M-O-M\equiv \ + \ RO-\underset{\underset{O}{\|}}{C}R' \quad (3)$$

$$2 \equiv M-O-\!\!\!\!\prec \ + \ PhCH_2OH \ \xrightarrow{-\mathit{i}PrOH} \ \equiv M-O-M\equiv \ + \ PhCH_2CH_2\overset{OH}{\underset{|}{C}}HCH_3 \quad (4)$$

$$2 \equiv M-OR \ + \ 2\ O=\!\!\!\prec \ \xrightarrow{-2\ ROH} \ \equiv M-O-M\equiv \ + \ O=\!\!\!\prec \quad (5)$$

Scheme 3.1 Selected reaction pathways to metal oxide formation in nonaqueous systems. Eq. (1): Alkyl halide elimination, Eq. (2): ether elimination, Eq. (3): ester elimination, Eq. (4): C–C bond formation between benzylic alcohols and alkoxides, Eq. (5): aldol-like condensation reactions.

Eq. (1) summarizes the condensation between metal halides and metal alkoxides, resulting from the reaction of metal halides with alcohols, with release of an alkyl halide. One of the first examples was the preparation of anatase nanocrystals from titanium isopropoxide and titanium chloride [23].

Ether elimination (Eq. (2)) leads to the formation of a M–O–M bond upon condensation of two metal alkoxides with elimination of an organic ether. This mechanism was reported for the formation of hafnium oxide nanoparticles [80].

The ester elimination process involves the reaction between metal carboxylates and metal alkoxides (Eq. (3)), as reported for zinc oxide [38]. Analogous to ester eliminations are amide eliminations. By reacting metal oleates with amines one can induce the controlled growth of titania nanorods [57].

In addition to these three main routes, more complex pathways have recently been reported for metal oxide nanoparticles. $BaTiO_3$ formation occurs via a much more elaborate mechanism involving a C–C bond formation between the isopropoxy ligand of titanium isopropoxide and the solvent benzyl alcohol (Eq. (4)), followed by the release of a hydroxy species from benzyl alcohol, which induces further condensation to the metal oxide [84]. Similar C–C coupling mechanisms were also found in the case of CeO_2 and Nb_2O_5 [31].

Ketones are scarcely used as solvents for the nonaqueous synthesis of metal oxides, probably due to their generally low boiling points. This problem can be circumvented by performing the reactions in autoclaves and also the condensation processes often take place at moderate temperatures. The release of oxygen usually involves aldol condensation, where two carbonyl compounds react with each other with elimination of water, which acts as an oxygen supplying agent for the metal oxide formation (Eq. (5)). Literature examples include the formation of ZnO [19] and TiO_2 [87] in acetone.

Detailed discussions of these and other reaction pathways along with an exhaustive overview of literature examples can be found elsewhere [31].

3.7
Summary and Outlook

Aqueous sol–gel procedures have been applied for the synthesis of bulk metal oxides with great success. However, on the nanoscale, the high complexity of aqueous systems constitutes a major problem, which makes it rather difficult to obtain full control over the structural and morphological features of nanoparticles. Synthesis procedures performed in organic solvents with exclusion of water drastically decrease the reaction rates, allowing better control over particle size, shape, size distribution, crystallinity and surface properties. The replacement of water by the more controllable reactivity of the oxygen–carbon bond in organic solvents leads to the situation where basic organic reaction principles supply the water or, more formally, the oxygen necessary for oxide formation. Many observations point to the fact that the organic side of the process plays a crucial role in influencing the final inorganic product. An illustrative example was reported by Li et al. for the synthesis of Fe_3O_4 nanoparticles using iron(III) chloride as precursor in 2-pyrrolidone [93]. These authors found that the solvent is unstable under reaction conditions and forms carbon monoxide, which is responsible for the partial reduction of Fe(III) to Fe(II). Accordingly, the organic solvent is not only crucial with respect to particle stabilization, but also with respect to the intrinsic composition.

In the last few years, intense research efforts on nonaqueous sol–gel procedures have produced a wide variety of nanocrystalline metal oxides, sometimes with fantastic particle morphologies and outstandingly small size distributions. Although investigations with respect to a tailoring of particle size and shape as well as the elucidation of potential reaction mechanisms represent important progress in nanoparticle synthesis, the ultimate goal of rational synthesis design is still far away. The big question of finding a relationship between a particular synthesis system and the final particle morphology still awaits an answer.

Acknowledgement

Financial support by the Max Planck Society is gratefully acknowledged.

References

1 J. L. G. Fierro, L. L. G. Fierro, *Metal Oxides: Chemistry and Applications*, CRC Press Taylor & Francis, Boca Raton, **2005**.
2 G. Eranna, B. C. Joshi, D. P. Runthala, R. P. Gupta, *Crit. Rev. Solid State Mater. Sci.* **2004**, *29*, 111.
3 M. Winter, R. J. Brodd, *Chem. Rev.* **2004**, *104*, 4245.
4 K. L. Chopra, P. D. Paulson, V. Dutta, *Prog. Photovolt: Res. Appl.* **2004**, *12*, 69.
5 A. J. Moulson, J. M. Herbert, *Electroceramics: Materials, Properties, Applications*, John Wiley & Sons, Chichester, **2003**.
6 B. L. Cushing, V. L. Kolesnichenko, C. J. O'Connor, *Chem. Rev.* **2004**, *104*, 3893.
7 L. L. Hench, J. K. West, *Chem. Rev.* **1990**, *90*, 33.
8 D. C. Bradley, R. C. Mehrotra, I. P. Rothwell, A. Singh, *Alkoxo and Aryloxo Derivatives of Metals*, Academic Press, London, **2001**.
9 N. Y. Turova, E. P. Turevskaya, *The Chemistry of Metal Alkoxides*, Kluwer Academic Publishers, Boston, **2002**.
10 J. Livage, M. Henry, C. Sanchez, *Prog. Solid State Chem.* **1988**, *18*, 259.
11 C. J. Brinker, G. W. Scherer, *Sol-Gel Science: The Physics and Chemistry of Sol-Gel Processing*, Academic Press, London, **1990**.
12 R. J. P. Corriu, D. Leclercq, *Angew. Chem. Int. Ed.* **1996**, *35*, 1420.
13 L. G. Hubert-Pfalzgraf, *J. Mater. Chem.* **2004**, *14*, 3113.
14 M. In, C. Sanchez, *J. Phys. Chem. B* **2005**, *109*, 23870.
15 A. W. Dearing, E. E. Reid, *J. Am. Chem. Soc.* **1928**, *50*, 3058.
16 A. Vioux, *Chem. Mater.* **1997**, *9*, 2292.
17 J. N. Hay, H. M. Raval, *Chem. Mater.* **2001**, *13*, 3396.
18 M. Inoue, *J. Phys.: Condens. Matter* **2004**, *16*, S1291.
19 S. C. Goel, M. Y. Chiang, P. C. Gibbons, W. E. Buhro, *Mater. Res. Soc. Symp. Proc.* **1992**, *271*, 3.
20 M. Inoue, H. Kominami, T. Inui, *Appl. Catal., A* **1993**, *97*, L25.
21 M. Inoue, H. Kominami, H. Otsu, T. Inui, *Nippon Kagaku Kaishi* **1991**, 1364.
22 H. Kominami, J. Kato, Y. Takada, Y. Doushi, B. Ohtani, S. Nishimoto, M. Inoue, T. Inui, Y. Kera, *Catal. Lett.* **1997**, *46*, 235.
23 T. J. Trentler, T. E. Denler, J. F. Bertone, A. Agrawal, V. L. Colvin, *J. Am. Chem. Soc.* **1999**, *121*, 1613.
24 M. Ivanda, S. Music, S. Popovic, M. Gotic, *J. Mol. Struct.* **1999**, *481*, 645.

25 C. de Mello Donega, P. Liljeroth, D. Vanmaekelbergh, *Small* **2005**, *1*, 1152.
26 Y. W. Jun, J. H. Lee, J. S. Choi, J. Cheon, *J. Phys. Chem. B* **2005**, *109*, 14795.
27 S. Kumar, T. Nann, *Small* **2006**, *2*, 316.
28 A. Nel, T. Xia, L. Mädler, N. Li, *Science* **2006**, *311*, 622.
29 T. Hyeon, *Chem. Commun.* **2003**, 927.
30 M. Niederberger, G. Garnweitner, N. Pinna, G. Neri, *Prog. Solid State Chem.* **2005**, *33*, 59.
31 M. Niederberger, G. Garnweitner, *Chem. Eur. J.* **2006**, *12*, 7282.
32 T. Hyeon, S. S. Lee, J. Park, Y. Chung, H. B. Na, *J. Am. Chem. Soc.* **2001**, *123*, 12798.
33 J. Park, K. An, Y. Hwang, J. G. Park, H. J. Noh, J. Y. Kim, J. H. Park, N. M. Hwang, T. Hyeon, *Nat. Mater.* **2004**, *3*, 891.
34 Q. Song, Z. J. Zhang, *J. Am. Chem. Soc.* **2004**, *126*, 6164.
35 F. X. Redl, K. S. Cho, C. B. Murray, S. O'Brien, *Nature* **2003**, *423*, 968.
36 E. V. Shevchenko, D. V. Talapin, C. B. Murray, S. O'Brien, *J. Am. Chem. Soc.* **2006**, *128*, 3620.
37 E. V. Shevchenko, D. V. Talapin, N. A. Kotov, S. O'Brien, C. B. Murray, *Nature* **2006**, *439*, 55.
38 J. Joo, S. G. Kwon, J. H. Yu, T. Hyeon, *Adv. Mater.* **2005**, *17*, 1873.
39 J.-W. Seo, Y.-W. Jun, S. J. Ko, J. Cheon, *J. Phys. Chem. B* **2005**, *109*, 5389.
40 D. Zitoun, N. Pinna, N. Frolet, C. Belin, *J. Am. Chem. Soc.* **2005**, *127*, 15034.
41 R. L. Penn, J. F. Banfield, *Am. Mineral.* **1998**, *83*, 1077.
42 X.-L. Li, Q. Peng, J.-X. Yi, X. Wang, Y. Li, *Chem. Eur. J.* **2006**, *12*, 2383.
43 T. Yu, J. Joo, J. Park, T. Hyeon, *Angew. Chem. Int. Ed.* **2005**, *44*, 7411.
44 N. R. Jana, Y. Chen, X. Peng, *Chem. Mater.* **2004**, *16*, 3931.
45 A. K. Boal, K. Das, M. Gray, V. M. Rotello, *Chem. Mater.* **2002**, *14*, 2628.
46 P. D. Cozzoli, A. Kornowski, H. Weller, *J. Am. Chem. Soc.* **2003**, *125*, 14539.
47 M. Yin, C. K. Wu, Y. Lou, C. Burda, J. T. Koberstein, Y. Zhu, S. O'Brien, *J. Am. Chem. Soc.* **2005**, *127*, 9506.
48 J. Rockenberger, E. C. Scher, A. P. Alivisatos, *J. Am. Chem. Soc.* **1999**, *121*, 11596.
49 Z. Li, H. Chen, H. B. Bao, M. Y. Gao, *Chem. Mater.* **2004**, *16*, 1391.
50 S. H. Sun, H. Zeng, D. B. Robinson, S. Raoux, P. M. Rice, S. X. Wang, G. X. Li, *J. Am. Chem. Soc.* **2004**, *126*, 273.
51 J. Tang, J. Fabbri, R. D. Robinson, Y. M. Zhu, I. P. Herman, M. L. Steigerwald, L. E. Brus, *Chem. Mater.* **2004**, *16*, 1336.
52 W. S. Seo, H. H. Jo, K. Lee, J. T. Park, *Adv. Mater.* **2003**, *15*, 795.
53 W. S. Seo, H. H. Jo, K. Lee, B. Kim, S. J. Oh, J. T. Park, *Angew. Chem. Int. Ed.* **2004**, *43*, 1115.
54 M. Epifani, J. Arbiol, R. Díaz, M. J. Perálvarez, P. Siciliano, J. R. Morante, *Chem. Mater.* **2005**, *17*, 6468.
55 Y. W. Jun, M. F. Casula, J. H. Sim, S. Y. Kim, J. Cheon, A. P. Alivisatos, *J. Am. Chem. Soc.* **2003**, *125*, 15981.
56 J. Tang, F. Redl, Y. Zhu, T. Siegrist, L. E. Brus, M. L. Steigerwald, *Nano Lett.* **2005**, *5*, 543.
57 Z. Zhang, X. Zhong, S. Liu, D. Li, M. Han, *Angew. Chem. Int. Ed.* **2005**, *44*, 3466.
58 K. Lee, W. S. Seo, J. T. Park, *J. Am. Chem. Soc.* **2003**, *125*, 3408.
59 M. Shim, P. Guyot-Sionnest, *J. Am. Chem. Soc.* **2001**, *123*, 11651.
60 P. D. Cozzoli, M. L. Curri, A. Agostiano, G. Leo, M. Lomascolo, *J. Phys. Chem. B* **2003**, *107*, 4756.
61 J. Joo, T. Yu, Y. W. Kim, H. M. Park, F. X. Wu, J. Z. Zhang, T. Hyeon, *J. Am. Chem. Soc.* **2003**, *125*, 6553.
62 C. Wang, Z. Deng, Y. Li, *Inorg. Chem.* **2001**, *40*, 5210.
63 C. Wang, Z. X. Deng, G. H. Zhang, S. S. Fan, Y. D. Li, *Powder Technol.* **2002**, *125*, 39.
64 G. Li, L. Li, J. Boerio-Goates, B. F. Woodfield, *J. Am. Chem. Soc.* **2005**, *127*, 8659.
65 A. A. Khaleel, *Chem. Eur. J.* **2004**, *10*, 925.
66 M. Niederberger, M. H. Bartl, G. D. Stucky, *J. Am. Chem. Soc.* **2002**, *124*, 13642.
67 Z. W. Zhao, Z. P. Guo, H. K. Liu, *J. Power Sources* **2005**, *147*, 264.

68 J. Ba, J. Polleux, M. Antonietti, M. Niederberger, *Adv. Mater.* **2005**, *17*, 2509.
69 J. Buha, I. Djerdj, M. Antonietti, M. Niederberger, unpublished.
70 H. G. Choi, Y. H. Jung, D. K. Kim, *J. Am. Ceram. Soc.* **2005**, *88*, 1684.
71 M. Niederberger, M. H. Bartl, G. D. Stucky, *Chem. Mater.* **2002**, *14*, 4364.
72 M. Niederberger, G. Garnweitner, F. Krumeich, R. Nesper, H. Cölfen, M. Antonietti, *Chem. Mater.* **2004**, *16*, 1202.
73 J. Polleux, N. Pinna, M. Antonietti, M. Niederberger, *Adv. Mater.* **2004**, *16*, 436.
74 J. Polleux, N. Pinna, M. Antonietti, C. Hess, U. Wild, R. Schlögl, M. Niederberger, *Chem. Eur. J.* **2005**, *11*, 3541.
75 J. Polleux, N. Pinna, M. Antonietti, M. Niederberger, *J. Am. Chem. Soc.* **2005**, *127*, 15595.
76 M. Niederberger, G. Garnweitner, J. Buha, J. Polleux, J. Ba, N. Pinna, *J. Sol-Gel Sci. Tech.* **2006**, *40*, 259.
77 J. Klongdee, W. Petchkroh, K. Phuempoonsathaporn, P. Praserthdam, A. S. Vangnai, V. Pavarajarn, *Sci. Technol. Adv. Mater.* **2005**, *6*, 290.
78 W. Payakgul, O. Mekasuwandumrong, V. Pavarajarn, P. Praserthdam, *Ceram. Int.* **2005**, *31*, 391.
79 N. Pinna, M. Antonietti, M. Niederberger, *Colloids Surf., A* **2004**, *250*, 211.
80 N. Pinna, G. Garnweitner, M. Antonietti, M. Niederberger, *Adv. Mater.* **2004**, *16*, 2196.
81 N. Pinna, G. Neri, M. Antonietti, M. Niederberger, *Angew. Chem. Int. Ed.* **2004**, *43*, 4345.
82 G. Garnweitner, M. Niederberger, *J. Am. Ceram. Soc.* **2006**, *89*, 1801.
83 M. Niederberger, N. Pinna, J. Polleux, M. Antonietti, *Angew. Chem. Int. Ed.* **2004**, *43*, 2270.
84 M. Niederberger, G. Garnweitner, N. Pinna, M. Antonietti, *J. Am. Chem. Soc.* **2004**, *126*, 9120.
85 N. Pinna, G. Garnweitner, P. Beato, M. Niederberger, M. Antonietti, *Small* **2005**, *1*, 112.
86 J. Polleux, A. Gurlo, M. Antonietti, M. Niederberger, *Angew. Chem. Int. Ed.* **2006**, *45*, 261.
87 G. Garnweitner, M. Antonietti, M. Niederberger, *Chem. Commun.* **2005**, 397.
88 B. C. Gaskins, J. J. Lannutti, *J. Mater. Res.* **1996**, *11*, 1953.
89 G. Garnweitner, J. Hentschel, M. Antonietti, M. Niederberger, *Chem. Mater.* **2005**, *17*, 4594.
90 N. Pinna, S. Grancharov, P. Beato, P. Bonville, M. Antonietti, M. Niederberger, *Chem. Mater.* **2005**, *17*, 3044.
91 N. Pinna, G. Garnweitner, M. Antonietti, M. Niederberger, *J. Am. Chem. Soc.* **2005**, *127*, 5608.
92 J. Ba, D. Fattakhova Rohlfing, A. Feldhoff, T. Brezesinski, I. Djerdj, M. Wark, M. Niederberger, *Chem. Mater.* **2006**, *18*, 2848.
93 Z. Li, Q. Sun, M. Y. Gao, *Angew. Chem. Int. Ed.* **2005**, *44*, 123.

4
Growth of Nanocrystals in Solution

R. Viswanatha and D. D. Sarma

4.1
Introduction

Tunability of various physical and chemical properties of materials by varying the size in the region of nanometers has opened up many new directions in several fields of current research and modern technologies [1, 2]. In particular, the study of systematic changes in the electronic structure of solids as a function of size in this nanometric regime has been intensively investigated in recent times (see for example Refs. [3–6]). Sustained efforts in this field have established interesting applications such as UV protection films [7], fluorescent sensors in biological applications [8], photocurrent generation in various devices [9], optical switches [10], catalytic reactions [11], nanotweezers [12] and other optoelectronic devices [13]. One of the major aspects necessary for the actual realization of these applications is the ability to synthesize nanocrystals of the required size with a controlled size distribution. The growing demand to obtain such nanocrystals is met largely by the solution route synthesis of nanocrystals [14, 15], due to its ease of implementation and high degree of flexibility. The main difficulty with this method is that the dependences of the average size and the size distribution of the generated particles on parameters of the reaction are not understood in detail and, therefore, the optimal reaction conditions are arrived at essentially in an empirical and intuitive manner [16]. The understanding of coarsening processes that lead to growth of the minority phase within a majority phase by combining smaller particles is important both scientifically and for technological considerations [17–21]. In recent times, there has been a renewed interest in understanding the growth process of a solid phase within another solid in the nanometric regime. Interestingly, the kinetics of the growth of a solid from a solution, which is the most popular chemical method to produce a wide variety of systems with dimensions in the nanometer region, has been relatively less investigated [22–26]. However, there is a popular belief that the growth in solution occurs via a diffusion limited Ostwald ripening process [14, 15].

In this chapter, we briefly discuss the theory of nucleation. We then review the various theoretical aspects of the different processes that influence the growth rate. Following the discussion of the theoretical framework, we review experimental progress made in investigating the growth of various nanocrystals.

4.2
Theoretical Aspects

The growth of nanocrystals in solution involves two important processes, the nucleation followed by the growth of the nanocrystals. We discuss these two processes in the following two subsections.

4.2.1
Theory of Nucleation

La Mer and coworkers [27] studied extensively nucleation and growth in sulfur sols, from which they developed an understanding of the mechanism for the formation of colloids or nanocrystals from a homogeneous, supersaturated medium. Their mechanism suggested that a synthesis of the colloid should be designed in such a way that the concentration increases rapidly, rising above the saturation concentration for a brief period, when a short burst of nucleation occurs with the formation of a large number of nuclei in a short space of time. These particles grow rapidly and lower the concentration below the nucleation level whilst allowing the particles to grow further at a rate determined by the slowest step in the growth process, thus separating the nucleation and growth in time. La Mer's mechanism is depicted schematically by means of the simple diagram shown in Fig. 4.1. The requirements for monodispersity, as evident from La Mer's diagram, are a high rate of nucleation leading to the burst of nuclei formation in a short period, an initial fast rate of growth of these nuclei to reduce the concentration below the nucleation concentration rapidly and an eventual slow rate of growth leading to a long growth period compared to the nucleation period. They also derived the rates of growth of sols prepared by the above-mentioned dilution method. The growth rates obtained were reproduced from previous theoretical considerations which allowed the estimation of the value of the diffusion coefficient of sulfur. It was also claimed that the application of this method to the estimation of diffusion coefficients was valid for any colloidal system that was characterized by a small particle size distribution at all stages of its growth. Further, a qualitative explanation was offered for the necessary conditions under which mono-dispersed colloids might be prepared by both the dilution and acid decomposition of sodium thiosulfate methods. This mechanism of La Mer was later widely applied in attempts to prepare various nearly-monodisperse particles in homogeneous solutions [28], yet success was achieved only after tedious trial-and-error attempts to tune the major parameters such as the concentration of reactants. So far it has not been possible to have a generalized approach to the

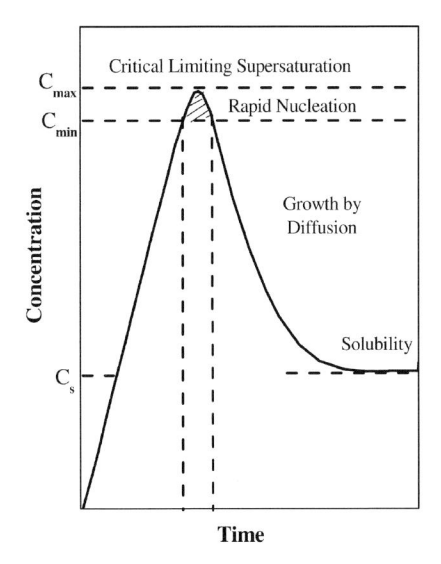

Fig. 4.1 Schematic diagram illustrating La Mer's condition for nucleation.

synthesis of different systems by following the necessary conditions of La Mer [29]. Generalizations between preparations have been few and it is now believed that La Mer's mechanism is rigorously appropriate only for the system for which it was developed, that is sulfur sols, and it may not have significance as a general approach to a wide variety of systems. It was later observed that La Mer's condition for nucleation is neither a necessary nor sufficient condition for monodispersity but that the specific growth mechanism also plays an important role in deciding the size and the size distribution.

Regardless of the rigorous validity of La Mer's prediction in the context of diverse systems, the key idea of separating the nucleation stage and growth process in time is often used to obtain nearly monodisperse particles. In most of the cases in recent times, synthesis has been carried out by mixing the reactants together, often by injecting one of the components into the remaining ones, in a very short time. This is to ensure that the entire nucleation takes place in that short time, followed by a much slower growth process, thereby attempting to separate the two stages temporally. In the next subsection, we discuss the various aspects of the mechanism of the growth of nanocrystals from solution.

4.2.2
Mechanism of Growth

Nucleation occurs over some time with constant monomer concentration. Eventually surface growth of clusters begins to occur which depletes the monomer supply. When the monomer concentration falls below the critical level for nucleation (critical supersaturation level), nucleation ends. A general analysis of the growth process is then important to understand nanocrystal synthesis. In general, the surface to volume ratio in smaller particles is quite high. As a result of the large surface area present, it is observed that surface excess energy becomes more important in very small particles, constituting a non-negligible percentage of the total energy. Hence, for a solution that is initially not in thermodynamic equilibrium, a mechanism that allows the formation of larger particles at the cost of smaller particles reduces the surface energy and hence plays a key role in the growth of nanocrystals. A colloidal particle grows by a sequence of monomer diffusion towards the surface followed by reaction of the monomers at the surface of the nanocrystal. Coarsening effects, controlled either by mass transport or diffusion, are often termed the Ostwald ripening process. This diffusion limited Ostwald ripening process is the most predominant growth mechanism and was first quantified by Lifshitz and Slyozov [30], followed by a related work by Wagner [31], known as the LSW theory.

The diffusion process is dominated by the surface energy of the nanoparticle. The interfacial energy is the energy associated with an interface due to differences between the chemical potential of atoms in an interfacial region and atoms in neighboring bulk phases. For a solid species present at a solid/liquid interface, the chemical potential of a particle increases with decreasing particle size, the equilibrium solute concentration for a small particle is much higher than for a large particle, as described by the Gibbs–Thompson equation. The resulting concentration gradients lead to transport of the solute from the small particles to the larger particles. The equilibrium concentration of the nanocrystal in the liquid phase is dependent on the local curvature of the solid phase. Differences in the local equilibrium concentrations, due to variations in curvature, set up concentration gradients and provide the driving force for the growth of larger particles at the expense of smaller particles [32].

Now, assume that the average radius of the particles is r. The bulk liquid phase is considered to have a uniform supersaturated monomer concentration, c_b, while the monomer concentration at the particle interface, c_i and the solubility of the particle with a radius r is given by c_r. The flux of monomers, J, passing through a spherical surface with radius x within the diffusion layer, is given by Fick's first law as

$$J = 4\pi x^2 D \, dC/dx \tag{4.1}$$

At steady state, where J is constant over the diffusion layer x, the above equation can be integrated approximately to obtain

$$J = \frac{4\pi D r (r + \delta)}{\delta} (c_b - c_i) \tag{4.2}$$

where, δ is the thickness of the diffusion layer.

This flux can be equated to the consumption rate of the monomer species at the surface of the particle. That is,

$$J = 4\pi r^2 k_d (c_i - c_r) \qquad (4.3)$$

where k_d is the rate constant of a simple first order deposition reaction. In solution, it is difficult to measure c_i and hence it is necessary to eliminate that variable from the two equations. Assuming that $dr/dt = JV_m/4\pi r^2$, we get

$$\frac{dr}{dt} = \frac{\frac{D}{r}\left(1 + \frac{r}{\delta}\right) V_m (c_b - c_r)}{1 + \frac{D}{k_d r \left(1 + \frac{r}{\delta}\right)}} \qquad (4.4)$$

The terms c_b and c_r are related to the particle radius, r, by the Gibbs–Thompson equation given by the expression

$$c_r = c_\infty \exp\left(\frac{2\sigma V_m}{rRT}\right) \approx c_\infty \left(1 + \frac{2\sigma V_m}{rRT}\right) \qquad (4.5)$$

where c_∞ is the concentration of a flat particle, σ is the interfacial energy, V_m is the molar volume and R is the universal gas constant. The approximate expression on the right is derived by retaining only the first two terms in the expansion of the exponential function under the assumption of a small value of the argument, $2\sigma V_m/rRT$.

Similarly, c_b can be written as

$$c_b = c_\infty \exp\left(\frac{2\sigma V_m}{r_b RT}\right) \approx c_\infty \left(1 + \frac{2\sigma V_m}{r_b RT}\right) \qquad (4.6)$$

Since diffusion layer thicknesses are typically of the order of microns, in the case of nanocrystals we assume that $r \ll \delta$. Substituting Eqs. (4.5) and (4.6) into Eq. (4.4) we obtain the equation

$$\frac{dr}{dt} = \frac{2\sigma V_m^2 c_\infty}{RT(1/D + 1/k_d r)} \frac{(1/r_b - 1/r)}{r} \qquad (4.7)$$

We shall now explore the different behaviors of growth arising from this differential equation in various limits.

4.2.2.1 Diffusion Limited Growth: Lifshitz–Slyozov–Wagner (LSW) Theory and Post-LSW Theory

Diffusion limited growth: When the diffusion is the slowest step in the growth process, characterized by $D \ll k_d r$ in Eq. (4.7), the particle growth is essentially controlled by the diffusion of the monomers to the surface. In this limit, Eq. (4.7) reduces to the form,

$$\frac{dr}{dt} = \frac{2\sigma D V_m^2 c_\infty}{RT} \frac{(r/r_b - 1)}{r^2} = K_D (r/r_b - 1)/r^2 \quad (4.8)$$

where K_D, given by $2\sigma D V_m^2 c_\infty / RT$, is a constant. If the total mass of the system is explicitly kept conserved, the LSW theory showed that the ratio r/r_b is a constant. Making this assumption in Eq. (4.8), we see that Eq. (4.8) reduces to the form

$$\frac{dr}{dt} = K_D * \text{const}/r^2 \quad (4.9)$$

that can be easily solved to obtain the dependence of the particle size on time. It is given by

$$r^3 - r_0^3 = Kt \quad (4.10)$$

where r_0 is the average radius of the particle at time $t = 0$. K is given by the expression

$$K = \frac{8\sigma D V_m^2 c_\infty}{9RT} \quad (4.11)$$

where D, the diffusion constant of the system at any given temperature, is given by the equation, $D = D_0 \exp(-E_a/k_b T)$.

LSW theory: The original idea of a growth mechanism driven by the reduction of the surface energy was proposed by Ostwald in 1901 [33]. It is interesting to note that a quantitative theory based on these ideas was derived [30, 31] nearly 50 years after Ostwald's discovery of the phenomenon. The discovery of the LSW theory marked one of the major advances in the theory of Ostwald ripening. The LSW theory, considering a diffusion limited growth, but following a different and more rigorous method than that described earlier, was able to make quantitative predictions on the long-time behavior of the coarsening process.

The LSW approach is based on the following basic assumptions.
- This theory examines the growth of spherical particles in a supersaturated medium.
- Particles are assumed to grow or shrink only in relation to the mean field concentration set at infinity.
- The total mass of the solute is conserved.
- The size distribution of the growing phase under the assumption of being spherical is characterized only by a radius distribution in terms of a continuous function, valid in the limit of a sufficiently large number of particles in the system to justify such a continuum description.
- Processes such as nucleation and aggregation that introduce new particles are negligible.

Using these assumptions, they arrived at a universal asymptotic solution. First, the concept of a critical radius, r_b, was introduced, which separated the smaller sized particles ($r < r_b$), shrinking in size, from the larger particles ($r > r_b$) that became larger with time. When the radius r is found to be equal to r_b, the growth is found to be zero. In the asymptotic limit, it was shown that the ratio of the mean radius, r, of the system of particles to the critical radius, r_b, is a constant.

It is interesting to note that the asymptotic state of the system is predicted to be independent of the initial conditions. This was obtained as a direct consequence of the mass conservation of the solute on the asymptotic solution of the continuity equation and the kinetic equation. These authors further showed that the time evolution of distribution of radii of the particles at any given time has the following functional form,

$$D(\xi) = \kappa \xi^2 \left(\frac{3}{3+\xi}\right)^{7/3} \left(\frac{1.5}{1.5-\xi}\right)^{11/3} \exp\left(\frac{-\xi}{1.5-\xi}\right) \tag{4.12}$$

where, $\xi = x/r$, x is the radius of various particles and r the average mean radius and κ depends only on time t by the relation $\kappa = \kappa_c/(1+t/\tau_D)^{4/3}$. τ_D is the time constant given by the expression

$$\tau_D = \frac{9r_0^3 RT}{64\sigma DC_\infty V_m^2} \tag{4.13}$$

We have illustrated a typical form of the distribution function, $D(\xi)$, in Fig. 4.2. This functional form shows a typical characteristic asymmetry at the higher particle diameter side cutting off sharply when the radius is 1.5 times the average radius. Further, integrating Eq. (4.12), they obtained an expression for the average radius, r, of the particle at time t, given by $r^3 - r_0^3 = Kt$, similar to Eq. (4.10).

Post-LSW theories: The coarsening phenomenon has been experimentally investigated in diverse systems, including detailed studies involving precipitate for-

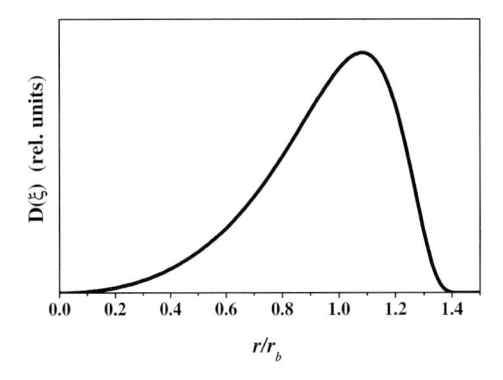

Fig. 4.2 Typical line shape of the size distribution of a purely diffusion-controlled reaction as predicted by the LSW theory.

mation in alloys, like Co in Cu–Co alloy [34] and Ni in Ni–Fe alloys [35]. It was observed that while the temporal power law for r, (i.e., $r^3 \propto t$) was confirmed by these experiments [36], the temporal evolution of the size distribution did not agree with that of the experimental results. The experiments, in general, showed a broader and more symmetric distribution. Further experiments also showed that there exists a dependence of the rate constant K on the volume fraction [37], where the volume fraction is defined as the ratio of the volume of the particular phase to that of the total volume of the system. While the LSW theory is a pioneering work in this field, it does not consider the effects of the finite volume fraction of the coarsening phase. We now take a look at the post-LSW theories, taking into account the finite volume fraction of the growing phase. Moreover, the LSW theory assumes that their rate equation is valid at a very low volume fraction, ϕ, of the coarsening phase, such that the coarsening rate is independent of the surroundings. This unspecified low volume fraction, being much lower than realistic values of ϕ encountered in typical experiments, led to disagreements between experimental results and theoretical predictions. These disagreements were found to be due the high strength of the diffusional interactions between neighboring particles that arose from the long-range coulomb interaction surrounding the particles. This can be taken into account by statistically averaging the diffusional interactions of the particle with its surroundings.

In order to improve the theoretical description beyond the LSW description, several groups addressed theoretically some of the above-mentioned shortcomings of the original theory. The prominent ones among them include the works by Brailsford and Wynblatt [38], Voorhees and Glickman [39], Marquse and Ross [40] and Tokuyama and Kawasaki [41]. It is surprising to note that each of the groups have used identical microscopic equations to perform the statistical averaging of the growth rate and arrive at qualitatively different results; this apparently intriguing situation arises from using different methods of averaging, such as chemical rate theory, computer simulation techniques, a multiple scattering method or a scaling expansion technique. However, all these theories agree on some of the points. For example, the temporal power law reported in the LSW theory is not dependent on the volume fraction, ϕ, validating this aspect of LSW theory. On the other hand, the particle size distribution function becomes broader and more symmetric than the LSW distribution with an increase in the volume fraction. The rate constant, K, rises rapidly at low volume fractions and is followed by a slower increase at higher volume fractions. The predictions of the different theoretical works mentioned above are almost identical till a volume fraction of 0.1, unlike the predictions of LSW theory that did not show any dependence of the rate constant K on the volume fraction. However, these theories do not agree with each other on several other important aspects. For example, the dependence of the rate constant K on the volume fraction ϕ was predicted to follow $K(\phi) - K(0) \sim \phi^{(1/2)}$ in Ref. [40] but $K(\phi) - K(0) \sim \phi^{(1/3)}$ in Ref. [39].

The above discussion shows that while the LSW theory represented a considerable advance in describing the diffusion-controlled Ostwald ripening quantitatively, it is an oversimplified approach in many respects in comparison to the complex and diverse situations encountered in actual practice. Though later works made significant progress by accounting for some of the concerns arising from the drastic assumptions in the LSW theory, these still left unanswered a large number of questions about the size distribution in the diffusion-controlled growth. For example, it is not known what particle size distribution results from nucleation followed by the growth process and if this influences the system's approach to the asymptotic state. Moreover, at $\phi \neq 0$, diffusional interactions between particles are present and thus the spatial distribution of the particles becomes important. However, all the theories described earlier assume random spatial distribution of particles and the role of spatial correlations between particles is largely ignored; it is not yet fully understood how these assumptions may affect the description of the growth kinetics.

4.2.2.2 Reaction-limited Growth

So far, we have only considered the diffusion-limited regime, characterized by $D \ll k_d r$ in Eq. (4.7). Besides the diffusion process, that is accounted for in the LSW theory, another important process in the growth of any particle is the reaction at the surface where the units of diffusing particles are assimilated into the growing nanocrystal. We now turn to this opposite limit of reaction-limited growth.

If $k_d r \ll D$ in Eq. (4.7), then the growth rate is limited by the surface reaction of the monomers. Then the rate law for the average radius of growing nanocrystals can be reduced from Eq. (4.7) to the form given by

$$\frac{dr}{dt} = \frac{2\sigma k_d V_m^2 c_\infty}{RT} \frac{(r/r_b - 1)}{r} = K_R (r/r_b - 1)/r \qquad (4.14)$$

If the conservation of mass is valid, similar to the diffusion limited case, the ratio of the average radius to the critical radius (r/r_b) is observed to be constant [32, 66]. Hence the above differential equation in the case of simple first order reaction can be integrated under this assumption to give rise to the equation of the form

$$r^2 \approx K_R t \qquad (4.15)$$

Thus, it can be seen that the mean particle size grows as a function of the square root of time. The size distribution in the case of the reaction controlled Ostwald ripening was established long back by Wagner [31] and is expected to follow the equation of the form

$$D(\xi) = \kappa_R \xi \left(\frac{2}{2-\xi}\right)^5 \exp\left(\frac{-3\xi}{2-\xi}\right) \qquad (4.16)$$

where, $\xi = x/r$ and κ_R depends on time t by the relation $k_R = k'/(1 + t/\tau_R)^2$. τ_R is the time constant given by the expression

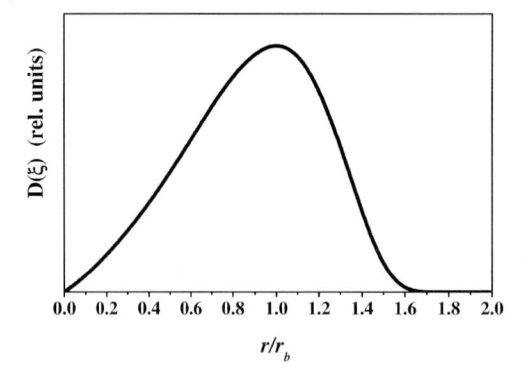

Fig. 4.3 Typical line shape of the size distribution of a purely reaction-controlled reaction as predicted by Wagner.

$$\tau_R = \frac{r_{b0}^2 RT}{\sigma k_d C_\infty V_m^2} \quad (4.17)$$

The typical form of the distribution function in the case of reaction controlled kinetics is shown in Fig. 4.3. The above expression is valid for radii less than two times the critical radius r_b after which the distribution goes to zero. In contrast, in the diffusion limited case, the distribution function was found to go to zero at 1.5 times the critical radius. It is observed that the size distribution in the reaction limited growth is broader and more symmetrical than that obtained from the diffusion-controlled growth (Fig. 4.2 and Eq. (4.12)).

4.2.2.3 Mixed Diffusion–Reaction Control

While in the preceding two sections, we have discussed the two limiting cases, in practice it is to be expected that both diffusion and surface reaction will contribute to the growth process in real experimental conditions, rendering the applicability of the limiting cases uncertain and of doubtful relevance in general. Specifically, several recent reports [42, 44, 66] have stressed that the growth in a variety of realistic systems does not belong to either of the two limits, namely the diffusion- or the reaction-limited regimes, but is controlled by a combination of diffusion and reaction at the surface.

It has been shown by LSW theory that, when the mass is conserved, the ratio r/r_b is a constant and equal to one in the diffusion-limited case [30]. Later, it was observed that, in the reaction-limited case, when the mass is kept conserved, r/r_b is a constant and equal to 8/9 [45]. Recently it was shown [42, 66] that if the conditions in limiting cases can be extended to the intermediate case, in other words if r/r_b is assumed to be a constant, K', Eq. (4.7) takes the form

$$\frac{dr}{dt} = \frac{2K'\sigma V_m^2 c_\infty}{RTr_2(1/D + 1/k_d r)} \quad (4.18)$$

This equation can be simplified and written in the following form

$$\frac{r^2\, dr}{D^*\text{const}} + \frac{r\, dr}{k_d^*\text{const}} = dt \tag{4.19}$$

which can be easily integrated to obtain

$$t = Ar^3 + Br^2 + C \tag{4.20}$$

where $A = RT/2DK'\sigma V_m^2 c_\infty$ and $B = RT/2k_d K' V_m^2 c_\infty$ and C is a constant. Thus, this expression, while separating the diffusion and the reaction terms, provides us with an analytical solution for the transient growth regime under the assumptions mentioned above. These expressions were found to explain satisfactorily some of the experimental observations [66] discussed in a later section.

Talapin et al. [44] have further suggested that the growth of nanocrystals could not be satisfactorily explained by approximating the Gibbs–Thompson equation (Eqs. (4.5) and (4.6)) to just the first two terms of the exponential expansion. We note that the size of the growing crystal, r, appears in the denominator of the argument for the exponential function in Eq. (4.5). This implies that the argument becomes increasingly larger for decreasing r, thereby making the finite polynomial expansion of the exponential function increasingly inaccurate. Thus, it is necessary to go beyond such an approximate expression, as used earlier (see Eq. (4.5)), particularly in the case of nanocrystal growth. Hence retaining the exponential term in the Gibbs–Thompson equation and assuming $r \ll \delta$ in Eq. (4.4) they obtain the equation

$$\frac{dr^*}{dt} = \frac{S - \exp(1/r^*)}{r^* + K \exp[\alpha/r^*]} \tag{4.21}$$

where r^* is the dimensionless average radius given by

$$r^* = \frac{RT}{2\sigma V_m} r \tag{4.22}$$

The other parameters include the rate constant, K, monomer oversaturation parameter, S, and the dimensionless time τ. They are given by the equations,

$$\tau = \frac{R^2 T^2 D c_\infty}{4\sigma^2 V_m} t \tag{4.23}$$

$$K = \frac{RT}{2\sigma V_m} \frac{D}{k_d} \tag{4.24}$$

$$S = c_b/c_\infty \tag{4.25}$$

In arriving at the above equations, the volume fractions were kept lower than 10^{-3}, so that the corrections due to the diffusional interactions between neighboring particles were not necessary. Monte-Carlo simulations were performed by calculating the remaining monomer concentration and the monomer oversaturation S in each cycle and substituting in Eq. (4.21) to calculate the growth of the next step. Different statistical parameters like oversaturation of monomers, nanocrystal concentration, average particle size, standard deviation in the size distribution were monitored at each step. These calculations included the nucleation step in the growth dynamics and assumed that the nucleation and growth processes were completely separated in time. Though detailed kinetics of nucleation is not known experimentally in real systems, these calculations suggested that the influence of initial conditions on the growth of nanocrystals was negligibly small as long as the rate of nucleation was much higher than the growth rate. This part of the result is in agreement with that of the LSW theory predicting a unique shape of the particle size distribution independent of the initial conditions of nucleation. In the case of the growth of a single nanocrystal, some amount of the monomer is consumed in the reaction. For growths with a finite volume of solution, the bulk concentration of monomer and the value of the oversaturation S decrease gradually. This gives rise to a shift in the critical radius towards larger sizes and the growth rate decreases.

In the case of nanocrystal ensembles with a diffusion-controlled growth, this approach predicts that an initially symmetric normal radius distribution, as well as the standard deviation, evolves in time toward the asymmetric negatively signed one. Initial conditions have significant influence on the size distribution only in a short transient period and only minor changes were observable at later stages of growth. When the reaction at the surface is much slower than the diffusion process, thereby becoming the rate determining step, the particle size distribution is found to be systematically broader than observed for diffusion-controlled growth. In the diffusion-controlled growth, in the early period, it is observed that the growth rate is much higher than predicted by the LSW theory. The narrowest size distribution is achieved when the particle growth is influenced more by the diffusion process than the reaction at the surface. They also established conditions for the evolution of ensembles leading to either "focusing" or "defocusing" of the particle size distribution. It was observed that the size distribution of the nanocrystal in the early stages of growth underwent a strong focussing effect if the particles had high initial oversaturation parameters, in other words, a large excess of the monomers. The excess of monomer affects strongly the evolution of the size distribution during the initial stages of nanocrystal growth. A fast increase in r accompanied by a strong narrowing in the size distribution is observed, followed by subsequent broadening without almost any change in r. However, the value of r and the standard deviation depend on the initial monomer concentration. The nanocrystals have initially positive growth rates and smaller nanocrystals grow faster than the larger ones. The number of particles remains nearly constant during the stage of "focusing" of the size distribution. During the focusing stage the size distribution remains nearly symmetric, and it is well described by a normal distribution. The defocusing is accompanied by a transition from symmetric to asymmetric stationary size distribution. During the focusing stage the oversaturation drops to some equilibrium value, and the number of nanocrystals starts to decrease due to the dissolution of the smallest particles.

4.3 Experimental Investigations

Having discussed the theoretical framework, we now review in this section the experimental situation in investigating the growth kinetics of various nanocrystals. Though the potential of solution phase synthesis of metals and metal oxide nanocrystals has long been realized, there were few mechanistic studies on these systems till about a decade ago. However, in this past decade, there has been a surge of interest [14, 15, 46–55, 66] in understanding the growth mechanism in nanocrystals. As already mentioned, the growth of nanocrystals is believed to take place predominantly via diffusion-limited coarsening, known as "Ostwald ripening". This has been reported so far for the cases of TiO_2 [46], InAs and CdSe [14] and ZnO [47] nanocrystal growth. Typically in any such experiment, the average size of the growing nanocrystal is monitored by measuring for example, the average diameter, d, of spherical particles as a function of the time, t and then analyzing whether d^3 is proportional to t in accordance with the prediction of the LSW theory (see Eq. (4.10)). The average size of nanocrystals can be measured either directly employing microscopic techniques or indirectly by monitoring a size-dependent physical property, for example absorption or emission energies and then deducing the size from the measured quantity with the use of a suitable calibration curve that relates the physical property to the size of the nanocrystals. For example, Oskam et al. [46] studied the growth kinetics of TiO_2 by transmission electron micrograph (TEM) recorded at various stages of growth, arriving at the conclusion that the growth can be described by the $d^3 \propto t$ prediction of the LSW theory.

Though TEM is the most direct tool to analyze the size and the size distribution in nanocrystals, it is a very time consuming process and it would be impossible to follow the size and size distribution of a fast growing particle during the synthesis by this technique. Moreover, *in situ* monitoring of the growth of nanocrystals in a solution is obviously beyond the scope of TEM. Therefore alternative methods have been devised and employed in recent times to probe *in situ* growth of nanocrystals in real time. Since the scattering of electromagnetic waves is a pronounced function of the particle size, this phenomenon is being used increasingly to study *in situ* growth of various nanocrystal systems. Considering the nanometer size of these particles in the region of interest, small angle X-ray scattering (SAXS), particularly using synchrotron radiation, becomes a powerful tool to probe the growth of nanocrystals in the size regime of typically 0.5–10 nm with a sub-second time resolution [54, 56].

Viswanatha et al. [54] used time-resolved SAXS at a third-generation synchrotron source to study the growth of CdS nanocrystals in the absence of any capping agents. They studied the dependence of the average diameter of growing nanocrystals and found it to follow the cube-root of time dependence, indicative of an Ostwald ripening mechanism. Further, it was established that the growth kinetics rigorously follows the LSW theory, not only in terms of the growth of the average diameter of the nanocrystals, but also in terms of the time dependence of the size distribution and the temperature dependence of the rate constant, establishing a remarkable adherence to the diffusion-limited growth or Ostwald ripening in the quantum confinement (<5 nm) regime.

Another alternate approach, based on the optical properties of nanocrystals, has also gained recent popularity as an effective tool to investigate the growth mechanism. It is well-known that the bandgap of nanocrystals in the small sized (<10 nm) regime is generally a strongly varying function of the size; this phenomenon is often termed the quantum size effect. It has been extensively probed both experimentally and theoretically. Based on a range of theoretical approaches to calculate the electronic structure of nanocrystals, such as *ab initio* methods [57], semi-empirical pseudo-potential methods [58, 59], and parametrized tight binding methods [60–67], it is now possible to quantitatively estimate the size of the semiconducting nanocrystal from the shift in bandgap obtained from UV absorption data. This has led to an extensive increase in the study of growth kinetics in recent times by monitoring the time-dependent UV absorption of *in situ* reaction mixtures. Recently, it has been shown that it is also possible to estimate the size distribution from UV absorption data [53] and also photoluminescence data [14]. This has led to a considerable advance in the study of growth kinetics in semiconductor nanocrystals.

Peng et al. [14] used UV absorption to determine the size of the nanocrystals and the width of the photoluminescence data to determine the size distribution to study the growth of CdSe and InAs nanocrystals. They observed a focusing and defocusing effect similar to that expected of Ostwald ripening behavior. More recently, Qu et al. [68] developed an *in situ* method of measuring the UV absorption of solutions at high temperatures and carried out real time measurements with a time resolution of a few milliseconds. They showed that in the case of CdSe the growth consists of a prolonged formation of relatively small particles (nucleation) followed by a focusing of the size distribution when the distribution changed from an asymmetric line shape to a symmetric one. This stage was followed by a stable phase which is most likely due to the monomer concentration in the solution being close to the solubility of the particles in the solution. The fourth stage showed the relatively large particles in the distribution growing even bigger while the relatively small ones in the solution disappeared. Hence the authors suggested this stage as the main course of the Ostwald ripening process.

Metal nanocrystals, like Au and Ag are known to show intense plasmon bands in the visible region, which can be easily monitored by optical absorption spectroscopy. Recently, it has been shown by Scaffardi et al. [69] that the size of the nanocrystal can be obtained from Mie theory [70] by analyzing the width of the plasmon band in the case of smaller particles (<10 nm) and by analyzing the position of the band for larger particles. This observation suggests the possibility of studying the growth of the metal nanocrystals by monitoring the time evolution of the spectral features of the plasmon band during the chemical reaction leading to the growth of metal nanocrystals.

In the following sections, we critically review typical growth kinetics with illustrative examples from the literature. Recently it was shown that an apparently linear dependence of d^3 on t, especially only in the asymptotic limit, does not rigorously establish the validity of the LSW theory, although this criterion has been used [46, 47] extensively in earlier studies of growth of such particles. In general, d^x as a function of t may appear linear within the experimental error limit for a wide range of x-values. Therefore, it becomes necessary to verify explicitly the expected dependences of the rate constant on temperature and the concentrations of the reactants, which provide more sensitive and critical testing grounds for the growth mechanism. More detailed studies on these systems suggest that, in most cases, the growth kinetics cannot be explained within the framework of diffusion-limited Ostwald ripening. We discuss in detail one example of the growth of metal nanocrystals, using Au as a test case [72], and one example of semiconducting nanocrystals, using ZnO [48, 53, 55, 66] as an illustration; these studies show that the growth kinetics often violates the over-simplified predictions of a simple diffusion-limited growth model, in contrast to the popularly held belief of the validity of diffusion-limited Ostwald ripening of nanocrystals in such cases.

4.3.1
Au Nanocrystals

Interest in the colloidal metal particles dates back to the time of Michael Faraday who recognized that the different colors of gold sols could be indicative of the different sizes or states of aggregation of the particles. An aspect of importance in the context of metal particles is the phenomenon of nucleation and growth in the nanometer regime. As early as 1951, a similar study was carried out by Turkevich et al. [71] on colloidal gold kinetics of gold sols in the micron length scales. In recent times Seshadri et al. [72] have studied the growth kinetics of gold in the nanometric regime quite extensively using transmission electron microscopy; we discuss this work in detail here to illustrate the complex growth mechanism of metallic nanocrystals. They studied the growth kinetics by investigating the particle size distribution at various times. It was observed that the growth was initially quick, slowing down at longer times. A plot of the mean diameter as a function of $t^{1/3}$ was found to be nearly a straight line, as expected in the case of Ostwald ripening. However, in contrast to the prediction of LSW theory in the context of the diffusion-limited Ostwald ripening process, the size distribution was nearly symmetric and it was found that it could be described well by a Gaussian profile, as shown in Fig. 4.4. While LSW and other theories of growth predict an asymmetric size distribution function (see Eqs. (4.12) and (4.16) and Figs. 4.2 and 4.3), attempts to use asymmetric profiles were not successful in describing the experimental results. It is possible to arrive at a more symmetric size distribution

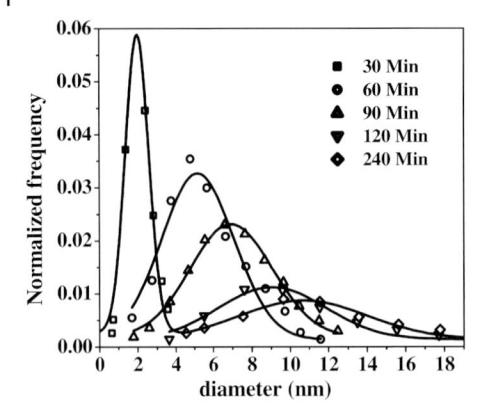

Fig. 4.4 Typical variation of the particle size distribution at various time intervals. The solid lines show the Gaussian fits to the experimental data. (Adapted from Ref. [72].)

within the LSW theory, but this happens only at high enough volume fractions. Considering that the volume fractions in these specific studies were below 10^{-4}, an exceptionally high volume fraction cannot be invoked to explain the experimentally observed symmetric size distribution. This report suggests that the growth process should be essentially stochastic and implies that the nucleation and growth is well separated. Further, it is observed that the average diameter and the standard deviation show the same time dependence. Normalized intensity as a function of x/r collapses into a single curve, thereby suggesting that it is a stochastic process. It was concluded that the growth was activated, since the growth kinetics showed a temperature dependence. This report provides a new growth law to fit the observed particle size distribution and its time dependence. However, further investigation is required to understand the growth mechanism completely in such cases.

4.3.2
ZnO Nanocrystals

Growth kinetics in the case of ZnO nanocrystals is one of the earliest [47] as well as the most extensively investigated systems [48–50, 52, 53, 55, 66, 73]. However, the study of the growth kinetics of ZnO nanocrystals from solutions has been fraught with contradictory claims till recently. It has been known for a long time that ZnO can be synthesized using zinc acetate $(Zn(OAc)_2)$ and a base such as NaOH. However, the intriguing aspect in the preparation of the samples has been the observation that the presence of a small or even trace amount of water in the synthesis of ZnO nanocrystals influences strongly the size of the nanocrystals [74, 75]. In order to understand the extraordinary sensitivity of ZnO nanocrystal size to the presence of any trace amount of water, we first review the growth kinetics of ZnO in the absence of any base and only in the presence of

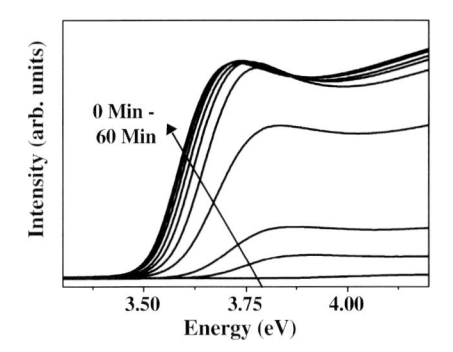

Fig. 4.5 UV absorption curves obtained at equal intervals of time for a typical reaction carried out at 318 K with 100 mM of water. (Adapted from Ref. [66].)

precisely defined quantities of water. One of the first reports of the growth kinetics of ZnO using water as a reactant was put forth recently by Hu et al. [55]. They studied the growth kinetics of ZnO by measuring the time-resolved UV absorption spectra and by analyzing the shift in the band edge using the effective mass approximation (EMA) [76]. However, it has long been shown that EMA overestimates the size of the nanocrystal and hence is not a very reliable tool to obtain the size of nanocrystals. Here we discuss the work carried out more recently in our own group, circumventing the earlier shortcomings [66].

A typical set of optical absorption spectra for a given concentration of reactants and at a fixed temperature (Fig. 4.5) shows a clear increase in the absorption intensity with increasing time. This suggests an increase in ZnO concentration with time. More significantly, there is a systematic shift in the absorption edge to lower energies with increasing time, indicating a steady growth of larger particles. The average diameter, d, of the nanocrystals was estimated from the dependence of the bandgap on particle size, based on a realistic tight binding modeling of first principle electronic structure calculations for ZnO [66], which is known to provide a realistic variation in the bandgap energy with size, unlike the earlier proposed EMA. We show typical variations of the cube of the diameter, d, vs. the time, t for several temperatures in Fig. 4.6(a). Since the size information from the shift in the absorption edge is derived indirectly via the dependence of the electronic structure on the size of the nanocrystals, the sizes obtained from the analysis of the absorption spectra were verified at a few specific points in time by monitoring the TEM intermittently.

Though the d^3 vs. t behavior deviates from the expected linearity based on the LSW theory (see Eq. (4.10)) in the small time regime, it indeed follows a linear relation at higher time-scales remarkably well, shown by thick solid lines, as if suggesting a dominantly diffusion-limited growth in the long time limit. However, it has also been noted that these results show acceptable linear behavior within experimental uncertainties for x-values ranging from 2.3 to 4, as illus-

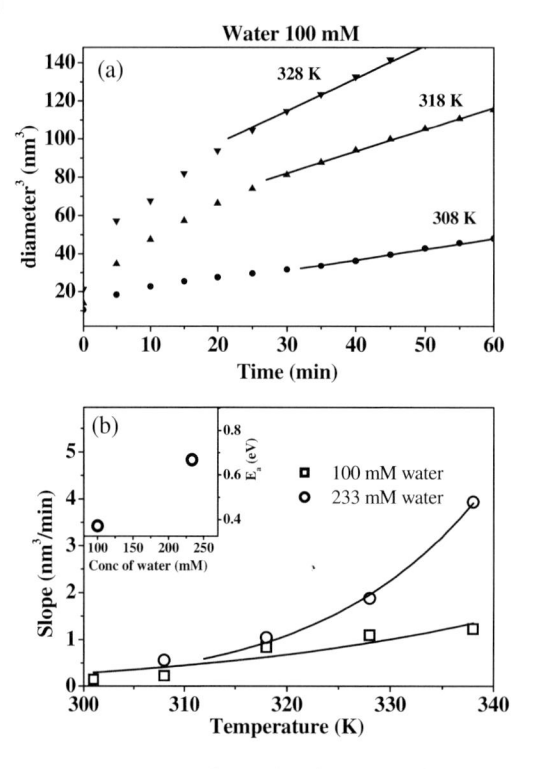

Fig. 4.6 (a) Variation of the cube of diameter of ZnO nanocrystals shown as a function of time for different temperatures at a fixed water concentration (233 mM). (b) The variation of the slopes of the linear part as a function of temperature for different water concentrations. The solid lines show the fits obtained from the function of the form given in Eq. (4.11). The inset shows the variation of the activation energies as a function of concentration of water. (Adapted from Ref. [66].)

trated by a plot of d^4 vs. t for the same set of $d(t)$ used for Fig. 4.7, indicating the difficulty in establishing the mechanism of growth kinetics only on the basis of the value of the exponent in $d^x \propto t$ plots. In order to probe the growth mechanism more rigorously, it is necessary to go beyond the simple verification of the d^3 vs. t plot being linear or not. It should be noted here that the slope (K) of the linear plot of d^3 vs. t has a well-defined dependence on the temperature, T, via the dependence of the diffusion constant, D, on T (see Eq. (4.11)). One can easily obtain the slopes, K, from the linear plots in Fig. 4.6(a) at different temperatures for various concentrations of the reactants, namely water in this case. Least-squared error fits of the K value to the form described in Eq. (4.11) show that though the observed dependence of K on T is reasonably well described by this functional form at higher concentration of water, the fit is far from satisfactory for lower water concentration, as shown in Fig. 4.6(b). Even more significantly, the activation energy, E_a, obtained from the best-fit curves and plotted

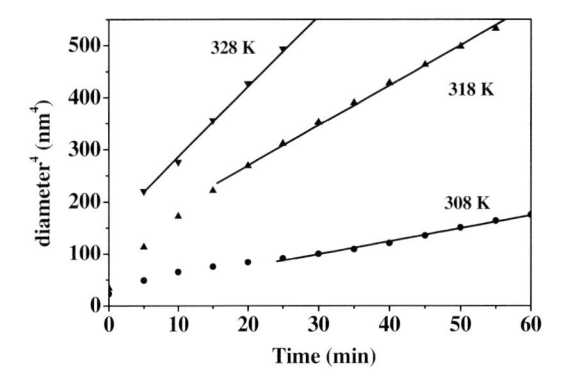

Fig. 4.7 Variation of the fourth power of the diameter of ZnO nanocrystals shown as a function of time for different temperatures at a fixed water concentration (233 mM).

(open circles) in the inset of Fig. 4.6(b) clearly shows a pronounced dependence on the concentration of water. This is clearly an unphysical situation, since the activation energy associated with the diffusion is clearly a function of the solvent and cannot depend on the solute concentration in the dilute limit (typically < 10 mmol of the reactant). This clearly suggests that a purely diffusion-controlled mechanism cannot explain the observed growth laws, in spite of the apparently linear behavior of the d^3 vs. t plots.

In the case of the present reaction of zinc acetate with water to form ZnO nanocrystals, the Zn^{2+} ions are obtained by the complete dissociation of zinc acetate into Zn^{2+} and OAc^- ions. The hydroxyl ions for the reaction are obtained from the dissociation of water. Ideally the nanocrystals of ZnO are assumed to comprise tetrahedrally coordinated Zn and O atoms and only the surface Zn atoms are terminated with a hydroxyl ion instead of the O ion. Small ZnO clusters are nucleated at time $t = 0$. The growth of the clusters occurs by the dehydration of terminating OH^- ion using the freely available dissociated OH^- ion in the solution. It is followed by the capturing of the Zn^{2+} ions available in the solution and brought near the surface of the cluster by the process of diffusion. The growth of the nanocrystal is thus further continued by the Zn^{2+} ion capturing the OH^- ion and so on. Hence the Zn^{2+} ions and the OH^- ions in solution may be considered as the pseudo-monomers of the growth process. In this present model the availability of the OH^- ions is governed by the dissociation of water. This is, however, controlled by the dissociation constant of water and hence the reaction cannot be assumed to be instantaneous. The presence of a prolonged reaction suggests that there should be an increase in the concentration of ZnO nanocrystals, at least for the lower concentrations of water. Further, it should be noted that the dissociation constant of water increases by a couple of orders of magnitude with increasing temperature providing a large number of OH^- ions at higher temperatures and thus increasing the rate of the reaction drastically. Thus the reaction mechanism to form ZnO can be written as $H_2O \leftrightarrow H^+ + OH^-$, $Zn^{2+} + 2OH^- \leftrightarrow Zn(OH)_2 \leftrightarrow ZnO + H_2O$.

A colloidal particle grows by a sequence of monomer diffusion towards the surface due to the concentration gradients set up by the differences in local curvatures and then the reaction of the monomers on the surface. Since the absorbance in the UV absorption spectrum increases with time (Fig. 4.5), it is clear that the reaction forming ZnO take place over time scales comparable to that of the growth of the nanocrystals, especially at lower temperatures and lower concentrations of water, suggesting that the nucleation is not separated from the growth. The reaction mechanism suggests that both the diffusion of Zn^{2+} ions and the rate at which the reactions take place at the surface have to be taken into consideration in the modeling of the growth process. Clearly then the growth in these nanocrystals belong to the transient regime of growth, discussed in the earlier section. Analyzing the reaction mechanism, using Eq. (4.7), it can be seen that the reaction term is relatively more important for small d, which is consistent with our observation (Fig. 4.6(a)) of more marked deviations from a purely diffusion-controlled growth at early times or the initial stages of the growth, when the nanocrystals are evidently the smallest in size. Replacing average radius r by the average diameter d in Eq. (4.20), we obtain the relation, $t = Bd^3 + Cd^2 + $ const, with $B = KT \exp(E_a/k_B T)$ and $K \propto 1/(D_0 \gamma V_m^2 c_\infty)$. The coefficient C is of the form $C \propto T/k_d \gamma V_m^2 c_\infty$. Using this expression relating d and t in the transient regime of growth, the experimentally observed variation $d(t)$ can be fitted for different temperatures, as shown in Fig. 4.8 by thick solid lines through experimental data obtained at different temperatures. The remarkable goodness of fits over

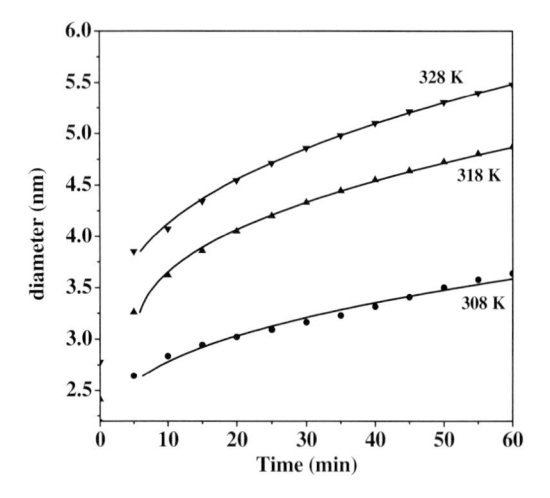

Fig. 4.8 Variation of the diameter of ZnO nanocrystals shown as a function of time for different temperatures at a fixed water concentration (100 mM). (Adapted from Ref. [66].)

the entire time regime, in contrast to those in Fig. 4.6(a), provides a conclusive validation of this description for the growth of ZnO nanocrystals. Further, the values of B for different temperatures and different concentrations of water can be extracted from the fits. The expected temperature dependence of the coefficient B was carried out and the activation energy, E_a, obtained from the least square fits to $B(T)$, can be plotted as a function of water concentration. While the results for the activation energy obtained earlier assuming only a diffusion-controlled growth (shown in the inset of Fig. 4.6(b)) exhibited a pronounced dependence on the water concentration, the new estimates of E_a using the transient growth model recovers a physically acceptable scenario of concentration-independent activation energy, 0.74 ± 0.01 eV. This firmly establishes the proposed growth mechanism.

We note here that the linear dependence of d^3 on t has often been used in the past literature to conclude about the growth model in a wide variety of systems, such as TiO_2 [46], CdSe [14], and ZnO [55]. However, in view of the results discussed here, it is clear that the cubic dependence of diameter on time alone is not enough to determine the mechanism of growth for nanocrystals and it is necessary to study the other dependences. One such example of growth studies was performed in the group of Searson to study the growth of ZnO in the presence of a base such as NaOH. The distinguishing feature of this case compared to the previous example is that a strong base (NaOH) providing a very large number of OH^- ions is used to carry out the reaction in contrast to the presence of only water, providing OH^- concentration of the order of 10^{-7}. We note here that most of the syntheses of ZnO nanocrystals, independent of the growth studies, indeed use such a base to carry out the reaction to form ZnO [47, 80].

The growth kinetics in the presence of a base, such as NaOH, as the reactant has been extensively studied by Searson et al. [47, 49, 51, 55]. They have measured the *in situ* UV absorption spectra for the reaction carried out at different temperatures. The bandgap thus obtained is converted to the size of the particle using EMA. They have shown that the cube of the diameter of the particles is linearly dependent on the time, as expected by the LSW theory. Further, they have obtained the values of σ and V_m from literature and determine c_∞ experimentally. Using these values, they have calculated the value of the diffusion coefficient D from the slopes of the linear curves. For the case of ionic diffusion, it is known that the diffusion constant can be obtained by the Stokes–Einstein relation given by

$$D = k_B T / 6\pi \eta a \qquad (4.26)$$

where η is the viscosity of the solvent and a is the hydrodynamic radius of the solute. The values obtained for the diffusion coefficient from the value of the slope K were compared with that of the Stokes–Einstein diffusion model and found to be consistent with the typical values for ions in solution at room temperature and in good agreement with that obtained from the Stokes–Einstein model.

These authors have also studied the effects of various anions like $Zn(OAc)_2$, $ZnBr_2$ and $Zn(ClO_4)_2$ on the growth kinetics. They found that that the particle growth at longer time is determined solely by diffusion in the case of $Zn(OAc)_2$ and $ZnBr_2$. However, in the case of $Zn(ClO_4)_2$, the radius increases more rapidly than for coarsening. The authors suggested that aggregation could be a competing mechanism of growth to the coarsening process and is dependent on surface chemistry resulting in either oriented or random attachment of particles. Random aggregation usually leads to the formation of porous clusters of particles whereas epitaxial attachment of particles leads to the formation of secondary particles with complex shapes and unique morphologies. They believe that the faster increase in the radius in the case of $Zn(ClO_4)_2$ arises due to epitaxial aggregations, resulting in much larger particles. The rate constants were obtained for all three different anions and plotted in the Arrhenius plot. From this plot, the activation energies were calculated to be 0.21 eV for $ZnBr_2$, 0.35 eV for $Zn(OAc)_2$ and 0.46 eV for $Zn(ClO_4)_2$. They also studied the effect of solvents on the kinetics by using ethanol, propanol, butanol, pentanol and hexanol as solvents in separate reactions. It is found that the coarsening rate constant increases not only with increasing temperature, but also with longer solvent chain length. The effect of solvent chain length on the rate constant is believed to arise from solvent viscosity, surface energy, and the bulk solubility of ZnO in these different solvents. These results illustrate that the solvent is an important parameter in controlling details of the growth kinetics.

A very recent work [53] has suggested a non-monotonic dependence of growth on the NaOH concentration, thus indicating a qualitative departure from the Ostwald ripening behavior. This suggests that the growth mechanism in ZnO is quite complex; while it has been extensively studied, more experiments are needed to understand completely the growth mechanism in these nanocrystals.

4.3.3
Effect of Capping Agents on Growth Kinetics

So far, we have discussed the growth of nanocrystals in the absence of any capping agent, which, when present, inhibits the growth of nanocrystals by effectively passivating their surfaces. However, the synthesis of almost all nanocrystals is in reality carried out in the presence of a capping agent to stabilize the desired size for a given application, thereby making the growth process complex and beyond the scope of LSW theory. The effect of the capping agent on the modification of the growth kinetics is very specific to the choice of capping agent, thereby defying any general theoretical approach; additionally very little is known about the growth process in such a complex reaction. In particular, there is hardly any theoretical approach to understanding such non-ideal reaction conditions. On the other hand, it is obvious that a detailed understanding of such growth processes is essential, if we are to employ rational syntheses rather than empirical ones. Thus, there have been a few attempts to probe experimentally different growth processes in the presence of a variety of passivating agents in order to understand the influence that each reaction parameter has on the average size and the size distribution. We present here a few selected examples of technologically important systems, like CdSe and ZnO, to illustrate some of the important points.

4.3.3.1 Effect of Oleic Acid on the Growth of CdSe Nanocrystals

CdSe nanocrystals have attracted the attention of many groups because of the possibility of tuning their emission wavelength through the entire visible spectrum by changing their size and the high quantum efficiency of such emission. Peng et al. [14] discovered a very facile reaction for the synthesis of high quality CdSe nanocrystals. However, a capping agent is inherently present in the synthesis, making it necessary to study the role of capping agents in such a reaction. This was recently discussed in detail by Bullen et al. [78] for the case of oleic acid and TOP as capping agents. As the nuclei grow, van der Waals interactions can cause rapid coalescence of nuclei and an unrestrained particle growth. However, capping agents may be added during the synthesis to adsorb and limit particle–particle aggregation, though such molecules may, in principle, also hinder monomer deposition. For example, ligands such as oleate and TOP are chemically bonded to both the precursor and the particles that form and hence hinder particle growth.

Bullen et al. carried out time-resolved UV absorption and fluorescence measurements. From the shift in the absorption edge due to quantum confinement effects, they obtained the sizes of the nanocrystals using the bandgap shift vs. size calibration data presented by Yu et al. [79]. The nucleation of CdSe in hot octadecene can provide an ideal environment for instantaneous nucleation, using up all the monomers; therefore, a subsequent reduction in the temperature allows only the growth of the particle, thus providing an ideal system to separate out nucleation and particle growth in solution. The number of nuclei formed during the nanocrystal synthesis and an estimation of the initial size of these nuclei prior to growth have been studied [78]. It was found that the nuclei concentrations were constant throughout the reaction, within experimental error, and that it is a very fast process that ceases almost immediately after monomer injection. It was observed that the rate is controlled by the reaction at the surface and a rate equation was obtained. It was found that the rate of growth of the nanocrystal is influenced by two effects. The surface area of each nanocrystal increases over time tending to make the reaction rate accelerate; on the other hand, decreasing concentrations of Cd and Se monomers cause the growth rate to slow down. As a consequence of these two opposite trends, the radius is observed to grow almost linearly at very early times. However, the growth rate proceeds toward saturation in the long time regime. These results can be understood in the following way using the chemical rate equation. The rate must obey an equation of the form

$$\frac{d[Cd]_t}{dt} = -kA(t)[Cd]_t N(t) \qquad (4.27)$$

where $[Cd]_t$ is the concentration of available Cd at time t, $A(t)$ is the surface area of each particle at time t, $N(t)$ is the number of particles at time t, and k is an interfacial rate constant that reflects the rate determining steps during deposition. Under steady-state growth conditions both Se and Cd should deposit at equal rates, so the particle growth can be described in terms of either species. From the experiments it is observed that the number of particles remains constant following nucleation, i.e., $N(t) = N_0$; furthermore, assuming spherical symmetry we get $A(t) = 4\pi r^2(t)$. Then the rate of increase in the volume of precipitated CdSe in the entire solution is given by

$$\frac{dV}{dt} = -V_m \frac{d[Cd]_t}{dt} \qquad (4.28)$$

where V_m is the molar volume of CdSe. From Eqs. (4.27) and (4.28), it is seen that

$$\frac{dr}{dt} = kV_m N_0 [Cd]_t = k[V_m([Cd]_0 - [Cd]_{eq}) - 4N_0 \pi r^3/3] \qquad (4.29)$$

since the effective concentration of free Cd, $[Cd]_t$, is equal to the initial concentration injected $[Cd]_0$, less the amount already deposited, $[Cd]_{dep} = 4N_0\pi r^3/3V_m$, less the amount remaining at equilibrium, $[Cd]_{eq}$.

At the maximum value of r, given by $r_{max} = (Y/Z)^{1/3}$ where $Y = V_m([Cd]_0 - [Cd]_{eq})$ and $Z = 4N_0\pi/3$, all the excess Cd in solution would have been consumed and hence, the radius was found to saturate after some time. On fitting the data, it was found that the best fits for the rate constant were at least 8 orders of magnitude below the diffusion-limited rate constant. This suggests that the growth occurs at a far slower rate and hence smaller particles are formed. The rate constant for growth did not change significantly with different amounts of capping agents. These results indicated that the oleic acid is not only an efficacious capping agent for CdSe nanocrystals in octadecene, but it markedly influences the primary nucleation steps in two distinct ways. First, the number of nuclei is reduced drastically as oleic acid is added since the nucleation is more difficult in the presence of oleic acid due to its complexation with Cd; the results also show that the initial nuclei in the presence of oleic acid are smaller than in its absence. Secondly, the complexing agent not only reduces the rate of nucleation by reducing the active monomer concentration, but it also rapidly caps the nuclei as they form. These two effects compete with each other. If there is too much capping agent, nucleation can be completely hindered, ultimately leading to indiscriminate growth of a small population of nuclei. However, because there are fewer nuclei formed in the presence of the ligand, larger clusters were unexpectedly found to form with increasing concentration of oleic acid. It is also noted that since the two processes, nucleation and particle growth, are decoupled in this system, there are two, well-defined activation energies. This shows clearly that the capping agents not only determine the rate of growth, but also play a major role in determining the number and size of the nuclei formed during injection; this, in turn, has a very pronounced influence on the subsequent growth rate.

4.3.3.2 PVP as a Capping Agent in the Growth of ZnO Nanocrystals

Various reports in the literature [66, 80] have now established that polyvinyl pyrollidone (PVP) is an effective capping agent to restrict the growth of ZnO nanocrystals. The temporal evolution of the average diameter, d, as a function of time, t, for the different ratios of Zn^{2+} to PVP is shown for a fixed NaOH concentrations in Fig. 4.9 [53]. The figure shows that there is a rapid and sustained growth of nanocrystals with time in every case, in spite of the presence of the capping agent. It was found that it was not possible to fit any of the curves to the form $d^x - d_0^x = Kt$, for any given value of x, even for the case without any PVP. This suggests that the growth process observed here is qualitatively different from the Ostwald ripening mechanism, even in the absence of any capping agent, as discussed in a previous section. The present example, involving the presence of a capping agent [53] is even less understood and is certainly beyond the scope of the LSW theory. In the absence of any guidelines provided by a theoretical understanding, it was noted that an empirical fit in terms of $(d - d_0)^x = Kt$ describes well the experimental results in every case, as shown by the best fit results with the solid lines overlapping the experimental data points in Fig. 4.9.

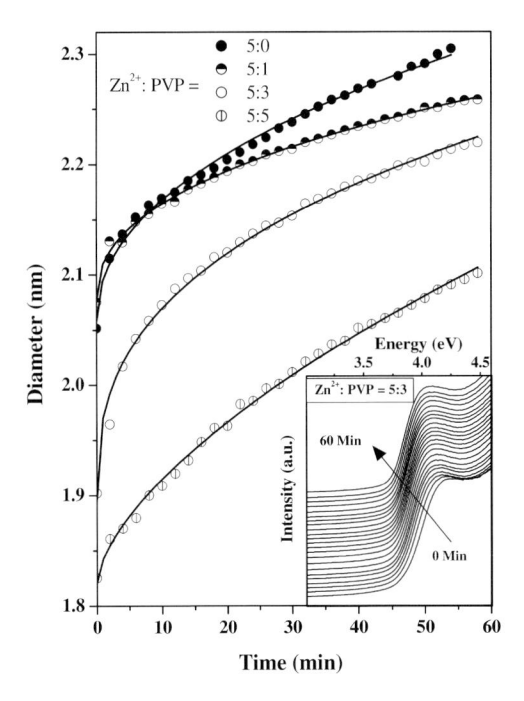

Fig. 4.9 Variation in the size of ZnO nanocrystals shown as a function of time for different ratios of Zn^{2+} to PVP at a fixed concentration (0.5 mmol) of NaOH. The inset shows the UV absorption curves for a typical system as a function of time ranging from 0 to 60 min. (Adapted from Ref. [53].)

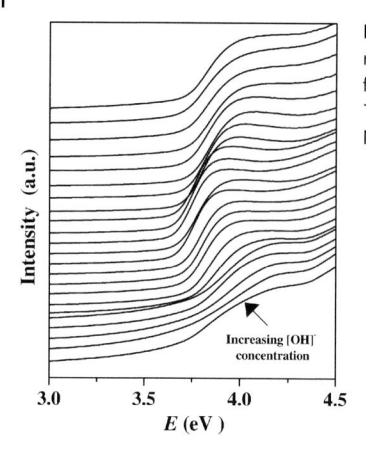

Fig. 4.10 The UV absorption curves recorded for ZnO nanocrystals with Zn^{2+} to PVP ratio of 5:3 as a function of NaOH concentration varying from 0.04 to 1 mmol at fixed time of 160 min after the addition of NaOH. (Adapted from Ref. [53].)

Further, it was observed that the absorption edge of the products after a fixed time of reaction appears at different energies for different concentrations of NaOH, as shown in Fig. 4.10. It is observed to vary non-monotonically, thereby indicating the complex dependence of the average nanocrystal size on NaOH concentration. Additionally, the sharpness of the absorption edge is found to vary significantly and non-monotonically as a function of NaOH concentration. These intriguing non-monotonic dependences of the size and size distribution on NaOH concentration after a fixed reaction time are shown in Fig. 4.11. Fig. 4.11(a) shows the expected result that the average particle size is largest in the absence of PVP (Zn^{2+} to PVP ratio = 5:0). A low Zn^{2+} to PVP ratio of 5:1 is also found to be inefficient in passivating the nanocrystals (Fig. 4.11(b)), minimal changes were observed compared to the uncapped system (Fig. 4.11(a)) in terms of both size and size distribution. However, the average size of the nanocrystal was found to decrease systematically as the PVP concentration was increased, as shown in Fig. 4.11(c) and (d). The results shown in Fig. 4.11 reveal that the dependences of d on the concentration of NaOH for all concentrations of PVP were essentially similar.

The insets to Fig. 4.11(a) and (b) revealed a slight monotonic increase in the relative distribution at low PVP concentration as the NaOH concentration increases. Interestingly, at higher concentrations of PVP (insets to Fig. 4.11(c) and (d)), we observed a striking decrease in the size distribution or a "narrowing" effect. This narrowing effect was understood in the following way. In the presence of high concentrations of PVP and the absence or low concentration of NaOH, the ZnO-like clusters are formed using the oxide ions from the PVP and isopropanol in the medium catalyzed by the slight basicity of the medium. In this case the rate determining step of the reaction is found to be the formation of the ZnO-like clusters and these clusters are immediately capped efficiently with PVP. Therefore, further ripening that leads to more uniform sized clusters is strongly suppressed at higher PVP concentration due to the effective passivation. This is also suggested by the fact that there is hardly any growth of the nanocrystal size

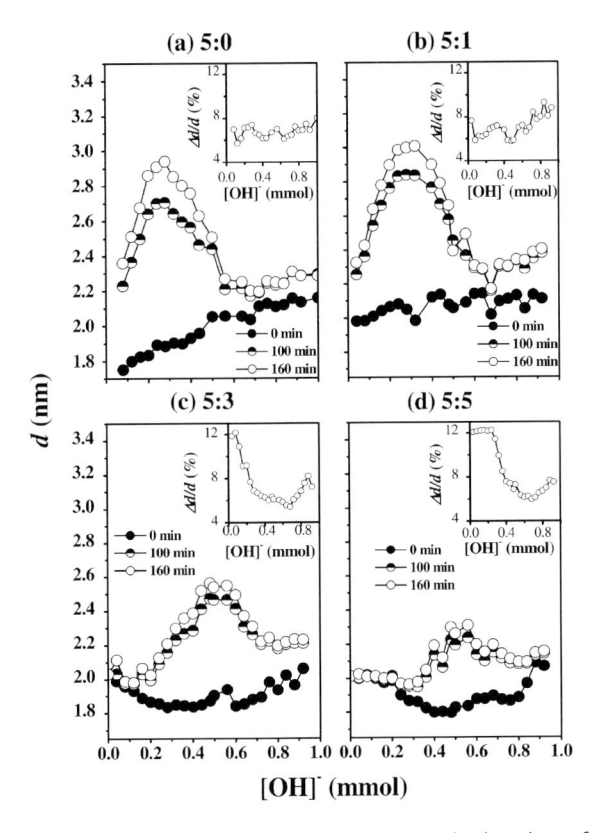

Fig. 4.11 Variation in the size of ZnO nanocrystals plotted as a function of NaOH concentration at times 0 min (closed circles), 100 min (half-open circles), and 160 min (open circles) for Zn^{2+} to PVP ratios of (a) 5:0, (b) 5:1, (c) 5:3, and (d) 5:5. The corresponding insets show the variation in size distribution at 160 min. (Adapted from Ref. [53].)

with time, indicated by the overlap of the three time (0, 100 and 160 min) plots, for the large PVP concentrations (Fig. 4.11(c) and (d)) in the low NaOH concentration regime. Hence, a broad size distribution was observed in such cases. Beyond a certain critical concentration of NaOH, namely about 0.1 to 0.2 mmol in Fig. 4.11(c) and (d) respectively, an evident growth of the nanocrystal size with time was observed, suggesting that, at these higher concentrations, NaOH is effective in reacting with the small pre-formed ZnO clusters in spite of the presence of PVP, thereby leading to the growth, though at a much slower rate than in the uncapped case. This removal of the capping agent at higher concentration of NaOH also allows a ripening process in competition with the passivating process by PVP, that gives rise to a higher degree of uniformity in size. Interestingly it was also noted that the minimum in the relative deviation corresponds to the same NaOH concentration that is required to just reach the low size regime in the range of the higher NaOH concentration regime. If the NaOH concentration is higher than this optimal value, the relative deviation becomes larger, leading to a poorer quality of the product, though the average particle size remains approximately the same. Thus, the present study establishes the necessity to use the optimal NaOH concentration to obtain small nanocrystals with a minimum size distribution.

4.3.3.3 Effect of Adsorption of Thiols on ZnO Growth Kinetics

Though the effect of thiols in capping the sulfides is well-known [81], it has been recently observed that the growth of ZnO nanocrystals can be altered by the addition of thiols. Pesika et al. [50] carried out reactions to form ZnO using zinc acetate and a base such as NaOH and studied the effect of the addition of thiol on the growth of ZnO. This was done by adding different amounts of octanethiol to a reaction of ZnO and carrying out an *in situ* UV absorption study. From the observed shifts in the band edge, they obtained the sizes of the nanocrystals using EMA and found that the growth of the nanocrystals is retarded by the addition of octanethiol. Further, it was observed that the retardation is more efficient on increasing the concentration of octanethiol.

We have studied the effect of adding a large amount of thiol on the growth of ZnO nanocrystals, when the capping agent was added after different intervals of time; the results are shown in Fig. 4.12 as plots of d vs. t. The vertical error bars on each data point represent the size distribution, while the average size is denoted by the data point itself. The data with filled circles show that the uncapped ZnO continues to grow with time, with the average diameter of the nanocrystals reaching up to ca. 5.5 nm after nearly 2 h. The size distribution also exhibits a

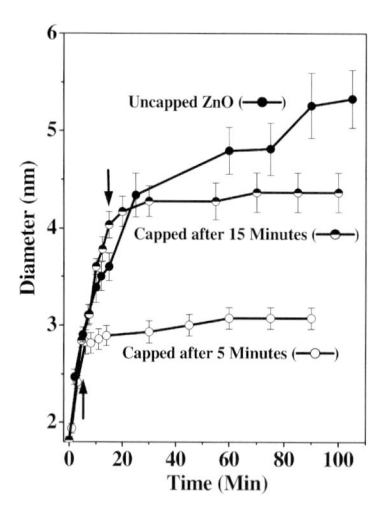

Fig. 4.12 Temporal evolution of the diameter of ZnO nanocrystals with thiol added at different intervals of time. The size distribution is shown as error bars in the graph.

monotonic increase in the uncapped case. The arrows on the other two sets of data mark the time when the thiol was added to the solution. It is interesting to note that with the addition of thiol, the growth continues for a short period of approximately 5 min and then the growth effectively stops. Additionally, the size distribution also does not change with time after a short time of adding the capping agent, as shown by the error bars in Fig. 4.12. Thus, it can be observed that thiol is a very efficient capping agent for ZnO.

The above examples make it clear that though the capping agents are expected to reduce the size and the size distribution, it is necessary to maintain the synthesis condition at the optimal level in order to prevent reversal of roles in the reaction. The study of the growth kinetics in the more realistic and complex cases of capped nanocrystals provide us with a handle to understand the details of growth under actual synthetic conditions and thereby help us to achieve the optimal conditions.

4.4
Concluding Remarks

Solution chemistry is the most versatile and highly flexible technique to tune the size and size distribution of nanocrystals. From a fundamental point of view, the growth kinetics of a solid in solution constitutes a very important field of study; however, there are relatively few such investigations reported in the literature. We have presented in this chapter the growth of nanocrystals from solutions in terms of some of the available theoretical models, followed by specific experimental results as illustrative examples of different growth models. We also indicate how such studies may open up new routes to the rational synthesis of nanocrystals.

It is observed that the growth of nanocrystals can be controlled either by diffusion or by the reaction at the surface. In the diffusion-controlled regime, we have discussed the assumptions as well as the results obtained from the LSW theory. We have also discussed models with specific improvements over the LSW theory in the diffusion-limited regime. We have presented the dependence of diameter and the form of the size distribution in the reaction-limited regime. Then these two limiting behaviors were integrated into a single model which accounts for the regime where both diffusion and reaction at the surface are important. We discussed the essential results obtained within this transient regime at the end of the section dealing with various theoretical models.

While there are several studies in the literature reporting diffusion-limited Ostwald ripening as the mechanism of growth, we present the cases of Au nanocrystals and ZnO nanocrystals, exhibiting remarkable departure from the expected Ostwald ripening process. In the case of Au nanocrystals, it is observed that though the diameter is found to increase as the cube-root of time, the size distributions are found to be very symmetric, unlike that expected for Ostwald ripening behavior. In the case of ZnO nanocrystals synthesized even in the absence of a base, it was found that though $d^3 \propto t$ is satisfied in the asymptotic limit, the activation energy is found to depend on the concentration of the reactant, if the experimental data are analyzed in the usual manner within the diffusion-limited ripening process. A better analysis suggests that ZnO growth kinetics belong to the transient Ostwald ripening regime.

We have then explored the effects of capping agents on the growth kinetics in three typical cases, namely the effect of oleic acid on the growth of CdSe nanocrystals, PVP on the growth of ZnO and thiol on the growth of ZnO. In all the cases, the capping agent is found to have multiple roles and affects the kinetics in a complicated and unique manner that is beyond the scope of the existing theories.

References

1 *Nanometer Scale Science and Technology*, M. Allegrini, N. Garci, O. Marti (Eds.), IOS Press, Amsterdam, 2001.
2 *Nanotechnology Research Directions: IWGN Workshop report*, M. C. Roco, B. S. Williams, A. P. Alivisatos (Eds.), Kluwer Academic Publishers, Dordrecht, 1999.
3 A. D. Yoffe, *Adv. Phys.* **50**, 1 (2001).
4 A. P. Alivisatos, *J. Phys. Chem.* **100**, 13226 (1996).
5 *Encyclopedia of Nanoscience and Nanotechnology*, H. S. Nalwa (Ed.), American Scientific Publishers, 2004.
6 *Chemistry of Materials*, C. N. R. Rao, A. K. Cheetham, A. Muller (Eds.), Wiley-VCH, Weinheim, 2004.
7 E. A. Meulenkamp, *J. Phys. Chem. B*, **102**, 5566 (1998).
8 M. Bruchez Jr., M. Moronne, P. Gin, S. Weiss, A. P. Alivisatos, *Science*, **281**, 2013 (1998).
9 J. Nanda, K. S. Nagaraj, B. A. Kuruvilla, G. L. Murthy, D. D. Sarma, *Appl. Phys. Lett.*, **72**, 1335–1337 (1998).
10 C. T. Tsai, D. S. Chuu, G. L. Chen, S. L. Yang, *J. Appl. Phys.*, **79**, 9105–9109 (1996).
11 A. T. Bell, *Science*, **299**, 1688 (2003).
12 P. Boggild, T. M. Hansen, C. Tanasa, F. Grey, *Nanotechnology*, **12**, 331 (2001).
13 H. Lin, S. Tzeng, P. Hsiau, W. Tsai, *Nanostruct. Mater.*, **10**, 465–477 (1998).
14 X. Peng, J. Wickham, A. P. Alivisatos, *J. Am. Chem. Soc.*, **120**, 5343 (1998).
15 W. W. Yu, X. Peng, *Angew. Chem. Int. Ed.*, **41**, 2368 (2002).
16 J. S. Bradley, in *Clusters and Colloids. From Theory to Applications*, G. Schmid (Ed.), VCH, New York, 1994.
17 A. Onuki, *Phase Transition Dynamics*, Cambridge University Press, Cambridge, 2002.
18 A. J. Bray, *Adv. Phys.*, **43**, 357 (1994).
19 K. Binder, P. Fratzl, in *Phase Transformations in Materials*, G. Kostorz (Ed.), Wiley-VCH, Weinheim, 2001.
20 G. R. Carlow, M. Zinke-Allmang, *Phys. Rev Lett.*, **78**, 4601 (1997).
21 J. Alkemper, V. A. Snyder, N. Akaiwa, P. W. Voorhees, *Phys. Rev. Lett.*, **82**, 2725 (1999).
22 C. A. Jeffrey, E. H. Conrad, R. Feng, M. Hupalo, C. Kim, P. J. Ryan, P. F. Miceli, M. C. Tringides, *Phys. Rev. Lett.*, **96**, 106105 (2006).
23 P. Gangophadhyay, R. Kesavamoorthy, S. Bera, P. Magudapathy, K. G. M. Nair, B. K. Panigrahi, S. V. Narasimhan, *Phys. Rev. Lett.*, **94**, 047403 (2005).
24 S. Takakusagu, K. Fukui, R. Tero, F. Nariyuki, Y. Iwasawa, *Phys. Rev. Lett.*, **91**, 066102 (2003).
25 A. F. Craievich, G. Kellermann, L. C. Barbosa, O. L. Alves, *Phys. Rev. Lett.*, **89**, 235503 (2002).
26 R. D. Averitt, D. Sarkar, N. J. Halas, *Phys. Rev. Lett.*, **78**, 4217 (1997).

27 V. K. La Mer, R. H. Dinegar, *J. Am. Chem. Soc.*, **72**, 4847 (1950); V. K. La Mer, *Ind. Eng. Chem.*, **44**, 1270 (1952).
28 E. Matijevic, *Chem. Mater.*, **5**, 412 (1993).
29 M. A. Watzky, R. G. Finke, *J. Am. Chem. Soc.*, **119**, 10382 (1997) and references therein.
30 I. M. Lifshitz, V. V. Slyozov, *J. Phys. Chem. Solids*, **19**, 35 (1961).
31 C. Wagner, *Z. Elektrochem.*, **65**, 581 (1961).
32 T. Sugimoto, *Adv. Colloid Interface Sci.*, **28**, 165 (1987).
33 W. Ostwald, *Z. Phys. Chem.*, **37**, 385 (1901).
34 J. D. Livingston, *Trans. Metall. Soc. A. I. M. E.*, **215**, 566 (1959).
35 A. J. Ardell, R. B. Nicholson, *J. Phys. Chem. Solids*, **27**, 1793 (1966).
36 C. H. Kang, D. N. Yoon, *Metall. Trans. A*, **12**, 65 (1981).
37 C. K. L. Davies, P. Nash, R. N. Stevens, *J. Mater. Sci.*, **15**, 1521 (1980).
38 A. D. Brailsford, P. Wynblatt, *Acta Metall.*, **27**, 489 (1979).
39 P. W. Voorhees, M. E. Glicksman, *Acta Metall.*, **32**, 2001 (1984).
40 J. A. Marqusee, J. Ross, *J. Chem. Phys.*, **80**, 536 (1984).
41 M. Tokuyama, K. Kawasaki, *Physica A*, **123**, 386 (1984).
42 R. Viswanatha, C. Dasgupta, D. D. Sarma, unpublished results.
43 R. Viswanatha, P. K. Santra, C. Dasgupta, D. D. Sarma, to be published.
44 D. V. Talapin, A. L. Rogach, M. Haase, H. Weller, *J. Phys. Chem. B*, **105**, 12278 (2001).
45 E. M. Lifshitz, L. P. Pitaevskii, *Landau and Lifshitz Course of Theoretical Physics*, Vol. 10, Ch. 12, Butterworth-Heinemann, 1981.
46 G. Oskam, A. Nellore, R. L. Penn, P. C. Searson, *J. Phys. Chem. B*, **107**, 1734 (2003).
47 E. M. Wong, J. E. Bonevich, P. C. Searson, *J. Phys. Chem. B*, **102**, 7770 (1998).
48 E. M. Wong, P. G. Hoertz, C. J. Liang, B. Shi, G. J. Meyer, P. C. Searson, *Langmuir*, **17**, 8362 (2001).
49 N. S. Pesika, K. J. Stebe, P. C. Searson, *J. Phys. Chem. B.*, **107**, 10412 (2003).
50 N. S. Pesika, Z. Hu, K. J. Stebe and P. C. Searson, *J. Phys. Chem. B*, **106**, 6985 (2002).
51 Z. Hu, G. Oskam and P. C. Searson, *J. Colloid Interface Sci.*, **263**, 454 (2003).
52 G. Oskam, Z. Hu, R. L. Penn, N. Pesika, P. C. Searson, *Phys. Rev. E*, **66**, 11403 (2002).
53 R. Viswanatha, D. D. Sarma, *Chem. Eur. J.*, **12**, 180 (2006).
54 R. Viswanatha, S. R. Santra, S. Sapra, S. S. Datar, H. Amenitisch, D. D. Sarma, to be published.
55 Z. Hu, D. J. Escamilla Ramirez, B. E. Heredia Cervera, G. Oskam, P. C. Searson, *J. Phys. Chem. B*, **109**, 11209 (2005).
56 G. Beaucage, *J. Appl. Crystallogr.*, **28**, 717–728 (1995).
57 F. Buda, J. Kohanoff and M. Parrinello, *Phys. Rev. Lett.*, **69**, 1272 (1992).
58 M. V. Ramakrishna, R. A. Friesner, *Phys. Rev. Lett.*, **67**, 629 (1991).
59 (a) H. Fu, A. Zunger, *Phys. Rev. B.*, **55**, 1642 (1997); (b) H. Fu, A. Zunger, *Phys. Rev. B*, **56**, 1496 (1997).
60 P. E. Lippens, M. Lannoo, *Phys. Rev. B*, **39**, 10935 (1989).
61 J. M. Jancu, R. Scholz, F. Beltram, P. Bassani, *Phys. Rev. B*, **57**, 6493 (1998).
62 G. Allan, Y. M. Niquet, C. Delerue, *Appl. Phys. Lett.*, **77**, 639 (2000).
63 S. Sapra, N. Shanthi, D. D. Sarma, *Phys. Rev. B*, **66**, 205202 (2002).
64 S. Sapra, R. Viswanatha, D. D. Sarma, *J. Phys. D*, **36**, 1595 (2003).
65 S. Sapra, D. D. Sarma, *Phys. Rev. B*, **69**, 125304 (2004).
66 R. Viswanatha, S. Sapra, B. Satpati, P. V. Satyam, B. N. Dev, D. D. Sarma, *J. Mater. Chem.*, **14**, 661 (2004).
67 R. Viswanatha, S. Sapra, T. Saha-Dasgupta, D. D. Sarma, *Phys. Rev. B*, **72**, 045333 (2005).
68 L. Qu, W. W. Yu, X. Peng, *Nano Lett.*, **4**, 465 (2004).
69 L. B. Scaffardi, N. Pellegri, O de Sanctis, J. O. Tocho, *Nanotechnology*, **16**, 158 (2005).

70 G. Mie, *Ann. Phys.*, **25**, 377 (1908).
71 J. Turkevich, P. C. Stevenson, J. Hillier, *Discuss. Faraday Soc.*, **11**, 55 (1951).
72 R. Seshadri, G. N. Subbanna, V. Vijayakrishnan, G. U. Kulkarni, G. Ananthakrishna, C. N. R. Rao, *J. Phys. Chem.*, **99**, 5639 (1995).
73 R. Viswanatha, H. Amenitsch, D. D. Sarma, *J. Am. Chem. Soc.*, in press.
74 E. A. Meulenkamp, *J. Phys. Chem. B*, **102**, 5566 (1998).
75 R. Viswanatha, S. Sapra, S. Sen Gupta, B. Satpati, P. V. Satyam, B. N. Dev, D. D. Sarma, *J. Phys. Chem. B*, **108**, 6303 (2004).
76 (a) A. L. Efros, A. L. Efros, *Sov. Phys. Semicond.*, **16**, 772 (1982); (b) L. E. Brus, *J. Chem. Phys.*, **79**, 5566 (1983).
77 T. Vossmeyer et al., *J. Phys. Chem.*, **98**, 7665 (1994).
78 C. R. Bullen, P. Mulvaney, *Nano Lett.*, **4**, 2303 (2004).
79 W. W. Yu, L. Qu, W. Guo, X. Peng, *Chem. Mater.*, **15**, 2854 (2003).
80 L. Guo, S. Yang, C. Yang, P. Yu, J. Wang, W. Ge, G. K. L. Wong, *Chem. Mater.*, **12**, 2268 (2000).
81 J. Nanda, S. Sapra, D. D. Sarma, N. Chandrasekharan, G. Hodes, *Chem. Mater.*, **12**, 1018 (2000).

5
Peptide Nanomaterials: Self-assembling Peptides as Building Blocks for Novel Materials

M. Reches and E. Gazit

5.1
Overview

Molecular self-assembly represents a central theme in modern nanotechnology. The molecular recognition and assembly of simple building blocks may lead to the formation of complex and ordered structures on the nano-scale. Specific biochemical recognition and biocompatibility are among some of the attractive features of biological building blocks. Specifically, peptides are among some of the most studied building blocks for biological and bio-inspired self-assembly. Peptides combine the usefulness of facile chemical synthesis and chemical flexibility, together with biocompatibility and relative stability. Various peptide building blocks have been used for the formation of nanometric assemblies. These building blocks include cyclic and linear peptides that have various functional groups that can mediate self-assembly processes. Peptide-based materials also include peptide conjugates such as aliphatic linkers as well as block copolymers composed of poly-amino acids. The self-assembly may be facilitated by hydrophobic, aromatic, and electrostatic interactions. The resulting nanostructures include nanotubes, nanospheres, nanotapes, nanoribbons, and hydrogels with nano-scale order. Some of these nanostructures have already been used in nanotechnological applications, for example as, novel antibacterial agents, templates for the fabrication of inorganic materials, elements in biosensors, and molecular scaffolds for tissue engineering and regeneration.

5.2
Introduction

Molecular self-assembly is a spontaneous process by which molecules interact with each other to form well-ordered supramolecular structures. The self-association between molecular building blocks is mediated through specific molecular recognition processes. In this way, extraordinarily complex assemblies are

Nanomaterials Chemistry. Edited by C. N. R. Rao, A. Müller, and A. K. Cheetham
Copyright © 2007 WILEY-VCH Verlag GmbH & Co. KGaA, Weinheim
ISBN: 978-3-527-31664-9

formed in various natural systems by simple biological molecules such as oligonucleotides, phospholipids, and peptides. This process has motivated many studies in the field of nanotechnology and material science to exploit biological elements and to design nanometric structures that will self-assemble using a bottom-up approach.

Numerous naturally occurring structures self-assemble by the interactions between protein building blocks. These include the main constituent of microtubules [1], the tubulin protein, which self-assembles into nanometric fibers [2, 3], the viral capsid of many viruses [4], amyloid fibrils, which are associated with numerous human disorders [5], and many other complex structures. These assemblies are studied for their application in nanotechnology as a scaffold for the fabrication of organic and inorganic nano-objects and as a physical support for macromolecular organization.

S-layers, the crystalline cell surface that exists in all archeial and many bacterial species, are one of the earliest self-assembled structures to be exploited for biotechnological and nanotechnological applications [6]. Their recrystallization onto silicon and other surfaces with oblique, square or hexagonal lattice symmetry has been exploited for nanofabrication [7]. This was followed by efforts to utilize other proteins and peptides as building blocks for the formation of well-ordered structures on the nano-scale. The major advantage that proteins and peptides have as building blocks for the fabrication of novel nanomaterials stems from their amino acid sequence, since each of the amino acids represents different physicochemical properties that have structural and functional properties. In this way, very short peptides can possess catalytic activity and can bind metal and other inorganic materials. While attempting to mimic self-assembly processes that occur in nature with large proteins, short peptides were suggested as simpler building blocks.

In this chapter we will discuss the various approaches suggested for the design of self-assembled peptides into nanostructures. We will describe the basis for the design of various building blocks and their molecular self-association properties. In addition, we will present the applications proposed for these nano-assemblies.

5.3
Cyclic Peptide-based Nanostructures

Numerous approaches for fabricating peptide-based self-assembled structures have been introduced during the last decade. The first demonstration of designed nanostructures with well-ordered morphology was reported more than a decade ago by Ghadiri and coworkers. Here, cyclic peptides were designed with an alternating even number of D- and L-amino acids that interact through hydrogen bonding into an array of self-assembled nanotubes (Fig. 5.1, [8]). The alternating chirality of the amino acid elements, a motif that was bio-inspired by a naturally occurring non-ribosomal bacterial peptide, resulted in a flat β-sheet conformation that allowed the stacking of the circular building blocks one on top of the other to

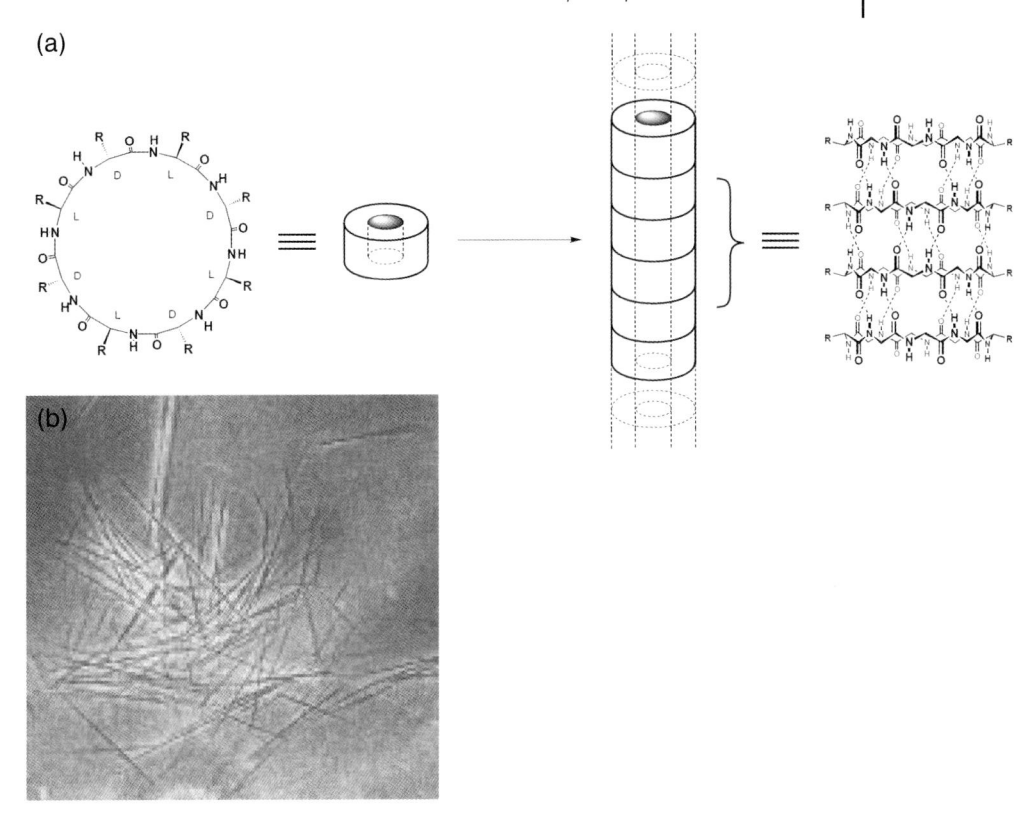

Fig. 5.1 Formation of tubular assemblies by cyclic D,L-peptides. (a) Schematic diagram of the nanotubes' formation. The cyclic rings of the peptides are stacked together to form hollow nanotubes, which then interact with each other to form microcrystals. (b) Phase-contrast microscopy images of cyclic D,L-peptide nanotube microcrystals. With permission from Wiley-VCH, courtesy of M. Reza Ghadiri.

form hollow supramolecular cylindrical assemblies. The internal diameter of the nanotubes ranges from 7 to 8 Å and can be controlled by changing the number of amino acids in the cyclic peptide sequence [8]. In addition, the nanotubes' internal and external interfaces can be modified by changing the composition of their amino acids. Several applications have been proposed for these peptide nanostructures. One of them involves their use as antibacterial agents, since they have the ability to destabilize both Gram-positive and -negative bacterial membranes [9]. Their use as nanocontainers for drug delivery applications was also suggested. Non-biological applications for these tubular structures involve their decoration with chemical groups, which enables their electrical activity and their further use as conductive wires in molecular electronics [10].

Another cyclic peptide that was later reported to self-assemble into tubular structures is the Lanreotide, a synthetic octapeptide that is synthesized as a growth hormone inhibitor [11]. This octapeptide self-assembles through segregation of aliphatic/aromatic residues into monodispersed nanotubes with a diameter of 244 Å and a wall thickness of 18 Å. It was suggested that these nanotubes could serve in nanofiltration of biological molecules and as a template for the fabrication of ordered mesoporous materials. At high concentrations this peptide forms hydrogel (Autogel), which is already being utilized in acromegaly treatment [11].

5.4
Linear Peptide-based Nanostructures

Linear peptides can also self-assemble into tubular structures, as was demonstrated by Matsui and coworkers. This class of peptides, termed bolaamphiphiles peptides, is composed of two hydrophilic peptides that are conjugated through a hydrophobic peptide linker. The driving force for the formation of these structures is most likely the intermolecular association of the hydrophobic moieties in aqueous solution to form ordered structures, similar to a micellization process.

Both X-ray scattering and Raman spectroscopy studies revealed that the bolaamphiphile peptide-based nanotubes contain carboxylic acid groups as well as free amide groups that intercalated metal ions such as copper and nickel. The peptide nanotubes were shown to serve as a template for metal nanowires fabrication by using an electroless deposition procedure [12].

The ability to control their position on gold electrodes was further studied in order to combine these metallic nanowires in nano-electronic applications [13]. Methods that were used involve molecular recognition between the free groups on the external surface of the nanotubes and chemical groups of self-assembled monolayers (SAMs) that were deposited onto a gold substrate, thus enabling the positioning of the tubes on a surface [13].

The free amide groups on the bolaamphiphiles' peptide nanotubes were also exploited to immobilize a histidine-rich peptide sequence that enables the formation of copper nanocrystals on the tubular structures [14].

A parallel molecular direction for the assembly of nano-scale peptide structures, which was developed independently by Stupp and coworkers, is based on the conjugation of a single hydrophilic peptide motif with a hydrophobic alkyl tail to form amphiphilic peptides. These peptide conjugates can self-assemble into well-ordered nanofibers. Their morphology, surface chemistry, and bioactivity can be modified by the sequence selection of hydrophilic components as well as by modifying the alkyl tail [15, 16]. The design of the peptide sequence enables the fabrication of composite materials by direct mineralization of the nanofibers with hydroxyapatite [15]. In addition, these peptide-amphipiles can be used as artificial three-dimensional scaffolds because the self-assembled nanofibers form a fibrilar network that can incorporate cell adhesion motifs such as the laminin epitope IKVAV [17].

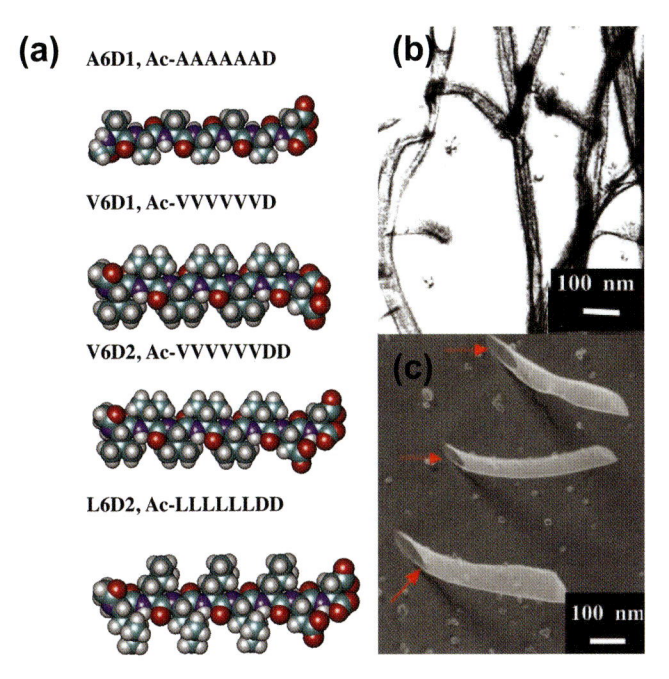

Fig. 5.2 Formation of nanotubes and nanovesicles by surfactant-like peptides. (a) Space-filling molecular models of surfactant-like peptides; each peptide is composed of a negatively charged amino acid and hydrophobic tails with increasing hydrophobicity. (b) Quick-freeze deep-etch TEM image of the self-assembled structures formed by a surfactant-like peptide. (c) Quick-freeze deep-etch TEM image of the vesicles and nanotubes; the open ends of the nanotubes are clearly observed using this techniques. Copyright 2002, National Academy of Sciences, U.S.A. Courtesy of Shugunag Zhang.

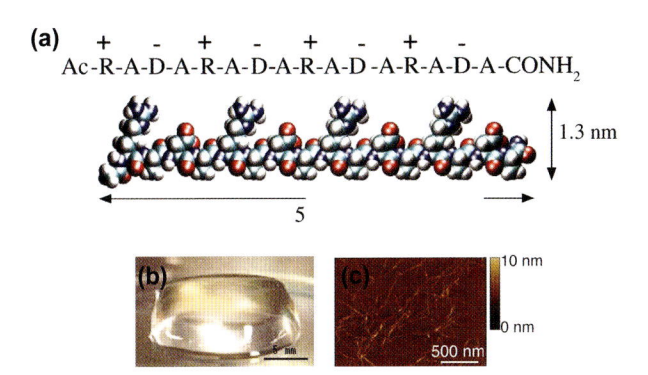

Fig. 5.3 Ionic self-complementary peptides for the fabrication of novel nanomaterials. (a) Amino acid sequence and molecular model of the ionic self-complementary RADA16-I. (b) Photograph of the RADA16-I hydrogel. (c) Atomic force microscope analysis of the RADA16-I nanofiber scaffold. Copyright 2005, National Academy of Sciences, U.S.A. Courtesy of Shugunag Zhang.

The use of non-conjugated peptides as building blocks to form ordered nanoscale structures was demonstrated later by Zhang and coworkers. The first class of linear peptides that was shown to self-assemble into nanometric structures was the self-complementary ionic peptides (Fig. 5.2). The first member of this class of peptides was discovered from a DNA binding protein in yeast called Zuotin. These peptides are highly (>50%) charged with opposing charges using alternately basic and acidic amino acids. Upon their transfer into aqueous solution they form two distinct surfaces, one hydrophilic and the other hydrophobic and spontaneously self-assemble as a 'molecular Lego' to form well-ordered nanometric structures [18]. The well-ordered nanofibers self-assembled by this class of peptides can form a macroscopic hydrogel scaffold that supports three-dimensional cell growth and therefore can be used for tissue engineering and regeneration (Fig. 5.2, [19]).

Another class of linear self-assembled peptides discovered later by Zhang and coworkers are surfactant-like peptides characterized by well-defined hydrophilic and hydrophobic residues (Fig. 5.3, [20]). The design of the peptides mimics the physical properties of biological phospholipids. The peptides are composed of one or two negatively charged amino acids at the C-termini of the peptide, six hydrophobic amino acids such as valine, alanine or leucine, and acetylated N-termini to eliminate the positive charge. These peptides self-assemble in water into nanotubes and nanovesicles that can be utilized as drug delivery agents and for solubilizing and stabilizing membrane proteins.

Linear oligopeptides were also shown to self-assemble into β-sheet tapes [21]. These peptides are related to a transmembrane domain of the IsK protein that forms β-sheet structures in lipid bilayers. Based on the suggestion that peptide fragments of this protein will form ordered structures in amphiphilic solvents, their self-assembly in methanol and chloroform was explored. Indeed, peptide fragments of 24 residues were able to assemble into transparent viscoelastic gels in these solvents. These gels are composed of β-sheet tapes with a diameter of about 8 nm and a length of about 1 µm. The nano-tapes are similar in their morphology to amyloid fibrils and facilitate the formation of biocompatible gels that possess the adequate mechanical strength needed for tissue engineering applications. Upon increasing the peptide concentration, these peptides can form a hierarchy of structures such as tapes, twisted ribbons, and fibrils, which comprise a class of soft-solid-like nanostructure materials [22]. In addition, the peptides' self-assembly can be reversibly controlled by introducing amine or carboxyl groups into the peptide sequence upon changing the pH of the solution [23]. The charge introduced into these peptides also facilitates a coulombic attraction between oppositely charged peptides and leads to spontaneous self-assembly of fibrillar structures [24].

5.5
Amyloid Fibrils as Bio-inspired Material: The Use of Natural Amyloid and Peptide Fragments

Amyloid fibrils are naturally occurring fibrilar structures that are formed by the self-assembly of various proteins associated with numerous diseases. These fibrils also appear in several non-pathological microbial systems such as biofilms and aerial hyphea, and it was suggested that they represent a generic structural form of aggregating proteins [25–27]. These fibrilar structures attract much attention from both the biomedical and the nanotechnological point of view. In dealing with the biomedical aspects of these fibrils, great effort is being invested to better understand the underlying molecular mechanism that leads to amyloid fibril formation [28–34]. From the technological point of view, these nano-objects were proposed as a scaffold for the fabrication of conducting metal wires (Fig. 5.4, [35]).

Peptides can also self-assemble into amyloid fibrils, and they are being widely used as a model for amyloid fibril formation since they are easy to design and synthesize. These peptide-based fibrils can also be utilized as an alternative to protein-based fibrils for the nucleation of metals and for the fabrication of conductive nanowires. One example is the formation of amyloid-like structures by peptide fragments from the fiber protein of adenovirus with a specific relation to the folds of the native structure of the protein. This viral natural β-fibrous protein, as well as other elongated proteins such as phage P22, folds into the β-helical conformation that was suggested as a model for amyloid fibril formation [36–38].

Another polypeptide that forms amyloid in non-pathological conditions is Medin, the main constituent of aortic medial amyloid, which is deposited as amy-

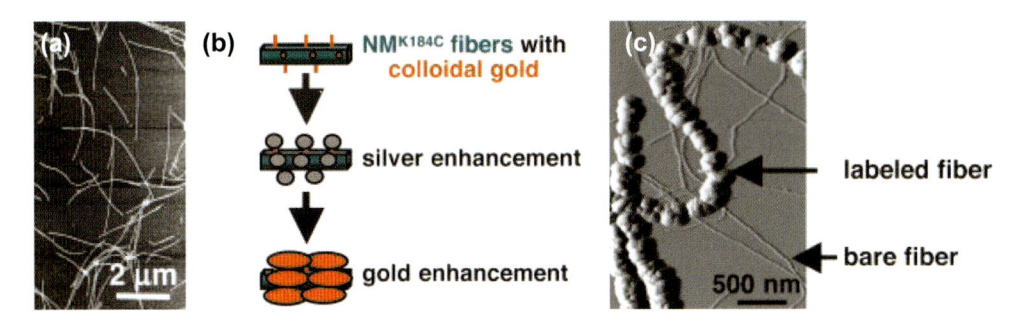

Fig. 5.4 Amyloid fibrils as bio-inspired material. (a) Atomic force microscope analysis of amyloid fibrils formed by the N-terminal and middle region (NM) of yeast *Saccharomyce scerevisiae* Sup35p. (b) These amyloid fibrils were used as a template for the fabrication of conductive nanowires, as illustrated in the schematic diagram. NM amyloid fibrils were genetically engineered to introduce cysteine amino acids that contain a thiol group. These thiols were then interacted with gold particles that were enhanced with silver or gold solution. (c) AFM analysis of the resulting gold-toned fibers. Copyright 2003, National Academy of Sciences, U.S.A. Courtesy of Susan L. Lindquist.

loid fibrils in the aorta of virtually all individuals older than 60 years. Short fragments of Medin, NH$_2$-NFGSVQFV-COOH, and its truncated analog, the NH$_2$-NFGSVQ-COOH peptide, can self-assemble into well-ordered amyloid fibrils and can be used for studying amyloid fibril formation [39, 40].

Short peptides corresponding to pathologically associated proteins are being widely used as a model for amyloid fibril formation. This includes the NFGAIL hexapeptide fragment of the islet amyloid polypeptide (IAPP), which is involved in type II diabetes disease [41]. This short peptide can form well-ordered amyloid fibrils *in vitro* with the same morphology and biophysical properties as the fibrils formed by the full-length protein [41].

The high occurrence of aromatic amino acids in these amyloidogenic peptides led us to suggest that aromatic interactions play a central role in the formation of amyloid fibrils [42]. Based on this hypothesis, we identified another minimal active amyloidogenic peptide of the hIAPP, the NH$_2$-NFLVH-COOH hexapeptide fragment [43]. Additionally, we were the first to report on an amyloidogenic pentapeptide, the NH$_2$-DFNKF-COOH fragment from the human hormone calcitonin, which self-assembled into amyloid fibrils with the same ultrastructural properties as much longer polypeptides. We also reported on a shorter truncated tetrapeptide, NH$_2$-DFNK-COOH, which can form amyloid fibrils with a slightly different morphology [44]. These findings are consistent with a recent research study on the ability of tripeptides to form amyloid fibrils. In this study it was revealed that three tripeptides: Boc-Ala-Aib-Val-OMe, Boc-Ala-Aib-Ile-Ome, and Boc-Ala-Gly-Val-OMe can self-associate to form supermolecular β-sheet structures and further aggregate into amyloid-like fibrils [45].

Another class of self-assembled peptides that are derived from amyloidogenic proteins are linear bis-conjugated peptides from the prion octapeptide [46, 47]. The first reported peptide of this class, a bis-scaffold containing the pentapeptide PHGGG from the prion octapeptide, can self-assemble into well-ordered fibrils on the nano-scale [46]. In addition, a dipeptide-conjugate derived from this octarepeat sequence can form fibrils that can nucleate copper in the same manner as the octapeptide repeat [47].

5.6
From Amyloid Structures to Peptide Nanostructures

As part of our attempts to get an insight at the molecular level into the interactions that lead to the formation of amyloid fibrils, we used a reductionist approach to identify the minimal determinate that would mediate amyloid fibril formation. Taking this approach, we identified several short amyloidogenic peptides. Based on the aromatic nature of these peptides, we proposed a novel mechanistic model for amyloid fibril formation and suggested that stacking interactions between the aromatic rings contribute order and directionality in this self-assembly process.

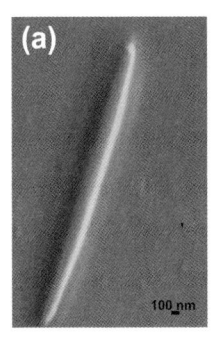

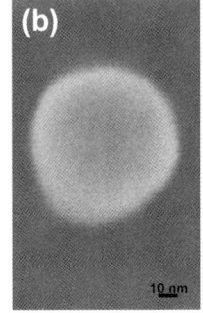

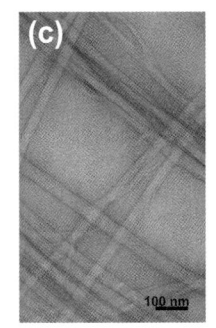

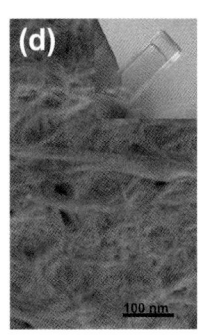

Fig. 5.5 Electron microscopy analysis of nano-assemblies formed by aromatic homo-dipeptides. (a) Scanning electron microscopy micrograph of the tubular structures self-assembled by the diphenylalanine peptide. (b) High-resolution scanning electron microscopy analysis of the closed-caged spherical structures formed by the diphenylglycine peptide. (c) Transmission electron microscope analysis of thin tubular structures self-assembled by the di-D-2-naphthylalanine peptide. (d) Environmental scanning electron microscopy analysis of the hydrogel formed by the Fmoc–Phe–Phe–COOH peptide, an image of the hydrogel in an inverted cuvette is in the inset.

Based on the suggested model and our motivation to reveal the minimal amyloid-forming fragment, we explored the diphenylalanine core recognition motif of the β-amyloid polypeptide, which is associated with Alzheimer's disease. Importantly, we revealed that this dipeptide can self-assemble into a novel type of peptide nanotubes [48]. These aromatic structures, also termed ADNT (aromatic dipeptide nanotubes), are formed as individual entities with a long persistence length (Fig. 5.5(a)). Their physical and chemical stability is remarkable as they are stable at high temperatures and when exposed to numerous organic solvents. In addition, their Young's modulus, indicating the mechanical stability, is the highest known for biological assemblies [49, 50].

These nanostructures were shown to serve as a casting mold for silver nanowires and as a scaffold for platinum nanoparticles [48, 51]. Their application in electrochemical biosensing devices was also shown and resulted in a more sensitive electrochemical signal when compared to standard electrodes [52, 53].

The discovery of highly ordered tubular structures formed by an aromatic dipeptide led to the development of the concept of using aromatic homo-dipeptides for designing nanometic structures (Fig. 5.5). It was further demonstrated that aromatic homo-dipeptides can self-assemble into closed-caged nanospheres, amyloid-like structures and plate-like structures (Fig 5.5) [54–56]. In addition, it was shown that one of these modified aromatic homo-dipeptides, Fmoc-Phe-Phe-COOH, can form a highly rigid and biocompatible hydrogel. It was suggested that this hydrogel be used in tissue engineering and drug delivery applications [57].

5.7
Bioinspired Peptide-based Composite Nanomaterials

Block copolypeptides were also suggested for the fabrication of peptide-based biomaterials [58]. Using these peptides, much effort was devoted to mimicking the hierarchal organization of inorganic materials with respect to proteins in biological systems. One example is the group of synthetic cysteine–lysine block copolypeptides reported by Deming and coworkers. These block copolypeptides can mimic the properties of silicatein protein, which can facilitate the formation of silica structures with controlled shapes [59]. The formation of hydrogels was also observed with block copolypeptide, upon their synthesis as amphiphiles containing charged and hydrophobic segments [60].

The ability of peptides to nucleate inorganic materials was further utilized when short peptides with a high affinity to metal were introduced into the coat protein of the M13 filamentous bacteriophage. The peptides were first selected through an evolutionary screening process of phage-display and then incorporated into the generic scaffold of the M13 coat structure. In this way, Belcher and coworkers demonstrated the direct synthesis of magnetic and semiconducting nanowires by templating CoPt, FePt, ZnS, or CdS on the virus surface [61, 62]. The synthesis of metal nanowires was further exploited for the fabrication of lithium ion battery electrodes [63]. This was done by templating cobalt oxide on the M13 virus followed by ordering of the viruses using a layer-by-layer technique [64].

5.8
Prospects

Peptide building blocks are currently being explored as a potential material for the design of various nanostructures. This class of nano-blocks has many advantages for this purpose since they are easy to synthesize, relatively inexpensive, and can be easily decorated with chemical and biological modifications. Their design as a nanomaterial is inspired from nature since larger polypeptides and proteins can spontaneously self-assemble under ambient conditions into complex architectures on the nano-scale. In addition, their ability to nucleate the mineralization of inorganic substances and facilitate the casting of their nano-scale organization can be exploited for the fabrication of composite materials. Peptides have already been shown to self-assemble into ordered structures on the nano-scale; moreover, their decoration with functional elements and inorganic materials was also demonstrated. The next stage of studies will be associated with the utilization of these structures for various applications that include tissue engineering and regeneration, contrast agents, and biosensors. Much effort should be devoted in order to achieve better control over the self-assembly processes. This may lead to the fabrication of uniform nanostructures with identical dimensions. In addition, the ability to control the spatial organization of the peptide-based nanostructures is also required for their integration into complex nano-assemblies and devices.

Acknowledgments

Support from the Israel Science Foundation (F.I.R.S.T. program) is gratefully acknowledged.

References

1. E. Nogales, M. Whittaker, R. A. Milligan, K. H. Downing, *Cell*, **1999**, *96*, 79–88.
2. R. A. B. Keates, R. H. Hall, *Nature*, **1975**, *257*, 418–421.
3. K. H. Downing, E. Nogales, *Curr. Opin. Cell. Biol.*, **1998**, *10*, 16–22.
4. A. Klug, *Angew. Chem. Int. Ed. Engl.*, **1983**, *22*, 565–636.
5. J. C. Rochet, P. T. Lansbury Jr., *Curr. Opin. Struc. Biol.*, **2000**, *10*, 60–68.
6. U. B. Sleytr, P. Messner, D. Pum, M. Sára, *Angew. Chem. Int. Ed.*, **1999**, *38*, 1034–1054.
7. U. B. Sleytr, M. Sara, D. Pum, B. Schuster, *Prog. Surf. Sci.*, **2001**, *68*, 231–278.
8. M. R. Ghadiri, J. R. Granja, R. A. Milligan, D. E. Mcree, N. Khazanovich, *Nature*, **1993**, *366*, 324–327.
9. S. Fernandez-Lopez, H. S. Kim, E. C. Choi, M. Delgado, J. R. Granja, A. Khasanov, K. Kraehenbuehl, G. Long, D. A. Weinberger, K. M. Wilcoxen, M. R. Ghadiri, *Nature*, **2001**, *412*, 452–455.
10. N. Ashkenasy, W. S. Horne, M. R. Ghadiri, *Small*, **2006**, *2*, 99–102.
11. C. Valery, M. Paternostre, B. Robert, T. Gulik-Krzywicki, T. Narayanan, J. C. Dedieu, G. Keller, M. l, Torres, R. Cherif-Cheikh, P. Calvo, F. Artzner, *Proc. Natl. Acad. Sci. U. S. A.*, **2003**, *100*, 10258–10262.
12. H. Matsui, P. Pan, B. Gologan, S. H. Jonas, *J. Phys. Chem. B.*, **2000**, *104*, 9576–9579.
13. H. Matsui, B. Gologan, S. Pan, G. E. Douberly, *Eur. Phys. J. D.*, **2001**, *16*, 403–406.
14. I. A. Banerjee, L. Yu, H. Matsui, *Proc. Natl. Acad. Sci. U. S. A.*, **2003**, *100*, 14678–14682.
15. J. D. Hartgerink, E. Beniash, S. I. Stupp, *Science*, **2001**, *294*, 1684–1688.
16. J. J. Hwang, S. N. Iyer, L. S. Li, R. Claussen, D. A. Harrington, S. I. Stupp, *Proc. Natl. Acad. Sci. U. S. A.*, **2002**, *99*, 9662–9667.
17. G. A. Silva, C. Czeisler, K. L. Niece, E. Beniash, D. A. Harrington, J. A. Kessler, S. I. Stupp, *Science*, **2004**, *303*, 1352–1355.
18. T. C. Holmes, S. de Lacalle, X. Su, G. S. Liu, A. Rich, S. Zhang, *Proc. Natl. Acad. Sci. U. S. A.*, **2000**, *97*, 6728–6733.
19. H. Yokoi, T. Kinoshita, S. Zhang, *Proc. Natl. Acad. Sci. U. S. A.*, **2005**, *102*, 8414–8419.
20. S. Vauthey, S. Santoso, H. Y. Gong, N. Watson, S. Zhang, *Proc. Natl. Acad. Sci. U. S. A.*, **2002**, *99*, 5355–5360.
21. A. Aggeli, M. Bell, N. Boden, J. N. Keen, P. F. Knowles, T. C. B. McLeish, M. Pitkeathly, S. E. Radford, *Nature*, **1997**, *386*, 259–262.
22. A. Aggeli, I. A. Nyrkova, M. Bell, R. Harding, L. Carrick, T. C. B. McLeish, A. N. Semenov, *Proc. Natl. Acad. Sci. U. S. A.*, **2001**, *98*, 11857–11862.
23. A. Aggeli, M. Bell, L. M. Carrick, C. W. G. Fishwick, R. Harding, P. J. Mawer, S. E. Radford, A. E. Strong, N. Boden, *J. Am. Chem. Soc.*, **2003**, *125*, 9619–9628.
24. A. Aggeli, M. Bell, N. Boden, L. M. Carrick, A. E. Strong, *Angew. Chem. Int. Edit.*, **2003**, *42*, 5603–5606.

25 M. R. Chapman, L. S. Robinson, J. S. Pinkner, R. Roth, J. Heuser, M. Hammar, S. Normark, S. J. Hultgren, *Science*, **2002**, *295*, 851–855.
26 D. Claessen, R. Rink, W. de Jong, J. Siebring, P. de Vreugd, F. G. Boersma, L. Dijkhuizen, H. A. Wosten, *Genes. Dev.* **2003**, *17*, 1714–1726.
27 M. F. Gebbink, D. Claessen, B. Bouma, L. Dijkhuizen, H. A. Wosten, *Nat. Rev. Microbiol.* **2005**, *3*, 333–341.
28 J. D. Harper, P. T. Lansbury Jr., *Annu. Rev. Biochem.*, **1997**, *66*, 385–407.
29 M. Sunde, C. C. F. Blake, *Q. Rev. Biophys.*, **1998**, *31*, 1–39.
30 C. M. Dobson, *Trends Biochem. Sci.*, **1999**, *24*, 329–332.
31 J. D. Sipe, A. S. Cohen, *J. Struct. Biol.*, **2000**, *130*, 88–98.
32 R. B. Wickner, K. L. Taylor, H. K. Edskes, M. L. Maddelein, H. Moriyama, B. T. Roberts, *J. Struct. Biol.*, **2000**, *130*, 310–322.
33 E. Gazit, *Angew. Chem. Int. Ed.*, **2002**, *41*, 257–259.
34 E. Gazit, *Curr. Med. Chem.*, **2002**, *9*, 1725–1735.
35 T. Scheibel, R. Parthasarathy, G. Sawicki, X. Lin, H. Jaeger, S. L. Lindquist, *Proc. Natl. Acad. Sci. U. S. A.*, **2003**, *100*, 4527–4532.
36 K. Papanikolopoulou, V. Forge, P. Goeltz, A. Mitraki, *J. Biol. Chem.*, **2004**, *279*, 8991–8998.
37 M. Luckey, J. Hernandez, G. Arlaud, V. T. Forsyth, R. W. Ruigrok, A. Mitraki, *FEBS Lett.*, **2000**, *468*, 23–27.
38 K. Papanikolopoulou, G. Schoehn, V. Forge, V. T. Forsyth, C. Riekel, J. F. Hernandez, R. W. Ruigrok, A. Mitraki, *J. Biol. Chem.*, **2005**, *280*, 2481–2490.
39 B. Häggqvist, J. Näslund, K. Sletten, G. T. Westermark, G. Mucchiano, L. O. Tjernberg, C. Nordstedt, U. Engström, P. Westermark, *Proc. Natl. Acad. Sci. U. S. A.*, **1999**, *96*, 8669–8674.
40 M. Reches, E. Gazit, *Amyloid*, **2004**, *11*, 81–89.
41 K. Tenidis, M. Waldner, J. Bernhagen, W. Fischle, M. Bergmann, M. Weber, M. L. Merkle, W. Voelter, H. Brunner, A. Kapurniotu, *J. Mol. Biol.*, **2000**, *295*, 1055–1071.
42 E. Gazit, *FASEB J.*, **2002**, *16*, 77–83.
43 Y. Mazor, S. Gilead, I. Benhar, E. Gazit, *J. Mol. Biol.*, **2002**, *322*, 1013–1024.
44 M. Reches, Y. Porat, E. Gazit, *J. Biol. Chem.*, **2002**, *277*, 35475–35480.
45 S. K. Maji, D. Haldar, M. G. B. Drew, A. Banerjee, A. K. Das, A. Banerjee, *Tetrahedron*, **2004**, *60*, 3251–3259.
46 C. Madhavaiah, S. Verma, *Chem. Commun.*, **2004**, 638–639.
47 C. Madhavaiah, K. K. Prasad, S. Verma, *Tetrahedron Lett.*, **2005**, *46*, 3745–3749.
48 M. Reches, E. Gazit, *Science*, **2003**, *300*, 625–627.
49 L. Adler-Abramovich, M. Reches, V. L. Sedman, S. Allen, S. J. B. Tendler, E. Gazit, *Langmuir*, **2006**, *22*, 1313–1320.
50 N. Kol, L. Adler-Abramovich, D. Barlam, R. Z. Shneck, E. Gazit, I. Rousso, *Nano Lett.*, **2005**, *5*, 1343–1346.
51 Y. J. Song, S. R. Challa, C. J. Medforth, Y. Qiu, R. K. Watt, D. Pena, J. E. Miller, F. van Swol, J. A. Shelnutt, *Chem. Commun.*, **2004**, *9*, 1044–1045.
52 M. Yemini, M. Reches, J. Rishpon, E. Gazit, *Nano Lett.*, **2005**, *5*, 183–186.
53 M. Yemini, M. Reches, E. Gazit, J. Rishpon, *Anal. Chem.*, **2005**, *77*, 5155–5159.
54 M. Reches, E. Gazit, *Nano Lett.*, **2004**, *4*, 581–585.
55 M. Reches, E. Gazit, *Isr. J. Chem.*, **2005**, *45*, 363–371.
56 M. Reches, E. Gazit, *Phys. Biol.*, **2006**, *3*, S10–19.
57 A. Mahler, M. Reches, M. Rechter, S. Cohen, E. Gazit, *Adv. Mater.*, **2006**, *18*, 1365–1370.
58 T. J. Deming, *Nature*, **1997**, *390*, 386–389.
59 J. N. Cha, G. D. Stucky, D. E. Morse, T. J. Deming, *Nature*, **2000**, *403*, 289–292.
60 A. P. Nowak, V. Breedveld, L. Pakstis, B. Ozbas, D. J. Pine, D. Pochan, T. J. Deming, *Nature*, **2002**, *417*, 424–428.
61 C. Mao, D. J. Solis, B. D. Reiss, S. T. Kottmann, R. Y. Sweeney, A.

Hayhurst, G. Georgiou, B. Iverson, A. M. Belcher, *Science*, **2004** *303*, 213–217.

62 Y. Huang, C. Y. Chiang, S. K. Lee, Y. Gao, E. L. Hu, J. De Yoreo, A. M. Belcher, *Nano. Lett.*, **2005**, *5*, 1429–1434.

63 K. T. Nam, D. W. Kim, P. J. Yoo, C. Y. Chiang, N. Meethong, P. T. Hammond, Y. M. Chiang, A. M. Belcher, *Science*, **2006**, *312*, 885–888.

64 P. J. Yoo, K. Nam, J. Qi, S. Lee, J. Park, A. M. Belcher, P. T. Hammond, *Nat. Mater.*, **2006**, *5*, 234–240.

6
Surface Plasmon Resonances in Nanostructured Materials

K. G. Thomas

6.1
Introduction to Surface Plasmons

Interaction of light with nanostructured metal films and nanoparticles can give rise to hybrid surface waves (light waves coupled to free electrons in a metal), broadly termed as surface plasmons (SPs) [1–7]. One of the exciting features of surface plasmons is their ability to confine (trap) light at a metal/dielectric interface, which can either localize or propagate depending upon the dimensionality, of the nanostructured material. The confinement of light at the interface results from the collective oscillation of conduction electrons at the metal surface. According to the Drude–Lorentz model, metal is denoted as plasma consisting of an equal number of positive ions (fixed at the crystal lattice) and conduction electrons which are free and mobile. In the presence of electromagnetic radiation, the free electrons are displaced by the electric vector and the columbic attraction between the electron cloud and nuclei is the main restoring force [2, 3, 6, 8, 9]. As a result of the oscillating nature of the electric field of light, the electron clouds coherently oscillate over the surface with a resonance frequency ω_p. For a bulk metal with infinite dimensions in all the three directions, the resonance frequency of plasma (ω_p) [8] can be expressed as

$$\omega_p = (Ne^2/\varepsilon_0 m_e)^{1/2} \tag{6.1}$$

where N is the number density of the electrons, ε_0 is the permittivity of vacuum, and e and m_e are the charge and effective mass of the electron, respectively. Depending on the dimensionality of the nanostructured materials, different boundary conditions can be imposed on the electron plasma and the quantized plasma oscillations are called plasmons. The surface plasmons are broadly classified as propagating surface plasmons (PSP) and localized surface plasmons (LSP) [9]. Nanostructured systems of noble metals, such as silver and gold, are of great interest since their localized surface plasmons resonate at the optical wavelength of electromagnetic radiation. This chapter provides a brief summary of the recent

---

Nanomaterials Chemistry. Edited by C.N.R. Rao, A. Müller, and A.K. Cheetham
Copyright © 2007 WILEY-VCH Verlag GmbH & Co. KGaA, Weinheim
ISBN: 978-3-527-31664-9

developments on surface plasmons, with special emphasis on the surface plasmon resonances of spherical and rod-shaped gold nanostructures.

6.1.1
Propagating Surface Plasmons

Propagating surface plasmons (often called surface plasmon polaritons) [1, 10] are associated with smooth, thin films of gold and silver wherein the plasmon feels a boundary due to the planar surface. The surface plasmon in such a metal/dielectric interface has a combined electromagnetic wave as well as surface charge character and is highly bound to the surface (Fig. 6.1(a)). As a result of the combined character, an enhanced field is observed near the surface in the perpendicular direction. The amplitude decays exponentially with distance from the surface and is said to be evanescent or near-field in nature. In the dielectric medium, perpendicular to the metal surface, the decay length of the field in the dielectric material (δ_d) is of the order of half the wavelength of light used, whereas the decay length into the metal (δ_m) is determined by the skin depth (Fig. 6.1(b)) [1]. Once SPs are generated on a flat metal surface they can propagate, however excitation of surface plasmons on such surfaces with freely propagating light is not possible. This is because the momentum of the SP mode is greater than that of the free-space photon of the same frequency due to the combined electromagnetic wave and surface charge character of SP. Such materials can be excited by matching the momenta by various techniques such as (i) the use of prism coupling which can enhance the momentum of incident light, (ii) involving scattering from topographic defects in the thin film (subwavelength protrusion or hole) which can generate local SPs and (iii) by making use of periodic corrugation on the surface of the metal [1].

Once the light is converted to SP mode on a metal surface it can propagate but is gradually attenuated owing to the loss arising from the absorption in the metal [1]. For example, in the case of a relatively absorbing metal such as aluminum the propagation length of SP (δ_{SP}) at a wavelength of 500 nm is ∼2 µm. In contrast, for metals such as silver which possess low loss in the visible range, δ_{SP} is typically in the range 10–100 µm which increases towards 1 mm by using the near-infrared telecom band of 1.55 µm. A representative example of surface plasmon propagation length scales is given in Fig. 6.2. More recently, Ebbesen and coworkers have developed subwavelength waveguide components (Y-splitters, interferometers and ring resonators), operating at telecom wavelengths, which can afford efficient large angle bending and splitting of radiation [11]. Another rich potential of surface plasmons was demonstrated earlier by the same group with the discovery of enhanced light transmission through subwavelength holes which is attributed to the activation of surface plasmons [1, 12–14]. The phenomenon of enhanced transmission is in contrast to the standard aperture theory which predicts that the transmission of light through a small hole is very weak. Interestingly, they have demonstrated the transmission spectra of hole arrays that can be tuned by adjusting the period and symmetry of the holes (Fig. 6.3).

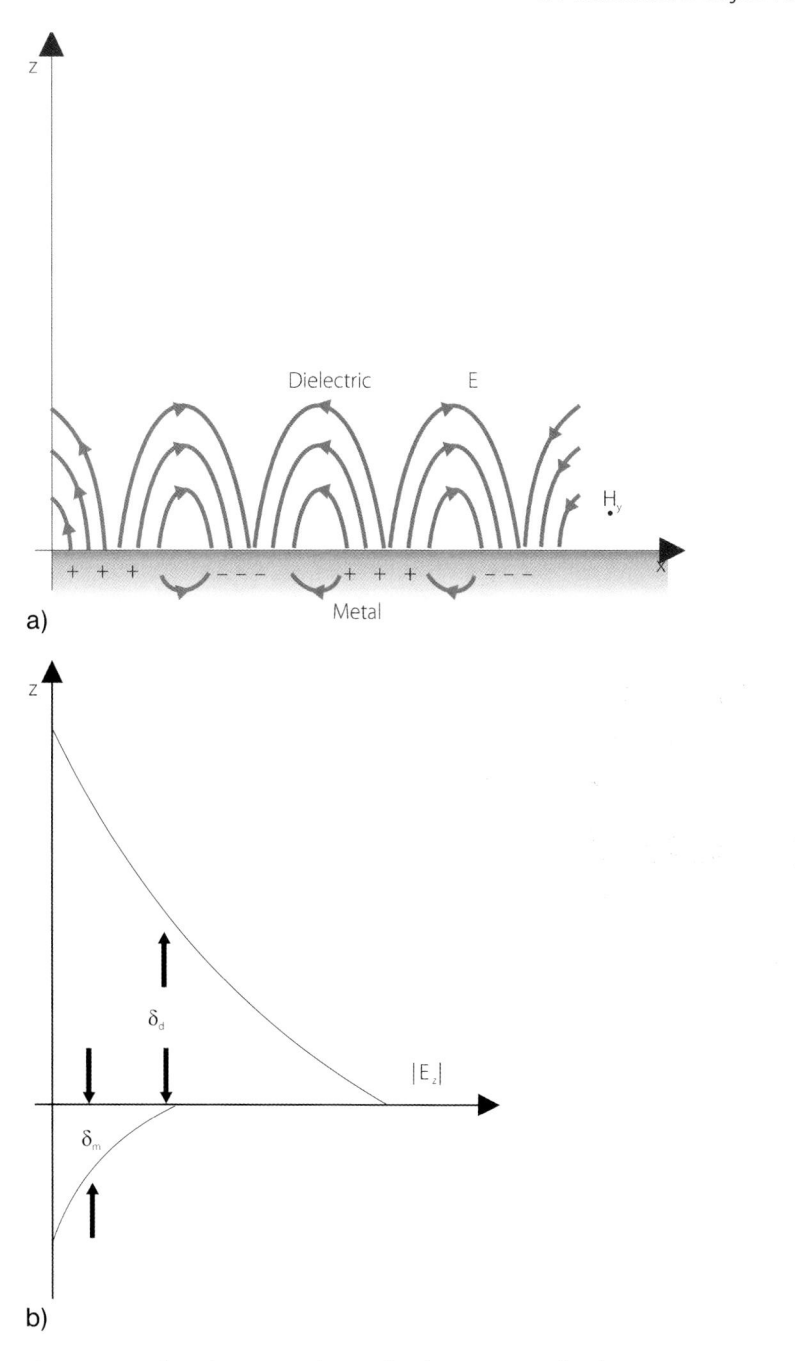

Fig. 6.1 (a) Surface plasmons at the interface between a metal and a dielectric material having a combined electromagnetic wave and surface charge character and (b) decay length of the field in various directions (reproduced with permission from Ref. [1], copyright 2003 Nature Publishing Group).

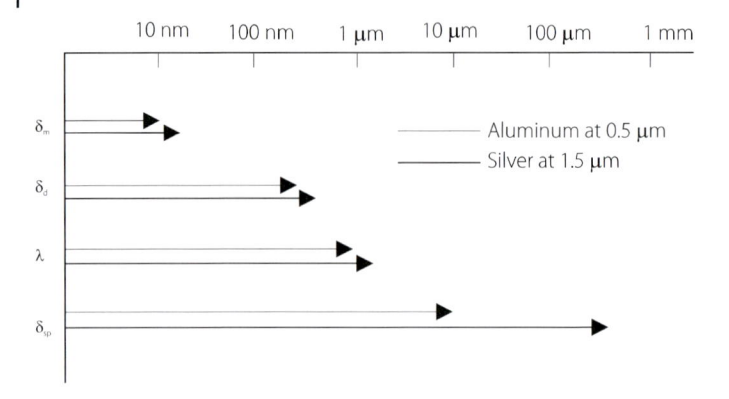

Fig. 6.2 Three characteristic length scales (δ_d, δ_m and δ_{SP}) for surface Plasmon-based photonics and the associated wavelength (λ) of the light; (δ_d, the decay length of the field in the dielectric material; δ_m, the decay length into the metal; δ_{SP}, the propagation length of SP) (reproduced with permission from Ref. [1], copyright 2003 Nature Publishing Group).

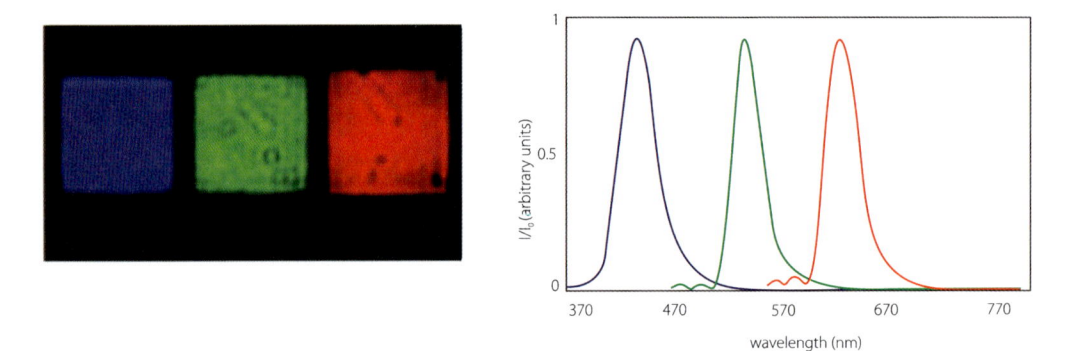

Fig. 6.3 Normal incidence transmission images (left) and spectra (right) for three square arrays of subwavelength holes. For the blue, green and red arrays, the periods were 300, 450 and 550 nm, respectively, the hole diameters were 155, 180 and 225 nm (reproduced with permission from Ref. [1], copyright 2003 Nature Publishing Group).

A variety of approaches have been demonstrated for the fabrication of nanostructures on solid supports for plasmonic applications. These include various lithographic methods such as electron-beam lithography, nanosphere lithography and dip-pen nanolithography [6, 15]. The knowledge of optical near fields around the metal nanostructure is essential for tailoring their properties for various device applications. Near-field techniques such as photon scanning tunneling microscopy are currently utilized for mapping surface plasmons on a metal surface [16, 17]. Fundamental investigations on surface plasmons have revealed several exciting properties and the most notable ones are their ability to trap, concentrate and channel light through subwavelength structures. Based on these studies several SP-based components have been proposed which may have potential applications in optical devices, chemical and bio-sensors and spectroscopy [1].

6.1.2
Localized Surface Plasmons

In contrast to the propagating nature of surface plasmons on a planar metal/dielectric interface, the electron plasma in metal nanoparticles is confined in a finite volume (spread in different spatial dimensions depending upon their shape) which is smaller than the wavelength of light [3–5, 8]. Localized surface plasmons (LSPs) originate from the collective oscillation of free electrons, confined at the surface of metal nanoparticles, when excited with electromagnetic radiation. Unlike bulk metals, the surface plasmons in metal nanoparticles, especially Ag, Au and Cu, can be directly excited by freely propagating electromagnetic radiation in the visible region. When a metal nanoparticle is irradiated by light, the oscillating electric field causes the conduction electron to oscillate coherently (Fig. 6.4) [18]. The absorption occurs when the incident photon frequency is in resonance with the collective oscillation of the conduction electrons, resulting in unique optical properties. The brilliant colors of silver and gold metal colloids were utilized, even centuries ago, by Egyptians (for e.g., stained glass windows) and Romans (for e.g., Lycurgus Cup) and there were several attempts to provide a scientific rationale for their optical properties. It was Michael Faraday who made the remarkable observation that colloidal gold exists in the reduced state as exceedingly fine particles and these divided metals are responsible for their colors. Later in 1908, Gustay Mie provided a quantitative description of the resonance by solving Maxwell's equations for spherical particles, with appropriate boundary condition [19]. According to Mie theory, the total cross-section consists of scattering and absorption (often termed extinction) and is given as summation

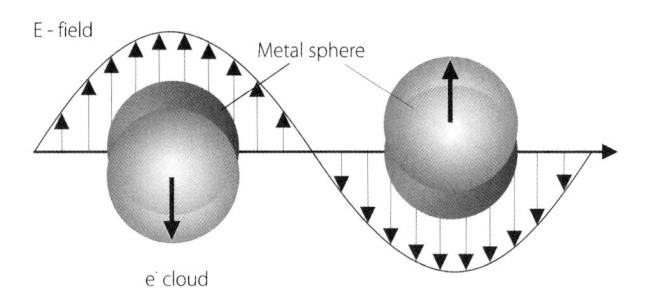

Fig. 6.4 Schematic of plasmon oscillation for a sphere, showing the displacement of the conduction electron charge cloud relative to the nuclei (reproduced with permission from Ref. [18], copyright 2003 American Chemical Society).

over all electric and magnetic oscillations [20]. The contribution of absorption and scattering mainly depends on the size and shape of the particles and these aspects are discussed in the subsequent sections.

6.2
Tuning the Surface Plasmon Oscillations

The frequency of oscillation of metal nanoparticles is determined mainly by four parameters: the density of the electron, the effective mass of the electron and the shape and size distribution of the charge [18]. This allows the tuning of the optical properties of noble metal nanoparticles (position of plasmon resonance band, the extinction cross-section, and the ratio of scattering to absorption) by varying the size, shape and dielectric environment. It is also possible to tune the optical properties of nanoparticles by functionalizing with photo- or electroactive molecules on the surface and these aspects have been reviewed [21, 22]. Several approaches, such as increasing the diameter of the Au nanoparticle or the aspect ratio of the Au nanorods, have been reported for tuning the plasmon resonance band to longer wavelengths. These aspects are discussed below.

6.2.1
Size of Nanoparticle

Typically, the surface plasmon band for Au nanoparticles having diameter of ~10 nm peaks around 520 nm [23]. By adopting the Turkevish method, Link and El-Sayed synthesized Au nanoparticles having diameters in the range 9–99 nm and observed a bathochromic shift in the plasmon resonance band with increase in diameter. For nanoparticles having diameters 9, 22, 48 and 99 nm, the extinction maxima (λ_{max}) were observed at around 517, 521, 533 and 575 nm, respectively (Fig. 6.5A) [23]. The same research group has more recently calculated the dependence of the nanoparticle diameter on the maximum of the plasmon resonance band by adopting Mie theory and the discrete dipole approximation (DDA) method (*vide infra*) [24]. A bathochromic shift in the λ_{max} from 520 to 550 nm was observed on increasing the nanosphere diameter from 20 to 80 nm, which is attributed to the electromagnetic retardation in larger nanoparticles. Alkenethiolate-stabilized Au nanoparticles having mean size in the range 1.5–5.2 nm were synthesized by Murray and coworkers and they observed a sharp decrease in the intensity of the surface plasmon band for particles having diameters below 3.2 nm [25]. This effect is attributed to the onset of quantum size effect and was further established by theoretical methods [26]. The absorption spectra of uncoated gold particles were calculated by Templeton et al. as a function of core diameter. The calculated spectra predict a pronounced SP band for particles having 5.2 nm diameter which undergo broadening and dampening with decrease in size (Fig. 6.5B) [26]. The surface plasmon band is absent for Au nanoparticles having core diameter less than 2 nm. The dampening of the SP mode is

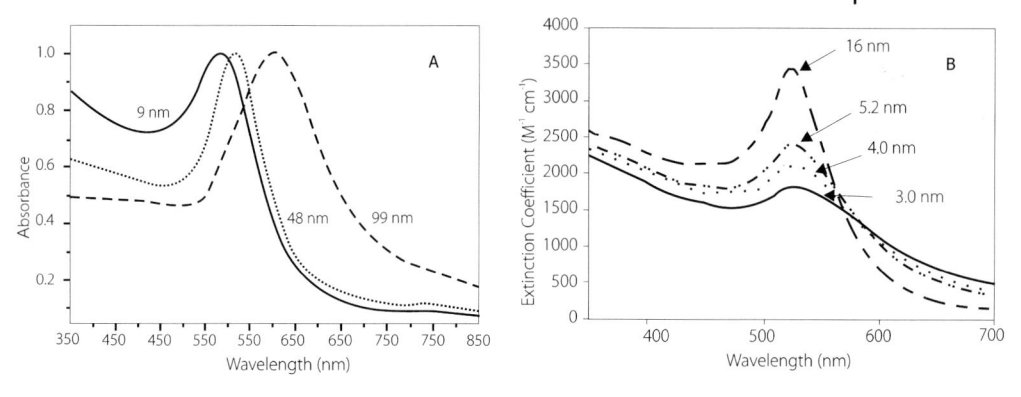

Fig. 6.5 (A) UV–vis absorption spectra of 9, 48, and 99 nm gold nanoparticles in water. (B) Calculated absorbance spectra of uncoated Au clusters, in water ($n_d^{20} = 1.3329$), as a function of nanoparticle diameter (reproduced with permission from Refs. [23, 26], copyright 1999 and 2000 American Chemical Society).

attributed to the surface scattering of conduction electrons which follows 1/radius dependence [3].

6.2.2
Shape of Nanoparticle

Recent developments in classical wet chemistry methods and lithographic techniques have enabled the synthesis of nonspherical nanostructures possessing well defined shapes (for example, ellipsoids, rods, cubes, disks, tetrahedra, cylinders, pyramids, triangular prisms, and multipods) and these aspects are discussed in extensive reviews [4, 5, 9, 27–30]. Among the various anisotropic shapes, nanorods have attracted much attention due to (i) their ease of preparation, (ii) the ability to fine tune their optical properties from the visible to the near-infrared region, by varying their aspect ratios and (iii) their potential use in the biomedical field and as "interconnectors" in nanoscale devices and waveguides [4, 28, 31, 32]. Recent progress on the synthesis, structural characterization, and potential applications of Au nanorods are described in excellent reviews [4, 28, 31]. These studies have shown that Au nanorods possess interesting physical properties and a brief summary of their optical properties is presented in this section.

The shape dependence of surface plasmon resonance for ellipsoids was predicted by Gans in 1912 by extending Mie theory for spherical particles, within the dipole approximation [33]. Although it was predicted that the surface plasmon resonance in small elliopsoids splits into two distinct bands, only recently was it proved experimentally. This is due to the difficulties associated with the preparation of ellipsoidal/cylindrical particles. Esumi and coworkers first demonstrated the formation of rod-like gold particles by UV irradiation (253.7 nm) of

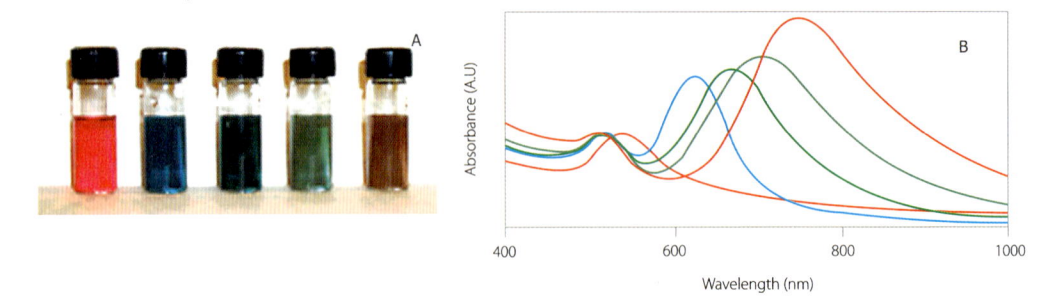

Fig. 6.6 (A) Image of photochemically prepared gold nanorods solution and (B) corresponding UV–vis spectrum. The leftmost solution was prepared with no silver ion addition. The other solutions were prepared with the addition of 15.8, 31.5, 23.7, 31.5 μL of 0.01 M AgNO$_3$ solution, respectively. The middle solution was prepared with longer irradiation time (54 h) than that for all other solutions (30 h), and the transformation into shorter rods can be seen (reproduced with permission from Ref. [40], copyright 2002 American Chemical Society).

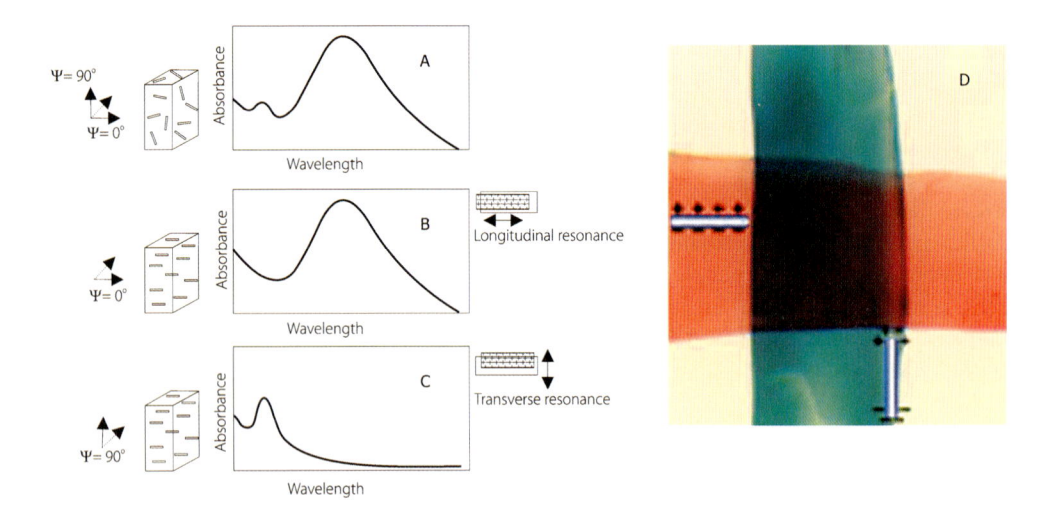

Fig. 6.7 Schematic representation of the absorbance of light by (A) randomly oriented, (B) and (C) aligned gold nanorods. The schematics in the absorbance spectra indicate the direction of the plasmon oscillation with respect to the main axis of the particle and the corresponding resonance. The dotted arrow indicates the wave vector of the incident light. (reproduced with permission from Ref. [41], copyright 1999 American Chemical Society). (D) Photograph of PVA containing gold nanorods aligned parallel (blue film) and perpendicular (red film) to the electric field of polarized incoming light (reproduced with permission from Ref. [42], copyright 2005 Wiley-VCH).

HAuCl$_4$ solutions in the presence of rodlike micelles of hexadecyltrimethylammonium chloride [34]. More recently, several approaches have been reported for the synthesis of gold nanorods of varying aspect ratios, in high yield and monodispersity. This includes hard template-directed synthesis using anodic aluminum oxide templates [35–37], electrochemical [38, 39] and photochemical methods [34, 40] and a seed mediated growth process [28]. Among these the photochemical method has received special attention due to the ease in synthesizing gold nanorods, of varying aspect ratios, by just changing the silver ion concentration [40]. The amount of silver ions added to the system dictates the aspect ratio of the rods produced (Fig. 6.6) [40].

Gold nanorods possess two distinct surface plasmon absorption bands associated with the transverse and longitudinal oscillations of electrons. Interestingly, the former band absorbs at 520 nm whereas the latter absorbs at a longer wavelength, which undergoes a bathochromic shift with an increase in aspect ratio. Electric field alignment studies by van der Zande et al. have shown that the short wavelength band ($\sim$520 nm) originates from the transverse mode of vibration and the long wavelength band from the longitudinal mode of vibration [41]. In the absence of an electric field the gold rods are randomly oriented and the absorbance spectra display both the transverse and the longitudinal resonances (Fig. 6.7A). Interestingly, in the presence of an electric field, the relative magnitude of the resonances is sensitive to the induced orientational order in the rod dispersion. When the incident light is polarized parallel to the applied electric field, the transverse mode of vibration disappears whilst the longitudinal mode is retained (Fig. 6.7B). In contrast, the transverse resonance survives at the expense of the longitudinal resonance in the presence of perpendicularly polarized light (Fig. 6.7C). The optical effect of poly(vinylalcohol) (PVA) films containing oriented gold nanorods placed parallel and perpendicular to the polarization direction is shown in Fig. 6.7D (parallel film appears blue and the perpendicular one red) [42]. The different colors arise from the selective excitation of the longitudinal and transverse plasmon modes, respectively.

Simulation of the optical absorption spectra of gold nanorods as a function of their aspect ratio and medium dielectric constant has been carried out by Link and El-Sayed (Fig. 6.8) [43, 44]. Both experimental and theoretical studies indicate a linear relationship between the absorption maximum of the longitudinal plasmon resonance and the mean aspect ratio (determined from TEM). The authors have also calculated the absorption spectra of Au nanorods as a function dielectric constant of the medium. For Au nanorods of a fixed aspect ratio, a bathochromic shift in the longitudinal plasmon resonance was observed with increase in dielectric constant of the medium and these aspects were further proved experimentally.

It is essential to know the concentration of Au nanorods in order to carry out any meaningful physical investigations and for this purpose it is important to determine the extinction coefficient of Au nanorods. The extinction coefficient of Au nanorods (synthesized electrochemically having an aspect ratio 2.7, length of 27 nm and width of 10 nm) was first reported by El-Sayed and coworkers as $1.9 \pm 0.4 \times 10^9$ M^{-1} cm^{-1} at 695 nm [45]. We have synthesized gold nanorods

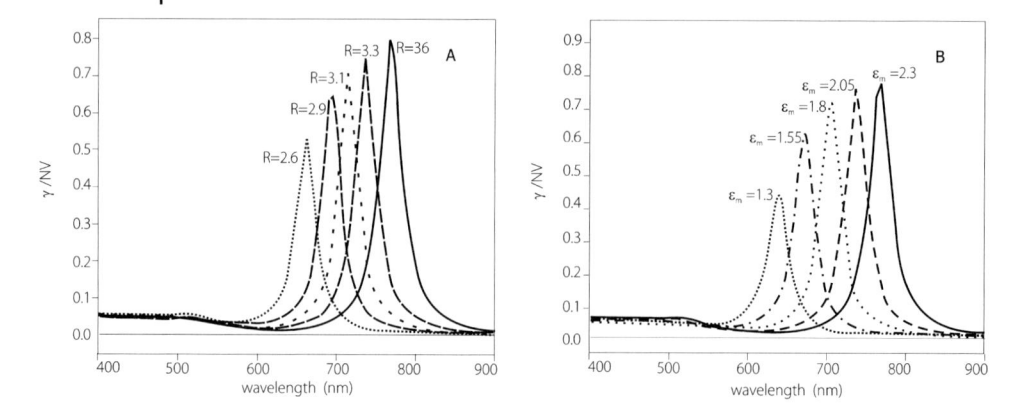

Fig. 6.8 Calculated absorption spectra of elongated ellipsoids with (A) varying aspect ratios, 'R' (dielectric constant of the medium = 2.05) and (B) varying dielectric constant of the medium, ε_m (R = 3.3) (reproduced with permission from Ref. [43], copyright 2005 American Chemical Society).

of different aspect ratios following a photochemical method and the extinction coefficients were determined in conjunction with transmission electron microscopy (TEM) and inductively coupled plasma (ICP) analysis [46, 47]. For Au nanorods of aspect ratio 2.9 (average length and diameter 47.3 nm and 20.4 nm, respectively), the extinction coefficients (ε) of the transverse and longitudinal plasmon absorption bands were estimated as 0.27×10^{10} M^{-1} cm^{-1} (515 nm) and 0.53×10^{10} M^{-1} cm^{-1} (700 nm), respectively. Comparable extinction coefficient values were reported for the longitudinal band of Au nanorods having an aspect ratio 2.9 by various other groups; Liao and Hafner reported 'ε' as 0.44×10^{10} M^{-1} cm^{-1} (796 nm) [48] and Orendorff and Murphy reported 'ε' as 0.39×10^{10} M^{-1} cm^{-1} (728 nm) [49]. Even though the aspect ratios are same, the lengths and diameters of the rods used by various groups are different and this may be the main reason for variations in the values of the extinction coefficients. Orendorff and Murphy have also reported an increase in the extinction coefficient value for the longitudinal plasmon band of Au nanorods from 2.5×10^9 to 5.5×10^9 M^{-1} cm^{-1} on increasing the aspect ratios from 2.0 to 4.5 [49].

6.2.3
Dielectric Environment

The surface plasmon peak of metal nanoparticles is greatly influenced by the local dielectric environment and extensive theoretical as well as experimental reports are available. The plasmon peak position depends on the refractive index of the surrounding medium (n) and can be determined by Eq. (6.2) [8],

$$\lambda^2 = \lambda_p^2(\varepsilon^\infty + 2\varepsilon_m) \tag{6.2}$$

where, λ_p is the bulk plasmon wavelength, ε^∞ is the high-frequency dielectric constant due to interband and core transitions, and ε_m is the dielectric constant of the medium (the refractive index of the medium is directly related to its dielectric constant, i.e., $n = (\varepsilon_m)^{1/2}$). In most cases, Au nanoparticles are protected with an organic capping agent to prevent agglomeration. Equation (6.2) does not account for the contribution of the dielectric shell, hence is applicable only for unprotected gold nanoparticles. By involving the contribution of the monolayer, Murray, Mulvaney and coworkers [26] modified Eq. (6.2) for alkanethiolate protected Au nanoparticles as

$$\lambda^2 = \lambda_p^2[(\varepsilon^\infty + 2\varepsilon_m) + 2g(\varepsilon_s - \varepsilon_m)/3] \tag{6.3}$$

where, ε_s is the dispersionless optical dielectric function of the shell, g is the volume fraction of the shell layer which increases with chain length of the organic capping agent. In the case of very small gold nanoparticles, the size of the core and organic shell becomes comparable and the g value is large. Hence the solvent refractive index effect is quite small for alkanethiolate-protected Au clusters having a diameter ~2 nm. However, for dodecanethiolate-protected Au nanoparticles having an average diameter of 5.2 nm, an 8 nm red shift in plasmon band is observed when the solvent refractive index is varied from 1.33 to 1.55 [26]. This result is in accord with the calculated spectral shift obtained using the modified Eq. (6.3) [26].

Recently, we have probed the changes in the surface plasmon band of tetraoctylammonium bromide (TOAB) capped spherical gold nanoparticles in solvents of varying refractive index [50]. Based on the surface plasmon band shifts, solvents were classified into two general categories: (i) solvents that do not complex with the Au nanoparticle surface and that can alter the refractive index and (ii) solvents that complex with the gold nanoparticle surface. Solvents such as cyclohexane, toluene, o-xylene, chlorobenzene, and o-dichlorobenzene fall into the first category, and the surface plasmon band gradually shifts toward longer wavelengths with increase in refractive index of the solvent. A plot of the square of the observed position of the surface plasmon band of TOAB-capped Au nanoparticles in these solvents as a function of the dielectric constant of the medium, showed a linear relationship (Fig. 6.9) [50]. The diameter of the TOAB-capped gold nanoparticles is in the range 5–6 nm and the contribution of the 'g' value becomes less with increase in the core size. Moreover, in the present case, TOAB-capped gold nanoparticles are stabilized through noncovalent interactions. Hence the bulk solvent molecules are able to penetrate through the shell and directly influence the surface plasmon band of the metal particle. Interestingly, the surface plasmon band position remains unaffected in polar solvents with nonbonding electrons (e.g., tetrahydrofuran, dimethylformamide, dimethyl sulfoxide, and acetone); this is attributed to the direct interaction of these solvents with the Au surface. Such complexation processes may override the effects of refractive index since they substantially alter the electron density of the Au nanoparticle surface.

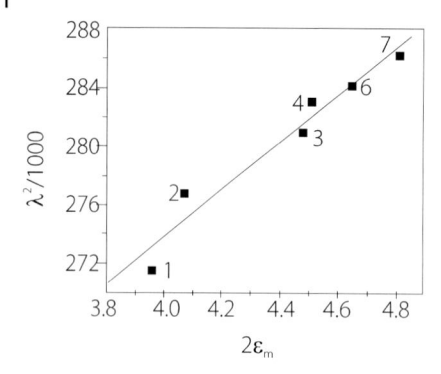

Fig. 6.9 Dependence of the square of the observed peak position of the surface plasmon band as a function of twice the dielectric constant of the medium (reproduced with permission from Ref. [50], copyright 2002 American Chemical Society).

Heterostructured nanoparticles are another interesting class of nanomaterials which consist of a dielectric core (for example, silica) and a noble metallic shell, and they possess very interesting optical properties [51, 52]. Plasmon resonances in these systems are shifted to much longer wavelengths than those in the corresponding solid metal nanoparticles. More interestingly, it is possible to tune the plasmon resonance from the visible to the far-infrared region by just varying the thickness of the metallic shell. Well established synthetic protocols are available for controlling the thickness of the metallic shell around a dielectric core [51, 52]. The nanoshell plasmon resonance is highly sensitive to the local dielectric environment and is useful in the design of sensors. The tunability of the plasmon resonance to the near-infrared region has been explored for various biomedical applications such as drug delivery, rapid whole-blood immunoassay and in cancer diagnosis and therapeutics.

6.3
Excitation of Localized Surface Plasmons

As discussed in previous sections, the high optical cross-section of the surface plasmon band of Au nanoparticles (typically 4–5 orders of magnitude higher than for conventional dyes) is one of the exciting features which provide excellent opportunities in the biomedical field for diagnosis, imaging and photothermal treatment of cancer cells [24]. The absorption cross-section of the surface plasmon band of an Au nanoparticle (product of the efficiency of absorption and the cross-sectional area of the nanoparticle) having a diameter of 40 nm is calculated as 2.93×10^{-15} m^2 at 528 nm, which corresponds to a molar extinction coefficient of 7.66×10^9 M^{-1} cm^{-1} [24]. The light absorbed by gold nanoparticles is efficiently converted into localized heat and this strategy has been successfully used for the laser photothermal destruction of cancer cells [53–55]. Other superior properties of Au nanoparticles, compared to conventional dyes, are facile immunotargeting ability, nonsusceptibility to photobleaching and chemical/thermal denaturation. Gold nanoparticles have been used *in vivo* as a radiotracer [56] since 1950s and recent studies have shown that these nanoparticles, with suitable surface modifiers, are nontoxic to human cells (noncytotoxic) [57]. However, for *in vivo* imaging and therapeutic applications the optical resonance of the nanoparticles is strongly desired to be in the near-infrared (NIR) region of the biological water window, where the optical penetration through tissue is highest [58]. Surface plasmon absorption of gold nanorods can be conveniently tuned to this region by varying their aspect ratio. Thus Au nanorods and gold–silica nanoshell systems have emerged as potential candidates for various biomedical applications [53, 55].

6.3.1
Multipole Resonances

The spectral features of the surface plasmon band are determined by the relative dimensions of the particle in relation to the wavelength of the electromagnetic radiation [18, 20, 30]. When the dimensions of nanoparticles are much smaller than the wavelength of light, the particle experiences a uniform electromagnetic field. As a result, all the conduction electrons move in-phase with the electromagnetic radiation producing only a dipole type of oscillation. The surface plasmon band of a spherical silver or gold nanoparticle having diameter of approximately 5–50 nm is dipolar in nature, resulting in a single narrow peak. However, with increase in size, the field across the particle becomes nonuniform and light cannot polarize the nanoparticles homogeneously. This results in a phase retardation of the applied field inside the material. As a consequence, broadening in the dipolar surface plasmon band is observed in larger particles along with the possibility of multipolar excitation (quadrupole, octupole etc.) [59–61].

Various extinction modes due to multipolar excitation have been classified by the multipolar order 'v' of the spherical particles [20]. They are divided into electric and magnetic modes which correspond to the surface plasmon and eddy current, respectively. The extinction loss due to each mode consists of true absorption (A) as well as a scattering (S) component and the total extinction spectrum is given by

$$E(\omega) = \Sigma_v (A_v + A'_v + S_v + S'_v) \tag{6.4}$$

where A and S corresponds to the absorption and scattering contribution from surface plasmons and A' and S' correspond to the absorption and scattering contribution from the eddy current. The contribution due to the magnetic mode is small for these particle sizes and spectral region and hence is not considered. The extinction losses due to electric field interaction result in distinct spectral bands for noble metal nanoparticles since they consist of nearly free electrons (*vide supra*). Thus the total extinction spectrum in the case of noble metal nanoparticles can be deduced as

$$E(\omega) = \Sigma_\nu (A_\nu + S_\nu) \qquad (6.5)$$

The contributions of the quadrupole surface plasmon absorption was first experimentally separated from the dipolar extinction by Kreibig et al. for spherical Ag nanoparticles (having a diameter ∼63 nm) using optical and photothermal spectroscopy [20]. More recently, high order multipole resonances have been observed by various groups in silver and gold nanomaterials possessing different shapes, such as spheres, rods and prisms [59–62]. To observe these bands, nanomaterials should be extremely pure and homogeneous and their dimensions should be sufficiently large.

High quality synthetic procedures were adopted by various groups for synthesizing chemically clean metal nanoparticles. More recently, Chumanov and coworkers prepared free standing naked Ag nanoparticles of varying dimensions by reducing Ag_2O using hydrogen gas [63]. These particles are chemically clean and do not possess any chemical species on the surface, except water and oxygen. The extinction spectra of small Ag particles showed an initial peak position around 420 nm which is attributed to the dipole component of the plasmon resonance. As the particle size increases, the intensity of this band increases with concomitant shifts in its position to longer wavelengths and the authors could also observe the successive formation of high-order multipole resonances (Fig. 6.10) [59–61]. On increasing the size of Ag nanoparticles to ∼90 nm, the dipole resonance band shifts to 490 nm, along with the formation of a new band corresponding to the quadrupole mode at 420 nm. On further increasing the size of the particles (∼170 nm), both the dipole and quadrupole resonances shift to the red spectral range (630 and 470 nm, respectively) and a new band emerges at around 430 nm corresponding to the octupole mode. The band corresponding to the octupole mode, with further increase in the particle size, gradually evolves into a distinct peak and shifts to 475 nm. Subsequently, a new shoulder corresponding to the hexadecapole mode was observed in the spectrum at 430 nm. For a particle of ∼215 nm size, the authors could observe quadrupole, octupole, and hexadecapole modes of the plasmon resonance at 543, 483, and 432 nm, respectively, with the broad dipole peak at the near-IR region of the spectrum.

Highly structured multipole plasmon resonances were reported in gold nanorods and nanoprisms by Schatz, Mirkin and coworkers [62, 64]. Usually, wet chemical methods have been adopted for the synthesis of gold nanorods of smaller dimensions and they possess two distinct surface plasmon absorption bands associated with the transverse and longitudinal oscillations of electrons. Due to polydispersity and the presence of other shapes, nanorods synthesized by this method are not useful for mapping out the multipole plasmon resonances. For investigating these aspects, authors have synthesized high purity Au nanorods of larger dimensions (average diameter 85 nm and length 96–1175 nm) by electrochemically depositing them in anodic aluminum oxide templates (hard

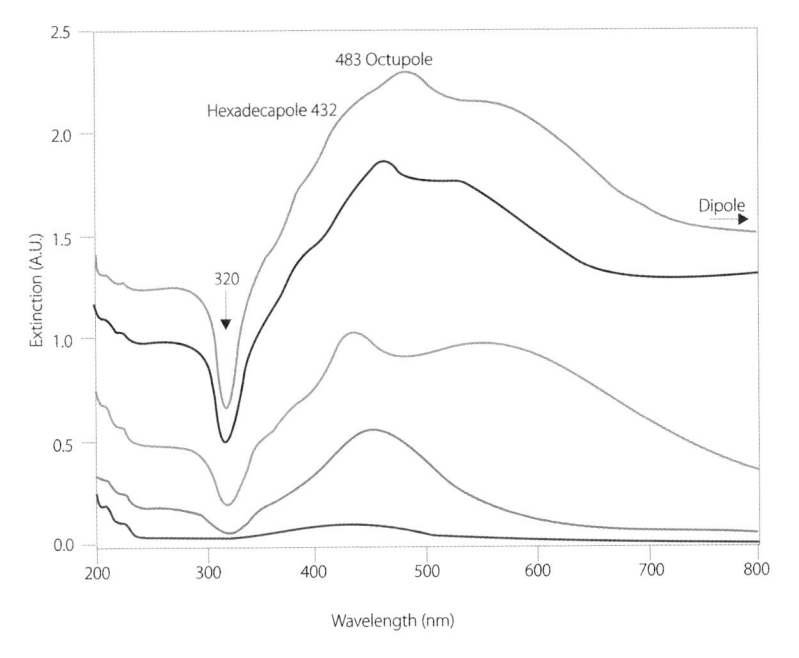

Fig. 6.10 Extinction spectra of different sized Ag nanoparticles synthesized by the hydrogen reduction method and the spectra corresponding to aliquots taken at different intervals (reproduced with permission from Ref. [60], copyright 2005 American Chemical Society).

template-directed synthesis). The higher order multipole resonances observed in the experimental spectra (Fig. 6.11) [62] match well with the theoretical extinction spectra obtained by the DDA method. When the length of the nanorod is 96 nm (diameter ~85 nm), one prominent broad peak is observed around 600 nm. Since the aspect ratio is close to 1, the longitudinal plasmon (labeled I) overlaps with the transverse dipole mode. Gans theory predicts that the transverse mode will shift to shorter wavelengths (blue shift) and the longitudinal mode to longer wavelengths (red shift) with increase in the aspect ratio and this trend was observed in gold nanorods synthesized via the hard template-directed synthesis. It is interesting to note that the authors could observe the successive formation of all the longitudinal resonance modes up to seventh order in Au nanorods. Each of the even (labeled as II, IV, and VI) and odd (labeled as I, III, V and VII) longitudinal modes shift to longer wavelengths with increase in aspect ratio.

Extremely pure Au nanoprisms were synthesized by the same group and they could observe a quadrupole plasmon resonance band in these systems ($\lambda_{max} \approx 800$ nm) along with an in-plane dipole band with $\lambda_{max} \approx 1300$ nm [64]. This band was not observed in previous reports, may be due to the inhomogeneity and impurities present.

It is believed that an understanding of the various multipole resonance modes can provide fundamentally important correlations on the structure and optical

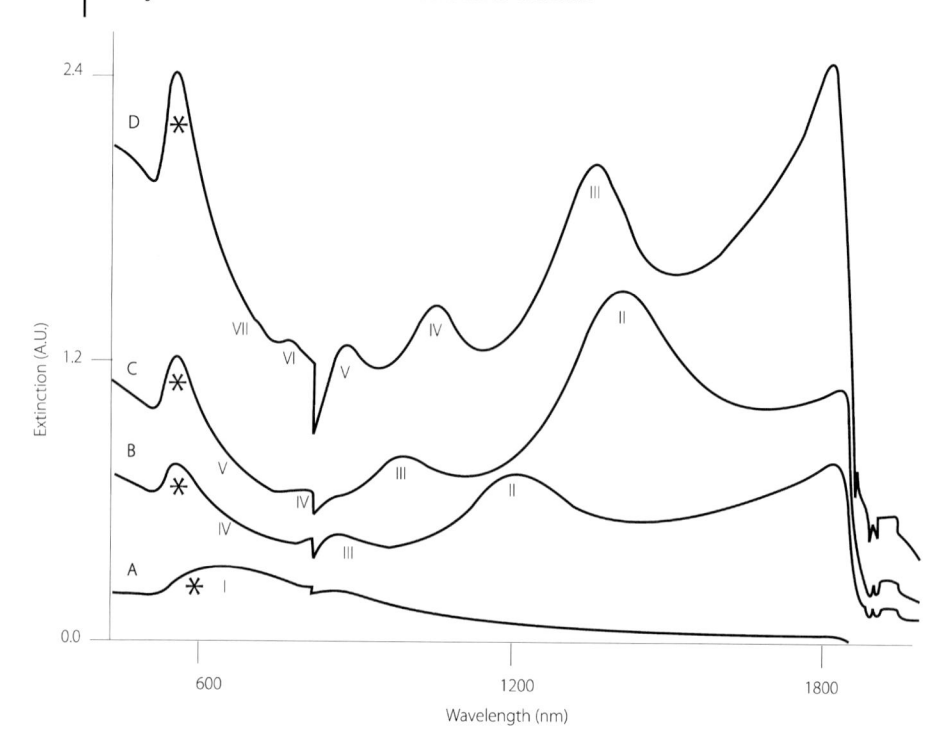

Fig. 6.11 UV–vis–NIR spectra of the (A) 96, (B) 641, (C) 735, and (D) 1175 nm in length gold rods in D_2O. The Roman numeral labels the multipole order associated with each plasmon resonance and the transverse dipole mode, labeled with an asterisk. Orders were assigned on the basis of theoretical calculations (reproduced with permission from Ref. [62], copyright 2006 American Chemical Society).

properties of nanomaterials [62]. It can also provide a benchmark spectrum for evaluating the quality as well as the homogeneity of various anisotropic nanomaterials. The near-field profiles and the far-field scattering patterns of multipolar harmonics may be potentially useful in applications pertaining to light signal routers, light manipulators, or multistep enhancers in processes such as second harmonic generation [65].

6.3.2
Absorption vs. Scattering

Surface plasmon excitation in Ag and Au nanoparticles results in strong enhancements of the absorption, scattering, and local electric field around the metal particles. These fundamental properties provide newer possibilities in the field of biodiagnostics, which surpass the selectivity, sensitivity and multiplexing capabilities of many conventional assays, biomedical imaging and therapeutics (for example, destruction of cancer cells using photothermal therapy) [24, 53–55]. For sensing and other diagnostic applications, it is desirable to have nanomaterials with a high absorption cross-section whereas particles having strong optical scattering are useful for imaging applications [24]. It is important to understand the contributions of absorption/scattering to the total extinction by varying the dimensions and to find methods to control their relative contributions.

Maxwell's equations for the optical response of a homogeneous, isotropic sphere to an electromagnetic field of light provide exact formulae for calculating both scattering and absorption cross-sections. However, this analytical method has a fundamental limitation that the exact solutions are restricted to highly symmetric particles such as spheres. Numerical approximate methods are generally used to calculate the optical properties of small particles of arbitrary shapes and several methods have been developed in the past decade [66]. These include (i) surface-based methods such as the transition matrix (T-matrix) method [67] and the generalized multipole technique (GMT) [68] where the particle's surface is discretized and solved numerically and (ii) volume-based methods such as the discrete dipole approximation (DDA) [69] and the finite different time domain (FDTD) methods [70] where the entire volume is discretized. Among the various methods, DDA has emerged as the most powerful electrodynamic method for calculating the scattering and absorption cross-section of small particles of any arbitrary shape. This method was originally developed by Purcel and Pennypacker for addressing the scattering problem of interstellar dust particles [69]. More recently, Schatz and coworkers have successfully demonstrated the use of the DDA method for studying the extinction spectrum and local electric field distribution of metallic systems having different geometries and environments [18, 30, 66]. In this method, a particle of an arbitrary shape is replaced in an assembly of finite polarizable cubical elements. Each of these cubes is small enough that one has to consider only the dipole interaction with an applied electromagnetic field and the induced fields in other elements. Thus the particle can be considered as a cubic array of dipoles, which reduces the solution of Maxwell's equation to an algebraic problem involving many coupled dipoles.

Several groups have compared the optical properties of noble metal nanostructures obtained from experimental and DDA simulation. Anisotropic noble metal nanoparticles such as Ag nanodisks, Ag triangular nanoprisms, Au nanoshells and Au multipods show intriguing optical properties and possess tunable surface plasmon bands depending on their shapes [30]. The extinction spectra of these particles were compared by experiment as well by the DDA method. Compared to spherical particles, the surface plasmon resonances of these anisotropic particles are red-shifted and in some cases observed as distinctive dipole and quadrupole plasmon modes. Sosa et al. have used the DDA method to study the extinction, absorption, and scattering efficiencies of nanoparticles of different sizes and shapes (spheres, ellipsoids, cubes, tetrahedra, cylinders, and pyramids), made out of silver and gold and identified the main optical signature associated with various geometries [71]. They observed that the spectra are more complex as the particle has less symmetry and more vertexes.

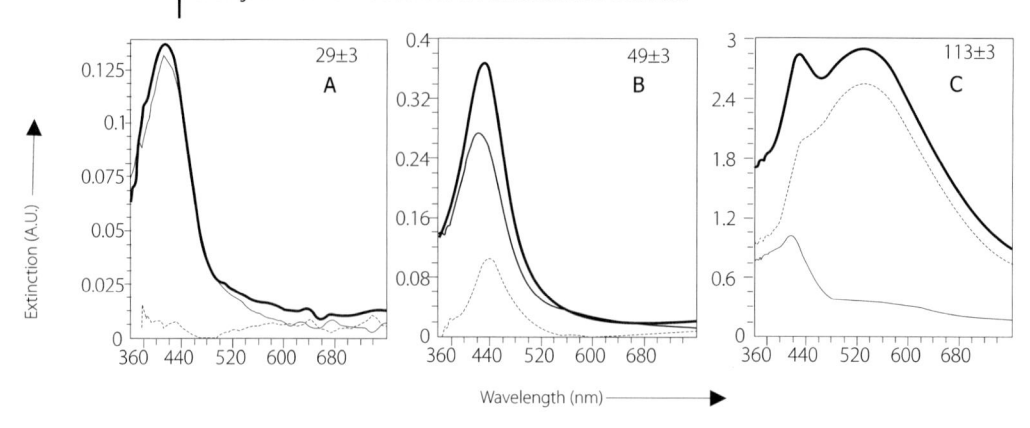

Fig. 6.12 Extinction (solid line), scattering (dotted line), and absorption (thin line) spectra of silver nanoparticle suspensions normalized per particle. The mean particle sizes are noted in each panel. The units on the y axis are multiplied by 10^{10} (reproduced with permission from Ref. [61], copyright 2005 Wiley-VCH).

In a recent article, Evsnoff and Chumanov have presented the experimental measurements of the extinction, scattering and absorption cross-sections and efficiencies for silver nanoparticles of 16 different sizes, having diameters ranging from 29 to 136 nm [59]. For this purpose they used highly pure Ag nanoparticles, free of extraneous species, having narrow size and shape distribution and a high degree of crystallinity. The extinction spectra obtained by standard UV–vis measurements represent transmission losses due to both true absorption and scattering which were separated using an integrating sphere. The extinction is dominated almost entirely by absorption for small particles with diameter around 30 nm (Fig. 6.12A) [59]. The contribution of scattering in the extinction spectrum increases with particle size and becomes equal to absorption at a particle diameter of ca. 52 nm (Fig. 6.12B) [59]. The effect of phase retardation can be observed for larger particles (with diameter > 76 nm) resulting in the appearance of additional plasmon modes (Fig. 6.12C) [59].

In an elegant theoretical study, El-Sayed and coworkers calculated the absorption and scattering properties of gold nanoparticles of different size, shape, and composition [24]. They observed that the magnitude of extinction, as well as the relative contribution of scattering to the extinction, increases rapidly with increase in the size of the metal nanospheres. For Au nanospheres and nanorods, increase in the diameter resulted in an increase in the extinction cross-section as well as in the relative contribution of scattering (Fig. 6.13). However, an increase in the nanorod aspect ratio at a constant radius does not result in any considerable effect on either the extinction cross-section or the ratio of scattering to absorption. These studies clearly indicate that the relative scattering to absorption contribution could be easily tuned by changing the dimensions of the rods.

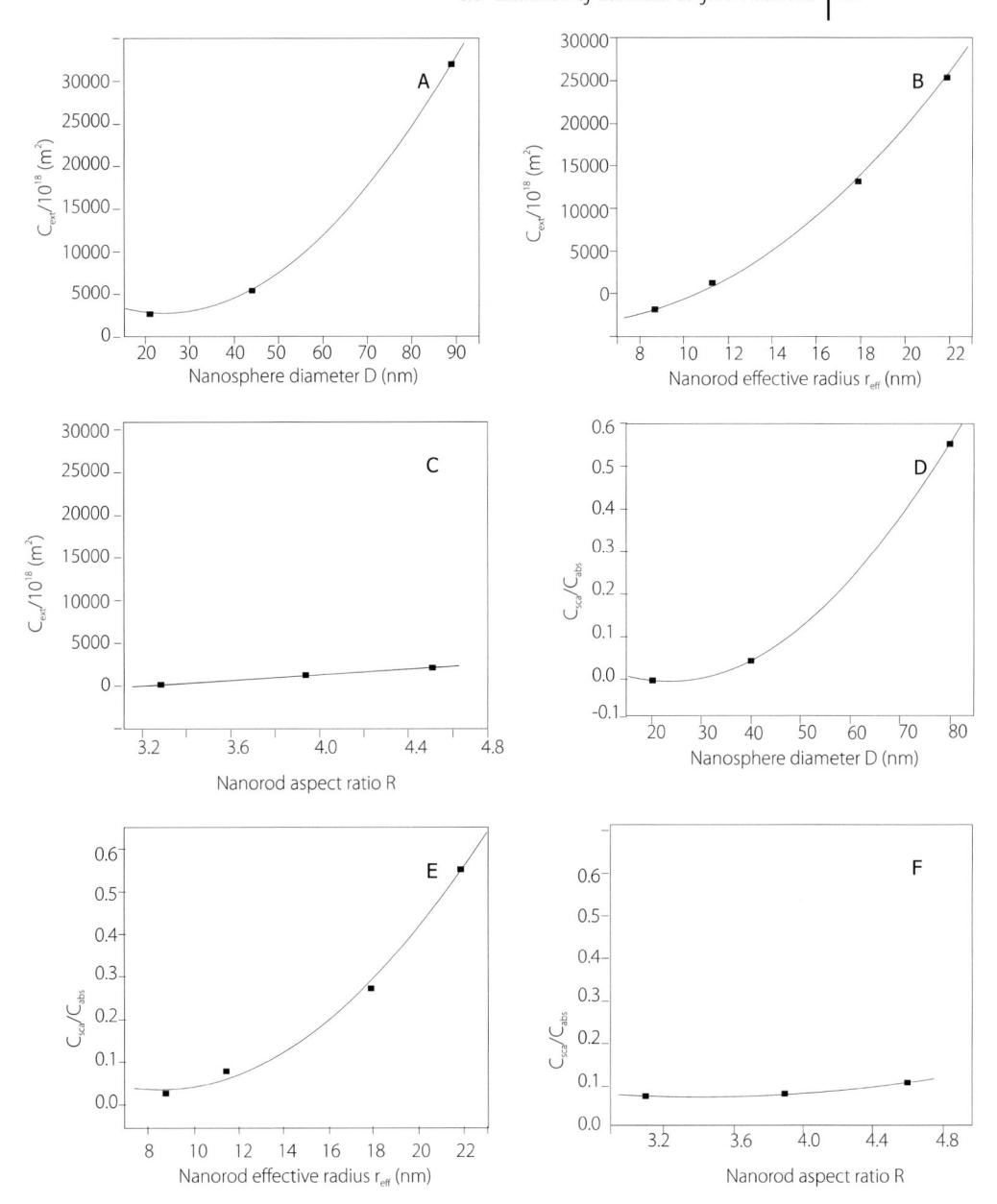

Fig. 6.13 (A)–(C) Tunability of the extinction cross-section of nanoparticles and (D)–(F) tunability of the ratio of scattering to absorption of nanoparticles. Variation of 'C_{ext}' with (A) nanosphere diameter 'D', (B) nanorod effective radius 'r_{eff}' at fixed aspect ratio ($R = 3.9$), and (C) nanorod aspect ratio 'R' at fixed r_{eff} of 11.43 nm. Variation of C_{sca}/C_{abs} with (D) nanosphere diameter 'D', (E) 'r_{eff}' at fixed aspect ratio $R = 3.9$, and (F) 'R' at fixed 'r_{eff}' of 11.43 nm (reproduced with permission from Ref. [24], copyright 2006 American Chemical Society).

6.4
Plasmon Coupling in Higher Order Nanostructures

The transport of optical energy using materials that are considerably smaller than the wavelength of light is one of the most challenging issues in the miniaturization of photonic components. The design of higher order nanostructured materials (for example, one-dimensional arrays of noble metal nanoparticles with defined particle spacing) is an essential requirement for achieving this goal. Lithographic methods such as electron beam lithography are more commonly used for the construction of higher order nanostructures and the details are summarized in recent reviews [6, 15]. Such nanostructures can convert photons into surface plasmons that are not diffraction limited (focused more tightly than $\lambda/2$). Within the propagation length, the surface plasmon modes can be decoupled to light and this possibility offers tremendous opportunities in the design of nanoscale optical and photonic devices such as metal nanoparticle plasmon waveguides. Maier et al. have recently demonstrated the transport of electromagnetic energy over a distance of 0.5 µm in plasmon waveguides consisting of closely spaced silver rods [32]. The waveguides were excited by the tip of a near-field scanning optical microscope and energy transport was probed by using fluorescent nanospheres.

Tailoring the optoelectronic properties of nanomaterials through the stepwise integration of nanoscale building blocks (nanoparticles, nanorods, nanotubes, etc.) is another major challenge in the area of nanotechnology. Recent studies have shown that it is possible to fine tune the optical properties of metallic nanoparticles by their controlled organization into periodic arrays [6, 15]. Two types of interactions exist in organized metal nanoparticles: near-field coupling and far-field dipolar coupling. Near-field coupling (evanescent coupling) is observed in an ensemble of closely packed particles wherein the particles nearly touch each other. In the latter case, the dipole field resulting from the plasmon oscillation of a metal nanoparticle induces an oscillation in a neighboring nanoparticle. A brief summary on the theoretical as well as the experimental investigations on plasmon resonances in nanoparticles arrays, with emphasis on 1D organization of spherical and rod shaped Au nanoparticles, is summarized in this section.

6.4.1
Assembly of Nanospheres

Assemblies of particles cannot be readily modeled by simple Mie theory due to the complexity involved in multiparticle scattering and interference effects. A convenient way to theoretically investigate electromagnetic interactions in arrays of spherical particles is to use high quality electrodynamic methods such as T-matrix or coupled multipole methods [72]. Recently, Schatz and coworkers have reported the influence of electromagnetic coupling between nanoparticles over a wide range of particle spacing and for a broad range of array structures by using these methods [72]. They have calculated the extinction spectra (dipole plasmon reso-

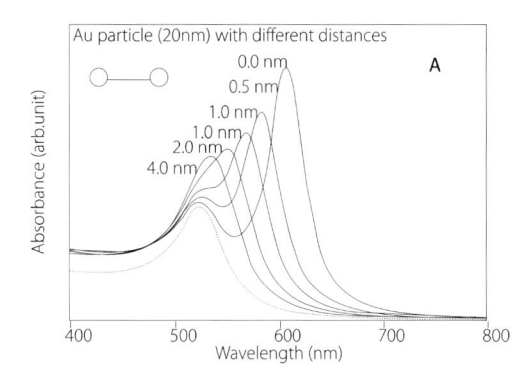

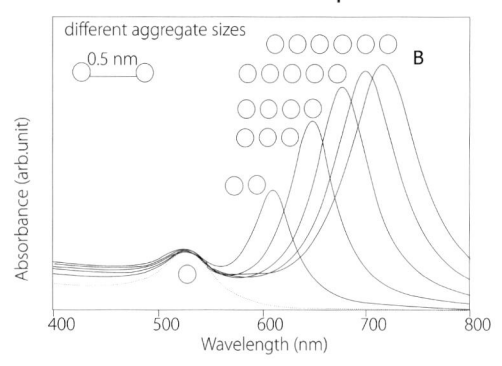

Fig. 6.14 Electrodynamic modeling calculations for Au nanoparticles. (A) Change of extinction spectra for 20 nm diameter particles with interparticle distance (s). (B) Extinction spectra of "line aggregates" of varying number ($d = 40$ nm, $s = 0.5$ nm) (reproduced with permission from Ref. [73], copyright 2004 American Chemical Society).

nance wavelength and plasmon width) of 1D chains and 2D arrays of silver nanoparticles by varying the particle size, array spacing, array symmetry, and polarization direction. Under both parallel and perpendicular polarization, the width of the plasmon band broadens in 1D chains as the spacing decreases. However, the plasmon resonance wavelength mostly shifts to the red region under parallel polarization and to the blue region under perpendicular polarization.

Zhong et al. have theoretically calculated the extinction spectra for clusters of two spherical Au nanoparticles (diameter 20 nm) at different separation distances by using generalized multiparticle Mie (GMM) solutions [73]. Isolated nanoparticles possess a single peak (Fig. 6.14A), while linked Au particle pairs (or larger aggregates) show two extinction maxima. With decrease in interparticle spacing, the short wavelength band becomes weaker whereas the intensity of the long wavelength band increases and shifts to longer wavelengths. The electrodynamic interaction between the nanoparticles is at a maximum when the particles touch each other and a maximum peak shift was observed. By varying the number of Au nanoparticles (diameter 40 nm), they calculated extinction spectra for line aggregates where the light polarization direction is oriented parallel (Fig. 6.14B) [73]. A bathochromic shift in λ_{max} was observed with increasing aggregate size. Various experimental methods such as electron beam lithography [74], laser trapping [75] and chemical functionalization [76] have been utilized by various groups for assembling metal nanoparticles.

By systematically varying the interparticle distances, Rechberger et al. have experimentally verified the surface plasmon interactions in pairs of identical nanoparticles having a particle diameter of 150 nm [74]. The particle pairs with varying interparticle distances of 150, 200, 300 and 450 nm were fabricated by electron beam lithography techniques. Representative scanning electron microscope (SEM) images of gold particle pairs are presented in Fig. 6.15A. The extinc-

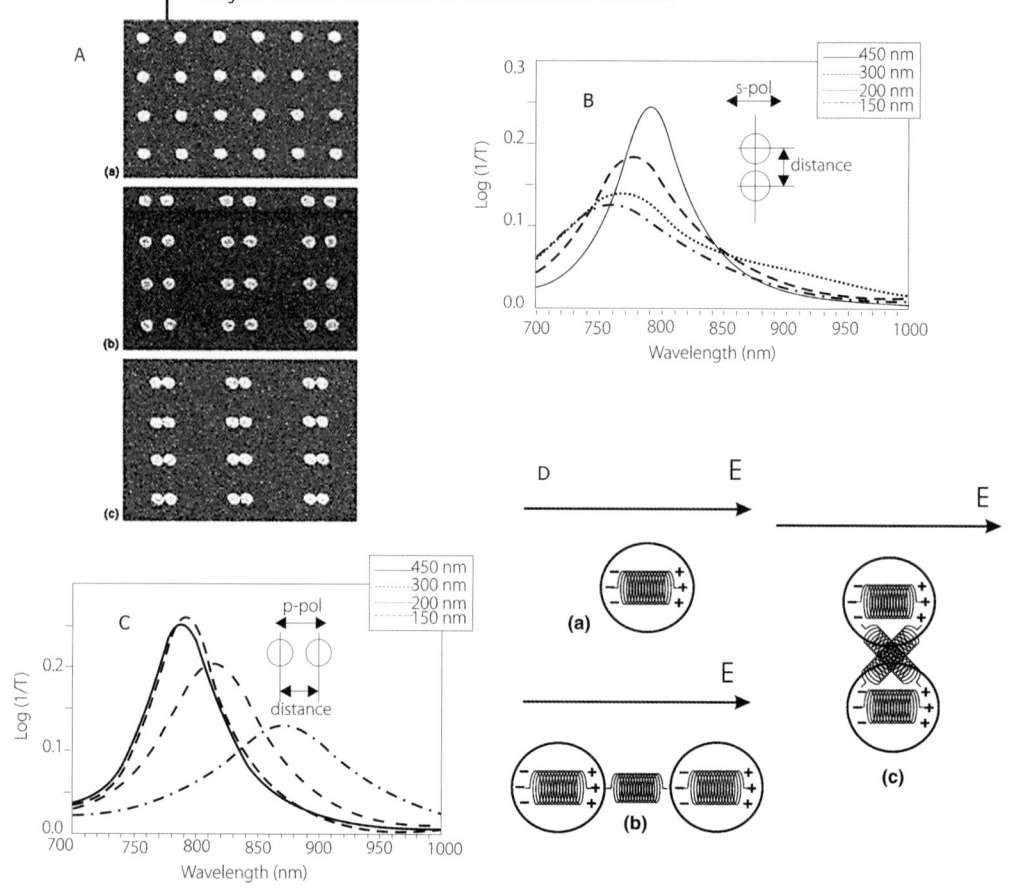

Fig. 6.15 (A) SEM images of particle pair samples with varying interparticle distance (center-to-center) of (a) 450 nm, (b) 300 nm and (c) 150 nm. (B)–(C) Extinction spectra of a 2D array of the Au nanoparticle pairs at different interparticle distances. The polarization direction of the exciting light is (B) parallel to the long particle pair axis and (C) orthogonal to it. (D) Illustratation of the electromagnetic interaction between closely spaced nanoparticles, (a) an isolated particle, (b) a pair of close particles with the polarization of the exciting field parallel to the long particle pair axis and (c) orthogonal to the long particle pair axis (reproduced with permission from Ref. [74], copyright 2003 Elsevier Science).

tion spectra at different interparticle distances were recorded using incident light, having polarization direction parallel and orthogonal to the long particle pair axis (Fig. 6.15B, C). When incident light having parallel polarization was used for excitation, a remarkable red shift of the surface plasmon extinction peak was observed with decrease in interparticle distance (Fig. 6.15B). In contrast, a distinct blue shift was observed with incident light having orthogonal polarization (Fig. 6.15C). The observed distance dependent peak shifts were qualitatively explained based on a simple dipole–dipole interaction model. When two particles are nearby, upon polarization, additional forces act on both particles, as illustrated in Fig. 6.15D. For example, if the driving field is parallel to the long particle pair axis, the repulsive forces within each particle are weakened due to attraction of two opposite charges on adjacent particles. As a result the resonance peak shifts to lower frequency. In contrast, when the driving field is normal to the long particle pair axis, the charge distributions of both particles act cooperatively to enhance the repulsive action in both particles, thus increasing the resonance frequency [74].

Another elegant approach to control the assembly of Au nanoparticles, in a reversible manner, has been demonstrated by Yoshikawa et al. [75]. They reported the optical trapping of Au nanoparticles by a focused laser beam which leads to the assembly of particles in the focal point. At lower laser power, the extinction spectrum possesses only a single band at around 530 nm corresponding to isolated Au nanoparticles. As the laser power increases, more compact gold nanoparticles assemblies were produced in the submicrometer optical cage. As a result an additional broad band was observed in the longer wavelength region (~660 nm) and its intensity increased with laser power. The spectral intensity ratio between the two bands (660 to 530 nm) was estimated for each spectrum and plotted against the laser power (Fig. 6.16) [75]. The spectral changes clearly indicate that the assembly of gold nanoparticles can be controlled reversibly and repeatedly by tuning the laser power.

Controlling the organization of metallic nanoparticles into 1D or 2D assemblies by adopting a chemical functionalizing route is more fascinating. Mann, Dujardin and coworkers have investigated the formation of a 1D array of Au nanoparticles by adopting a ligand exchange method [76]. The citrate stabilized spherical Au nanoparticles were exchanged with the ditopic molecule, namely 2-mercaptoethanol [$HS(CH_2)_2OH$]; MEA], resulting in the formation of chain-like superstructures. Time-dependent optical spectra were recorded at a fixed MEA/Au nanoparticle molar ratio (r = 5000:1) to investigate the mechanism of chain

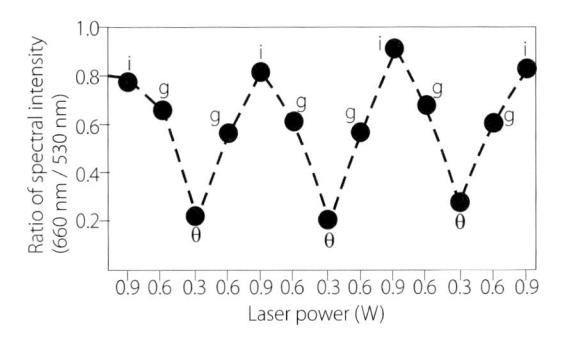

Fig. 6.16 Ratios of two bands (660 ± 5.6 to 530 ± 5.6 nm) plotted as a function of laser power. This procedure was repeated 12 times between 300 and 900 mW in steps of 300 mW (reproduced with permission from Ref. [75], copyright 2004 The American Physical Society).

self-assembly. A decrease in the intensity of the 520 nm band was observed along with the formation of a long wavelength band which progressively shifted to the red region with time (Fig. 6.17A) [76]. The corresponding TEM image, obtained immediately after addition of MEA, shows isolated nanoparticles. As reaction proceeds a variety of nanostructures, from short nanoparticle chains to extended networks of interconnected chains, were observed (Fig. 6.17B) [76]. These superstructures are quite stable and remained unchanged for prolonged incubation times (such as two weeks). The progressive formation of a low-energy longitudinal surface plasmon band is attributed to strong uniaxial coupling along the linear array.

6.4.2
Assembly of Nanorods

As mentioned in previous sections, closely packed 1D arrays of Au nanoparticles can, in principle, function as (i) guides of electromagnetic radiation (waveguides) allowing miniaturization of devices below the diffraction limit and (ii) interconnectors in optical and photonic devices. However, the isotropic nature of spherical Au nanoparticles prevents the selective binding of molecules on surfaces which restricts the possibility of designing 1D nanomaterials by chemical functionalization methods [21]. In contrast, the anisotropic features of Au nanorods allow their assembly in various orientations, and several attempts have been made to organize Au nanorods using electrostatic/supramolecular/covalent approaches [46, 47, 77–85]. This includes (i) the linear organization of Au nanorods using biotin-streptavidine connectors and lateral organization through electrostatic interactions by varying the pH of the medium [28, 79, 81]; (ii) longitudinal assembly through covalent functionalization by using α,ω-alkanedithiols [47], cooperative intermolecular hydrogen bonding by using 3-mercaptopropionic acid [77] and electrostatic interaction by using cysteine and glutathione [46]; (iii) end-to-end electrostatic assembly of Au nanorods on multiwall carbon nanotubes [78].

Using Au nanorods as specific examples, Gluodenis and Foss have predicated the effect of the interaction of nanorod pairs on the plasmon resonance spectra at different distances and orientation by using a simple quasistatic treatment [86]. According to this model, the axial interaction of Au nanorods may result in the formation of a red-shifted absorption band and lateral interactions lead to a blue shift of the longitudinal plasmon band. We have experimentally verified plasmon coupling in metal nanorods, by integrating them into ID assemblies, using chemical functionalization methods and these aspects are discussed below.

Interaction of bifunctional molecules such as 3-mercaptoproponic acid (MPA) and 11-mercaptoundecanoic acid (MUA) with Au nanorods (synthesized by a photochemical method) was investigated using absorption spectral studies and transmission electron microscopy [77]. As described in Section 6.2.2, Au nanorods possess two plasmon absorption bands; a short wavelength band at around 520 nm corresponding to the transverse mode of plasmon oscillation and a long wavelength band corresponding to the longitudinal mode of plasmon oscillation.

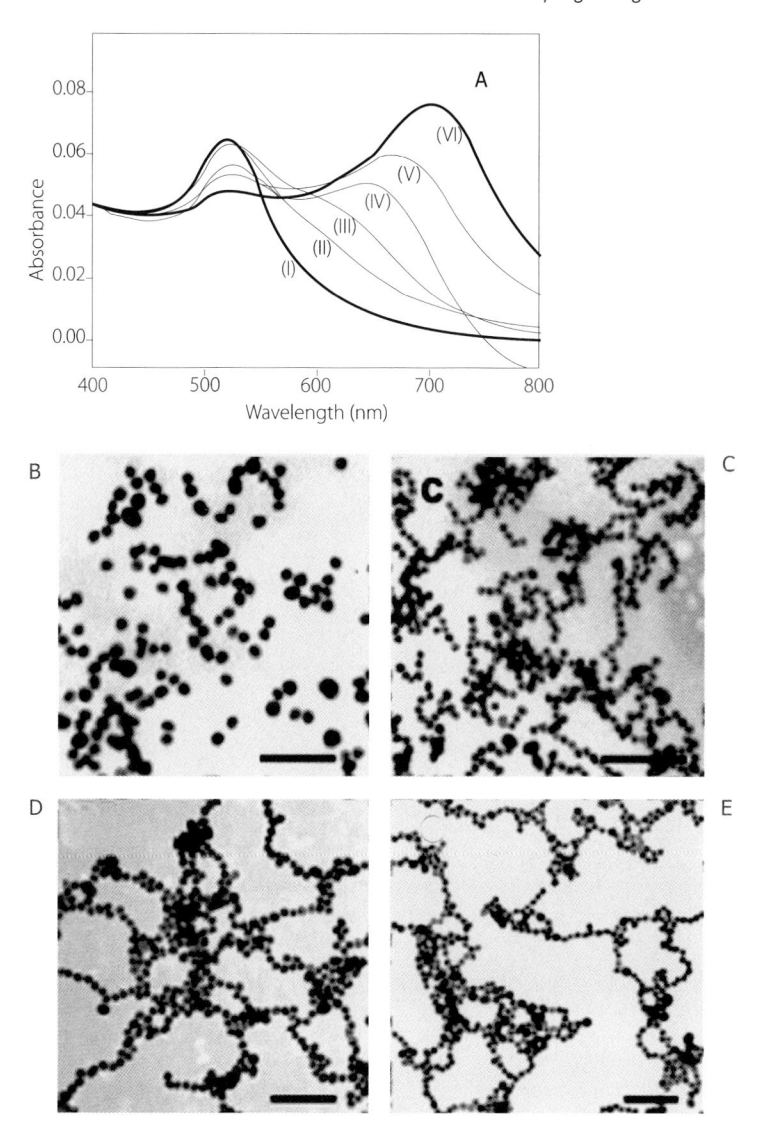

Fig. 6.17 (A) Time-dependent UV–vis spectra of Au nanoparticles recorded at various time intervals after addition of MEA at $r = 5000:1$. (i) 0, (ii) 3, (iii) 7, (iv) 24, (v) 48, and (vi) 72 h. (B)–(E) Corresponding TEM images taken after various time intervals, (B) 0 h, (C) 3 h, (D) 24 h, and (E) 14 days. Scale bars are 100 nm. (Reproduced with permission from Ref. [76], copyright 2005 Wiley-VCH).

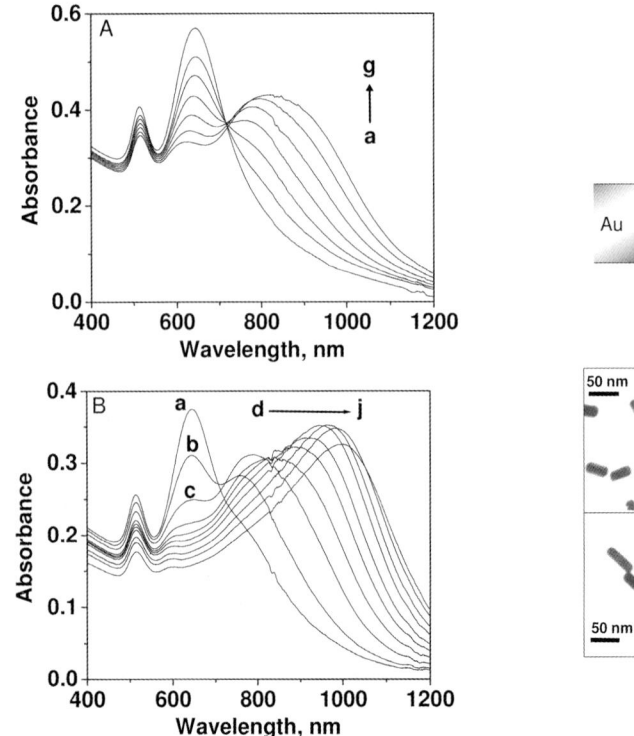

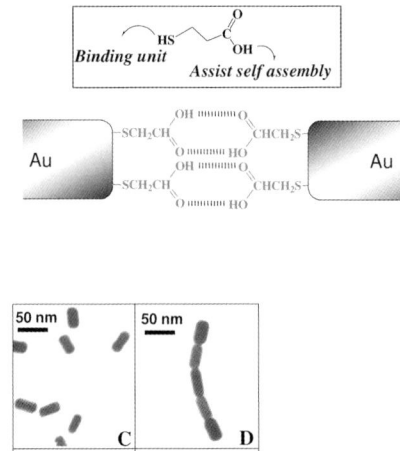

Fig. 6.18 (A) Absorption spectra of gold nanorods (aspect ratio 3) in acetonitrile–water (4:1) recorded immediately after the addition of MPA. [MPA]: (a) 0, (b) 3.6, (c) 4.5, (d) 5.4, (e) 6.3, (f) 7.2, and (g) 8.1 μM. (B) Absorption spectra of gold nanorods in acetonitrile–water (4:1) at different time intervals after addition of 15 μM MPA; (a) 0, (b) 4, (c) 8, (d) 12, (e) 16, (f) 20, (g) 30, (h) 45, (i) 60, and (j) 120 min. (C)–(F) TEM images of (C) Au nanorods in the absence of MPA. (D)–(F) are three separate examples of linearly assembled Au nanorods in the presence of MPA (reproduced with permission from Ref. [77], copyright 2004 American Chemical Society).

Addition of varying amounts of MPA/MUA led to a decrease in the intensity of the long wavelength band, accompanied by the formation of a new absorption band in the near-infrared region (Fig. 6.18A). The spectral interconversions observed through a clear isosbestic point indicate the existence of two different forms of nanorods in suspension. The newly formed absorption band at 800 nm slowly shifted toward longer wavelength with time (traces d to j in Fig. 6.18B) and this time-dependent bathochromic shift in the absorption band is more pronounced at higher concentrations of MPA/MUA. Interestingly, the short wavelength band of Au nanorods remains unaffected upon addition of MPA/MUA. TEM studies (Fig. 6.18C–F) showed that the Au nanorods were randomly distributed in the absence of MPA whereas in the presence of MPA, the rods were aligned in an end to end fashion. On the basis of high resolution transmission electron microscopy (HRTEM) studies, it has been recently reported that the edges of gold nanorods are dominated by [111] facets and the lateral sides by [100] and [110] facets [4, 28, 31]. Similar alignment of rods in an end to end fashion was earlier observed in the biotin-capped Au nanorod–streptavidin system [81]. As indicated in this study, the thiol groups preferentially bind to the [111] facets at the edges of the rod thus exposing the carboxylic group for further interactions (Fig. 6.18). The carboxyl groups appended at the ends of the Au nanorods dimerize through intermolecular hydrogen bonding, assisting the linear organization of Au nanorods. The linear assembly of Au nanorods results in the formation of a new red shifted band originating from a dipolar interaction. In the presence of external electromagnetic radiation, the charge distribution on the assembled Au nanorods get perturbed, resulting in an additional attractive force between the polarized negative and positive charges on adjacent rods. This additional attractive interaction in turn decreases the frequency of the longitudinal plasmon oscillation.

The aniosotropic surface binding properties of Au nanorods were further exploited for the selective detection of micromolar concentrations of cysteine and glutathione (γ-Glu-Cys-Gly) in the presence of other α-amino acids [46]. Selective detection of various α-amino acids/peptides is one of the major challenges in the field of chemical and biomedical sciences, mainly due to the structural similarity, incorporating both carboxylic and amino groups. Glutathione has very important biological functions: (i) keeping the cysteine thiol group in proteins in the reduced state and (ii) protecting the cells from oxidative stress (it traps the free radicals that damage DNA and RNA). Selective detection of glutathione using chemosensors is difficult due to the presence of several functional groups. Plasmon coupling of Au nanorods was observed in the presence of cysteine and glutathione through a two-point electrostatic interaction. The associated spectral changes in the intensity of the longitudinal surface plasmon absorption band were utilized for the selective detection of cysteine/glutathione (Fig. 6.19). One

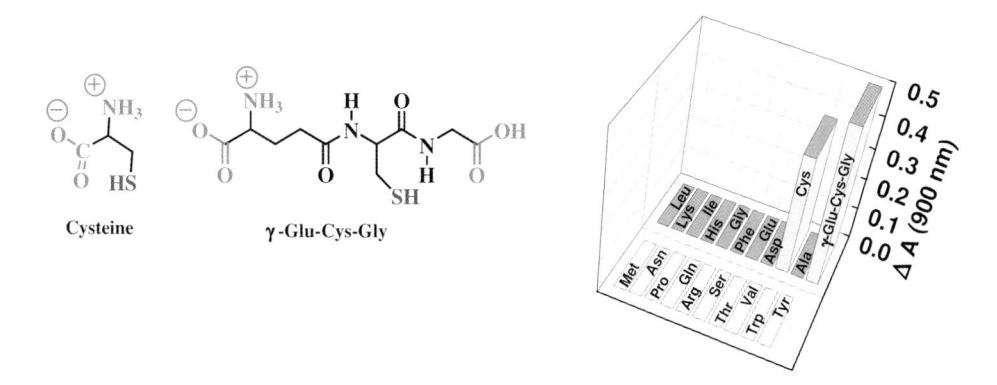

Fig. 6.19 3D plot showing the selectivity of cysteine (3 µM), glutathione (12 µM), and other α-amino acids (10 µM) (reproduced with permission from Ref. [46], copyright 2005 American Chemical Society).

of the significant features of the present system is the unique ability of Au nanorods to detect cysteine/glutathione in the presence of various other α-amino acids.

Mechanistic understanding of nanochain formation allows the incorporation of a desired number of Au nanorods in a nanochain, thereby tuning the optoelectronic properties of higher order nanostructures. With this viewpoint, the interaction of Au nanorods with a series of α,ω-alkanedithiols namely, 1,3-propanedithiol (**C3-DT**), 1,5-pentanedithiol (**C5-DT**), 1,6-hexanedithiol (**C6-DT**), 1,8-octanedithiol (**C8-DT**), and 1,9-nonanedithiol (**C9-DT**) were investigated. A decrease in the longitudinal plasmon absorption was observed along with a concomitant formation of a new red-shifted band above a critical concentration of dithiol (~0.6 µM for 0.12 nM Au nanorods), which is attributed to the plasmon coupling in assembled nanorods (Fig. 6.20). However, no noticeable spectral changes were observed below the critical concentration (Fig. 6.20A), and the TEM studies indicated that the nanorods remain isolated and randomly distributed. Two steps (Fig. 6.21) are involved in nanochain formation: an incubation step followed by the interlinking

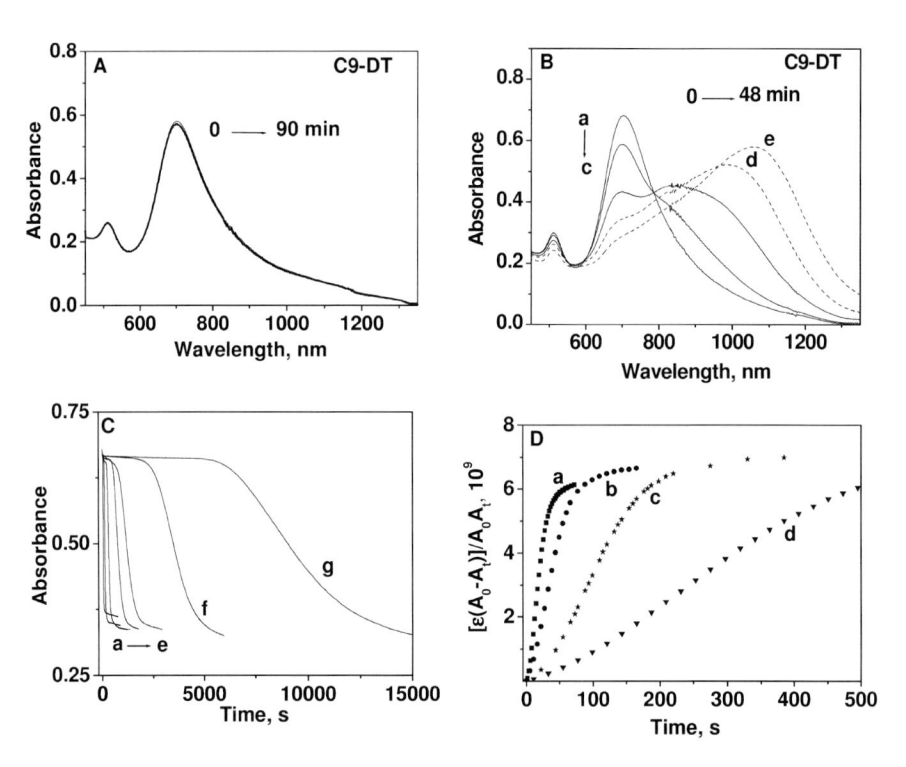

Fig. 6.20 Absorption spectral changes (A), (B) of Au nanorods having aspect ratio 2.9 [~0.12 nM] upon addition of **C9-DT**: (A) [0.5 µM], (B) [2.0 µM] at different time intervals. (C) absorption–time changes in the longitudinal absorption band of Au nanorods on addition of 2 µM of **C9-DT** at (a) 55, (b) 45, (c) 35, (d) 25, (e) 20, (f) 15 and (g) 10 °C, and (D) the kinetic plot of $[\varepsilon(A_0 - A_t)]/A_0 A_t$ versus time (reproduced with permission from Ref. [47], copyright 2006 American Chemical Society).

of nanorods. During the incubation period, one of the thiol groups of α,ω-alkanedithiols preferentially binds onto the edges of the nanorods leaving the other thiol group free. No spectral changes were observed during the incubation period which is analogous to the reaction between 1-alkylmercaptans and gold nanorods. Above the critical concentration, a chain up process proceeds through the interlocking of nanorods, initially to dimers and subsequently to oligomers, which results in longitudinal interplasmon coupling. The clear isosbestic point observed in the time-dependent absorption spectrum and dimers observed in the TEM monographs confirm the involvement of the dimerization step in the chain up process (Fig. 6.21). Time-dependent changes in the longitudinal absorption band of Au nanorods were investigated over a wide range of temperature (10–55 °C) and were analyzed for a second-order kinetic process (Fig. 6.20C, D). The linearity observed in the initial period further supports the dimerization mechanism,

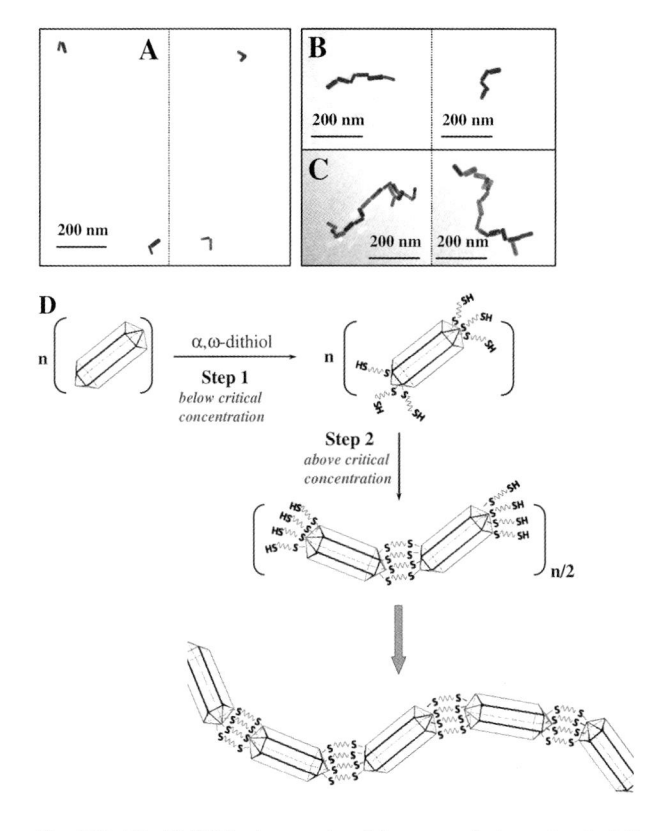

Fig. 6.21 (A)–(C) TEM micrographs of Au nanorods (aspect ratio 2.9) recorded in the presence of **C9-DT** [2.0 μM] at time intervals of (A) 500, (B) 1000 and (C) 1500 s. (D) Generalized scheme showing the stepwise formation of nanochains (reproduced with permission from Ref. [47], copyright 2006 American Chemical Society).

which deviates with time due to the contribution of other complex processes (Fig. 6.20D). The energies of activation for the dimerization of nanorods possessing **C9DT**, **C8DT**, and **C5DT** were estimated from the slope as 85.9, 80.2, and 33.8 kJ mol^{-1}, respectively. These results indicate that dimerization is more favored in the case of nanorods possessing **C5DT**. The large activation energy for dimerization further confirms that this step is not diffusion but activation controlled. To further characterize the surface interactions of α,ω-alkanedithiols on Au nanorods, Raman spectroscopic investigations were carried out. The FT-Raman spectrum of Au nanorods, recorded in the presence of **C9DT**, showed a new band at 475 cm^{-1} which is characteristic of a v_{Au-S} stretching mode.

Liz-Marzan and coworkers have reported the formation of strings of Au nanorods by using multiwall carbon nanotubes (MWNTs) as templates based on a polyelectrolyte layer-by-layer (LBL) approach [78]. The assembly is extremely uniform over long distances. At higher magnification, they observed a high degree of alignment of the nanorods, more or less in the form of stripes on opposite sides of the MWNTs, rather than uniformly covering the surface. It has been shown more recently that the potential decays more rapidly in areas with larger curvature (near the tips) for nanorods. This allows an end to end organization of nanorods rather than side-wise, resulting in a stringlike conformation. UV–vis spectra of the isolated nanorods with three different aspect ratios and the corresponding MWNT assembled nanorods are presented in Fig. 6.22. In all three cases the spectra measured after the assembly showed broader and red-shifted bands resulting from the uniaxial plasmon coupling between neighboring nanorods.

Gold nanoparticles have been used for the development of a wide range of biomolecular detection and extensive reviews on nanoparticle-based DNA and pro-

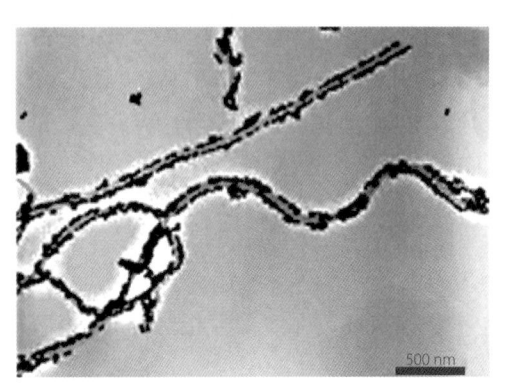

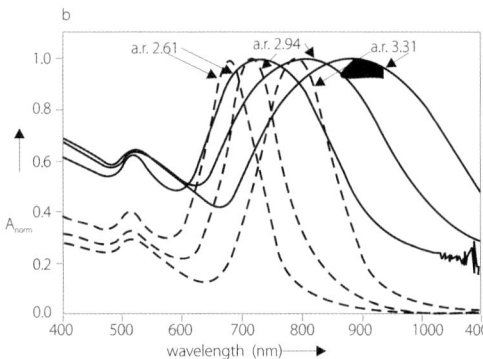

Fig. 6.22 (a) TEM images of Au nanorods (average aspect ratio 2.94), assembled on MWNTs (average diameter 30 nm) (b) UV–vis spectra of aqueous dispersions of individual Au nanorods (dashed lines) and nanorods attached on MWNTs (solid lines). The average aspect ratios (a.r.) of the nanorods are indicated (reproduced with permission from Ref. [78], copyright 2005 Wiley-VCH).

tein detection systems are available [87, 88]. Molecular rulers based on plasmon coupling of gold and silver nanoparticles have been reported recently by Alivisatos and coworkers [89, 90]. Coupling of localized surface plasmon resonances in noble metal nanoparticles offers a potential alternative to Forster resonance energy transfer for measuring nanometer-scale distances in biomolecular systems [91].

6.5 Summary and Outlook

Recent theoretical as well as experimental advances in the understanding of the surface plasmon behavior of nanostructured metal systems have provided several promising solutions for the confinement of light below the diffraction limit. These developments have led to several technological breakthroughs. The most noticeable of these are the surface plasmon-based photonic devices such as nanoscale waveguides and biosensors based on surface enhanced spectroscopy which allows the detection of single molecules. Newer synthetic strategies allowed the realization of numerous nanostructures having well defined size and shape. However our understanding of the mechanism of the growth process is still in its infancy and more studies are required. Such studies can provide guidance in the design of nanostructured materials with highly tunable optical properties.

Excitation of surface plasmons results in strong enhancements of the absorption, scattering, and local electric field around the metal particles. Also the magnitude of extinction and the relative contribution of scattering to the extinction can be fine tuned by structural modifications. Such unusual optical properties of surface plasmons may provide excellent opportunities in the biomedical field for diagnosis, imaging and therapeutics such as photothermal treatment of cancer cells. Over the past several years, exciting progress has been made in elucidating multipole resonance modes in metal nanoparticles. These results can provide fundamentally important correlations on the structure and optical properties of nanomaterials and possibly a benchmark spectrum for evaluating the quality as well as homogeneity of various nanomaterials [62].

One of the convenient methods for designing higher order nanostructured materials is through the stepwise integration of nanoscale building blocks. A brief summary of the recent developments in the hierarchical integration of nanospheres/nanorods into ID arrays is presented in Section 6.4. In general, such assemblies with well defined particle spacing result in selective plasmon coupling which can be further tuned by varying the distances and orientations. Approaches presented here for connecting nanorods into 1D arrays of desired chain length and thereby tuning their optoelectronic properties may have a wide-range of application in nanotechnology, particularly in nanoelectronics and plasmonics.

Acknowledgments

The author thanks the Department of Science and Technology, Government of India (SP/S5/NM-75/2002) and the Council of Scientific and Industrial Research (CMM 220239) for financial support and all his coworkers, whose names appear in the references cited in this article for their contributions. Special thanks are due to Professor M. V. George and Dr. Suresh Das (both at RRL, Trivandrum), Professor Prashant V. Kamat and Professor Dan Meisel (both at University of Notre Dame), Dr. Binil Itty Ipe and Dr. P. K. Sudeep (both former graduate students at RRL), and Mr. K. Yoosaf and Mr. P. Pramod (both graduate students at RRL) for their stimulating discussions on this topic at various stages. This is contribution No. RRLT-PPD-225 from the Regional Research Laboratory (Council of Scientific and Industrial Research), Trivandrum, India.

References

1 Barnes, W. L.; Dereux, A.; Ebbesen, T. W. *Nature* **2003**, *424*, 824.
2 Liz-Marzan, L. M. *Langmuir* **2006**, *22*, 32.
3 Kreibig, U.; Vollmer, M. *Optical Properties of Metal Clusters*, Springer-Verlag, New York, **1995**.
4 Burda, C.; Chen, X.; Narayanan, R.; El-Sayed, M. A. *Chem. Rev.* **2005**, *105*, 1025.
5 Daniel, M.-C.; Astruc, D. *Chem. Rev.* **2004**, *104*, 293.
6 Hutter, E.; Fendler, J. H. *Adv. Mater.* **2004**, *16*, 1685.
7 Rao, C. N. R.; Kulkarni, G. U.; Thomas, P. J.; Edwards, P. P. *Chem. Eur. J.* **2002**, *8*, 29.
8 Mulvaney, P. *Langmuir* **1996**, *12*, 788.
9 Xia, Y.; Halas, N. J. *MRS Bull.* **2005**, *30*, 338.
10 Ritchie, R. H. *Phys. Rev.* **1957**, *106*, 874.
11 Bozhevolnyi, S. I.; Volkov, V. S.; Devaux, E. S.; Laluet, J.-Y.; Ebbesen, T. W. *Nature* **2006**, *440*, 508.
12 Barnes, W. L.; Murray, W. A.; Dintinger, J.; Devaux, E.; Ebbesen, T. W. *Phys. Rev. Lett.* **2004**, *92*, 107401.
13 Ebbesen, T. W.; Lezec, H. J.; Ghaemi, H. F.; Thio, T.; Wolff, P. A. *Nature* **1998**, *391*, 667.
14 Dintinger, J.; Degiron, A.; Ebbesen, T. W. *MRS Bull.* **2005**, *30*, 381.
15 Girard, C.; Dujardin, E. *J. Opt. A: Pure Appl. Opt.* **2006**, *8*, S73.
16 Lim, J. K.; Imura, K.; Nagahara, T.; Kim, S. K.; Okamoto, H. *Chem. Phys. Lett.* **2005**, *412*, 41.
17 Krenn, J. R.; Dereux, A.; Weeber, J. C.; Bourillot, E.; Lacroute, Y.; Goudonnet, J. P.; Schider, G.; Gotschy, W.; Leitner, A.; Aussenegg, F. R.; Girard, C. *Phys. Rev. Lett.* **1999**, *82*, 2590.
18 Kelly, K. L.; Coronado, E.; Zhao, L. L.; Schatz, G. C. *J. Phys. Chem. B* **2003**, *107*, 668.
19 Mie, G. *Ann. Phys.* **1908**, *25*, 377.
20 Kreibig, U.; Schmitz, B.; Breuer, H. D. *Phys. Rev. B* **1987**, *36*, 5027.
21 Thomas, K. G.; Kamat, P. V. *Acc. Chem. Res.* **2003**, *36*, 888.
22 Shenhar, R.; Rotello, V. M. *Acc. Chem. Res.* **2003**, *36*, 549.
23 Link, S.; El-Sayed, M. A. *J. Phys. Chem. B* **1999**, *103*, 4212.
24 Jain, P. K.; Lee, K. S.; El-Sayed, I. H.; El-Sayed, M. A. *J. Phys. Chem. B* **2006**, *110*, 7238.
25 Hostetler, M. J.; Wingate, J. E.; Zhong, C.-J.; Harris, J. E.; Vachet, R. W.; Clark, M. R.; Londono, J. D.; Green, S. J.; Stokes, J. J.; Wignall, G. D.; Glish, G. L.; Porter, M. D.; Evans, N. D.; Murray, R. W. *Langmuir* **1998**, *14*, 17.

26 Templeton, A. C.; Pietron, J. J.; Murray, R. W.; Mulvaney, P. *J. Phys. Chem. B* **2000**, *104*, 564.
27 Wiley, B.; Sun, Y. G.; Mayers, B.; Xia, Y. N. *Chem. Eur. J.* **2005**, *11*, 454.
28 Murphy, C. J.; San, T. K.; Gole, A. M.; Orendorff, C. J.; Gao, J. X.; Gou, L.; Hunyadi, S. E.; Li, T. *J. Phys. Chem. B* **2005**, *109*, 13857.
29 Wiley, B.; Sun, Y. G.; Chen, J. Y.; Cang, H.; Li, Z. Y.; Li, X. D.; Xia, Y. N. *MRS Bull.* **2005**, *30*, 356.
30 Hao, E.; Schatz, G. C.; Hupp, J. T. *J. Fluorescence* **2004**, *14*, 331.
31 Perez-Juste, J.; Pastoriza-Santos, I.; Liz-Marzan, L. M.; Mulvaney, P. *Coord. Chem. Rev.* **2005**, *249*, 1870.
32 Maier, S. A.; Kik, P. G.; Atwater, H. A.; Meltzer, S.; Harel, E.; Koel, B. E.; Requicha, A. A. G. *Nat. Mater.* **2003**, *2*, 229.
33 Gans, R. *Ann. Phys.* **1912**, *37*, 881.
34 Esumi, K.; Matsuhisa, K.; Torigoe, K. *Langmuir* **1995**, *11*, 3285.
35 Foss Jr., C. A.; Hornyak, G. L.; Stockert, J. A.; Martin, C. R. *J. Phys. Chem.* **1992**, *96*, 7497.
36 Martin, C. A. *Chem. Mater.* **1996**, *8*, 1739.
37 van der Zande, B. M. I.; Böhmer, M. R.; Fokkink, L. G. J.; Schönenberger, C. *J. Phys. Chem. B* **1997**, *101*, 852.
38 Yu, Y. Y.; Chang, S. S.; Lee, C. L.; Wang, C. R. C. *J. Phys. Chem. B* **1997**, *101*, 6661.
39 Chang, S. S.; Shih, C. W.; Chen, C. D.; Lai, W. C.; Wang, C. R. C. *Langmuir* **1999**, *15*, 701.
40 Kim, F.; Song, J. H.; Yang, P. D. *J. Am. Chem. Soc.* **2002**, *124*, 14316.
41 van der Zande, B. M. I.; Koper, G. J. M.; Lekkerkerker, H. N. W. *J. Phys. Chem. B* **1999**, *103*, 5754.
42 Perez-Juste, J.; Rodriguez-Gonzalez, B.; Mulvaney, P.; Liz-Marzan, L. M. *Adv. Funct. Mater.* **2005**, *15*, 1065.
43 Link, S.; El-Sayed, M. A.; Mohamed, M. B. *J. Phys. Chem. B* **2005**, *109*, 10531.
44 Link, S.; Mohamed, M. B.; El-Sayed, M. A. *J. Phys. Chem. B* **1999**, *103*, 3073.
45 Nikoobakht, B.; Wang, J. P.; El-Sayed, M. A. *Chem. Phys. Lett.* **2002**, *366*, 17.
46 Sudeep, P. K.; Joseph, S. T. S.; Thomas, K. G. *J. Am. Chem. Soc* **2005**, *127*, 6516.
47 Joseph, S. T. S.; Ipe, B. I.; Pramod, P.; Thomas, K. G. *J. Phys. Chem. B* **2006**, *110*, 150.
48 Liao, H. W.; Hafner, J. H. *Chem. Mater.* **2005**, *17*, 4636.
49 Orendorff, C. J.; Murphy, C. J. *J. Phys. Chem. B* **2006**, *110*, 3990.
50 Thomas, K. G.; Zajicek, J.; Kamat, P. V. *Langmuir* **2002**, *18*, 3722.
51 Loo, C.; Lowery, A.; Halas, N.; West, J.; Drezek, R. *Nano Lett.* **2005**, *5*, 709.
52 Lal, S.; Westcott, S. L.; Taylor, R. N.; Jackson, J. B.; Nordlander, P.; Halas, N. J. *J. Phys. Chem. B.* **2002**, *106*, 5609.
53 Huang, X.; El-Sayed, I. H.; Qian, W.; El-Sayed, M. A. *J. Am. Chem. Soc.* **2006**, *128*, 2115.
54 El-Sayed, I. H.; Huang, X.; El-Sayed, M. A. *Cancer Lett.* **2006**, *239*, 129.
55 O'Neal, D. P.; Hirsch, L. R.; Halas, N. J.; Payne, J. D.; West, J. L. *Cancer Lett.* **2004**, *209*, 171.
56 Sherman, A. I.; Ter-Pogossian, M. *Cancer* **1953**, *6*, 1238.
57 Connor, E. E.; Mwamuka, J.; Gole, A.; Murphy, C. J.; Wyatt, M. D. *Small* **2005**, *1*, 325.
58 Weissleder, R. *Nat. Biotechnol.* **2001**, *19*, 316.
59 Evanoff, D. D.; Chumanov, G. *J. Phys. Chem. B* **2004**, *108*, 13957.
60 Kumbhar, A. S.; Kinnan, M. K.; Chumanov, G. *J. Am. Chem. Soc.* **2005**, *127*, 12444.
61 Evanoff Jr., D. D.; Chumanov, G. *ChemPhysChem* **2005**, *6*, 1221.
62 Payne, E. K.; Shuford, K. L.; Park, S.; Schatz, G. C.; Mirkin, C. A. *J. Phys. Chem. B* **2006**, *110*, 2150.
63 Evanoff, D. D.; Chumanov, G. *J. Phys. Chem. B* **2004**, *108*, 13948.
64 Millstone, J. E.; Park, S.; Shuford, K. L.; Qin, L.; Schatz, G. C.; Mirkin, C. A. *J. Am. Chem. Soc.* **2005**, *127*, 5312.
65 Oldenburg, S. J.; Jackson, J. B.; Westcott, S. L.; Halas, N. J. *Appl. Phys. Lett.* **1999**, *75*, 2897.

66 Yang, W.-H.; Schatz, G. C.; Van Duyne, R. P. *J. Chem. Phys.* **1995**, *103*, 869.
67 Mishchenko, M. I.; Travis, L. D.; Mackowski, D. W. *J. Quant. Spectrosc. Radiat. Transfer* **1996**, *55*, 535.
68 Ludwig, A. C. *Comput. Phys. Commun.* **1991**, *68*, 306.
69 Purcell, E. M.; Pennypacker, C. R. *Astrophys. J.* **1973**, *186*, 705.
70 Kottman, J. P.; Martin, O. J. F.; Smith, D. R.; Schultz, S. *Opt. Express* **2000**, *6*, 213.
71 Sosa, I. O.; Noguez, C.; Barrera, R. G. *J. Phys. Chem. B* **2003**, *107*, 6269.
72 Zhao, L.; Kelly, K. L.; Schatz, G. C. *J. Phys. Chem. B* **2003**, *107*, 7343.
73 Zhong, Z. Y.; Patskovskyy, S.; Bouvrette, P.; Luong, J. H. T.; Gedanken, A. *J. Phys. Chem. B* **2004**, *108*, 4046.
74 Rechberger, W.; Hohenau, A.; Leitner, A.; Krenn, J. R.; Lamprecht, B.; Aussenegg, F. R. *Opt. Commun.* **2003**, *220*, 137.
75 Yoshikawa, H.; Matsui, T.; Masuhara, H. *Phys. Rev. E* **2004**, *70*, 061406.
76 Lin, S.; Li, M.; Dujardin, E.; Girard, C.; Mann, S. *Adv. Mater.* **2005**, *17*, 2553.
77 Thomas, K. G.; Barazzouk, S.; Ipe, B. I.; Joseph, S. T. S.; Kamat, P. V. *J. Phys. Chem. B* **2004**, *108*, 13066.
78 Correa-Duarte, M. A.; Perez-Juste, J.; Sanchez-Iglesias, A.; Giersig, M.; Liz-Marzan, L. M. *Angew. Chem., Int. Ed.* **2005**, *44*, 4375.
79 Orendorff, C. J.; Hankins, P. L.; Murphy, C. J. *Langmuir* **2005**, *21*, 2022.
80 Sau, T. K.; Murphy, C. J. *Langmuir* **2005**, *21*, 2923.
81 Caswell, K. K.; Wilson, J. N.; Bunz, U. H. F.; Murphy, C. J. *J. Am. Chem. Soc.* **2003**, *125*, 13914.
82 Nikoobakht, B.; Wang, Z. L.; El-Sayed, M. A. *J. Phys. Chem. B* **2000**, *104*, 8635.
83 Mbindyo, J. K. N.; Mallouk, T. E.; Mattzela, J. B.; Kratochvilova, I.; Razavi, B.; Jackson, T. N.; Mayer, T. S. *J. Am. Chem. Soc.* **2002**, *124*, 4020.
84 Srivastava, S.; Frankamp, B. L.; Rotello, V. M. *Chem. Mater.* **2005**, *17*, 487.
85 Aizpurua, J.; Bryant, G. W.; Richter, L. J.; de Abajo, F. J. G.; Kelley, B. K.; Mallouk, T. *Phys. Rev. B* **2005**, *71*, 235420.
86 Gluodenis, M.; Foss, C. A. *J. Phys. Chem. B* **2002**, *106*, 9484.
87 Rosi, N. L.; Mirkin, C. A. *Chem. Rev.* **2005**, *105*, 1547.
88 Niemeyer, C. M. *Angew. Chem., Int. Ed.* **2001**, *40*, 4128.
89 Sonnichsen, C.; Reinhard, B. M.; Liphard, J.; Alivisatos, A. P. *Nat. Biotechnol.* **2005**, *23*, 741.
90 Reinhard, B. M.; Siu, M.; Agarwal, H.; Alivisatos, A. P.; Liphardt, J. *Nano Lett.* **2005**, *5*, 2246.
91 Thaxton, C. S.; Mirkin, C. A. *Nat. Biotechnol.* **2005**, *23*, 681.

7
Applications of Nanostructured Hybrid Materials for Supercapacitors

A. V. Murugan and K. Vijayamohanan

7.1
Overview

Nanostructured hybrid materials with both organic and inorganic components have attracted much attention recently due to the possibility of tailoring their dimensionality to facilitate a change in their fundamental properties including redox potential, conductivity and charge storage, in comparison with those of their bulk analogs. In the past, several nanostructured materials like carbon nanotubes (single-walled as well as multiwalled), ruthenium oxide functionalized carbon nanotubes and conducting polymer based nanostructured composites have been used to prepare excellent electrodes for electrochemical capacitors. Nevertheless, limited cycle life, high self-discharge and a large equivalent series resistance undermine their full utilization and new hybrid nanostructured materials are needed to solve some of these challenges. As a part of our continuing quest for preparing hybrid materials with novel or enhanced properties, we discuss here the present state of the rapidly emerging area of organic–inorganic nanostructured hybrids with regards to their synthesis, characterization, properties and applications as electrodes and electrolytes for supercapacitors.

7.2
Introduction

The emergence of nanotechnology as a unique and powerful interdisciplinary research activity with significant societal impact has affected almost all areas of science and technology, including electrochemical energy storage. For example, the uncanny ability to manipulate matter on the nanometer scale, has resulted in many remarkable materials with novel and significantly improved physical, chemical, mechanical and electrical properties compared to those of their bulk analogs [1]. Since, nanotechnology implies the control of matter at the atomic and molecular level, which requires working across the boundaries of classical disciplines, it is essential to work within a very basic scheme of the interdisciplinary character of nanotechnology realizing the overlap between nanotechnology and its related fields. This is especially significant since a strong hope is now put on nanotechnology, resembling a sort of dream in which we can, in principle, solve most of the daunting technological problems in diverse areas ranging from electronics, optoelectronics and photonics to energy storage, medicine and biology [1–4].

Nanostructures, defined as structures having at least one dimension in the range 1–100 nm, have attracted steadily growing interest for energy storage applications since our ability to generate such small structures has helped us to face many challenges in the design and development of materials for electrochemical energy storage, including batteries, fuel cells and supercapacitors. In recent years physicists and chemists have devoted increasing attention to these materials, and this interest is expected to increase further in the near future due to new properties acquired at this length scale depending on the size and shape [1–4].

The past few decades have witnessed an exponential growth in these materials and consequently it is almost impossible to provide a complete overview of the different types of nanostructured hybrid materials in any one chapter. Hence here we discuss the synthesis and characterization of nanostructured hybrid materials specially tailored for supercapacitor applications. We start with a broad definition of nanostrucured hybrid materials based on the dimensionality and electronic confinement effects, followed by the importance of these materials for electrochemical energy storage applications. Next we proceed with an elementary discussion of key criteria for selecting supercapacitor electrodes and we will also discuss the recent developments for both electrodes and electrolytes. Specific applications of these hybrid materials for rechargeable lithium batteries and fuel cells will not be extensively covered since excellent recent articles and reviews are available [5–8]. However, due to strong similarities of all these electrochemical systems, some overlap may be inevitable (for example, ionic liquids) since a few of the material parameters are fundamentally the same, even when we use them as conventional materials. The major limitations of currently used materials are highlighted with a view to suggesting the direction of research in this area to generate novel hybrid materials. We will conclude by pointing out the essential requirements for designing hybrid nanostructured materials for supercapacitor applications with enhanced performance.

7.3
Nanostructured Hybrid Materials

As commonly known there are different types of nanostructured materials ranging from zero-dimensional atomic clusters (quantum dots) to three-dimensional structures where, at least in one dimension, there is spatial quantum confinement facilitating size-dependent electronic properties. Consequently atoms, clusters, quantum dots and similar spatially confined molecular systems are called *zero-dimensionally modulated* (0D or more correctly quasi-zero dimensional) systems, which are spherical clusters with several thousands of atoms. If spatial confinement exists preferably in one dimension, wires, rods, belts and tubes in the nanometer range could be classified as *one-dimensionally modulated* (1D) systems (nanowires and nanotubes) and this classification provides a good method to investigate the dependence of electrical, thermal, optical and mechanical properties on the aspect ratio (that is, length divided by diameter) of these materials. Some of the established examples include size-dependent excitation or emission, quantized or ballistic conductance, coulomb blockade or single electron tunneling (SET) and metal–insulator transition and many of these properties play an important role when interconnects and active components are fabricated in nanoscale electronic, optoelectronic, electrochemical, and electromechanical devices [9–11].

If spatial confinement exists in a plane of nanometer thickness, these are called *two-dimensional layers* (2D), represented typically by many insertion materials such as graphite and layered-type dichalcogenides and oxides of the transition metals of Ti, Nb, Ta, Mo, W and V. Some of these materials are of great interest as positive and negative insertion electrode materials for electrochemical power sources. Several organic molecules and even some polymers can be organized in the interlayer gap to form many hybrid nanostructured materials, as discussed in detail later. Some of these materials, like MoS_2, could also function as solid lubricants because of their layered character [12]. The last class consists of *3D modulated* microstructures or a nanophase framework where cross-linked channels allow ion insertion if the size of the channels is sufficiently large to accommodate the ions. However, due to size-tunability, these materials are commonly considered as better model systems for investigating the dependence of electronic/ionic transport on dimensionality [1–4].

The above classification of nanostructured materials can be further broadened to encompass a large variety of organic and inorganic systems, either amorphous or crystalline, made of distinctly dissimilar components mixed at the nanometer scale. The term nanocomposite is thus used to indicate a distinct form of material which involves nanosized or molecular domain-sized particles embedded in an organic, polymeric, metallic or ceramic matrix. In all these cases, irrespective of the nature of the matrix, it is perceived that the intimate inclusion of these nanoparticles could completely change the intrinsic properties of these materials. In other words, the nanoparticles can serve as matrix reinforcement in order to change the physical properties of these base materials. With such a small amount of nanostructured material, a large amount of available interfacial area enables a complete transformation of the material's chemical, mechanical and morphological domain structure. Consequently much of today's research activity in the field of organic–inorganic hybrid nanocomposites is directed towards demonstrating mechanical and electrical properties superior to those of their separate components due to size confinement and dimensionality [13–23].

According to the nature of the organic–inorganic interface (that is, the nature of association between the inorganic and organic components) these nanocomposites can be classified into two types: one in which the inorganic nanoparticles are embedded in an organic matrix is called *"inorganic–organic nanocomposite"* (*IO materials*, focusing more on the inorganic materials) to denote hybrids where the

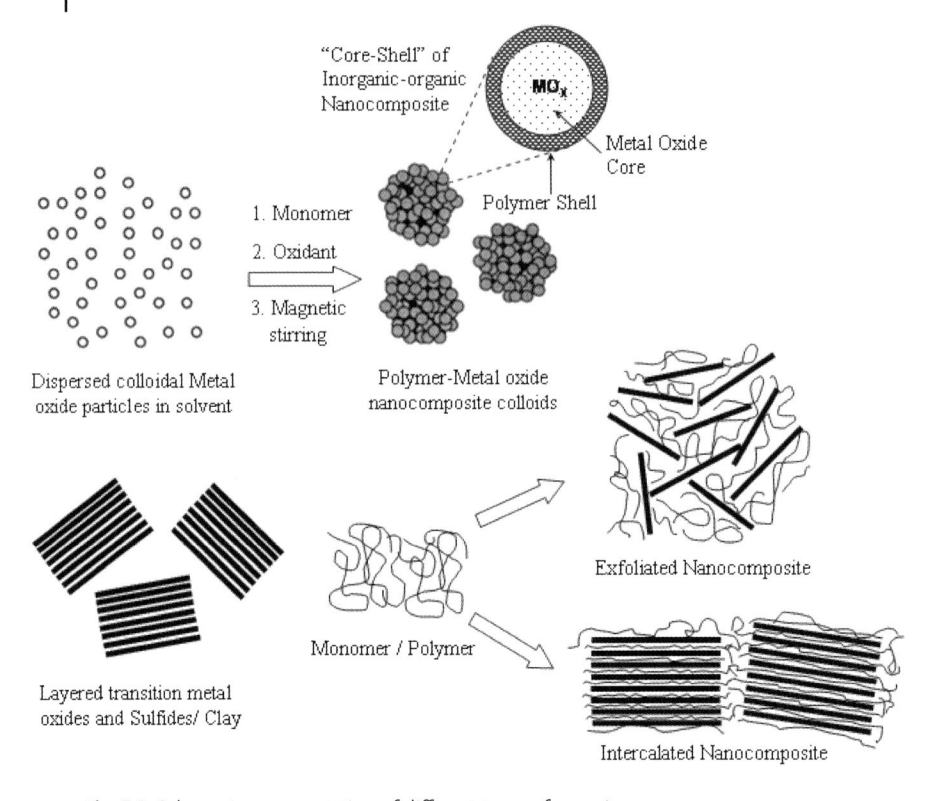

Fig. 7.1 Schematic representation of different types of organic–inorganic and inorganic–organic nanocomposites formation.

organic phase is host to an inorganic guest. The other type is *"Organic–inorganic nanocomposite"* (*OI materials*, focusing more on the organic materials) where the guest organic molecules are confined in inorganic host layers. However, in both cases the composite formation demands seamless blending or entrapment with compositional control at the interphase rather than simple blending or mixing and we will use the term hybrid nanomaterials to indicate useful properties arising from the presence of neither purely organic nor inorganic materials. Figure 7.1 provides a schematic representation of the different types of nanocomposites formed in this way. The synergistic behavior arising due to both anisotropy and the hierarchical nano-level organization are the primary reasons for their remarkable behavior [13–23].

7.4
Electrochemical Energy Storage

Electrochemical methods of energy storage possess several advantages over other modes of energy storage. For example, they are more efficient (since they are not subject to Carnot limitations), flexible and modular, as illustrated by many recent versions of microfuel cells and rechargeable lithium batteries [5–8]. Further, electrochemical energy generation and storage has become extremely important with increasing environmental pollution as it can decrease our dependence on limited fossil fuel reserves. As worldwide concern grows over fossil fuel usage, in terms of global warming, resource depletion, and other related issues there will be a progressive swing to the effective use of renewable energy sources. This will necessitate the development of improved methods of generating and storing electricity, from abundant resources without causing any environmental contamination. The rapid development of modern electronic and information technologies has created a strong demand for miniaturization, coupled with consumer need for lightweight portable electrochemical power sources, for a range of applications including communication devices, electric vehicles (EV), spaceships and pacemakers. Consequently, during the last few years, many developments have led to an entirely new class of nanostructured hybrid materials especially useful for electrochemical power sources [13–23].

Electrochemical power sources can be broadly classified in to three types: fuel cells, batteries and supercapacitors. Although many excellent hybrid nanostructured materials have been developed in the last decade for all these types of electrochemical power sources, we will focus only on supercapacitors since many innovative materials have been recently found to cause many dramatic performance related benefits. There have been many applications of hybrid nanomaterials in the area of both rechargeable lithium batteries and fuel cells, as electrocatalysts and other components to solve various difficulties and interested readers can look elsewhere, as already mentioned in the Introduction [13–27]. In the following section we discuss some essential features of supercapacitors before the actual survey of currently employed materials.

7.5
Electrochemical Capacitors

Electrochemical capacitors (ultracapacitors or supercapacitors), can act like static and passive electrical energy storage devices. However, they differ from conventional capacitors primarily in the mechanism by which they store energy. Supercapacitors hold their stored electrical charges in an ionic layer that forms at the interface between each of the two electrodes and a common electrolyte. For example, supercapacitors typically have energy densities that range from 300 times that of the largest conventional capacitors, to approximately one tenth of that of the lowest density batteries. However, their power densities are typically 10 times above those of most batteries. A comparison of power density of a supercapacitor with other different types of commonly used electrochemical power sources with respect to their ability to deliver energy and power is generally effected using a Ragone plot, as shown in Fig. 7.2 [27]. This clearly demonstrates that by offering high power and energy density coupled with low equivalent series resistance (ESR), a supercapacitor acts as a remarkable intermediate power source (a bridge

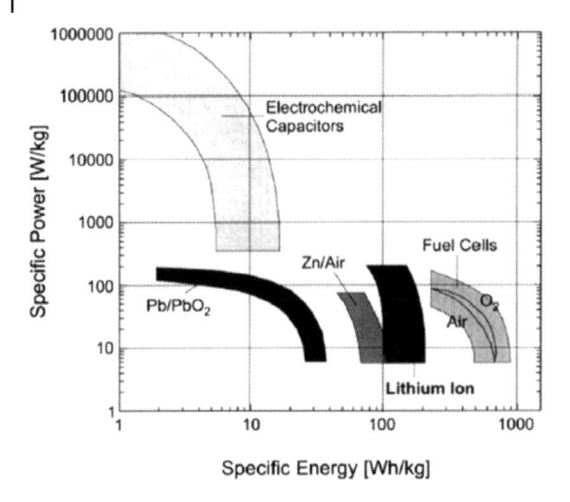

Fig. 7.2 Ragone plot of some of the electrochemical power sources such as fuel cells, batteries and supercapacitors.

between batteries and conventional capacitors) that features the best characteristics of both of these. Thus a strategically placed supercapacitor can prevent battery stress by supplying the peak power demands for specific applications, as illustrated in a schematic comparison of the charge–discharge profiles of a battery and a supercapacitor in Fig. 7.3. Due to the complementary aspects of power storage/retrieval, their combined application often allows a smaller capacity battery to be used, thus saving valuable weight and volume for a given application whilst extending the overall life span. In other words, together with a fuel cell or battery, a supercapacitor pack can be used as a hybrid power device for many critical applications. For instance, in the case of automotive applications, this can

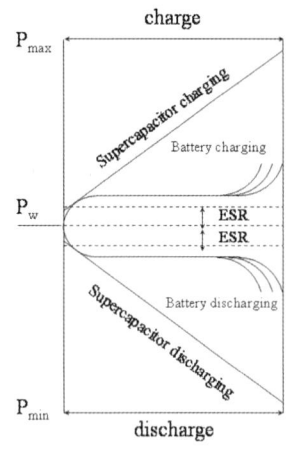

Fig. 7.3 A schematic comparison of the galvanostatic charge–discharge profiles of a supercapacitor and a battery; P_w represents the operating voltage of the supercapacitor akin to the open-circuit voltage of a storage battery, P_{max} and P_{min} represent the end-of-charge and end-of-discharge, respectively, and ESR is the equivalent series resistance.

supply the extra power needed for acceleration, concomitantly storing the energy during braking, which obviously could be reused in the next acceleration process, thus enabling fast regeneration for hybrid electric vehicle applications. Since their power density is higher than that of batteries and fuel cells due to their light weight and use of less space, important benefits are apparent for load leveling applications in addition to their excellent cold weather starting and increased cycle life [28–32]. There are also other important applications as back-up memory for consumer electronics, defibrillators for biomedical applications, power beaming in space, electric guns and high power lasers in military operations, where different types of design criteria are employed to match the required performance specifications. Consequently, in recent years, the practical realization of all these advantages have led to the development of many supercapacitors which can complement the power generation and storage characteristics of many advanced rechargeable batteries and fuel cells [33–41]. An even higher demand for the development of high specific energy supercapacitors or ultracapacitors is expected with technologies emerging today at a rapid pace.

7.5.1
Electrochemical Double Layer Capacitor vs. Conventional Capacitor

Electric charge is normally stored by keeping many free electrons (negative charge) in close proximity to materials that lack electrons (positive charge) as per the principles of electrostatics. In a conventional capacitor, this is accomplished by placing an insulator (called a dielectric) between two conductive surfaces or "plates". When a difference of electrical potential is applied to the plates, electrons will flow from the negative polarity side of the sources onto the plate to which it is connected. On the other side of the dielectric, electrons are repelled by the accumulated electrons on the negative plate and are driven toward the positive polarity side of the source until the full potential of the sources exists across the plates of the capacitor. Thus conventional capacitors store electrical energy by charge separation in a thin layer of dielectric material that is supported by metal plates that can also act as the terminals for the device. The energy stored in a capacitor is given by $\frac{1}{2}CV^2$, where C is its capacitance (in F) and V is the voltage between the terminal plates. The maximum voltage of the capacitor is dependent on the breakdown characteristics of the dielectric material. The charge Q (in Coulombs) stored in the capacitor is given by CV and the capacitance depends critically on the dielectric constant (ε), the thickness (t) of the dielectric and its geometric area (A) as expressed by, $C = \varepsilon A/4\pi t$ [28, 30]. Based on the preparation methods and materials used, conventional capacitors can be further classified into multilayered ceramic capacitors and electrolytic capacitors. Both are exquisite marvels of modern design and materials engineering although the former depends more on the modulation of ceramic composition to accomplish high values of dielectric constant [41].

An electrochemical double layer capacitor (EDLC) or supercapacitor comprises two "electrodes" much like those of a battery, which are immersed in or impreg-

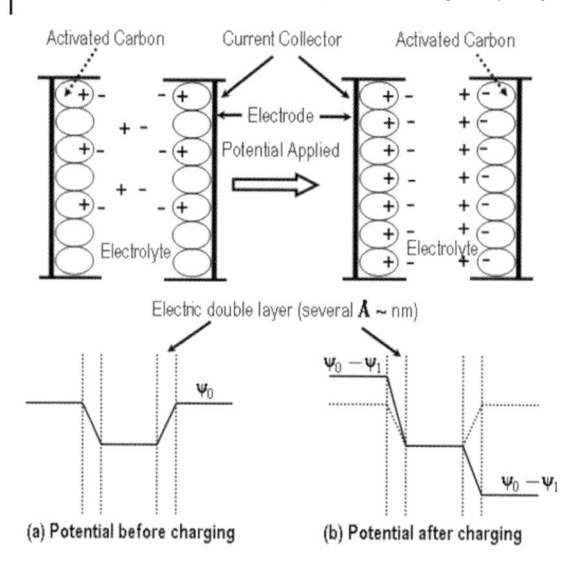

Fig. 7.4 A simple model of an electrochemical double-layer capacitor, where the charge storage occurs when a difference of electrical potential is applied to the two electrodes, essentially through the changes at the electrode/electrolyte interface.

nated with an electrolyte. However, a supercapacitor can have identical electrodes with respect to chemical composition, unlike those of batteries, charge storage occurs when a difference of electrical potential is applied to the two electrodes, essentially through the changes at the electrode/electrolyte interface (Fig. 7.4). Although a comprehensive discussion on their further classification and the mechanism of charge storage at the interface with sufficient electrochemistry background is available in a recent book by Conwey [28], it is essential to discuss the origin of enhanced capacitance since the organic part of many hybrid materials contributes especially to enhanced charge storage by virtue of its strong adsorption tendency.

7.5.2
Origin of Enhanced Capacitance

When an electronic conductor (metal) is brought into contact with a solid or liquid electrolyte, charge accumulation is achieved electrostatically on either side of the interface, leading to the development of an electrical double-layer which is essentially a molecular dielectric. When no charge transfer takes place across the interface, the current observed during this process is essentially a displacement current (non-faradaic current) due to the rearrangement of charges (conventionally described as an ideally polarized electrode). However, in some materials, charge storage could also occur by electron transfer that produces tangible oxidation state changes in the materials at particular potentials (the so-called ideally reversible electrode) as dictated by Faraday's laws (faradaic processes). Accordingly, electrochemical capacitors (ECCs) can be classified as electric double layer capacitors (EDLCs) or pseudocapacitors (or redox capacitors) depending on the nature of these interfacial processes as represented schematically in Fig. 7.5 [14]. The former has a capacitance associated with the non-faradaic charging (generally 10–30 µF cm^{-2} for Pt, Au electrodes) and discharging of the electrical double layer at the electrode–electrolyte interface and is hence called an electrical double layer capacitor (EDLCs), while the latter has faradaic current through surface redox reactions or adsorption-induced pseudocapacitance as the origin [31]. In pseudocapacitors, the non-faradaic double-layer charging process is usually accompanied by a faradaic charge transfer. Accordingly, the capacitance (C) of supercapacitance is given by

$$C = C_{dl} + C_\phi \tag{7.1}$$

where C_{dl} is the electrical double-layer capacitance and C_ϕ denotes pseudocapacitance. Double-layer capacitance (C_{dl}) is expressed as

$$C_{dl} = \varepsilon A / 4\pi t \tag{7.2}$$

where, ε is the dielectric constant 'A' is the surface area of the electrode, and 't' is the electrical double-layer thickness. It is obvious from Eq. (7.2) that for a large C_{dl} it would be mandatory to produce a thin, high surface-area electrical double-layer and it is this unique combination of high surface area (e.g. 2000 m^2 g^{-1}) with extremely small charge separation ($t \sim 10$ Å) which is responsible for the origin of the high value of the double layer capacitance for many hybrid materials. In contrast, the best type of ceramic multilayer capacitors use a combination of high ε (6500 to 10000) and low t (few µm), but has relatively lower capacitance values.

The pseudocapacitance (C_ϕ) is brought about by a surface redox-reaction of the type

$$O_{ad} + ne \rightarrow R_{ad} \tag{7.3}$$

where O_{ad} and R_{ad} are the adsorbed oxidants and reductants respectively, and n refers to the number of electrons (e) involved in the process. The special aspects of pseudocapacitance arising due to the process of the adsorption–desorption phenomenon of O_{ad}/R_{ad} on the electrode process can be understood by considering a simple model where the potential varies systematically with the coverage. According to Frumkin, for an electrical double-layer, the charge on the metal surface (q_m) could be express as:

$$q_m = q_0(1 - \theta) + q_1\theta \tag{7.4}$$

where q_1 is the charge associated with the coverage (θ) of the adsorbed species, and q_0 is the charge associated with the bare surface $(1 - \theta)$ [31]. It is estimated that the energy ($E = \frac{1}{2}CV^2$) stored in a 1 V electrical double-layer capacitor with porous carbon electrodes having surface area of 1000 m² g^{-1} is ∼28 W h kg^{-1}. For comparison, a 1 V supercapacitor with a pseudo-capacitance ($\theta = 0.2$, $n = 1$ and $g = 0$) held between carbon electrodes of a similar surface area could store as large an energy as ∼1876 W h kg^{-1}. In practice, however, it has not been possible to achieve these energy densities with any of the supercapacitors developed so far due to various practical limitations associated with adsorption such as adorbate–adsorbate interactions, solvation, local variation of ε etc.

Electrons involved in the electrical double-layer charging are the itinerant conduction-band electrons of the metal or carbon electrode, while the electrons involved in the faradaic processes are transferred to or from the valence-electron states (orbitals) of the redox active cathode or anode materials. The electrons may, however, arrive in or depart from the conduction-band states of the electronically-conducting support material, depending on whether the Fermi level in the electronically-conducting support lies below the highest occupied state (HOMO) of the reductant or above the lowest unoccupied state (LUMO) of the oxidant. This is the most important parameter which could be profitably used for designing hybrid material electrodes since the structure can be controlled in relation to the potential window of the electrolytes used. The HOMO–LUMO gap can indeed be tuned by doping the polymer matrix with a variety of ions and this

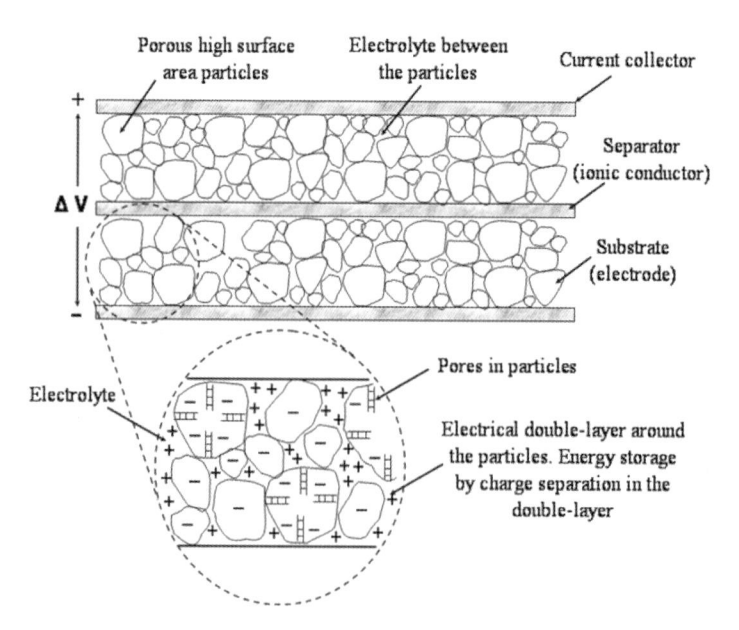

Fig. 7.5 A schematic representation of a supercapacitor where the energy storage occurs by charge separation in the double layer formed at the porous high surface area electrode/electrolyte interface.

approach has been demonstrated to be very valuable for designing many novel hybrid materials.

7.6 Electrode Materials for Supercapacitors

Like many other electrochemical devices, a supercapacitor mainly consists of three components: electrodes, electrolyte and the separator. We will not discuss the materials for separators since their selection is important only for liquid electrolytes to prevent their tendency for shorting. Electrolyte on the other hand, is an important component where one has to select from a diverse range of materials including liquid electrolytes, solid electrolytes and polymer electrolytes. More significant is the choice of electrodes as this plays a critical role in controlling cycle life, long-term stability, high surface areas, and resistance to electrochemical oxidation/reduction. The focus seems to be, however, on achieving high surface areas with low 'matrix' resistivity. Accordingly, electrochemical supercapacitors make use of three main classes of materials: (i) transition metal oxides (ii) electronically conducting polymers (iii) carbon nanotubes and related carbonaceous materials.

7.6.1 Nanostructured Transition Metal Oxides

The supercapacitive behavior of several transition metal oxides, such as RuO_2, IrO_2, NiO_x and manganese oxide has already been evaluated [42–46] several decades ago. Ruthenium oxide has been found to be one of the best materials for making supercapacitor electrodes and making nanoparticles of hydrous ruthenium oxide ($RuO_2 \cdot xH_2O$) with different water contents using the sol–gel method after annealing at different temperatures has provided many advantages. For example, hydrous ruthenium oxide $RuO_x(OH)_y$ has been recognized as one of the most promising candidates for electrodes as it can store charge by reversibly accepting and donating protons from an aqueous electrolyte. This process is governed by a potential-dependent equilibrium, as represented by:

$$RuO_x(OH)_y + \delta H^+ + \delta e^- \rightarrow RuO_x - \delta(OH)_{y+\delta} \qquad (7.5)$$

where $RuO_x(OH)_y$ is a mixed electronic–protonic conductor. Interestingly, its electrochemical properties depend on the amount of water incorporated in its structure and the change in oxidation state (Ru^{4+}/Ru^{3+}) of ruthenium, which is actually responsible for the capacitance.

Apart from providing the maximum specific capacitance of 770 F g^{-1}, use of ruthenium oxide also allows one to tune the electrochemical and physical properties, morphologies, crystalline structures and proton dynamics in $RuO_2 \cdot xH_2O$ for high-performance ultra-capacitors [28]. However, hydrous ruthenium oxide is a very expensive material and hence much effort is being made worldwide to replace ruthenium oxide by other nanostructured transition metal oxides. For example, many low-cost nanostructured transition metal oxides such as MnO_2, Fe_3O_4 and V_2O_5 have been recently developed as active electrode materials for aqueous electrochemical supercapacitors by many groups [47]. In particular, MnO_2 appears to be a promising material, compared to other materials like Fe_3O_4 and V_2O_5, in electrolytes such as 0.1 M K_2SO_4 due to its abundance and superior electrochemical properties, although the specific capacitance of the V_2O_5 electrode varies significantly after a few hundred cycles. For example, amorphous nanostructured manganese oxide, electrochemically deposited onto a stainless-steel electrode shows about 410 F g^{-1} specific capacitance along with a specific power of ca. 54 kW kg^{-1} at 400 mV s^{-1} sweep rate under potentiodynamic conditions [48]. Cycle life data for this also shows remarkable stability for the specific capacitance, even up to 10 000 cycles. Similarly, amorphous nanostructured tin oxide (SnO_2) prepared by potentiodynamic deposition onto inexpensive stainless-steel electrodes shows a maximum specific capacitance of 285 F g^{-1} at a scan rate of 10 mV s^{-1} [49]. Rutile-type $Ru_{1-x}V_xO_2$ nanoparticles with high surface area also exhibit extremely high capacitance compared to that of pure RuO_2 [50]. Other oxides like IrO_2 and NiO_x have also been prepared in nanostructured form by many other strategies including the commonly employed method of dispersion of transition metal oxides in other less expensive oxide/carbon matrices and in conducting polymers.

7.6.2
Nanostructured Conducting Polymers

The use of conducting polymers as active materials for supercapacitors, as illustrated by electrodes made up of polyaniline, polypyrrole and polythiophenes, demonstrates many advantages including fast doping–dedoping during charge–discharge, high charge density, easy chemical/electrochemical synthesis and low cost as compared to that of many noble metal oxides [51–55]. According to Rudge et al., there are three possible schemes by which conducting polymers can be used as electrodes in redox supercapacitors [53, 56]. In type I capacitors, two identical electrodes comprising symmetrical p-doped conducting polymer films are used, whereas two different p-doped conducting polymers are used in type II capacitors. Type II capacitors can be charged to a comparatively higher voltage by virtue of the difference in the potential range over which conducting polymers can be doped and de-doped. A mixture of both n- and p-doped conducting polymers can be used in type III capacitors and this can achieve the highest possible voltage (typically about 3 V) in the charged state. Since conducting polymers are generally permeable to small molecules [57] every volume element of the electrode material at the molecular level is in contact with the electrolyte solution, resulting in an extremely high effective surface area. However, due to their low compatibility with the electrolyte phase, the separation of the polymer chains in conducting polymers is generally small relative to the double layer thickness, causing a high electrolyte resistance in the polymer matrix [57] and a low effec-

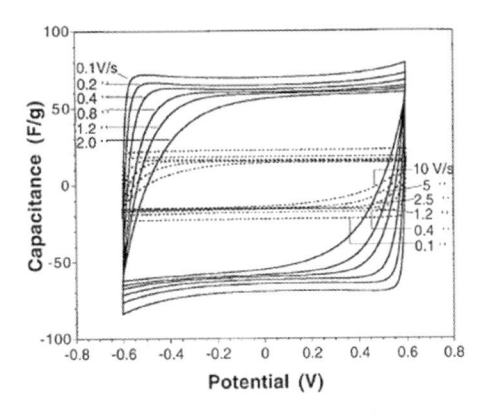

Fig. 7.6 Cyclic voltammograms at different scan rates for symmetrical two-electrode supercapacitors fabricated by using PEDOT–PSS/PPy electrodes and PEDOT–PSS electrodes in 1 M Na_2SO_4 aqueous electrolyte as per reference (adapted from Ref. [33]).

tive surface area. Compatibility of conducting polymers has, however, been improved in recent times using dispersions/solutions [51, 58, 59]. However, most of these conducting polymers lack adequate chemical stability due to their high propensity for degradation and consequently the cycle life of supercapacitor electrodes made from them is not very high [60]. Many of the conducting polymers used for fabricating supercapacitor electrodes are not nanostructured materials and hence we will not discuss them separately here. However, there are many exceptions, like the recent case of poly(3,4-ethylenedioxythiophene)-poly (styrenesulfonate) (PEDOT-PSS), which has been ionically crosslinked into a nanometer-scale conducting hydrogel with a diameter of 40–80 nm that shows high power density [33]. For example cyclic voltammograms at different scan rates for symmetrical two electrode supercapacitor cells with PEDOT-PSS/PPy electrodes and PEDOT-PSS electrodes in 1 M Na_2SO_4 aqueous electrolyte as shown in Fig. 7.6 clearly illustrate the superior performance. Similarly, polyaniline nanowires electrochemically deposited on stainless-steel electrodes at a potential of 0.75 V vs. SCE in 1 M H_2SO_4 electrolyte have been found recently to show 775 F g^{-1} at 10 mV s^{-1} along with good cycle life, suggesting the potential implications of these nanostructured conducting polymers for fabricating high performance supercapacitor electrodes [58].

7.6.3
Carbon Nanotubes and Related Carbonaceous Materials

Nanostructured carbononaceous materials including carbon nanotubes (CNT) represent an interesting class of materials for supercapacitor applications and many research groups have recently generated exciting results using these materials. For example, utilization of a matrix of vertically aligned CNTs as electrode structure by the MIT group has been found to generate power densities greater

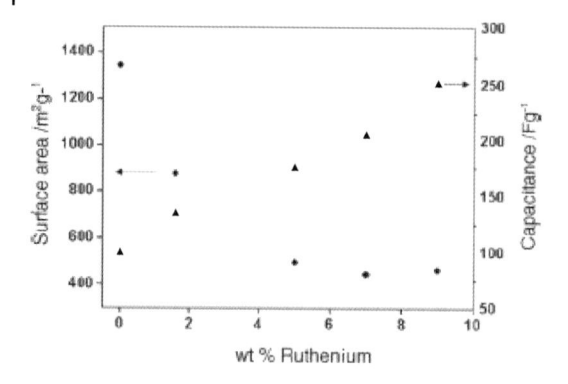

Fig. 7.7 Variation of surface area vs. specific capacitance with respect to wt% of ruthenium in coconut-shell-based activated carbon and [$RuO_x(OH)_y$] nanocomposite samples, adapted from Ref. [59].

than 100 kW kg^{-1}, a life time longer than 300 000 cycles and an energy density higher than 60 W h kg^{-1} [61].

Carbonaceous materials have been particularly popular for making electrodes for many types of batteries, fuel cells and ultracapacitors owing to their large surface area and high conductivity. Most of these carbonaceous materials are derived from precursors such as coconut shells, wood powders, coal tar, resins, resorcinol-formaldehyde and related polymers which yield active electrode materials with surface areas ranging between 1000 and 2000 m^2 g^{-1}, resulting in capacities as high as 500 F g^{-1} in alkaline electrolytes [62, 64]. More significantly, the surface of these materials can be functionalized by a variety of methods to tailor their properties and some of these surface-functionalized carbons shows surface areas up to even 3000 m^2 g^{-1}. Some, like activated carbon, carbon-black, carbon aerogel and carbon nanofibers are however, inexpensive and provide improved performance as supercapacitor electrodes. However their capacitance may not be very high as high surface area does not always mean high capacitance due to the intricate involvement of pore-size distribution [28, 65]. Figure 7.7 for example, shows the pore-size distribution with respect to capacitance variation for carbonaceous materials [65]. The variation of both specific surface area (m^2 g^{-1}) and specific capacitance of the composite with various percentages of ruthenium is also given in Fig. 7.7. Although activated carbon has a surface area of 1340 m^2 g^{-1}, the composite shows only 460 m^2 g^{-1} after impregnation with 9 wt.% ruthenium. This decreasing trend is not surprising as several pores are likely to be filled by hydrous ruthenium oxide, as illustrated by the data listed in Table 7.1 describing the variation in surface area, pore volume and pore-size distribution in activated carbon and the composite samples as a function of the amount of ruthenium.

Carbon nanotubes (CNTs) are attractive materials for electrodes of electrochemical supercapacitors due to their unique characteristics of chemical stability, low mass density, low resistivity and large surface area, as demonstrated in the SEM images of CNTs (Fig. 7.8). [66–68]. For example, CNT electrodes with a surface

Table 7.1 A comparison of surface area, porosity and electrochemical capacitance data for various amounts of $RuO_x.(OH)_y$ incorporated in activated carbon (adapted from Ref. [59]).

Amount of ruthenium (wt.%)	BET surface area ($m^2\ g^{-1}$)	Mesoporous area ($m^2\ g^{-1}$)	Pore volume ($cm^3\ g^{-1}$)	Average pore diameter (Å)	Capacitance ($F\ g^{-1}$)
0	1340	78	0.52	18	100
1.6	878	92	0.42	19	135
5	498	42	0.27	21	176
7	446	18	0.19	17	205
9	460	29	0.20	17	250

Fig. 7.8 Scanning electron microscopy images of carbon nanotubes grown directly on the graphite foil before using as supercapacitor electrodes. (a) low magnification to show the uniformity with respect to large area and length; (b) high magnification to show the diameter.

Table 7.2 Specific capacitance of different types of carbonaceous materials including single-walled carbon nanotube (SWCNT) and multi-walled carbon nanotube (MWCNT) as supercapacitor electrodes in various electrolytes.

Carbonaceous materials	Electrolyte	Capacitance/F g^{-1}	References
carbon cloth	KOH organic	200 100	28
coconut shells, activated carbon	H_2SO_4	100	59
carbon nanofibers	H_2SO_4	100	76
carbon aerogel	KOH	140	28
carbon black	KOH	95	28
single-walled carbon nanotube (SWCNT)	H_2SO_4	40–180	61, 83
Multi-walled carbon nanotube (MWCNT)	H_2SO_4	10–100	83, 84

area of 430 m^2 g^{-1} show a specific capacitance of 110 F g^{-1} and a power density of 8 kW kg^{-1} (at an energy density of 0.56 W h kg^{-1}) in a solution of 38 wt.% H_2SO_4 [66]. Interestingly, heat treatment at high temperature improves the specific capacitance of single-walled CNT to 180 F g^{-1} with a larger power density of 20 kW kg^{-1} (and energy density of 6.5 W h kg^{-1}). Although CNTs are potential candidates for supercapacitors, commercial application is yet to be realized because of difficulties in mass production and purification. Furthermore, the properties of CNTs depend on their prehistory and hence their reproducibility is often questionable with respect to their diameter and conductivity distribution. Other double layer capacitor materials such as activated carbon, carbon black, carbon aerogel and carbon nanofibers are, however, inexpensive and some of them provide improved performance as supercapacitor electrodes. A comparison of different types of carbonaceous materials including SWCNT and MWCNT as supercapacitor electrodes is provided in Table 7.2, where carbon cloth and aerogel materials show better performance even than carbon nanotubes.

7.7
Hybrid Nanostructured Materials

Recently, many new classes of nanostructured hybrid electrodes for supercapacitors have been developed by the unique combination of organic species with various inorganic materials [14, 65–72]. For example, electrochemical double-layer capacitors with larger storage abilities have been achieved by employing high surface area carbonaceous powders and carbon nanotubes with nanostructured transition metal oxides or conducting polymers (Table 7.3). The large capacitance exhibited by some of these systems has been demonstrated to originate from a combination of the double-layer capacitance and pseudocapacitance associated with the participation of adsorbed intermediates in the surface redox-type reac-

Table 7.3 A comparison of electrochemical double-layer capacitors prepared using various types of hybrid composites in different electrolytes.

Nanostructured hybrid materials	Electrolyte	Capacitance/F g^{-1}	References
NiO$_x$/carbon nanotube, CNT	KOH	160	28
RuO$_x$(OH)$_y$/coconut shells activated carbon	H$_2$SO$_4$	250	59
carbon nanofibers/PANI	H$_2$SO$_4$	264	76
RuO$_2$/carbon aerogel	H$_2$SO$_4$	270	35
PEDOT/CNT	H$_2$SO$_4$	160	75
RuO$_2$/activated carbon	H$_2$SO$_4$	470	65
RuO$_2$/CNT	H$_2$SO$_4$	295	66
PEDOT/RuO$_2$	H$_2$SO$_4$	420	56
MWCNT/PPy	H$_2$SO$_4$	160	69
PEDOT/MoO$_3$	EC/DMC	300	12
PANI/MWCNT	H$_2$SO$_4$	360	74

tions. For example, redox pseudo-capacitive materials (electronically conducting polymers with oxides of multivalent metals such as ruthenium and iridium) and double layer capacitive materials (nanostructured activated carbon, carbon black, carbon aerogel and carbon nanotubes as discussed above) are two classes of materials with improved performance and hence a composite using them will facilitate extensive applications in developing supercapacitor electrodes.

7.7.1
Conducting Polymer–Transition Metal Oxide Nanohybrids

Integration of a nanostructured transition metal oxide into a conducting polymer matrix to form a hybrid material is an effective way to harness the electrochemical activity of nanosized oxide clusters. By anchoring them into conducting polymers like polyaniline, the reversible redox chemistry of the otherwise soluble polyoxometalate clusters can be combined with that of the conducting polymer to bring benefit for energy storage applications [23]. The resulting hybrid polymer displays the combined activity of its organic and inorganic components to store and release charge in a solid state electrochemical capacitor. We discuss a few examples since this is an emerging area with many interesting applications in electrochemical energy storage and accordingly many conducting polymers with various nanostructured transition metal oxides hybrids have been reported recently for electrochemical supercapacitors [14, 69–72].

Let us consider a typical example of a transition metal oxide–conducting polymer composite: polypyrrole (PPy) mixed with nanosized colloidal Fe$_2$O$_3$. This shows an increased capacitance of about 420 F g^{-1} compared to that of the individual components, coupled with an enhanced stability of retaining 97% even after one thousand cycles [63]. Similarly, poly (3,4-ethylenedioxythiophene) PEDOT, based organo-inorganic hybrids combined at "nanoscale" level, show enhanced electrochemical storage ability due to synergistic effects [13–22]. For example, PEDOT nanocomposites with various transition metal oxides/sulfides have been prepared by using different methods [13–22], and specific systems like PEDOT–MoO_3 nanocomposite show remarkable supercapacitor behavior. More specifically, when PEDOT is intercalated between the layers of MoO_3 by microwave irradiation, an unusually enhanced double-layer capacitance for this nanocomposite ($\sim$300 F g^{-1}) is observed compared to that ($\sim$40 mF g^{-1}) of pristine MoO_3, which could be explained on the basis of higher electronic conductivity, enhanced bi-dimensionality and increased surface area facilitating this improved performance after polymer intercalation [14]. Comparative cyclic voltammograms of (a) crystalline MoO_3 and (b) layered PEDOT–MoO_3 nanocomposite, at several scan rates using 1 M $LiClO_4$ in a (1:1 v/v) mixture of ethylene and dimethyl carbonate (EC/DMC) are shown in Fig. 7.9, illustrating a drastic change in electrochemical prop-

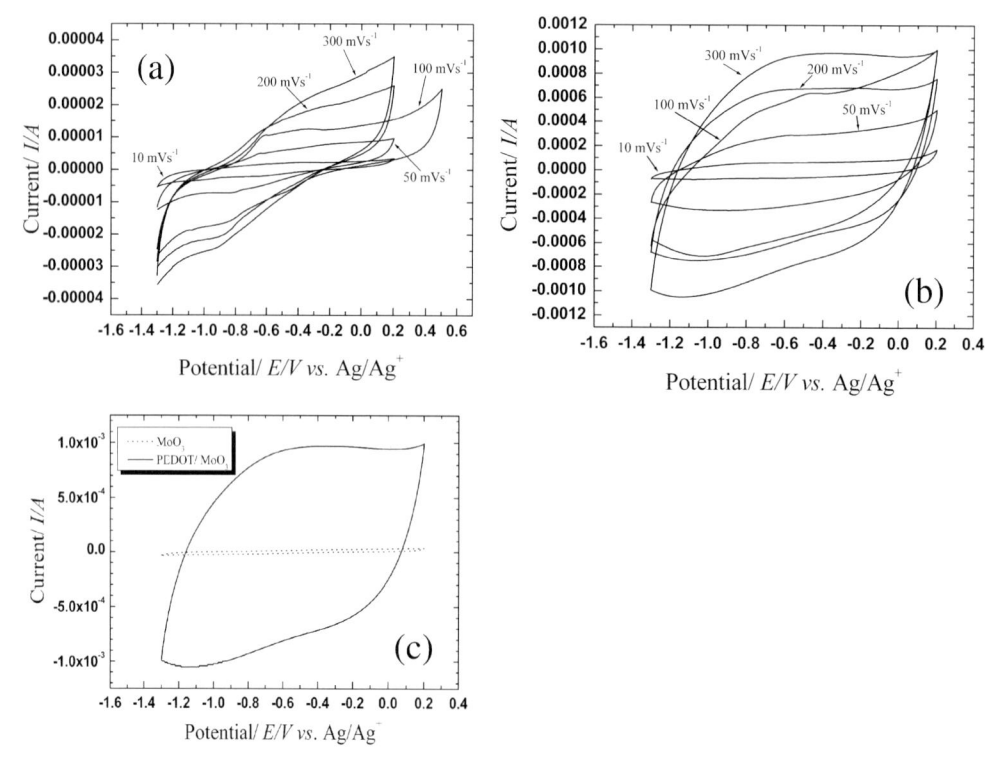

Fig. 7.9 Cyclic voltammograms of (a) crystalline MoO_3 and (b) layered PEDOT–MoO_3 nanocomposite, at several scan rates; (10, 50, 100, 200 and 300 m s^{-1}) (c) superimposed voltammograms of MoO_3 and the composite at 300 mV s^{-1} illustrating the supercapacitor behavior using 1 M $LiClO_4$ in EC/DMC (1:1 v/v).

erties induced by the polymer insertion. The observed linear increase in current with scan rate is expected for a strongly adsorbed electrochemical species. However, comparative cyclic voltammograms of crystalline MoO_3 and layered PEDOT–MoO_3 nanocomposite shown in Fig. 7.9(c) reveal the electrochemical features after nanocomposite formation [14]. Similarly, PEDOT, when mixed with nanosize RuO_2, produces a nanohybrid material which shows a capacitance of 420 F g^{-1} in aqueous electrolytes compared to about 930 F g^{-1} observed for the pristine oxide [60].

7.7.2
Conducting Polymer–Carbon Nanotube Hybrids

Recently, three types of electrically conducting polymers: polyaniline (PANI), polypyrrole (PPy) and PEDOT, have been used as supercapacitor electrode materials in the form of composites with multiwalled carbon nanotubes (CNTs). The energy storage in such a type of composite combines both faradaic and pseudocapacitance components and CNT has been shown to play the role of a perfect backbone for a homogenous distribution of conducting polymers in the composite [74–76]. It is well known that pure conducting polymers are mechanically weak, and hence, the carbon nanotubes can effectively preserve the active material from mechanical changes (shrinkage and breaking) during long cycling. Apart from providing excellent conducting and mechanical properties, the presence of nanotubes also improves the charge transfer that enables a high charge–discharge rate [75, 76]. Many research groups have reported the optimum use of CNTs in electrochemical capacitors as illustrated by a typical electrode composition with ca. 20 wt.% of CNTs although a careful selection of the potential range is necessary to get useful capacitance values [75–77]. Capacitance values ranging from 100 to 330 F g^{-1} could be reached for different asymmetric configurations with a capacitor voltage from 0.6 to 1.8 V. Since CNTs can play the role of an excellent support for conducting polymers, their composites can have unique microtextural, mechanical and conducting properties, allowing efficient charge propagation. This advanced type of CNT-conducting polymer composite in the doped state allows formation of a three-dimensional supercapacitor using a synergistic effect of complementary properties of both these components. The development of hybrid systems, with two different electrodes working in their optimal potential range, is also very promising. In all these cases, the applied potential is found to be one of the key factors influencing the specific capacitance of supercapacitor electrodes. For the operation of each electrode in its optimal potential range, asymmetric capacitors have been built with MWNT/PPy and MWNT/PANI as negative and positive electrodes, respectively. Perhaps this new morphology is responsible for the higher capacitance (320 F g^{-1}) of this electrode material and many similar supercapacitor electrodes have also been fabricated using activated carbon as a negative electrode and conducting polymer/CNTs composites as positive electrodes [74].

Supercapacitor behavior from multiwalled carbon nanotubes (MWCNTs) coated with polypyrrole (PPy) by electrochemical polymerization, shows enhanced specific capacitance from 50 F g^{-1} to 160 F g^{-1} [75]. The electrochemical deposition of PPy on MWCNT shows homogeneous nanocomposite formation, which applied to supercapacitor electrodes demonstrates a synergistic effect between the components. Similar studies [74] of the nanoporous composite of CNT and polypyrrole and also the graphite/polypyrrole composite [78] suggest a specific capacitance of about 400 F g^{-1} along with a Coulombic efficiency of 96–99%. Recently, determination of the specific capacitance of conducting polymer/nanotubes composite electrodes using different cell configurations of composite materials containing 20 wt.% of MWNT and 80 wt.% of chemically formed conducting polymers like PANI and PPy has highlighted the importance of the conducting properties of MWNT and their available mesoporosity allowing good charge propagation and the SEM images of these nanocomposites shown in Fig. 7.10 reveal interesting morphological changes and cyclic voltammograms of an asymmetric capacitor based on PANI–MWCNT positive electrode and PPy–MWCNT as negative electrode [79]. In the case of three electrode cells, extremely high values of capacitance can be found ranging from 250 to 1100 F g^{-1} while similar two electrode cells, give only 190 F g^{-1} for PPy/MWNT and 360 F g^{-1} for PANI/MWNT composites, respectively. Similarly many recent studies of MWCNT deposited with poly(3-methylthiophene) in 1 M LiClO$_4$/acetonitrile shows interesting pseudocapacitance effects [80].

Another nanocomposite prepared from PEDOT and MWCNTs by chemical or electrochemical polymerization of EDOT directly on the nanotubes or from a homogenous mixture of PEDOT and CNTs in acidic (1 M H$_2$SO$_4$), alkaline (6 M KOH) and organic (1 M TEABF$_4$ in acetonitrile) electrolytic solutions shows improved performance [81]. Due to the open mesoporous network of nanotubes, the easily accessible electrode/electrolyte interface allows quick charge propagation in the composite material along with an efficient reversible storage of energy in PEDOT during and subsequent to charging/discharging cycles. The values of capacitance for PEDOT/carbon composites range from 60 to 160 F g^{-1} and such a material also has a good cycling performance with good stability in all electrolytes. Polyaniline-coated carbon nanofiber fabricated using one-step vapor deposition polymerization techniques exhibits a maximum value of 264 F g^{-1} when the thickness of the polyaniline layer is ∼20 nm and detailed investigations show that uniform and ultrathin conducting polymer layers are formed on the carbon nanofiber surfaces regardless of the coating thickness [82].

7.7.3
Transition Metal Oxides–Carbon Nanotube Hybrids

Several transition metal oxide–CNT hybrids have been prepared by various researchers for fabricating excellent supercapacitor electrodes by mixing fine oxides/hydroxides with MWNTs using many methods like ultrasonic vibration in ethanol [68]. For example, a nonaqueous hybrid supercapacitor using manganese oxide/MWNTs composite and pure MWNTs as positive and negative electrodes, respectively, has been investigated by constant current charge/discharge methods.

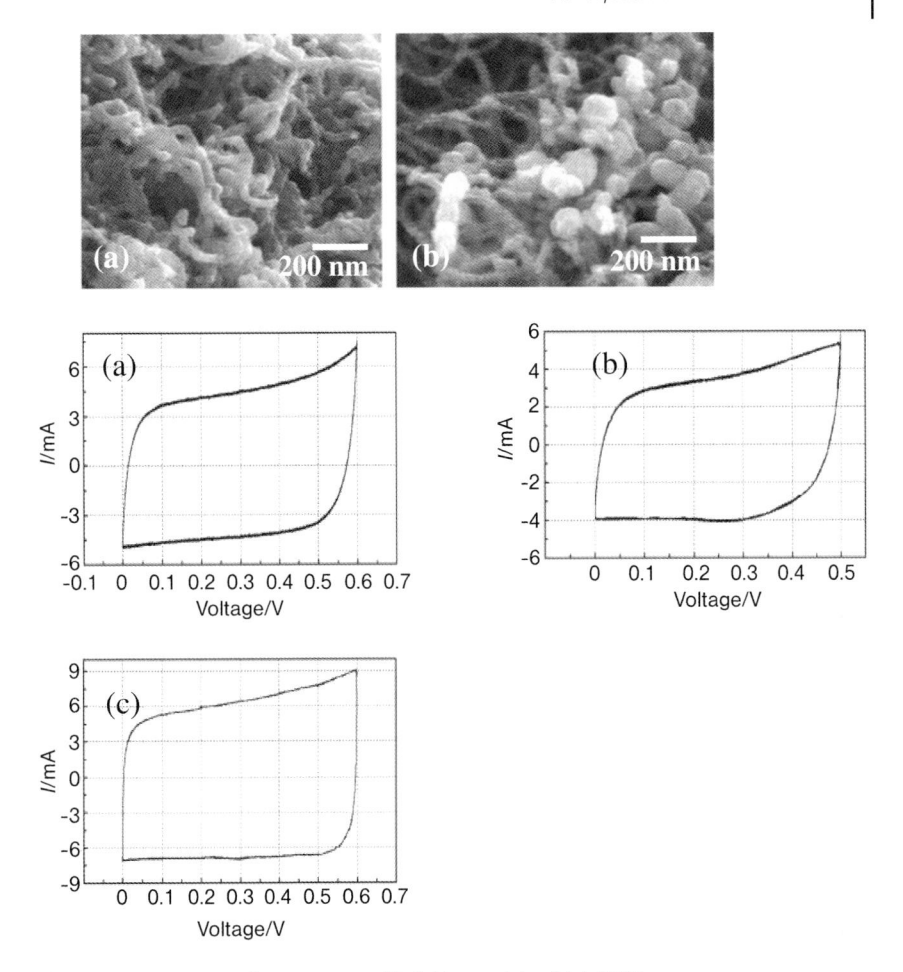

Fig. 7.10 SEM images of nanostructured hybrid materials of (a) PANI–MWCNT and (b) PPy–MWCNT; cyclic voltammogam of a symmetric capacitor based on (a) PPy–MWCNT and (b) PANI–MWCNT nanocomposite electrode respectively; (c) cyclic voltammogram of an asymmetric capacitor based on PANI–MWCNT as the positive electrode and PPy–MWCNT as the negative electrode [73].

This asymmetric hybrid capacitor shows better capacitance and energy characteristics than those of the symmetric ones based on individual manganese oxide/MWNT composite and MWNTs electrodes. The energy density of the hybrid capacitor can reach 33 W h kg^{-1} even at a current density of 10 mA cm^{-2} in 1.0 M LiClO$_4$ electrolyte, which is comparable to that of a manganese oxide/activated carbon hybrid capacitor [69].

Another hybrid system based on nickel oxide/CNT nanocomposite, prepared by a simple chemical precipitation method generates a CNT network in NiO, significantly improves the electrical conductivity of the host NiO by forming a network of CNT along with generating the active sites for the redox reaction of the metal oxide by increasing its specific surface area [70]. This increases the specific capacitance by 34% at a percolation limit of 10 wt.% of CNTs in addition to the improvement in the power density and cycle life. The enhanced electrochemical double layer capacitance of NiO/CNT composite arising from comparative cyclic voltammetry and SEM images of pristine NiO and NiO/CNT nanocomposite is revealed in Fig. 7.11.

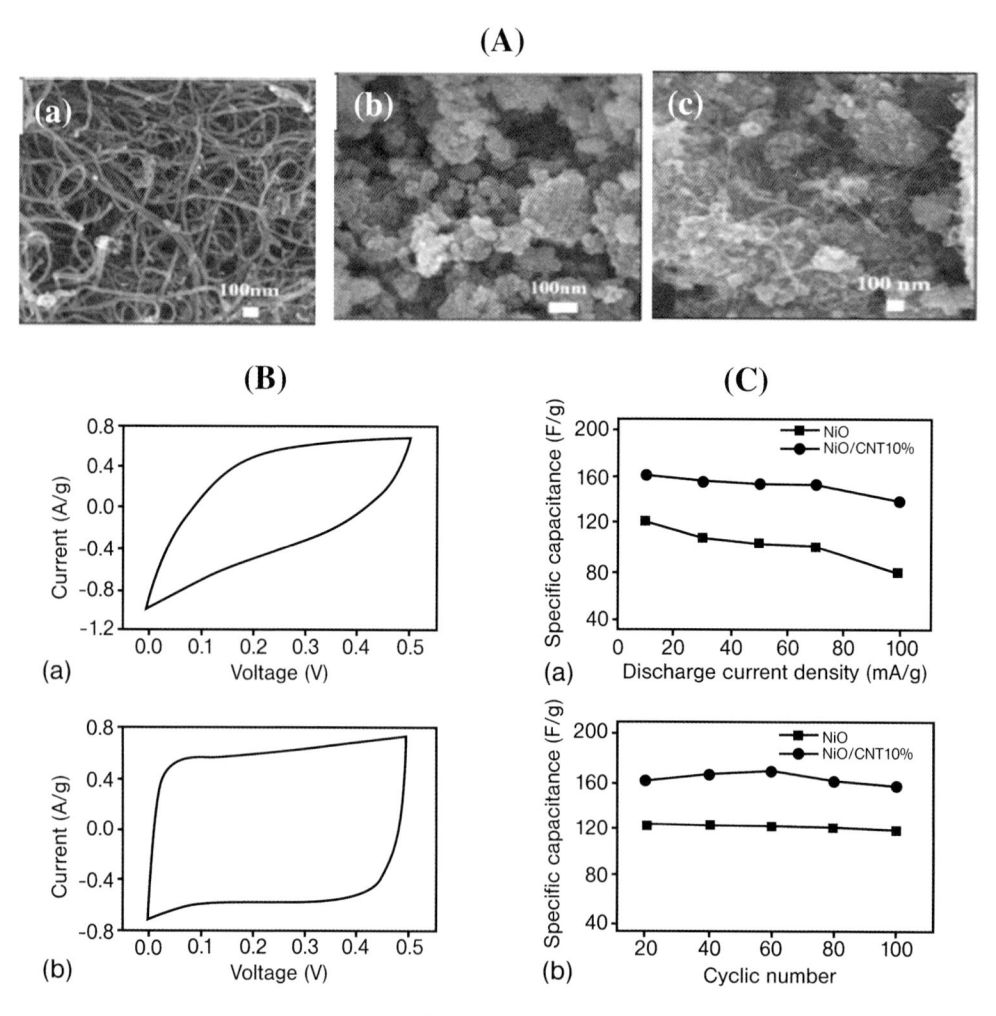

Fig. 7.11 (A) The SEM image of (a) the CNT, (b) the pristine NiO, (c) the NiO/CNT composite. (B) Electrochemical properties of supercapacitors using the bare NiO and NiO/CNT (10%) composite electrodes. The cyclic voltammetry behavior of (a) the bare NiO electrodes and (b) the NiO/CNT (10%) composite electrodes in 2 M KOH aqueous solution (sweep rate, 10 mV s^{-1}). (C) (a) Specific capacitance of the bare NiO and NiO/CNT (10%) composite as a function of the discharge current density. (b) Cycle test of the specific capacitance of the bare NiO and NiO/CNT (10%) composite [64].

Electrochemical characterization of activated carbon–ruthenium oxide nanoparticles (denoted as RuOx) composites for supercapacitors prepared by a modified sol–gel method shows high specific capacitance with a change depending on annealing in air [70]. The specific capacitance of activated carbon (denoted as AC) measured at 5 mA cm^{-2} is significantly increased from 27 to 40 F g^{-1} by the adsorption of RuOx nanoparticles with ultrasonic agitation in 1 M NaOH for 30 min. The total specific capacitance of a composite composed of 90 wt.% AC and 10 wt.% RuOx measured at 25 mV s^{-1} is about 63 F g^{-1}, which is increased to ca. 112 F g^{-1} when the RuOx has been previously annealed in air at 200 °C for 2 h. The specific capacitance of RuOx nanoparticles can be promoted from 470 to 980 F g^{-1} by annealing in air at 200 °C for 2 h [71]. Similar enhancement of the capacitance of multiwalled carbon nanotubes functionalized with ruthenium oxide has been observed by many others and attributed to the entangled network of nanotubes which forms open mesopores and their chemical stability with a basal geometry makes them suitable for such applications [72]. However, oxidative treatment to generate oxygenated functional groups on the tube-ends and along the sidewalls enables facile derivatization by hydrous ruthenium oxide to enhance the inherent capacitance. A specific capacitance of 80 F g^{-1} is obtained after ruthenium oxide functionalization, which is significantly greater than that of pristine MWNTs (30 F g^{-1}) in the same medium.

7.8
Hybrid Nanostructured Materials as Electrolytes for Super Capacitors

Commonly available supercapacitors are double-layer carbon supercapacitors (DLCS), which make use of electrolyte solutions consisting of a salt dissolved in an aprotic organic solvent capable of relatively high operating voltages. The main drawback of these supercapacitors is that the organic solvents often do not fulfill the requirements of environmental compatibility and safety for vapor generation, flammability and possible explosions. This is the case for DLCSs with acetonitrile-based electrolytes, which are some of the most common commercially available high-voltage DLCSs. The high vapor pressure of acetonitrile-based electrolytes requires a careful and expensive thermal control of the DLCSs when used in combination with fuel cells. Recently, aqueous and various nonaqueous liquid electrolytes as well as solid polymer electrolytes have been used for double-layer electrochemical capacitors [83]. Proton conducting properties for a large number of inorganic and polymer materials have been investigated because of the technological potential for applications in electrochemical capacitors [83]. However, their use in devices is limited by numerous materials requirements. For example, low temperature protonic conductors are stable only in a narrow temperature window (especially for polymer electrolytes in the temperature range 100–200 °C) and some of these polymer electrolytes have other disadvantages such as limited thermal stability and the possibility of many phase transitions linked with irreversible resistivity changes [84]. Besides these, there are also many other concerns related to low cost preparation, easy processing and environmental safety. Recently, several new synthetic routes have been developed for designing many organic–inorganic nanohybrid materials as realistic ion conducting electrolyte membranes for electrochemical capacitors [85]. In these materials, the structure of the hybrid has been designed at the molecular nanoscale to possess fast proton as well as lithium ion transport, mostly through modified organic ligands with inorganic surfaces.

7.8.1
Nanostructured Polymer Composite Electrolytes

Recently a family of nanostructured organic–inorganic composite polymer electrolytes has been developed which appears to be highly promising for resolution of the problems listed above for conventional solid polymer electrolytes (SPEs). This new family consists of transition metal oxides in a polymer matrix, such as nanosize SiO_2/polymer (polyethylene oxides (PEO); polypropylene oxide (PPO); polytetramethylene oxide (PTMO)) hybrids synthesized through sol–gel processes, forming the basis for a series of lithium ion conductors with superior properties as compared to conventional salt-in-polymer electrolytes [86]. Similarly, another interesting family of solid-state nanocomposite electrolytes using poly(ethylene oxide) (PEO) containing nanoscale fillers of layered double hydroxides (LDHs) act as electrolytes with high ionic conductivity [87].

7.8.2
Ionic Liquids as Supercapacitor Electrolytes

Although DLCSs with liquid electrolytes based on propylene carbonate are available in various sizes, attempts to further improve the performance at high temperature, especially the safety and the cell voltage, have resulted in the strategy of using ionic liquids as novel electrolytes [88–89]. Ionic liquids are a new type of solvents, which are actually molten salts at relatively low temperatures, consisting of entirely ionic species. Ionic liquids have no measurable vapor pressure, and possess high chemical stability at high temperatures. Furthermore, ionic liquids display wide electrochemical stability windows and good conductivities at the temperatures of interest so that they can be used as solvent-free "green" electrolytes for high voltage supercapacitors. The use of ionic liquids has been investigated both in DLCSs and in hybrid supercapacitors, although their cycling stability over many thousands of cycles is yet to be proven. Until now, ionic liquids have been applied in capacitors as solutions in molecular liquids (cyclic carbonates), making the system similar to conventional solutions of organic salts in nonaqueous solvents [89].

Let us consider some commonly used ionic liquid examples specially useful for supercapacitors. These include 1-ethyl-3-methyl imidazolium (EMImþ), 1-butyl-3-methyl imidazolium (BMImþ) and 1-methyl-1-propyl pyrrolidinium (BMPyþ) cations, as well as tetrafluoroborate, hexafluorophosphate and bis((trifluoromethyl)sulfonyl) imide anions [88]. The capacity of the double layer formed at the interface between the activated carbon and the ionic liquid (ca. 180 $F\,g^{-1}$), is comparable to the values for other electrolyte solutions of organic salts such as $(C_2H_5)_4NBF_4$, in acetonitrile or propylene carbonate. However, the energy stored in the capacitor containing ionic liquids is high, perhaps due to an improved electrochemical stability window. In addition, since ionic liquids are characterized by negligible vapor pressure, their use can lead to systems containing no volatile component, thus making these devices more environmentally friendly.

7.9
Possible Limitations of Hybrid Materials for Supercapacitors

Some of the significant concerns for nanotechnology in general are also valid for the application of hybrid materials for electrochemical energy storage. These include difficulties in producing high quality nanostructured materials of complex composition in high volume with consistent properties, preferably at low cost, in such a way that they can be conveniently processed into useful products. This is especially significant for nanostructured hybrid materials as they are often prepared using a combination of standard processes used for organic materials and also for inorganic materials with sometimes conflicting requirements of calcination, annealing and separation methods. Other major limitations of nanotechnology such as difficulty in scale-up, problems with reproducing properties as a function of size, contamination due to impurity accumulation and so forth also deserve special attention. Some of these hybrid materials have not been tested rigorously to establish their long term stability (both thermal and oxidative) and this could be exacerbated by applications like rechargeable batteries/supercapacitors where the continuous presence of an electric field during charging–discharging can cause permanent changes.

Cost is another critical issue to be considered for the application of these hybrid materials for supercapacitors although it may not be very important for strategic uses involving space and defence applications, where reliability in performance is more important. Since supercapacitors have EV as an important application, where a supercapacitor is coupled with a battery/fuel cell to meet high power requirements, the cost of the supercapacitor must be competitive with the use of more batteries or fuel cells to meet the power requirements of the system. Since the most critical factor in the cost of the supercapacitor is the cost of the electrode material, which in many cases includes preparation of high surface area, specialty carbon particulate or cloth coated with the hybrid materials, it is not possible to provide a realistic estimate of the device cost. CNT is presently expensive in relation to other carbonaceous materials although some of the hybrids may be cheap since processing is easier and large area electrodes can be fabricated to make supercapacitor banks. The use of ionic liquids and organic electrolytes make these costs rather high compared to organo-inorganic hybrid electrolytes although there are tangible advantages in terms of performance. Ultracapacitor development is continuing worldwide and the projected improvement in energy density should open new markets for ultracapacitors if their price can be significantly reduced by using these hybrid materials. Key factors in reducing the price are to utilize lower cost electrode materials and to develop assembly processes that can be automated at reasonable investment. The use of carbon blacks and low cost metal oxides in making hybrid materials could result in lower material and fabrication costs.

Generally the voltage of a supercapacitor is dictated by the available potential window prior to the commencement of any irreversible anodic and cathodic faradaic processes. Although the potential in aqueous electrolytes is only about 1.23 V, it can be extended even up to 3 V by a judicious choice of the supporting electrolyte and/or a solvent with high anodic and cathodic overpotentials. Consequently, the development of nonaqueous systems at present appears to be gaining momentum due to the increased energy density of such systems. For example, PEDOT–MoO_3 nanohybrid electrode material shows enhanced double-layer capacitance behavior in nonaqueous electrolytes and this hybrid strategy could provide an efficient way to fabricate cheaper supercapacitor electrodes. Similarly, the use of gel electrolytes holds the promise of combining the advantages of solid-polymer electrolytes with the increased voltage of a nonaqueous based system. Nevertheless, only carbon and RuO_x systems have so far been commercialized, although other systems are under various stages of development. Several characteristic features of carbon such as high corrosion resistance, acceptable electronic conductivity, high surface area after activation, wide availability in different structural forms and reasonable cost are especially significant for its use in supercapacitors and this coupled with hydrated ruthenium oxide in the form of a composite gives one of the best values for capacitance (700 F g^{-1}). Nevertheless, the total charge stored in a porous carbon-based electrode cannot be abstracted at short times and different types of mesoporous carbonaceous materials and CNT–polymer hybrids to ameliorate these limitations are at various stages of development.

7.10
Conclusions and Perspectives

Nanostructured hybrid materials derived from a seamless blending of both inorganic and organic building blocks offer great promise for obtaining unprecedented optical, electrical, electrochemical and mechanical properties due to their diverse methods of preparation and more significantly due to the possibility of dimensionality control. For example, nanostructured hybrid materials obtained by the insertion of interesting organic molecules and conducting polymers within the structure of an extended solid phase of 2D-transition metal oxides, sulfides, oxychlorides and phosphates typically represented by V_2O_5, MoO_3, FeOCl, $VOPO_4$, TaS_2, TiS_2, VS_2 and MoS_2 show remarkable synergistic properties, as illustrated by their potential for electrochemical energy storage.

Although, conductive polymers have high electronic conductivity and flexible processability, their poor long term stability, inferior fabrication capability, and structural instability during oxidation–reduction processes (that is, charge–discharge) coupled with poor capacity retention upon cycling (capacity fading) undermine the attempts to make efficient electrodes for electrochemical power sources like rechargeable lithium batteries and supercapacitor electrodes. One of the important strategies to overcome these limitations is the use of organo-inorganic conducting polymer-based intercalative nanocomposites where the same conducting polymer is prepared, often *in situ*, in the interlayer space of layered inorganic materials. Consequently, many of these limitations of conducting polymer electrodes can be alleviated by confining them with in the well-defined interlayer gap of layered inorganic structures, thus using these interesting hybrid nanocomposite materials as electrodes. Several such nanostructured hybrid materials have shown clear potential for their use as superior electrodes for electrochemical storage although a rigorous testing, validation and evaluation of their life in the actual device is still missing. The last few years have generated many excellent supercapacitor electrodes using aqueous, nonaqueous, polymeric and ionic liquid electrolytes by tuning the composition of many of these hybrid materials although a realistic assessment for successful commercial applications needs to surmount many barriers including cost and scale-up issues which are inherent in many nanotechnology processes. Nevertheless, a fascinating variety of materials with dimensional control and structures with interesting properties facilitating diverse applications are available for nanostructured hybrid materials which will continue to motivate interdisciplinary research in this area for a few more decades.

References

1 C. N. R. Rao, A. Müller, A. K. Cheetham (Eds.), *The Chemistry of Nanomaterials, Synthesis, Properties, and Applications*, Wiley-VCH, Weinheim, 1st Edn., **2004**.
2 J. T. Hupp, S. T. Nguyen, Synthesis, functionalization, and surface treatment of naonoparticles, *Interface* **2001**, *10*, 28–32.
3 M. L. Merlau, M. D. P. Mejia, S. T. Nguyen, J. T. Hupp, *Angew. Chem. Int. Ed. Engl.* **2001**, *40*, 4239–4242.
4 M. E. Williams, J. T. Hupp, Electrochemistry in Nanostructured Inorganic Molecular Materials, *Proc. Mater. Res. Soc.* **2001**, *676*, Y1.5.1.
5 K. V. Kordesch, G. R. Simader, *Chem. Rev.* **1995**, *95*, 191–208.
6 D. Linden, *Handbook of Batteries and Fuel Cells*, McGraw-Hill, New York, **1984**.
7 B. C. H. Steele, A. Heinzel, *Nature* **2001**, *414*, 345–348.
8 M. Winter, J. O. Besenhard, M. E. Spahr, Petr Novak, *Adv. Mater.* **1998**, *10*, 725–763.
9 K. K. Likharev, T. Claeson, *Sci. Am.* **1992**, 80–82.
10 K. K. Likharev, *IBM, J. Res. Dev* **1998**, *32*, 144–146.
11 G. Markovich, C. P. Collier, S. E. Henrichs, F. Remacle, R. D. Levine, J. R. Heath. *Acc. Chem. Res.* **1999**, *32*, 415–418.
12 J. A. Wilson, A. D. Yoffe, *Adv. Phys.* **1969**, *18*, 193–196.

13 A. Vadivel Murugan, M. Quintin, M. H. Delville, G. Campet, C. S. Gopinath, K. Vijayamohanan, *J. Mater. Res.* **2006**, *21*, 112–118.

14 A. Vadivel Murugan, A. K. Viswanath, G. Campet, C. S. Gopinath, K. Vijayamohanan, *Appl. Phys. Lett.* **2005**, *87*, 243511–243513.

15 A. Vadivel Murugan, A. K. Viswanath, C. S. Gopinath, K. Vijayamohanan, *J. Appl. Phys.* **2006**, *100*, 743191–743195.

16 A. Vadivel Murugan, C. W. Kwon, G. Campet, B. B. Kale, A. B. Mandale, S. R. Sainker, C. S. Gopinath, K. Vijayamohanan, *J. Phys. Chem B.* **2004**, *108*, 10736–10742.

17 A. Vadivel Murugan, M. Quintin, M. H. Delville, G. Campet, K. Vijayamohanan, *J. Mater. Chem.* **2005**, *15*, 902–909.

18 A. Vadivel Murugan, C. S. Gopinath, K. Vijayamohanan, *Electrochem. Commun* **2005**, *7*, 213–218.

19 A. Vadivel Murugan, C. W. Kwon, G. Campet, B. B. Kale, M. Tirupti, K. Vijayamohanan, *J. Power Sources* **2002**, *105*, 1–5.

20 A. Vadivel Murugan, B. B. Kale, C. W. Kwon, G. Campet, K. Vijayamohanan, *J. Mater. Chem.* **2001**, *11*, 2470–2475.

21 C. W. Kwon, A. Vadivel Murugan, G. Campet, J. Portier, B. B. Kale, K. Vijayamohanan, J.-Ho. Choy, *Electrochem. Commun.* **2002**, *4*, 384–387.

22 A. Vadivel Murugan, M. Quintin, M. H. Delville, G. Campet, C. S. Gopinath, K. Vijayamohanan, *J. Power Sources* **2006**, *156*, 615–619.

23 P. G. Romero, *Adv. Mater.* **2001**, *13*, 163–174.

24 A. K. C. Gallegos, M. L. Cantu, N. C. Pastor, P. G. Romero, *Adv. Funct. Mater.* **2005**, *15*, 1125–1133.

25 C. T. Hable, M. S. Wrighton, *Langmuir.* **1991**, *7*, 1305–1309.

26 B. Rajesh, K. R. Thampi, J.-M. Bonard, N. Xanthopoulos, H. J. Mathieu and B. Viswanathan, *Electrochem. Solid-State Lett.* **2002**, *5*, E71–E74.

27 M. Winter, J. O. Besenhard, M. E. Spahr and Petr Novak, *Adv. Mater.* **1998**, *10*, 725–763.

28 B. E. Conway, *Electrochemical Supercapacitors: Scientific Fundamentals and Technological Applications*, Kluwer Academic/Plenum, New York, **1999**.

29 B. E. Conway, *J. Electrochem. Soc.* **1991**, *138*, 1539–1342

30 S. Sarangapani, B. V. Tilak, C. P. Chen, *J. Electrochem. Soc.* **1996**, *143*, 3791–3799

31 A. K. Shukla, S. Sampath, K. Vijayamohanan, *Curr. Sci.* **2000**, *79*, 1656–1661.

32 A. Burke, *J. Power Sources* **2000**, *91*, 37–40.

33 S. Ghosh, O. Inganäs, *Adv. Mater.* **1999**, *11*, 1214–1218.

34 C. Downs, J. Nugent, P. M. Ajayan, D. J. Duquette, K. S. V. Santhanam, *Adv. Mater.* **1999**, *11*, 1028–1031.

35 K. Jurewicz, S. Delpeux, V. Bertagna, F. Béguin, E. Frackowiak, *Chem. Phys. Lett.* **2001**, *347*, 36–38.

36 W. Sugimoto, K. Yokoshima, Y. Murakami, Y. Takasu, *Electrochim. Acta* **2006**, *52(4)*, 1742–1748.

37 Y. Takasu, T. Nakamura, Y. Murakami, *Chem. Lett.* 1998, 1215–1216.

38 G. Z. Chen, M. S. P. Shaffer, D. Coleby, G. Dixon, W. Zhou, D. J. Fray, A. H. Windle, *Adv. Mater.* **2000**, *12*, 522–526.

39 J. M. Miller, B. Dunn, *Langmuir.* **1999**, *15*, 799–802.

40 M. Hughes, G. Z. Chen, M. S. P. Shaffer, D. J. Fray, A. H. Windle, *Adv. Mater.* **2002**, *14*, 1613–1615.

41 W. Hutcheson, Technology and Economics in the Semiconductor Industry, *Sci. Am.*, January, **1996**.

42 W. Chen, C. Hu, C. Wang, C. Min, *J. Power Sources* **2004**, *125*, 292–298.

43 V. H. Radosevic, K. Kvastek, M. Vukovic, D. Cukman, *J. Electroanal. Chem.* **2000**, *482*, 188–201.

44 C. Hu, K. Chang, *J. Power Sources* **2002**, *112(2)*, 401–409.

45 F. Zhang, Y. Zhou, H. Li, *Mater. Chem. Phys.* **2004**, *83*, 260–264.

46 M. Wu, G. A. Snook, G. Z. Chen, D. J. Fray, *Electrochem. Commun.* **2004**, *6*, 499–504.

47 T. Cottineau, M. Toupin, T. Delahaye, T. Brousse, D. Belanger, *Appl. Phys. A* **2006**, *82*, 599–606.
48 T. Shinomiya, V. Gupta, N. Miura, *Electrochim. Acta* **2006**, *51*, 4412–4419.
49 K. R. Prasad, N. Miura, *Electrochem. Commun.* **2004**, *6*, 849–852.
50 K. Yokoshima, T. Shibutani, M. Hirota, W. Sugimoto, Y. Murakami, Y. Takasu, *J. Power Sources* **2006**, in press.
51 T. Skotheim, R. L. Elsenbaumer, J. R. Reynolds (Eds.), *Handbook of Conducting Polymers*, Marcel Dekker, New York 1998.
52 C. Arbizzani, M. Mastragostino, L. Meneghello, R. Paraventi, *Adv. Mater.* **1996**, *8*, 331.
53 A. Rudge, J. Davey, I. Raistrick, S. Gottesfeld, J. P. Ferraris, *J. Power. Sources* **1994**, *47*, 89.
54 C. Arbizzani, M. Catellani, M. Mastragostino, C. Mingazzini, *Electrochim. Acta*. **1995**, *40*, 1871–1874.
55 C. Arbizzani, M. Mastragostino, L. Meneghello, *Electrochim. Acta*. **1996**, *41*, 4121–4124.
56 A. Rudge, J. Davey, I. Raistrick, S. Gottesfeld, *Electrochim. Acta*. **1994**, *39*, 273–287.
57 M. E. G. Lyons, *Electroactive Polymer Electrochemistry*, Plenum Press, New York, **1994**.
58 P. Novuk, O. Inganäs, R. Bjorklund, *J. Electrochem. Soc.* **1987**, *134*, 1341–1345.
59 V. Gupta, N. Miura, *Mater. Lett* **2006**, *60*, 1466–1469.
60 J.-I. Hong, I.-H. Yeo, W.-K. Paik, *J. Electrochem. Soc.* **2001**, *148*, A156–A163.
61 R. Signorelli, J. Schindall, J. Kassakian, http://less.mit.edu/less_research.htm
62 P. V. Adhyapak, T. Maddanimath, S. Pethkar, A. J. Chandwadkar, Y. S. Negi, K. Vijayamohanan, *J. Power. Sources* **2002**, *109*, 105–110.
63 R. Gangopadhyay, A. De, *Chem. Mater* **2000**, *12*, 608–622.
64 W. Li, G. Reichenauer, J. Fricke, *Carbon* **2002**, *40*, 2955–2959.
65 M. S. Dandekar, G. Arabale, K. Vijayamohanan, *J. Power. Sources* **2005**, *141*, 198–203.
66 C. Niu, E. K. Sichel, R. Hoch, D. Moy, H. Tennent, *Appl. Phys. Lett* **1997**, *70*, 1480–1482.
67 K. H. An, K. K. Jeon, W. S. Kim, Y. S. Park, S. C. Lim, D. J. Bae, Y. H. Lee, *J. Korean Phys. Soc.* **2001**, *39*, S511–S517.
68 E. Frackowiak, F. Béguin, *Carbon* **2002**, *40*, 1775–1787.
69 G.-X. Wang, B.-L. Zhang, Z.-L. Yu, M.-Z. Qu, *Solid State Ionics* **2005**, *176*, 1169–1174.
70 J. Y. Lee, K. Liang, K. H. An, Y. H. Lee, *Synth. Met.* **2005**, *150*, 153–157.
71 W.-C. Chen, C.-C. Hu, C.-C. Wang, C.-K. Minet, *J. Power Sources* **2004**, *125*, 292–298.
72 X. Qin, S. Durbach, G. T. Wu, *Carbon* **2004**, *42*, 423–460.
73 G. Arabale, D. Wagh, M. Kulkarni, I. S. Mulla, S. P. Vernekar, K. Vijayamohanan, A. M. Rao, *Chem. Phys. Lett.* **2003**, *376*, 207–213.
74 M. Hughes, G. Z. Chen, M. S. P. Shaffer, D. J. Fray, A. H. Windle, *Chem. Mater.* **2002**, *14*, 1610–1613.
75 K. Jurewicz, S. Delpeux, V. Bertagna, F. Béguin, E. Frackowiak, *Chem. Phys. Lett.* **2001**, *347*, 36–40.
76 E. Frackowiak, K. Jurewicz, S. Delpeux, F. Béguin, *J. Power Sources* **2001**, *97–98*, 822–825.
77 C. Downs, J. Nugent, P. M. Ajayan, D. J. Duquette, K. S. V. Santhanam, *Adv. Mater.* **1999**, *11*, 1028–1030.
78 J. H. Park, J. M. Ko, O. O. Park, D.-W. Kim, *J. Power Sources* **2002**, *105*, 20–25.
79 V. Khomenko, E. Frackowiak, F. Béguin, *Electrochim. Acta*. **2005**, *50*, 2499–2506.
80 Q. Xiao, X. Zhou, *Electrochim. Acta*. **2003**, *48*, 575–580.
81 K. Lota, V. Khomenko, E. Frackowiak, *J. Phys. Chem. Solids*. **2004**, *65*, 295–301.
82 J. Jang, J. Bae, M. Choi, S.-Ho. Yoon, *Carbon* **2005**, *43*, 2730–2736.
83 K. D. Kreuer, *Chem. Mater.* **1996**, *8*, 610–641.

84 I. G. Lineau, A. Denoyelle, J. Y. Sanchez, C. Poinsignon, *Electrochim. Acta.* **1992**, *37*, 1615–1618.
85 I. Honma, Y. Takeda, J. M. Bae, *Solid State Ionics* **1999**, *120*, 255–264.
86 I. Honma, S. Hirakawa, K. Yamada, J. M. Bae, *Solid State Ionics* **1999**, *118*, 29–36.
87 C.-S. Liao and W.-B. Ye, *J. Polym. Res.* **2003**, *10*, 241–246.
88 A. Lewandowski, M. Galin'ski, *J. Phys. Chem. Solids.* **2004**, *65*, 281–286.
89 C. Niu, E. K. Sichel, R. Hoch, D. Moy, H. Tennet, *Appl. Phys. Lett.* **1997**, *70*, 1480–1482.

8
Dendrimers and Their Use as Nanoscale Sensors

N. Jayaraman

8.1
Introduction

Among various macromolecular and polymeric architectures, the recently evolved dendritic architecture has become prominent in a wide variety of studies. An early report on the synthesis of cascade molecules [1] and the synthesis of a series of well-defined poly(amido amine) (PAMAM) dendrimers [2] initiated the evolution of the distinct class of macromolecules called dendrimers. Some of the unique architectural features of dendrimers are their uniform branching pattern throughout the structure, the increasing number of peripheral groups during their growth, their progressively increasing dense peripheries, their nanometric size and the monodispersity of their constitution. The dense peripheries that evolve during the growth of dendrimers result in relatively less dense interiors within the structure and both these dense peripheries and the less dense interiors become sites for variegated studies. The dendritic architecture does not seem to be known in nature to the extent that it is accomplished in chemical synthesis, yet highly branched polysaccharides, such as glycogen [3], might be considered as the naturally devised dendritic architectures. The salient features of a dendritic architecture are presented in Fig. 8.1. The variety of studies to which dendrimers have been subjected originates primarily from their architectural and structural features listed in part above. Structurally and architecturally driven effects are observed consistently from a number of targeted studies and these effects provide sufficient basis to engage dendrimers in a variety of chemical, biological and materials studies. Exploring the properties that emanate from the core, or from the internal structural components, or from the peripheries have become important in the sustained efforts to understand and apply the dendritic architectures and principles in as many studies as possible [4]. This chapter focuses on an introduction to various dendritic structures, followed by a summary of the recently evolved interest in interfacing the nanoscale dendrimers in the particular area of chemical and biological sensing.

Nanomaterials Chemistry. Edited by C. N. R. Rao, A. Müller, and A. K. Cheetham
Copyright © 2007 WILEY-VCH Verlag GmbH & Co. KGaA, Weinheim
ISBN: 978-3-527-31664-9

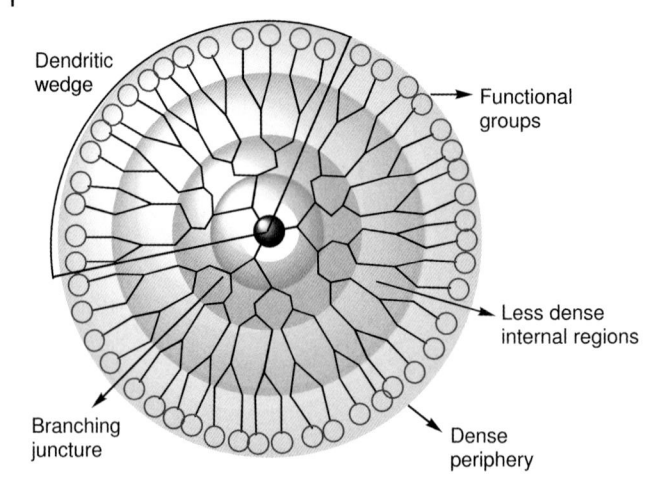

Fig. 8.1 A representation of a dendrimer and its structural features.

8.2
Synthetic Methods

Important structural features of dendrimers are the presence of (i) a core containing multiple reactive sites; (ii) linkers that functionalize the core unit; (iii) further functional groups at the termini of the linkers, presented in a branched fashion and (iv) a large number of chain ends, acting as the functionalized peripheries of the dendritic structure. It is important that every functional group at the core and branch locations undergoes reaction, so as to afford complete branching throughout the structure. There exist two established methods that allow accomplishment of the strict requirements of synthesis and the accompanying structural homogeneities. In one method, termed divergent growth [2], synthesis of the dendrimer begins with the core and continues by assembling the branching components at the growing peripheries of the dendrimer (Fig. 8.2). In the other method, termed convergent growth [5], synthesis begin with the peripheral component and the growth is directed inward, so as to realize wedges or dendrons. Multifold reaction of the core with the dendron completes the assembly of a dendrimer (Fig. 8.2). These two methods, over a period of time, have become the preferred methods of synthesis for a large variety of dendrimers. Other methods that are closely related to the above have also gained importance [6]. The divergent growth involves conducting a large number of reactions at the growing peripheries. Steric hindrance is not a serious constraint at the peripheral reactive sites and the high molecular weight differences between the growing dendrimer and the constituent monomer allow facile purification. The preferred step-wise synthetic sequence, iterated several times, needs to be efficient in order to accomplish the synthesis with complete control over the dendritic structure. The convergent growth relies

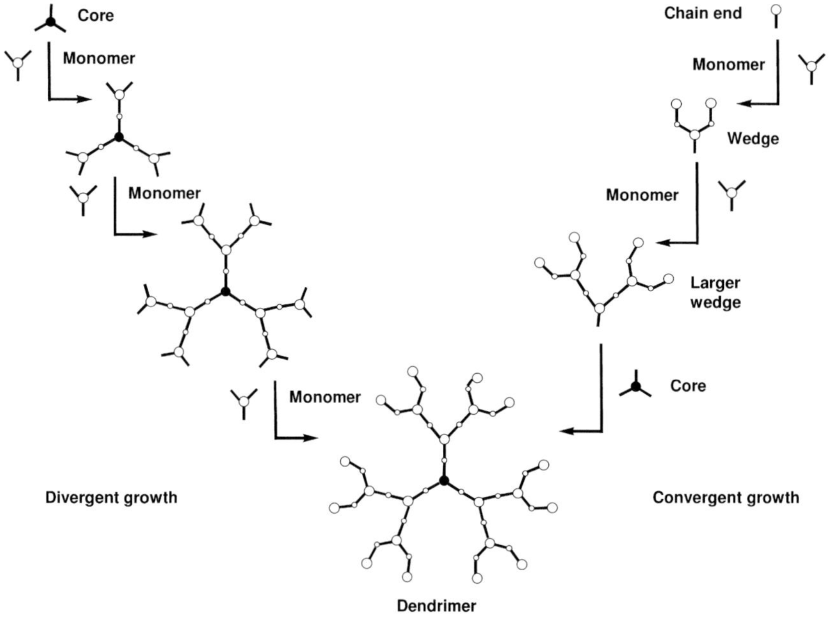

Fig. 8.2 Divergent and convergent approaches to the synthesis of a dendritic structure.

on synthesizing dendrons initially, by which the peripheral moieties of a dendrimer form the starting materials for the step-wise synthesis. The focal points of the branched monomers are reacted with the branched functionalities of the monomer, so as to form a dendron. Iterative reaction of the focal point functional group with the branched functionalities leads to the formation of a larger dendron. Such larger dendrons are reacted with a core containing a number of reactive sites, so as to complete the synthesis of a dendimer. The preferred reactions need to be efficient and should be amenable for iterative synthetic procedures. Both divergent and convergent synthetic methods are employed successfully in the synthesis of a large variety of generic dendrimers, employing new types of monomers in their constitution. Monomers, constituting the dendrimers, generally carry 3 to 5 functional groups, within these one functional group is coupled to the growing dendrimer and the rest of the functional groups act to generate a number of branches to the growing dendron or dendrimer. Every branch point of a dendrimer forms a generation and many generic dendrimers are known to be 4 to 6 generations in their largest generation. Generations of the order of 10 or more are also known for a few dendrimers, namely PAMAM dendrimers [2], polyallylated dendrimers [7] and polyphosphane dendrimers [8]. The divergent growth is more suitable for the synthesis of larger generations, although truncation in the structure could be a potential difficulty in the divergent synthesis of

Table 8.1 A compilation of dendrimers, constituted either with a distinct monomer or a linkage.

No.	Dendrimer constitution at a particular generation	Molecular weight of the largest dendrimer[e] (g mol^{-1})	Ref.
1		2661626[a]	7
2		934720[a]	2
3		139750[a]	15
4		94146[a]	8
5		73912[a]	14c

8.2 Synthetic Methods | 253

6	16	42731[b]
7	17	30799[a]
8	18	25681[a]
9	19	~25000[b]
10	20	24291[b]

Table 8.1 (continued)

No.	Dendrimer constitution at a particular generation	Molecular weight of the largest dendrimer[e] (g mol^{-1})	Ref.
11		20061[b]	21
12		13192[a]	22
13		13044[b]	23
14		11660[a]	24

8.2 Synthetic Methods | 255

15	25	11354[c]
16	26	11073[b]
17	27	10715[a]
18	28a	9720[a]
19	29	9337[a]

Table 8.1 (continued)

No.	Dendrimer constitution at a particular generation	Molecular weight of the largest dendrimer[e] (g mol^{-1})	Ref.
20		9158[b]	30
21		8824[b]	31
22		7928[b]	28b

8.2 Synthetic Methods

23	36 PF₆ — SiiPr₃ / SiiPr₃, 4
24	7606[a] 32
25	7549[b] 33
26	7508[b] 34
	7167[a] 12

Table 8.1 (continued)

No.	Dendrimer constitution at a particular generation	Molecular weight of the largest dendrimer[e] (g mol^{-1})	Ref.
27		6941[b]	35
28		6600[b]	36
29		6468[b]	37
30		5614[b]	38

8.2 Synthetic Methods | 259

Table 8.1 (continued)

No.	Dendrimer constitution at a particular generation	Molecular weight of the largest dendrimer[e] (g mol^{-1})	Ref.
36		4513[b]	13
37		3809[b]	42
38		3589[a]	43
39		3502[b]	44
40		3546[b]	45

8.2 Synthetic Methods | 261

41		46
		2461[a]
42		47
		2168[a]
43		48
		1918[b]
44		49
		1828[a]
45		50
		1740[a]
46		51
		1185[b]

[a] Divergent growth, [b] convergent growth, [c] double exponential growth, [d] orthogonal coupling method, [e] calculated molecular weight.

very large generations, essentially due to steric limitations on access to the reactive sites. The past decade has witnessed the identification of a number of new types of monomers, leading to new types of dendrimers.

Covalent bond formation is most important in dendrimer formation, followed by metal–ligand coordination bond formation [9] and non-covalent bond formation [10]. The large variety of dendrimers described in the literature also include several that were prepared for a particular type of study [11]. On the other hand, only a few dendrimers, namely, PAMAM [2], poly(propylene imine) (PPI) [12], polyaryl ether [5], polyamide [13], polyphosphane [8] and polysilane [14] have been utilized to conduct a number of further modifications and studies [4]. A compilation of dendrimers, constituted either by distinct monomers or the nature of linkage, is presented in Table 8.1. Note that the table presents most of the dendrimers conforming to the above classification.

8.3
Macromolecular Properties

8.3.1
Molecular Modeling and Intrinsic Viscosity Studies

The branches-upon branches architecture, nearly chain entaglement-free macromolecular structure, perfect structural homogeneities, nanometric dimensions and ability to present *endo*- and *exo*-receptor properties are inherent structural features of a dendrimer. It is still a challenging task, however, to characterize a dendrimer with respect to its solid state structural properties. Most of the physical methods of characterization are performed routinely to establish the constitution. The inability to determine the molecular structural details in the solid state led to investigations assessing the structural details through systematic molecular simulation methods. Gross structural features could still be obtained through small angle X-ray scattering (SAXS) and small angle neutron scattering (SANS) techniques in solution [52]. Molecular dynamics simulations [53] were useful to determine parameters such as radius of gyration, asphericity, end group distribution functions etc. The theoretical results from molecular dynamics simulation often compare favorably with results of experiments. Table 8.2 gives the data of radius of gyration, expressing the extent to which the dendritic molecules have expanded into space, for two prominent dendrimers, namely PAMAM [54] and PPI [55], and the series of PETIM dendrimers [56], reported recently. The table provides data from molecular modeling and the experimental data from SAXS and/or SANS.

Within the studied dendrimers, it is reasonably clear that the dimension of dendrimers increases with the generation number and is independent of the nature of the end group in the dendrimer. Molecular modeling studies have become more indispensable to gain many other molecular and supramolecular properties of dendrimers.

Table 8.2 A comparison of the radius of gyration from molecular modeling and SAXS/SANS experiments for a few dendrimers.

No. of terminal functional groups	PAMAM R_g (Å) [54]		PPI R_g (Å) [55]		PETIM R_g (Å) [56]	
	Modeling	SAXS	Modeling	SAXS	Modeling	SAXS
4	4.93		4.9	4.4	5.65	
8	7.46		6.0	6.9	8.90	7.86
16	9.17		7.4	7.3	14.14	10.74
32	11.23	15.8	10.0	11.6	17.73	14.88
64	14.5	17.1	12.5	13.1	21.13	
128	18.34	24.1	–		26.63	

Studies of intrinsic viscosity properties have opened a new dimension in understanding the structural properties of dendrimers. The intrinsic viscosities of linear and star polymers increase monotonically as the molecular weight increases. Experiments involving dendrimers have revealed that the observed intrinsic viscosities of a homologous series of dendrimers rise to a maximum with increasing generations before falling as the generations advance further (Fig. 8.3) [57]. The rise and fall in the intrinsic viscosities of increasing generations of dendrimers does not appear to be general. It was reported that the PPI series of dendrimers showed [55] relatively small changes in the intrinsic viscosities and no maximum could be found for up to five generations. The dependence of hydrodynamic vol-

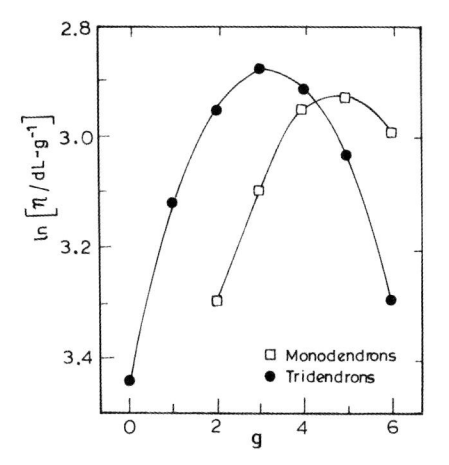

Fig. 8.3 A plot of intrinsic viscosities vs. generation number of dendrons. (Adapted from Ref. [57a].)

8.3.2
Fluorescence Properties

Dendritic architectures induce anomalous properties for the molecules. A recently reported observation on dendritic molecules is their fluorescence emission behavior. The carboxylic acid terminated PAMAM dendrimers were shown by Larson and Tucker to exhibit intrinsic fluorescence (λ_{ex} 380 nm; λ_{em} 440 nm) (Fig. 8.4) [58]. An excitation–emission matrix technique showed an overall increase in the relative fluorescence emission with increasing generations. The fluorescence lifetimes were found to lengthen with increasing generations. It was observed further by Bard and coworkers that ammonium persulfate treated hydroxy group functionalized PAMAM dendrimers (generations 0, 2 and 4) produced strong blue luminescence, with an emission band at 450 nm, upon excitation at 380 nm [59]. The formation of luminescence was ascribed to the presence of the hydroxy groups at the peripheries of the dendrimer, rather than the backbone of the dendrimer. A similar luminescence property was also observed for Au^{3+}-treated dendrimers, although the intensities were low, when compared to the persulfate-treated dendrimer. Following these reports, a strong fluorescence of PAMAM dendrimers (generations 2 and 4) was reported under acidic conditions [60a] and in ionic liquids [60b]. The fluorescence intensity maximum for generation 4 was at pH $\sim$ 2–5, whereas for the generation 2 dendrimer, the maximum intensity was observed at pH $\sim$ 4. It was reasoned that protonation of ter-

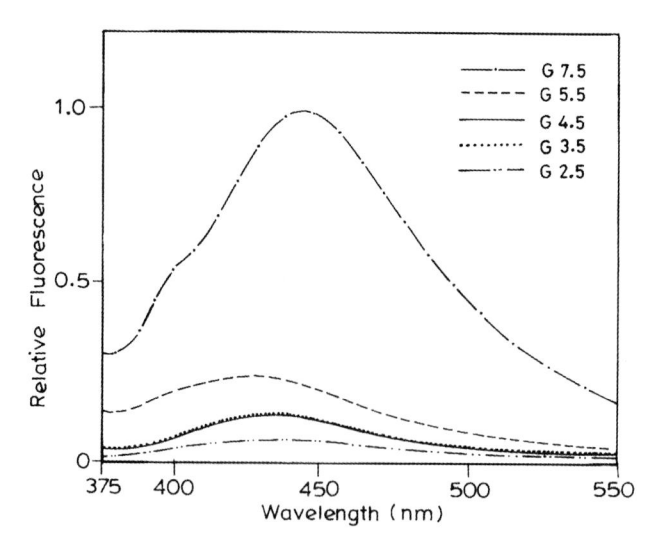

Fig. 8.4 Fluorescence emission profiles of a carboxylic acid-terminated PAMAM series of dendrimers (λ_{ex} = 380 nm). (Adapted from Ref. [58].)

tiary amines of the dendrimer enhanced the hydrogen bonding strengths at acidic pH and possible chemical reactions of the functional groups along the dendrimer branches for the observed fluorescence emission [60a]. The recently reported PETIM dendrimers have also been observed to exhibit fluorescence emission behavior (λ_{ex} 275 nm; λ_{em} 310 nm; λ_{ex} 330 nm; λ_{em} 390 nm) [61]. Quantum yields were measured to be in the order of 1 to 4.4%. pH-dependent emission was also observed for the PETIM dendrimers. These observations on different series of dendrimers indicate that the anomalous photophysical property could result from the unique dendritic architecture itself.

8.3.3
Endo- and Exo-Receptor Properties

The structurally homogeneous dendrimers are expected to have diameters in the range 2–7 nm, depending on the generation number, linker lengths and the nature of the solvent. The role of the solvent and its association with the constituents of dendrimers leads to either extended conformations or reduced conformations. Backfolding of the branches and chain ends towards the core may result in a dense core environment. Notwithstanding the constant dynamics associated with the fragments and the solvent association–dissociation, one of the recognized properties of dendrimers is their ability to act as hosts for small molecules [62]. An illustration of the encapsulation properties was the report by Meijer and coworkers on identifying the PPI dendrimers as the so-called 'dendritic boxes' [63]. Such dendritic boxes could encapsulate varied sizes of small organic molecules at sites of appropriate sizes within the dendritic structure. A box resulted when the peripheries of the dendrimer contained bulky protecting groups, by which the guest molecules were entrapped within the dendritic interiors. The guest release could, in turn, be augmented through removal of the protecting groups, thereby allowing the dendritic peripheries to become more open. The concept of small molecule encapsulation has been developed in chemical synthesis. A recent example of utilizing the inner cavities of phloroglucinol-based poly(alkyl aryl ether) dendrimers exemplified that selective photochemical changes could be accomplished on organic substrates that were encapsulated inside the dendrimer structure [64]. That the dendritic interior offers selective product formation could be compared to well-known hosts such as micelles and cyclodextrins. Encapsulation of functional moieties that form an integral part of the dendrimer constitution has also been studied in several instances [62a,b]. The covalent functionalization of a series of dendrimers with porphyrin as a core (Fig. 8.5) is one of the earliest examples of attempts to modulate the properties of such functional groups, as a result of their covalent dendritic encapsulation [65]. The effect of such encapsulation was found to be a large reduction in the reduction potentials of up to 300 mV.

Dendrimer-encapsulated nanoparticles is a thoroughly studied area of investigation [66]. The uniform composition, structure, nanometric dimensions, possibilities for encapsulation and reduction in agglomeration properties of encapsu-

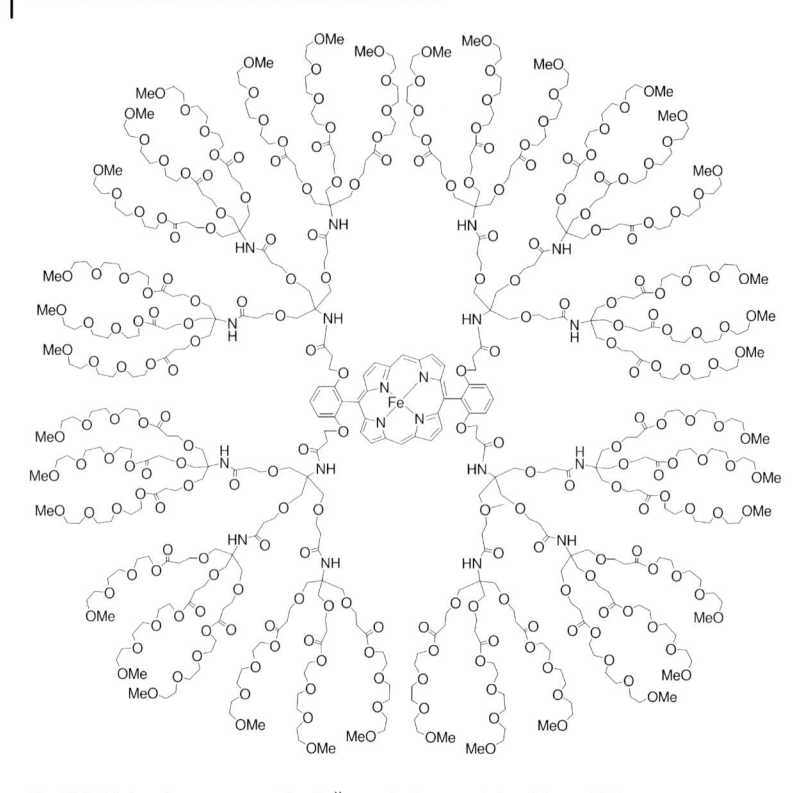

Fig. 8.5 Molecular structure of a Fe^{II}-porphyrin-cored dendrimer [65].

lated metals, are some of the features of dendrimers that form the basis of developing metal encapsulation by dendrimers. A number of catalytic and materials studies have been conducted with the dendrimer-encapsulated metals [66].

While studies targeting the utility of dendritic internal regions represent the *endo*-receptor properties, a large variety of studies has also been conducted utilizing the dendrimer peripheries as sites for functional group attachments. A homologous series of dendrimers, fully functionalized at the peripheries, provide a gradation of the number of such functional groups in the series. The associated functional group properties would then match according to the number of the groups. From this point of view, dendrimers are probably the only type of material where a precise and periodic number of functional groups can be installed. A number of studies concerning the functional and interfacial properties of different types of moieties have been conducted [67]. Observations ranging from the independent and unconnected nature of the functional groups [68] to the modulation and cooperativity [69] among the functional groups have been reported. Primary among the functional moieties that benefit from the dendritic arrangement are the organometallic catalysts [70] and small drug molecules for drug delivery purposes [71].

8.4
Chemical Sensors with Dendrimers

The unique architectural features of dendrimers have been applied systematically in chemical sensing. The sensors are often based on the formation of either nanoparticles or self-assembled mono- or multilayers or selective deposition of dendrimers as layers on sensors or dendrimer–polymer composites. Solution phase sensor studies have also been carried out in several instances. Vapor and gas sensing have been studied in detail, followed by organic substances, inorganic cations and anions. The following discusses the developments in the use of dendrimers as chemical sensors. The investigations are fairly recent and most of the developments have occurred within the last few years.

8.4.1
Vapor Sensing

Layer-by-layer assembly has been utilized to generate gold nanoparticle/ dendrimer composite films by Vossmeyer and coworkers [72]. Polyphenylene dendrimer was chosen, the hydrophobicity of this class of dendrimer being anticipated to enhance the possibilities for the detection of volatile organic compounds. The dendrimer was functionalized with thioctic acid at its peripheries and the dendrimer was then ligand exchanged with dodecylamine-stabilized gold nanoparticles, resulting in the disulfide-stabilized gold nanoparticles. Composite films were assembled layer-by-layer via alternated and repeated exposure to dodecylamine-stabilized gold nanoparticles and the dendrimer. The vapor sensing properties of the nanoparticles were measured by assessing the relative change in electrical resistance while exposing to various solvents. The response time upon exposure to vapors was within 5 s. Switching back to the initial state without vapor was possible upon exposure to air. The electrical resistivity response was excellent for organic solvent vapors and poor for hydroxylic solvents. The response was nearly nil for water (Fig. 8.6).

The electrical response of nanoparticles due to exposure to organic vapors was attributed to the swelling of the nanoparticle film, and also to changes in the dielectric properties of the film induced by sorption of the analyte. Further, the high sensitivities of dendrimer/gold nanoparticle were believed to be due to the open and shape persistent dendrimer structure that enhanced efficient uptake of the analyte molecules. The long chain thioctic acid was presumed to allow analyte-induced swelling of the composite material.

In addition to polyphenylene dendrimer, PAMAM and PPI dendrimers were also subjected to layer-by-layer assembly, so as to produce Au-nanoparticle/ dendrimer composite films [73]. Unlike in the case of polyphenylene dendrimers, which required further derivatization with thioctic acid, both PAMAM and PPI dendrimers were presented with amine terminal groups that acted as polyfunctional, chelating [74] and cross-linking ligands. These nanoparticle/dendrimer composites showed varied vapor sensing properties. The polyphenylene den-

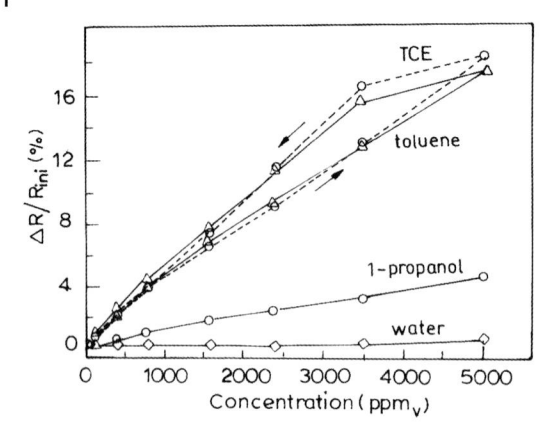

Fig. 8.6 Vapor sensing properties of a polyphenylene dendrimer gold nanoparticle, assessed through measurement of electrical resistance response vs. concentration of different solvent vapors. Arrows indicate measurement taken during increasing and decreasing concentrations of the vapor. (Adapted from Ref. [72].)

drimer containing Au nanoparticles showed strong response to toluene and no response to water. On the other hand, the PAMAM dendrimer containing nanoparticles showed strong response to water and little response to toluene. The PPI dendrimer containing nanoparticles showed intermediate behavior. These observations suggested that film sensitivities to different vapors were dependent on the nature of the dendrimer used.

The rigid conformational features of the polyphenylene dendrimers provide advantage for the vapor sensing properties of this class of dendrimers over the more flexible type dendrimers. Bargon, Müllen and coworkers studied [73d] the vapor sensing properties of a series of polyphenylene dendrimers, with and without further peripheral functionalization (for example, see Fig. 8.7), with the aid of the quartz crystal microbalance (QCM) technique. The sensor response, derived from QCM measurements for various organic vapors, was such that only polar aromatic compound vapors, such as acetophenone, aniline, benzaldehyde, benzonitrile, flourobenzene, nitrobenzene and 2-methylnitrobenzene were detected selectively. Chlorinated and unsubstituted aliphatic hydrocarbons, alcohols, amines, aldehydes and carbonyl compounds and their vapors were not detected. The observed selectivity was attributed to the formation of $\pi-\pi$ electron donor–acceptor complexes. Upon functionalizing the peripheries of dendrimers with groups such as cyano-, carboxyl-, or imino-substituents, the sensitivity to vapors of acetophenone, benzaldehyde or 1-methyl-2-pyrolidine could be enhanced. The acid-terminated dendrimers showed recognition of guest molecules carrying amine substituents and non-aromatic solvents such as acetone, acetonitrile, isopropyl methyl ketone and nitromethane. Such responses, selectivities and reproducibil-

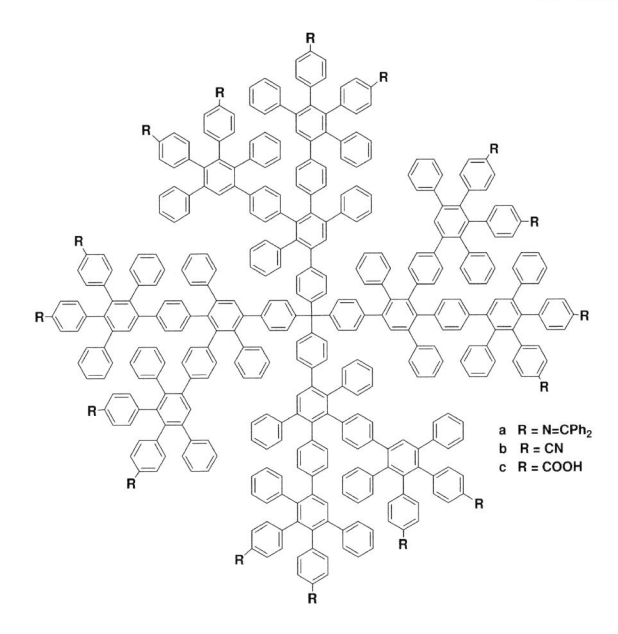

Fig. 8.7 Molecular structure of a functionalized polyphenylene dendrimer, used to detect polar aromatic molecules [73d].

a R = N=CPh$_2$
b R = CN
c R = COOH

ities were not observed for the hyperbranched polyphenylenes or for PAMAM dendrimers.

In one of the early studies on surface confinement of dendrimers suitable for purposes such as chemical sensing, Crooks and coworkers devised methods to chemically immobilize PAMAM dendrimers on to self-assembled monolayers (SAMs) [75]. Two strategies were developed. In one, the mercapto undecanoic acid SAM was coupled with amine terminated PAMAM dendrimers, followed by capping of the unreacted amine groups at the peripheries of the dendrimers with acid chlorides. In the second method, the dendrimer was functionalized with acid chlorides first, followed by their attachment to the SAM surface, by utilizing the remaining unreacted amine groups on dendrimers. Each of the above preparations provided different functional characteristics. A benzamido-functionalized dendrimer surface was produced by the above methods and these dendrimers were tested as chemically sensitive surfaces for detection of volatile organic compounds (VOC). It was found that higher sensor response, as measured through a surface acoustic wave (SAW) mass balance, was obtained for acid-terminated VOCs. Also, it was found that fourth generation dendrimer provided the maximum response among the dendrimers tested, namely, zero, second, fourth, sixth and eighth generations. It was reasoned that the fourth generation dendrimer was the smallest of the spheroidal dendrimers and the dendritic interiors were most accessible. Also, the surface prepared by the second method offered enhanced sensitivities.

8.4.2
Sensing Organic Amines and Acids

Quantification of organic amines and carboxylic acids was performed by Lewis and coworkers, using dendrimer–carbon composite materials [76]. In a study, PAMAM and PPI dendrimers were involved as insulating components of the carbon conductor composite during the vapor detection. The vapor detection was assessed through electrical resistance measurements by appropriately designed detectors comprised of dendrimer–carbon composites. The dendrimers were maintained in either the protonated or non-protonated free amine state. The protonated dendrimer thus formed good detectors for amine vapors through chemisorption, whereas, the non-protonated dendrimers provided the possibility to detect carboxylic acid vapors through a chemisorption process. It was possible to grade the type of amines, depending on the composition of dendrimer–carbon composite and the generation of the dendrimer involved. The dendrimer containing composite further showed ∼5000 times larger response to an amine, namely, butylamine, than non-dendrimeric carbon black polymer composites. The response time to reach maximum resistance change was 3 min and a duration of 50 min was required to purge the detectors to regenerate the surface devoid of the volatile organics. The non-protonated dendrimer–carbon composite detectors showed large response to different organic acids, such as, acetic, butyric, valeric and hexanoic acids. The dendrimer–carbon composite detectors were found to be ∼10^3–10^4 times more sensitive than non-dendrimeric insulating polymer–carbon black composites.

Modification of the electrodes to study analytes in aqueous solutions is an important requirement to detect, for example, organic contaminants in water. In order to develop an electoranalytical method, Godinez and coworkers utilized electrostatic aggregation of SAM layers, with either carboxylic acid or amine termini [77]. The corresponding oppositely charged PAMAM dendrimer was used for the electrostatic aggregation. The SAM–dendrimer aggregates, deposited on a gold surface, were tested as sensors for halogenated alkanoic acids, with the aid of electrochemical detection. It was found that the peak current ratio increased linearly with the concentration of the acids. The sensitivity of detection is much higher for the thiol-dendrimer modified gold surface than for base gold electrode. It was assumed that the improved performance of the thiol-dendrimer modified gold electrode was a consequence of the hydrophobic character of the alkanoic acids and the internal environment of the dendrimer molecules.

8.4.3
Vapoconductivity

Electrochemical methods seem to be more suitable for the detection of vapors than other detection techniques. The changing conductivities form the basis for observing the vapor sensing. "Vapoconductivity" was a term applied to denote the large increase in conductivities with conducting films. A pentaerythritol-cored *tris*-branched dendrimer with 36 carboxylic acid groups was functionalized with bi- and tetrathiophene moieties by Newkome and coworkers [78]. Oxidation of the dendritic thiophene moieties provided the corresponding cation radicals and such cation radical exposed dendrimer was used as a sensor for various volatile organic chemical vapors. Vapoconductivity experiments, using iodine oxidized film on glass with four probes, showed that the conductivity increased significantly for various organic vapors. The conductivity increase was nearly 800 times the original conductivity, in the case of polar organic compound vapors, namely, acetone vapors. It was presumed that plasticization of the conducting dendrimer film allowed faster electron transport between small stacks formed on individual thiophene containing dendrimer. The large increase in conductivity with dendrimer-containing films was found to be anomalous, since the corresponding polymeric thiophene films reduced the conductivity upon exposure to organic vapors.

8.4.4
Sensing CO and CO_2

Detection of gases such as carbon monoxide and carbon dioxide, that pose environmental, occupational and health hazards, has been explored by dendrimer film deposition on gold electrodes and electrical conductivity measurements. Ferrocenyl functionalized dendrimers and phthalocyanin-based dendrimers were utilized as substrates for the detection of gases. Kim and coworkers have studied carbon monoxide sensing by involving carbosilane dendrimer, with 48 peripheral ferrocenyl groups (Fig. 8.8) [79]. The dendrimer was deposited onto a silicon wafer, followed by deposition of Au for source and drain, and the electrical conductivities were measured. The current variation measurement, performed at various CO gas concentrations, showed that dendrimer film could detect up to 40% CO gas, in a mixture of CO and N_2 gases. Beyond this concentration, the conductance saturated. The transient response time was 50 s and the rising time was 150 s, during maximal absorption of the CO gas. The falling time was 420 s, after turning off the CO gas.

Dendritic phthalocyanins, in which three phthalocyanin moieties (Fig. 8.9) are incorporated in the dendrimer framework, were tested for the detection of CO_2 gas [80]. Phthalocyanin with Zn^{II} metal exhibited favorable CO_2 gas detection, in comparison to metal-free and Co^{II} metal containing phthalocyanins. The detections were performed by measuring the electrical conductivity of Zn^{II}-phthalocyanin dendrimer-coated interdigital transducers. The current increased sharply upon exposure to CO_2 gas. The transducer surface was recovered back upon purging with N_2 gas. The effect of various concentrations (500–8000 ppm) of CO_2 gas on the conductivity was assessed. The impedance spectroscopy was also studied and the impedance response changed significantly with exposure to high concentrations of CO_2.

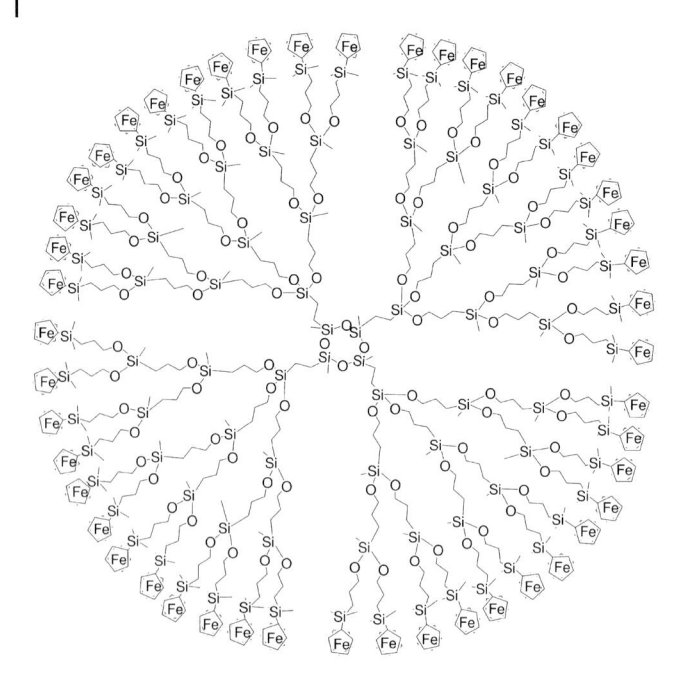

Fig. 8.8 Molecular structure of a ferrocenylated carbosilane dendrimer, used in carbon monoxide sensing [79].

8.4.5
Gas and Vapor Sensing in Solution

Studies of the gas and vapor sensing properties originating from dendrimers have been performed more often in the solution phase. Visible color changes, fluorescence, phosphorescence emissions and photoinduced electron transfer processes are a few of the detection methods that have been utilized to sense gases and vapors in solutions containing dendrimers.

From a series of studies, van Koten and coworkers have established the sulfur dioxide (SO_2) gas sensing abilities of aryl platinum^{II} complex dendrimers [81]. The so-called pincer ligands with NCN environment form square planar Pt^{II} complexes. Such Pt^{II} complexes, in organic solvents such as CH_2Cl_2, readily absorb SO_2 gas to form pentacoordinate complexes. The complexation process is accompanied by visible color change, i.e. change from colorless to bright orange on complexation to SO_2. Dendrimers were advanced as suitable supports to install several such active ligand–metal complexes. Figure 8.10 shows the molecular structure of a nanosize dendrimer containing six peripheral sensor units. When the dendrimer was exposed to atmospheric pressure of SO_2, an instantaneous color change occurred. Upon exposure to air or to solvents free of SO_2 or heating or reducing the pressure, the square planar Pt^{II} complex, devoid of SO_2 gas, was

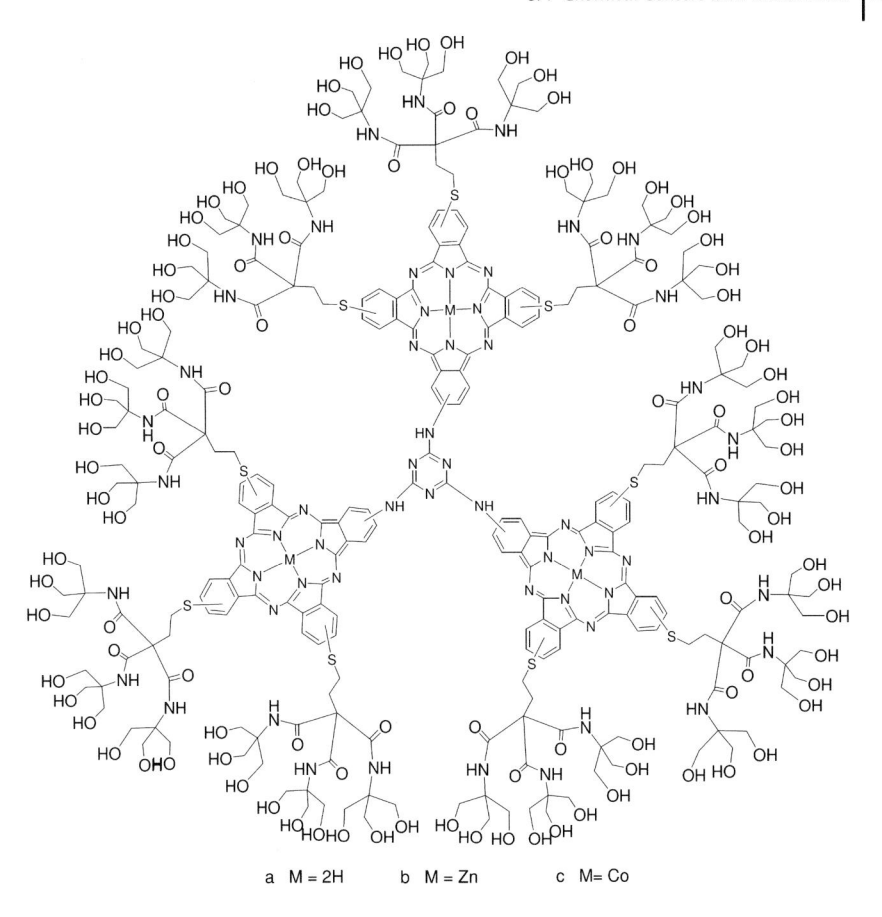

Fig. 8.9 Molecular structure of a branched porphyrin containing dendrimer, utilized as a CO$_2$ sensor [80].

regenerated in a few minutes. The shape persistent nature of the dendrimer allowed it to be recovered from solution by a nanofiltration technique.

Oxygen is an important molecule in the biological milieu and the detection of oxygen using dendrimer was studied by Vinogradov, Wilson and coworkers [82]. The detection was conducted through the ability of oxygen to quench phosphorescence. Pd or Pt complexes of porphyrins were studied due to their excellent phosphorescence properties. In order to enhance the two-photon absorption cross section, the porphyrin complexes were coupled with multiple copies of antenna chromophores, such that the two-photon excitation energy of the chromophore is channeled via energy transfer to the two-photon absorbing porphyrin complex. The antenna chormophore-porphyrin metal complex was further appended with dendritic moieties, so as to protect the complex from interactions with macromolecules of the medium and to control the oxygen diffusion to the complex.

Fig. 8.10 Changes in the coordination mode to the aryl-platinum complex containing dendrimer, upon absorption–desorption of SO_2 gas [81].

The molecular structure of a dendritic antenna-prophyrin metal complex is shown in Fig. 8.11. Systematic studies of the antenna-Pt porphyrin, modified with dendritic branches, demonstrated dramatic effects in oxygen quenching constants, relating directly to the effective protection of the phosphorescent core by the dendritic branching. The oxyen quenching constant of the dendritic com-

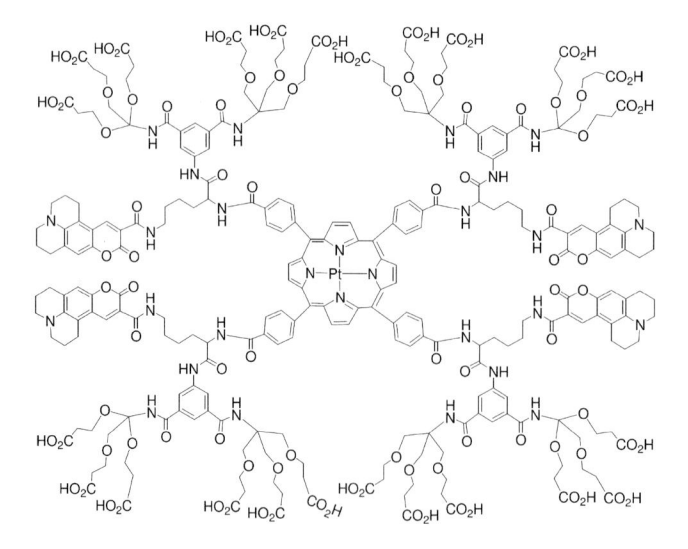

Fig. 8.11 Molecular structure of coumarin-containing Pt-*meso*-(4-aminophenyl) porphyrin dendritic antenna, designed for studies of oxygen sensing [82].

8.4.6
Chiral Sensing of Asymmetric Molecules

Site-selective positioning of functional entities within the dendritic architecture has the potential to bring new types of functions to the dendrimer. With an aim to explore new properties evolving from the dendritic architecture, Jiang, Aida and coworkers investigated the chiral sensing of asymmetric molecules by functional groups incorporated with the dendritic backbone [83]. Specifically, zinc porphyrins were incorporated as a structural element of dendrimers (for example, Fig. 8.12). A series of dendrimers, exhibiting two branches and three branches at the branch junctures were tested for chiral sensing of *RR*-, *SS*- and meso guest

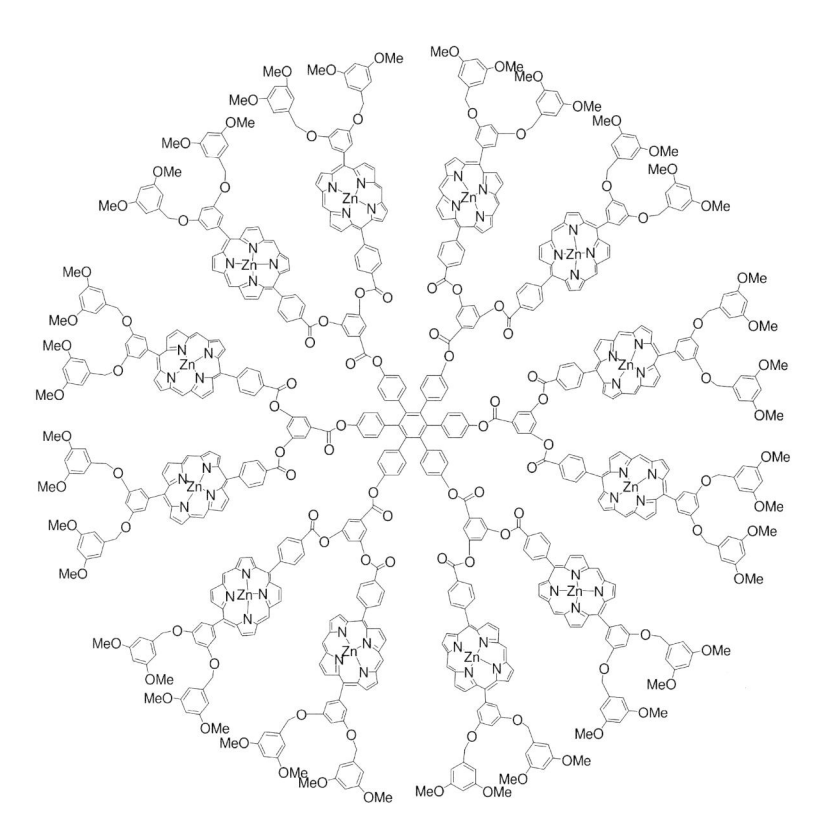

Fig. 8.12 Molecular structure of a branched Zn-porphyrin-containing dendrimer, which acted as a better chirooptical sensor of asymmetric compounds [83].

molecules. It was observed that the dendrimer possessing 12 zinc porphyrins at the intermediate layer of the dendrimer exhibited chiral sensing of the enantiomers. The sensing was monitored by induced circular dichroism and guest binding affinity measured by absorption spectroscopy. While the doubly branched dendrimer exhibited the sensing properties, the triply branched dendrimer did not show any cooperative behavior in chiral sensing. For both guest binding and chiro-optical sensing, the doubly branched zinc porphyrin was identified to be optimal. It was presumed that the doubly branched dendrimers possessed an efficient H-type aggregate of the zinc porphyrin units upon interaction with the chiral guest molecules and such H-type aggregate might have flexibility to twist the conformation to accommodate the chiral molecules, resulting in large induced circular dichroism responses.

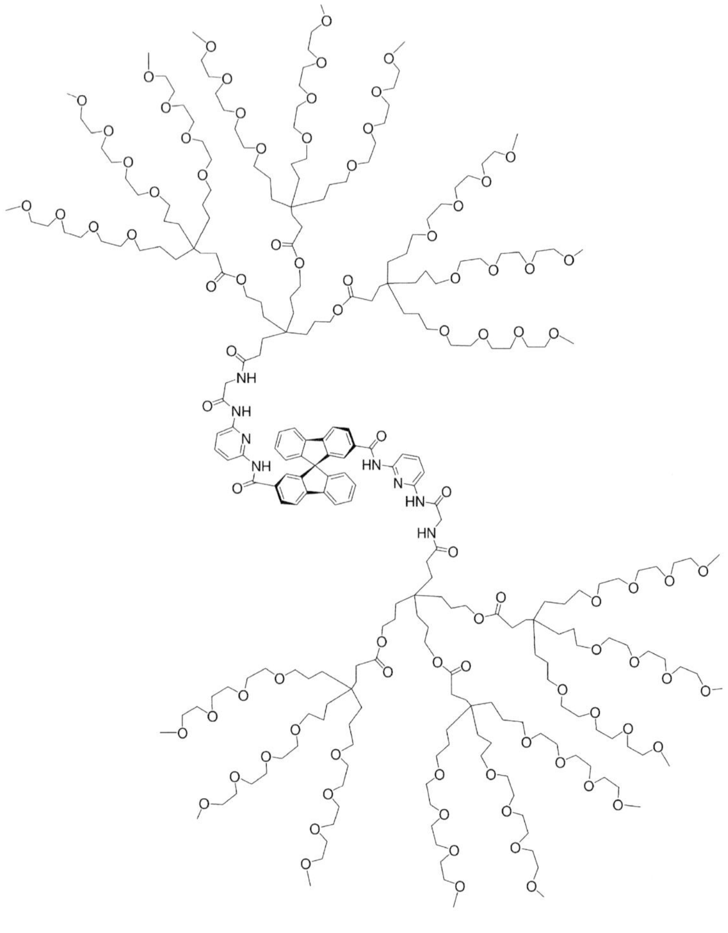

Fig. 8.13 A second generation 9,9′-spiro[9H-fluorene] cored dendrimer, which showed enantio- and diastereoselective complexation of sugars [84].

Enantio- and diastereomeric selection in chiral guest molecule sensing was investigated ealier by Diederich and Smith [84]. Rigid, optically pure 9,9′-spirobi[9H-fluorene]-cored first and second generation dendrimers were synthesized. These dendrimers were incorporated with hydrogen bonding cleft, at the interior segment of the structure and polyethylene glycol formed the peripheries (Fig. 8.13). Chiral sensing of D- and L-octylglucosides, as assessed by ^{1}H NMR chemical shifts of an amide group, had shown that while the association constants were similar to the sugars tested, there was reduction in the enantioselectivities and enhancements in the diastereoselectivities in the complexation in the series of generations 0 to 2. It was presumed that steric demands might disfavor certain complexation and also that the hydrogen bonding network with the guest might change the binding selectivities.

8.4.7
Fluorescence Labeled Dendrimers and Detection of Metal Cations

The fluorescence method is one of the frequently practiced detection methods in chemical and biological systems. In order to generate fluorescent dendrimers, one of the widely studied strategies is to incorporate fluorescence response moieties to pre-formed dendrimers [85]. A polylysine dendrimer functionalized with 24 dansyl units at its periphery (Fig. 8.14) was synthesized and its photochemical

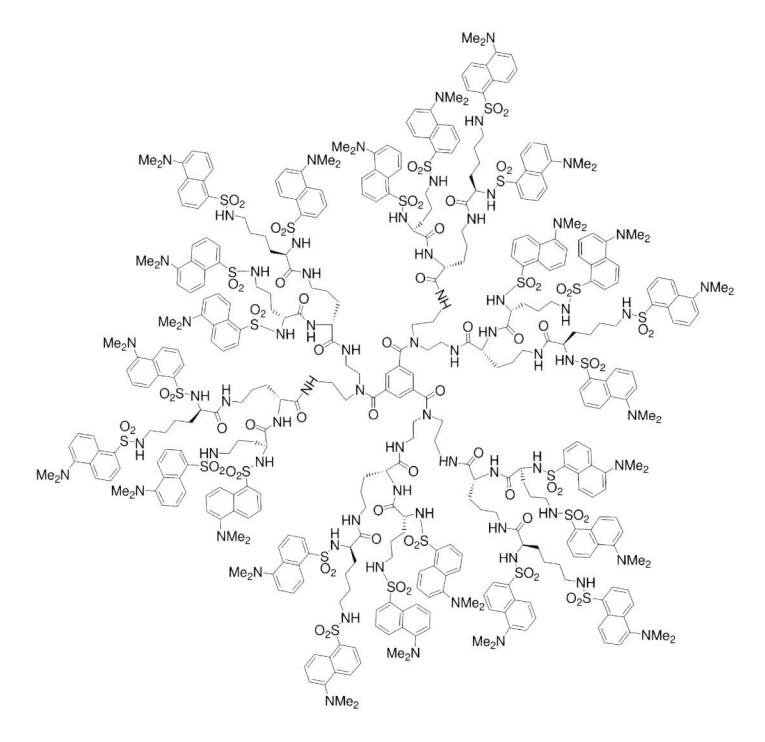

Fig. 8.14 Molecular structure of a multiple dansyl group containing dendrimer, utilized in metal ion sensing studies [86].

and photophysical properties investigated by Balzani, Vögtle and coworkers [86]. The dansyl chromophore exhibited intense absorption bands in the near-UV spectral region and a strong fluorescence in the visible region. Addition of a metal ion (Ni^{2+}, Co^{2+} and Zn^{2+}) and a base was observed to quench the fluorescence intensity of the dansyl group in the dendrimer and such quenching was absent for the monomeric dansyl group. It was rationalized that the metal ions coordinated to several amide units of the dendrimer, in a cooperative manner. It was observed that every metal ion (Ni^{2+} and Co^{2+}) quenched the fluorescence of about nine dansyl units and this effect was termed as sensing signal amplification in the quenching of a fluorescent dendrimer. Further Zn^{2+} ions displaced the coordinated Ni^{2+} and Co^{2+} metal ions. It was proposed that signal amplification effects in a dendritic structure should be useful in the design of sensitive fluorescent chemosensors.

In addition to polylysine dendrimers, PPI dendrimers were also functionalized with dansyl units at their peripheries and similar fluorescence quenching upon addition of Co^{2+} was demonstrated [87].

Functionalization of dendrimers with photofunctional moieties allows one to devise chemical sensors. Grabchev and coworkers designed and tested several 1,8-napthalimide derivatives functionalized dendrimers (Fig. 8.15) [88]. Inhibition of a photoinduced electron transfer (PET) process between the electron-

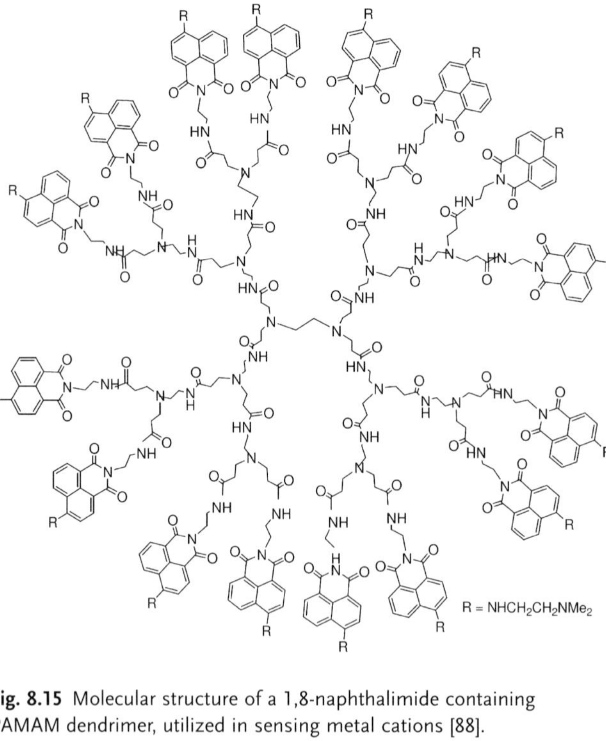

Fig. 8.15 Molecular structure of a 1,8-naphthalimide containing PAMAM dendrimer, utilized in sensing metal cations [88].

donating amino group and the electron-accepting naphthalimide was utilized as a means of sensing analyte molecules. In the absence of any analyte, a PET process operates, such that no fluorescence could be detected for the naphthalimide functionalized dendrimers. On the other hand, deactivation of the donor moiety through complexation with an analyte inhibited the PET process, allowing evolution of fluorescence. By this strategy, proton and a number of transition metal ions were tested as analytes of the sensor. The best performance of the sensor was observed with Cu^{2+} and Fe^{3+} ions.

A good sensor response was also found for Cu^{2+} and Co^{2+} ions, in a closely related study [89]. In this study, a PET process was not involved as a source of sensing, rather, efficient fluorescence quenching through electron or energy transfer reaction between the metal complex formed by the dendrimer core and 1,8-naphthalimide fluorophore units at the dendrimer peripheries was used. Metal ion detections were performed at submillimolar to micromolar concentrations. The PET-based principle to detect rare earth metal cations (Er^{3+}, Eu^{3+}, Gd^{3+}, Nd^{3+}, Tb^{3+}, Yb^{3+}) has also been tested [90].

8.4.8
Anion Sensing

Molecular recognition moieties present at the peripheries of dendrimers act as *exo*-receptors. In a series of studies, Astruc and coworkers have developed metal-

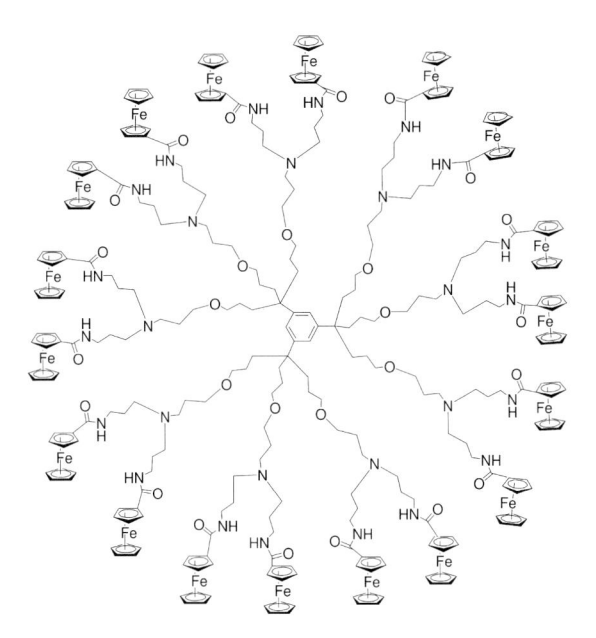

Fig. 8.16 Molecular structure of a ferrocene containing *nona*-branching cored dendrimer, used in anion sensing [92].

locene and Fe redox-active cluster functionalized dendrimers as hosts for recognition of various anions [91]. The ferrocene functionalized dendrimers of various generations, presenting 1, 3, 9 and 18 ferrocene moieties (for example, see Fig. 8.16) at their peripheries, were subjected to sensor studies with various anions, such as, HSO_4^-, Cl^-, NO_3^- and $H_2PO_4^-$. The molecular recognition studies were conducted through cyclic voltammetry, through the formation of ferricinium cation. The recognition of anions followed as a result of electrostatic interaction and hydrogen bonding between the amide linkage of the dendrimer and the anion. Better selectivity and sensing was observed for the larger generation dendrimer and with $H_2PO_4^-$ anion. This sensitivity and selectivity associated with increasing generations of dendrimer was termed the 'dendritic effect' in this study [92].

In a following study, the above authors utilized Fe_4 cluster as the peripheral functional groups (for example, see Fig. 8.17) and it was found that ATP^{2-} recognition occurred better than other anions, including $H_2PO_4^-$ [93]. This specificity was rationalized on the basis of the mutual nanosize of the Fe_4 clusters and ATP^{2-}, facilitating their interaction. Here too, a 'dendritic effect' was observed for higher generation dendrimer compared to the lower generation dendrimer, for selective ATP recognition. Further, the corresponding cobaltocenium functionalized dendrimers were also found to be sensors for small inorganic anions,

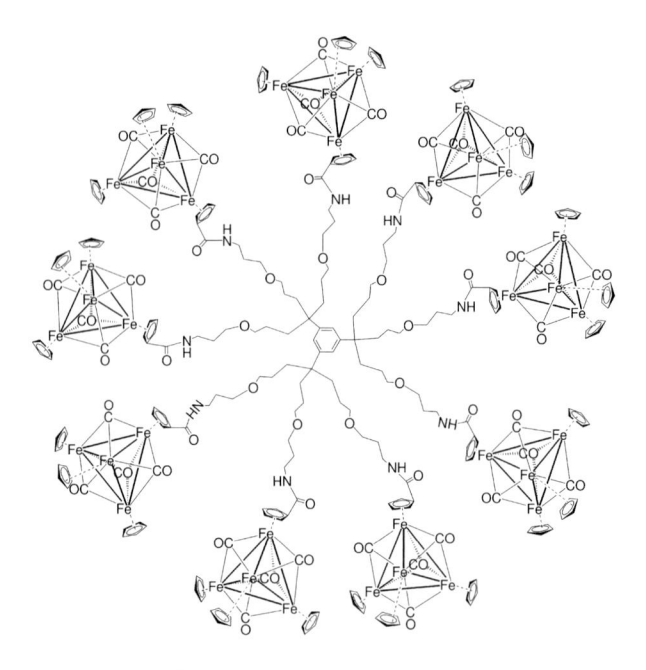

Fig. 8.17 Molecular structure of a redox-active [{cpFe(μ_3-CO)}$_4$] cluster containing dendrimer, which was found to exhibit selective binding to ATP^{2-} [93].

such as, $H_2PO_4^-$, HSO_4^- and Cl^-, as monitored by cyclic voltammetry and ^1H NMR spectroscopy [94].

Dendrimers with silicon linkages were subjected to ferrocene functionalization at the peripheries of the dendrimers, by Losada and coworkers [95]. The presence of –NH groups within the dendritic structure was established to be important for anion recognition, since in the absence of the –NH group the sensing of anions, as assessed by cyclic voltammetric experiments, did not occur. Anion recognition studies have also been performed by immobilizing the dendrimer onto electrodes and the feasibility of preparing an electrochemical sensor for recognition of anions was demonstrated.

Li^+-selective [96] and Ba^{2+}-selective dendrimers [97], silicon-based and TTF functionalized phosphorothioate dendrimers, were studied by electrochemical techniques, so as to evolve dendrimer-based sensors for these cations.

8.5
Dendrimer-based Biosensors

Biocompatible and non-toxic water-soluble dendrimers are useful in biological studies [98]. Many dendrimers have been tested as to their extent of acceptance in the biological context. For example, PAMAM and PPI dendrimers were studied for hemolytic cleavage of red blood cells. It was found that PAMAM and PPI dendrimers exhibited toxicity, but their toxicity was brought down to acceptable levels by modification of the surface functionalities with poly(ethylene glycol) or a carboxylic acid group [99]. In the series of PETIM dendrimers, no measurable cytotoxicity was observed and the results implied that these dendrimers having biocompatible functionalities would be suitable for *in vitro* biological studies [100]. In this section, the use of dendrimers in developing biosensors is discussed.

A biosensor is a device used to monitor and quantify biological reactions. The design and development of robust biosensors is an active area of investigation. The nanometric dendritic macromolecules have been subjected to biosensor studies in several instances in recent years. The stepwise chemical synthesis of a dendrimer provides the required flexibility to modify the constituent nature of the macromolecules. In addition, the extent of functional group loading on to dendrimer peripheries, periodic changes in the densities of the active moieties with the aid of different generations, and the possibilities for encapsulation of small molecules are some of the benefits of dendrimers upon which biological studies involving these macromolecules are undertaken. Major advances in the development of biosensors that utilize dendrimers are discussed below.

8.5.1
Acetylcholinesterase Biosensor

Biosensors based on acetylcholinesterase (AChE) are used extensively to detect low concentrations of pesticides. Organocarbamates and organophosphates,

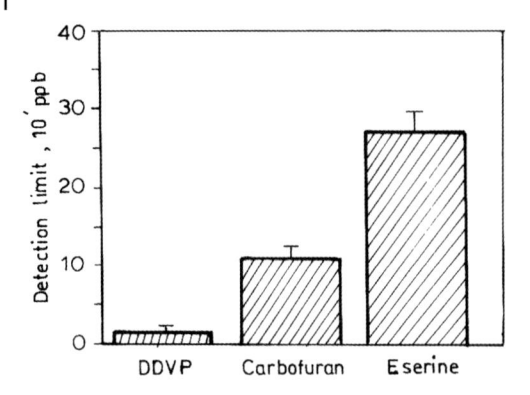

Fig. 8.18 Plot of detection limits of various pesticides. (Adapted from Ref. [101].)

major constituents of pesticides, irreversibly inhibit the AChE activity, concerned with the hydrolysis of acetylcholine to choline and acetic acid. PAMAM dendrimers have been used to design a biosensor for detection of such pesticides [101]. Fourth generation PAMAM dendrimer mixed with 1-hexadecanethiol was deposited onto a gold surface, into which enzymes AChE and choline oxidase were incorporated. The activities of the enzymes were measured through oxidation of H_2O_2, generated *in situ*, by the enzymes in the process of acetyl choline hydrolysis and oxidation of choline to betaine. The electrochemically measured biosensor assay was assessed for inhibition by pesticides, such as dimethyl-2,2-dichlorovinyl phosphate (DDVP), carboform and drug eserine. Figure 8.18 shows the inhibitory effect as a function of various pesticide concentrations. The most sensitivity was achieved for a concentration up to 1.3×10^{-3} ppb, which turned out to be ~1000 times better than potentiometric sensors. This high sensitivity was credited to the architectural features of dendrimers, providing higher surface densities of enzymes, and the ability to diffuse H_2O_2 efficiently to the electrodes, so as to enhance the sensor sensitivity. Detections through potentiometry and amperometry have also been demonstrated on electrodes modified with PAMAM dendrimers [102].

8.5.2
Dendrimers as Cell Capture Agents

Dendrimers serve as highly efficient capture agents of bacterial strains. Tabacco and coworkers have examined the PAMAM dendrimers as cell capture agents in an effort to develop real-time detection of bacterial contamination in dynamic aqueous environments [103]. The technique adopted was (i) deposition of dye encapsulated dendrimer on to the silanized glass plates (sensor plates); (ii) injection of contaminated aqueous samples passing through the sensor plate and (iii) detection of the dye with a blue laser light. The physically encapsulated dye (FAST DiA) in fourth generation PAMAM dendrimer contained quarternary ammonium groups at the peripheries. These ammonium groups were coupled covalently to the glutaraldehyde-coated glass plates. The dendrimers showed excellent physical stability and cell capture ability. The sensor detected 1×10^4 cells/mL of bacterial strain *E. coli* within 1 min of injection and the operational lifetime of the sensor was at least 64 h. A good selectivity for a bacterial strain was also demonstrated for the dendrimer-based biosensor. The major attributes of the dendrimer component within the sensor assembly were (i) covalent attachment of the quarternary ammonium functionality at the sensor surface; (ii) entrapment of the dye molecule within the dendrimer interiors; (iii) high affinity for the bacterial cells, even under the flow condition of the bacterial cell containing analyte. Current operational and storage durations were claimed to be beneficial for routine use in the field.

8.5.3
Dendrimers as a Surface Plasmon Resonance Sensor Surface

The macromolecular structures of the dendrimers were investigated to generate ultrathin film surfaces, in place of well-known polymers. Mark and coworkers employed an amine-terminated fourth generation PAMAM dendrimer as a macromolecular support on a gold surface, so as to evolve a surface plasmon resonance sensor surface [104]. Traditionally, macromolecules such as carboxymethylated dextrans have been used as the support in order to generate a sensor assembly. In order to minimize background effects, originating from such surfaces, the dendrimer was explored as a useful alternative. In this work, the dendrimer was linked covalently to an aminoundecane thiol functionalized surface. The surface was modified further with desired ligands, proteins and nucleic acids, through covalent bond formation. Various analytical methods were applied to establish the layer-by-layer architecture of the assembly. The extent of covalent functionalization of the ligands over the dendrimer supported sensor surface was much higher and this increased amount of loading was accounted for by the net increase in the total surface-accessible area that accompanied the binding of the dendrimer molecules and the large number of amine surface groups available for covalent coupling of the ligands. This particular study opens up the possibility for totally integrated optical biosensors and ultrathin films, employing dendrimer-based surface chemistries.

8.5.4
Layer-by-Layer Assembly Using Dendrimers and Electrocatalysis

Layer-by-layer (LBL) assembly is a powerful technique to assemble nanometric film fabrications, with precise control of film thickness. In an effort to address pertinent issues associated with surface enhanced Raman scattering intensities (SERS) of colloidal metallic solutions, dendrimers were involved to produce dendrimer–metal nanocomposite films by the layer-by-layer technique [105]. Increasing generations of amine-terminated PPI dendrimers utilized in the study showed higher Ag adsorption characteristics, reflecting the effect of the large number of peripheral groups available for the larger generation dendrimers, when compared to lower generation dendrimers. The layer-by-layer assembly of the dendrimer–Ag nanocomposite films was prepared by successive alternating immersions of glass slides in solutions of dendrimers and the colloidal Ag particles. Strong SERS enhancements were observed for analyte 2-naphthalenethiol cast on to the LBL films and such enhancements were significantly higher for the films prepared with higher concentration and higher generations of dendrimers. These results were attributed to the exceptional characteristics of dendrimers and the optical properties of the Ag nanoparticles.

The LBL assembly involving dendrimer was studied in detail in combination with heme protein and the catalytic efficiencies of the assembly by electrocatalysis was investigated [106]. LBL assembly of fourth generation PAMAM dendrimer and the heme proteins was developed, by employing the electrostatic interaction between these two macromolecules. The control of the film composition and thickness was assessed by UV spectroscopy, quartz crystal microbalance and cyclic voltammetry studies. The catalytic activities of the heme proteins to reduce substrates such as O_2, H_2O_2 and trichloroacetic acid were tested and the dendrimer–heme films were shown to provide a favorable environment for the heme proteins to transfer electrons with underlying electrodes. The favorable electrochemical and electrocatalytic properties of PAMAM/protein films were anticipated to have potential applicability in developing new types of electrochemical biosensors or bioreactors, without using any mediators.

8.5.5
SAM–Dendrimer Conjugates for Biomolecular Sensing

Efficient analytical tools assume significance in monitoring biologically important recognition processes. Foremost of the analytical tools are the electrochemical assays, relying on cyclic voltammetric, potentiometric and amperometric methods. Using these methods, the biomolecular recognition processes are eventually transduced as electrochemical signals. In one strategy, Yoon and coworkers applied dendrimer-coated electrode surfaces as the ligand component for their interaction with receptor containing analytes [107]. The biospecific interaction was initiated upon functionalizing the dendrimer-coated electrode surface with the ligand, namely biotin, in a particular study. Covalent attachment of the components was carried out. Thus, activated ester-terminated thiol SAM on the electrode surface was coupled with fourth generation PAMAM dendrimer. The presence of a dense amine group remains available for further reaction, after initial amide bond formation with the SAM surface. The free amine groups were functionalized as amides, linking the biotinyl groups. This constellation formed the biospecific ligand-presenting electrode surface. The recognition of antibiotin antibodies, which were labeled previously, was monitored through an enzyme catalyzed oxidation of a substrate, namely, 4-chloro-1-naphthol. The oxidized product of this substrate, benzo-4-chloro cyclohexadienone, was deposited on the elec-

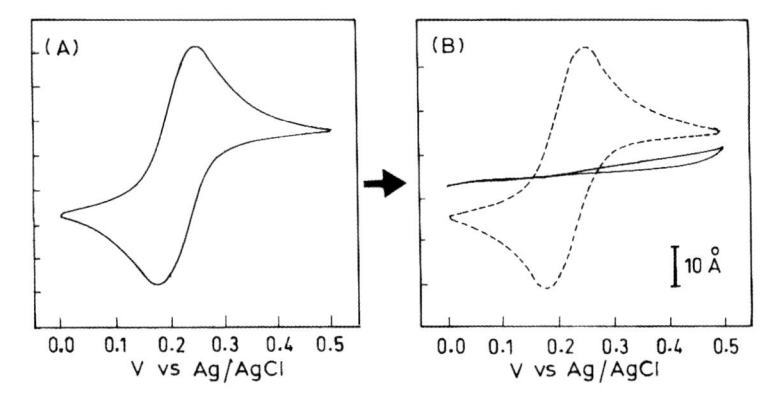

Fig. 8.19 Cyclic voltammograms representing the ferrocene/ferricenium couple, which acted as an electroactive signal tracer of a biocatalytic reaction [107].

trode surfaces. The extent of electrode surface deposition was, in turn, assessed through ferrocene methanol as the electroactive signal tracer. The dotted trace in the cyclic voltammogram represents the ferrocene/ferricenium couple (Fig. 8.19). The continuous trace represents the attempted redox reaction of the ferrocene/ferricenium couple, after the biospecific recognition process. With no distinguishable redox wave, the biospecific recognition blocked the electrode surface nearly completely, such that the redox reaction of ferrocene methanol was completely inhibited. The advantages of dendrimers as a platform, in comparison to polymers, includes the unique characteristics of dendrimer architectures, monodispersity, nanometric dimension and the presence of a large number of peripheral functional groups, onto which further functionalization with ligands of choice could be carried out.

In a related work, Kim and coworkers demonstrated surface coverage of dendrimer–ligand functionalized electrode films by antibody molecules as a measure of changes in the signal transduction [108]. Specifically, activated ester-terminated SAM was functionalized with amine-terminated fourth generation PAMAM dendrimer. The unreacted amino group in the dendrimer was double functionalized with a biotin derivative and ferrocene molecules. The affinity interaction of biotin molecules by anti-biotin IgG was followed by measuring the extent of surface coverage, as a result of steric inhibition of the biotin–antibody complex, through glucose oxidase mediated oxidation of glucose. When antibody recognition did not occur, the ferrocene moieties, present in the dendrimer molecule, mediated the electron transfer onto the electrode, thereby generating the signal. On the other hand, the antibody recognition resulted in an antigen–antibody complex at the electrode surface, and this, in turn, blocked ferrocene moieties as electron transfer mediators, resulting eventually in the loss of signal transduction. Modification through activation of external ferrocenylated dendrimer was also required, in order to minimize the nonspecific adsorption of glu-

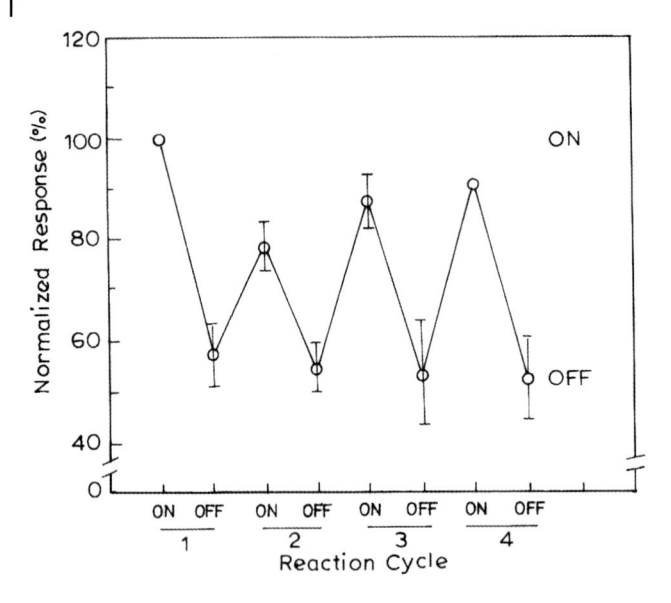

Fig. 8.20 The 'on' and 'off' peak current changes, resulting from antigen–antibody complexation–decomplexation reaction cycles. (Adapted from Ref. [108].)

cose oxidase on to the electrode surface. The antigen–antibody decomplexation was also achieved through addition of free biotin, which resulted in regeneration of the signal transduction. The 'on' and 'off', relating to association and dissociation, respectively, peak current changes in the cyclic voltammogram were also demonstrated (Fig. 8.20).

Partially ferrocene tethered fourth generation amine-terminated dendrimer was utilized to couple with enzyme glucose oxidase (periodate oxidized), deposited on to a cystamine-modified gold surface [109]. With the available unreacted amino groups in dendrimers, the second layer of the enzyme was coupled, which, in turn, could be subjected to react with another layer of ferrocene-tethered dendrimer. In repeating the process, multiple layers of ferrocenylated dendrimer–enzyme could be obtained (Fig. 8.21). The catalytic efficiencies of the enzyme/dendrimer-supported electrodes were tested and the bioelectrocatalytic signals from multilayered electrodes could reach 90% of the steady state values within 6 s. Further, the multilayered electrodes were found to be stable over a period of several days, thereby, demonstrating the advantages of dendrimers as supports for the biosensing purpose.

The enhanced capacity to immobilize ligands at the peripheries of dendrimers was utilized for electrochemical detection of DNA molecules [110]. A constellation of covalent coupling of ferrocenylated dendrimer onto carboxylate functionalized SAM, attachment of a capture probe at the peripheries of dendrimers, hybridization of oligomer DNA, complementary to the capture probe and further hybridization with an avidin conjugated alkaline phosphatase labeled DNA

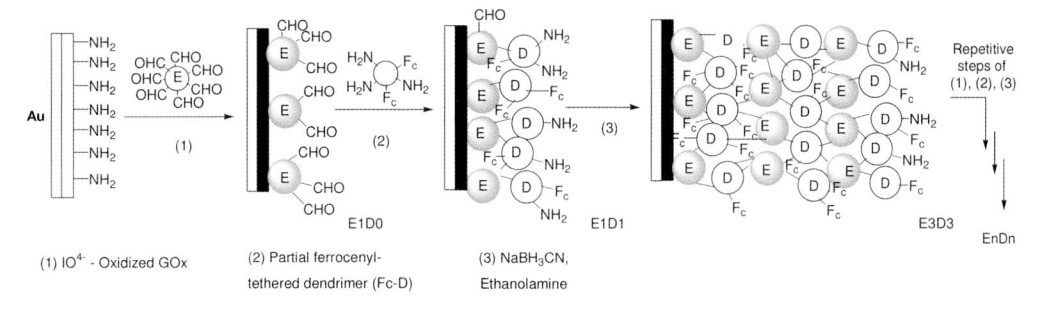

Fig. 8.21 Assembly of multiple layers of ferrocenylated PAMAM dendrimer–enzyme conjugate performed to construct reagentless enzyme electrodes. (Adapted from Ref. [109].)

oligomer formed the sensor assembly. The detection of target DNA was facilitated through alkaline phosphatase mediated hydrolysis of p-aminophenyl phosphate to p-aminophenol which was, in turn, oxidized to the corresponding quinonoid form, by a ferrocene/ferrocenium couple, thereby generating an electrochemical signal. A detection limit of 20 fmol was observed, in addition to achieving sequence selectivity in base-pairs during hybridization. The role of the ferrocenylated dendrimer here was first, to allow efficient attachment of the probe DNA on the sensor surface, through free amine moieties of the dendrimer and secondly, to act as an electrocatalyst to mediate electron transfer between the enzymatic reaction product and the electrode by means of a ferrocene/ferrocenium couple. The stepwise assembly of the DNA sensing electrode is depicted in Fig. 8.22.

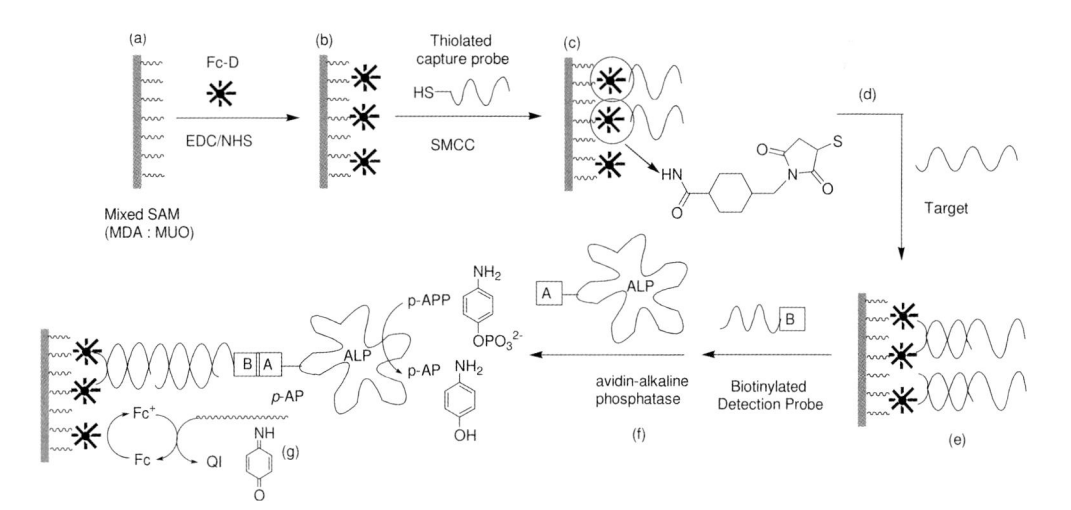

Fig. 8.22 The step-wise assembly of a DNA sensing electrode, is conjugated with a dendrimer component. (Adapted from Ref. [110].)

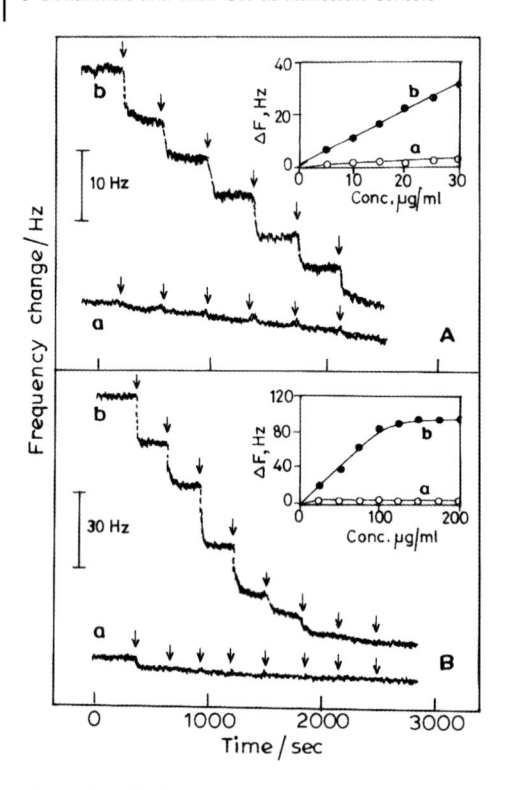

Fig. 8.23 QCM frequency change vs. time plots of conventional (a) and dendritic (b) DNA biosensor, with responses representing 5 (A) and 25 (B) µg mL^{-1} of a target 38-mer DNA (Adapted from Ref. [111].)

Unlike in the case of covalent attachment of probe DNA to SAMs, functionalized dendrimers such as PAMAM dendrimers constituted with single-stranded DNA oligomers have been explored as DNA biosensors [111]. The DNA dendrimers, containing 30 single-stranded arms, specific to a waterborne pathogen, namely, *Cryptosporidium parvum*, was immobilized on to a gold-coated quartz crystal and multiple layers of the dendrimer were deposited. The molecular weight of the fourth generation dendrimer was 8×10^6 Da, with a diameter of 120 nm. The quartz crystal microbalance technique was applied to monitor the oligonucleotide hybridization and the minimum detection limit was 1 µg/mL of target DNA. Hybridization reversibility was not, however, assessed, due to concerns of thermal melting and QCM resonant frequency. Figure 8.23 illustrates the detections observed as a function of QCM frequency change vs. time.

8.5.6
Dendrimer-based Calorimetric Biosensors

Calorimetric biosensors based on polydiacetylene and dendrimers have been developed, by utilizing polydiacetylenes as a visual color detector [112]. Mixed amine and hydroxy groups-terminated PAMAM dendrimers were used to produce diacetylene and biotin ligand functionalized dendrimers. The diacetylenes were further polymerized to polydiacetylenes to obtain highly networked polymers. The presence of several ligands in each dendrimer molecule allowed sensing receptors with fast response times and sensitivities. The advantage associated with polydiacetylenes was the visible color change as a function of recognition of specific analytes, or change in temperature.

8.5.7
Dendrimer-based Glucose Sensors

Developing a continuous monitoring method for glucose levels in blood samples is an active area of research, due to the utmost significance relating to blood sugar levels and the associated diabetes disease. One of the potentially applicable methods of continuous blood glucose level monitoring is an implantable chemical assay system, which relies on competitive binding of glucose to a protein–polysaccharide complex. The competitive binding of glucose is detected through fluorescence measurement. The protein of choice is a labeled lectin concanvalin A and the polysaccharide is a labeled dextran. Due to the proximity between the donor and acceptor fluorophore labels in the protein and polysaccharide, a fluorescence resonance energy transfer (FRET) occurs, and the fluorescence emission is monitored. Upon competitive binding of glucose, the efficiency of FRET reduces dramatically and an altered fluorescence emission occurs. The ratio of these two emission peaks provides a quantitative estimate of glucose concentration. A drawback identified in this assay was that the competitive binding of glucose to the protein–polysaccharide complex might be hindered due to incomplete dissociation of the complex, restricting the dynamic response of the assay. In an illustrative series of work, Coté and coworkers applied the dendrimer technology in the assay system [113]. Thus, glycosylated PAMAM dendrimers were applied in place of the dextran to produce the glycodendrimer–Con A complex. Competitive inhibition of this binding process by glucose was found to be superior. The wavelength of fluorescence emission of the labels was found to be non-detrimental. A high dynamic response, representing an effective inhibition of glycodendrimer–Con A binding by glucose, was also observed. The stability of the biosensor system was found to be only up to 4 days and the competitive inhibition was lost after this period. A schematic representation, as proposed by the authors, is presented in Fig. 8.24.

In addition to covalent coupling of dendrimers on carboxylic acid alkane thiol-coated SAMs, strong physisorption was also undertaken to produce a glucose biosensor, involving dendrimers. Hianik and coworkers developed a glucose biosensor based on the formation of dendrimer monolayers [114]. Specifically, a mixed monolayer of hexylmercaptane and PAMAM dendrimers, generations zero, one and four, was prepared on a gold support. The glucose-oxidizing flavin enzyme, namely, glucose oxidase, was adsorbed over the monolayer. The biosensor consisting of repeated dendrimer–glucose oxidase monolayers was also tested. Amperometry was used to monitor the oxidation of glucose by dendrimer glucose

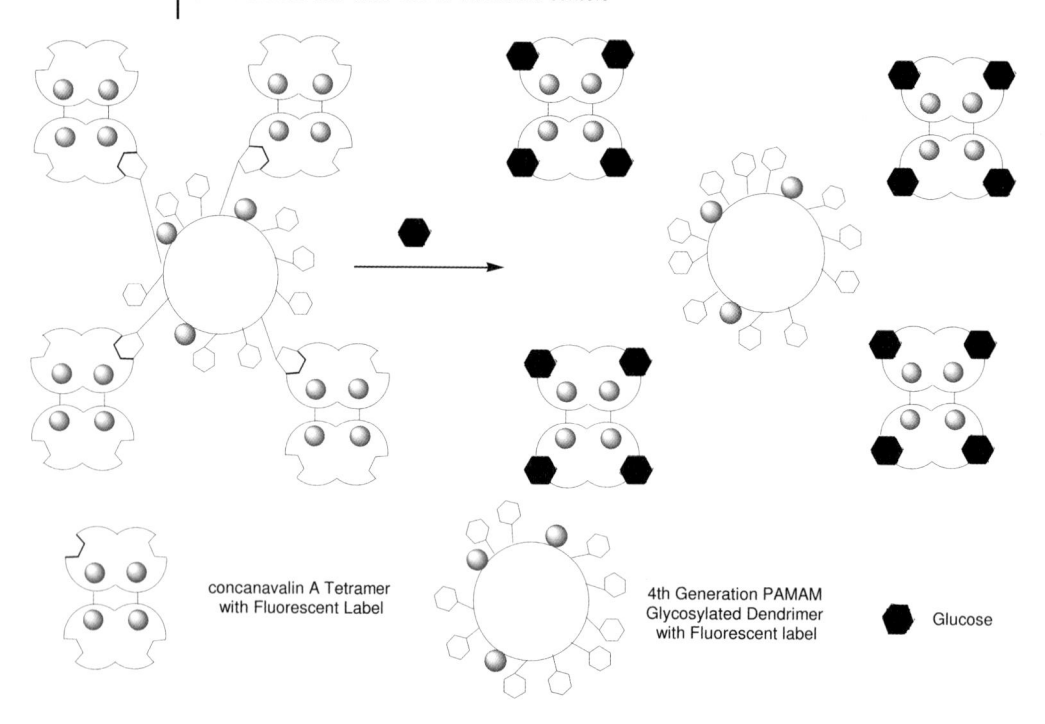

Fig. 8.24 A schematic representation of competitive binding of glucose to a glycodendrimer-con A complex. (Adapted from Ref. [113].)

oxidase monolayers, and it was observed that a low concentration of glucose provided a large current flow, with subsequent saturation at higher glucose concentration. Also, larger generation dendrimer increased the amperometric response of the sensor. This increase in response was attributed to (i) the increased volume of the dendrimer interior, which facilitates diffusion of small molecules, such as H_2O_2, to the electrode surface and (ii) the increased number of binding sites for the enzyme. The dendrimer–enzyme-containing biosensor was found to be stable for several days. Further modifications, involving cross linking of dendrimers with the enzyme, with the aid of glutaraldehyde, improved the sensitivity, response time, detection limit, turn over and stability of the biosensor [115]. The authors have also studied the extent of surface coverage by a mixed layer of alkane thiol–dendrimer and its influence on the activities of the enzyme [116].

Dendrimer encapsulated platinum particles have been utilized in biosensing applications. Carbon nanotubes functionalized with Pt-containing dendrimer was deposited with enzyme glucose oxidase onto the carbon nanotube, as well as onto the dendrimer scaffold [117]. This assembly was tested as a glucose biosensor, the response time and the detection limits were found to be 5 s and 2.5 µM of glucose, respectively. The encapsulation of Pt nanoparticles in the interior of the dendrimer facilitated the electron transfer, during oxidation, onto the electrodes.

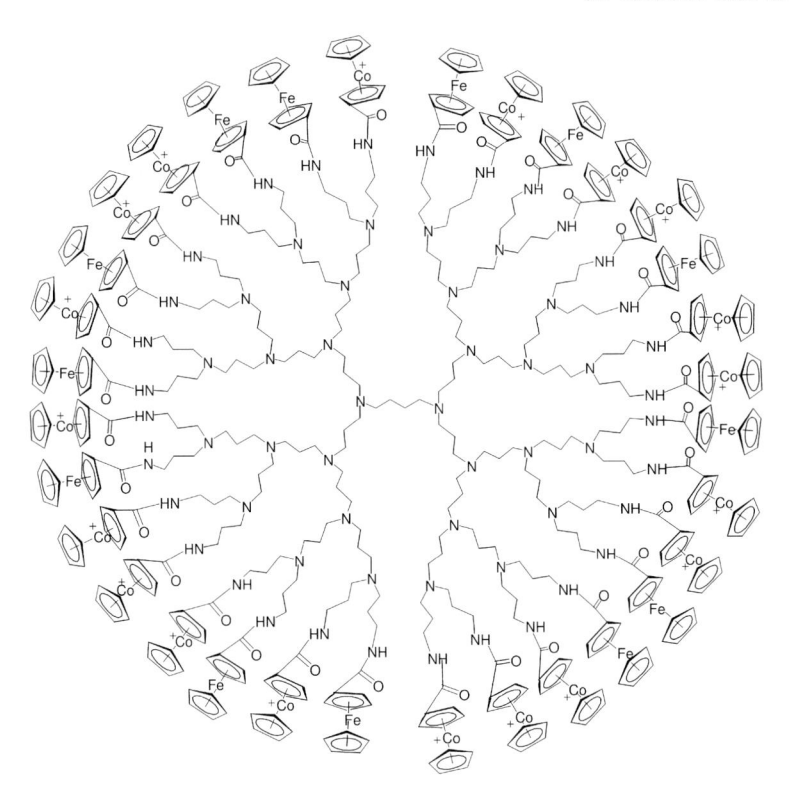

Fig. 8.25 Molecular structure of a ferrocene and cobaltocene-containing PPI dendrimer, utilized for glucose sensing under aerobic and anaerobic conditions [118].

Dendrimer-based glucose sensors have been approached by different methods. Dendrimers with two different electroactive moieties, namely ferrocene and cobaltocene (Fig. 8.25), were prepared and tested as glucose sensors, by Alonso and coworkers [118]. Dendrimers from 1 to 4 generations, coated with varied amounts of ferrocene and cobaltocene units, were deposited onto platinum or glassy carbon electrodes. The dendrimer-functionalized electrodes were further treated with enzyme glucose oxidase, such that dendrimer–enzyme complexation occurred through electrostatic interaction. The advantage of the mixed dendrimer deposition was demonstrated through glucose oxidation analysis under (a) aerobic and (b) anaerobic conditions. Under anaerobic conditions, the ferrocene/ferrocenium moiety mediated the electron transfer and a significant increase in the currents was observed. On the other hand, under aerobic conditions, the cobaltocene moieties acted as the mediator of the electron transfer, and again, significant changes in the volatmmograms were observed. The stability of the biosensor was also observed to be good. Under the anaerobic conditions, the higher generation dendrimers were more efficient as electron transfer mediators. In the presence of the mediator, i.e. oxygen, the sensitivity of the sensors increased with decreasing dendrimer generation. The authors have also previously reported ferrocenyl silicon-based dendrimers as mediators of glucose oxidation [119].

8.6
Conclusion and Outlook

From the forgoing compilation, it is evident that dendritic macromolecules have firmly established their importance in a variety of studies. Comparing the number of studies involving dendrimers, syntheses of new types of dendrimers themselves are lagging far behind. The perfect monodispersed constitution of dendrimers demands the synthesis to be multi-step in nature, which, in turn, demands that the chosen monomers and the reactions be efficient. With increasing avenues for application, synthesis of dendrimers tailored for a desired purpose will become imminent and the knowledge available currently should guide the efforts in synthesis. Similarly, molecular modeling studies have become essential to understanding the structure–property relationship of studies involving dendrimers, especially in the absence of atomic level details from solid state analysis. Anomalous properties of dendritic structures, such as intrinsic viscosities and fluorescence properties, will be useful when evolving new types of studies with dendrimers. With particular reference to chemical and biosensor investigations, the structural and architectural features of nanometric dendrimers have already been realized and the exploitation of dendrimers for the purpose of sensors has been phenomenal within the last few years. In this respect, real time applications, reaching the field, are distinct possibilities in the near future.

Acknowledgment

The author thanks the Department of Science and Technology, New Delhi and the Council of Scientific and Industrial Research, New Delhi, for financial support for the work in the area of dendrimers. Thanks are due to Mr. J. Nithyanandhan, for his help with the preparation of this chapter.

References

1 E. Buhleier, W. Wehner, F. Vögtle, *Synthesis* **1978**, 155.
2 D. A. Tomalia, H. Baker, J. Dewald, M. Hall, G. Kallos, S. Martin, J. Roeck, J. Ryder, P. Smith, *Polym. J.* **1985**, *17*, 117.
3 E. Goldsmith, S. Sprang, R. Fletterick, *J. Mol. Biol.* **1982**, *156*, 411.
4 (a) *Top. Curr. Chem.* **1998**, *197*; (b) *Top. Curr. Chem.* **2000**, *210*; (c) *Top. Curr. Chem.* **2001**, *212*; (d) *Top. Curr. Chem.* **2001**, *217*; (e) *Top. Curr. Chem.* **2003**, *228*.
5 C. J. Hawker, J. M. J. Fréchet, *J. Am. Chem. Soc.* **1990**, *112*, 7638.

6 (a) K. E. Uhrich, S. Boegeman, J. M. J. Fréchet, S. R. Turner, *Polym. Bull.* **1991**, *25*, 551; (b) R. Splider, J. M. J. Fréchet, *J. Chem. Soc., Perkin Trans. 1* **1993**, 913; (c) T. Kawaguchi, L. Walker, C. L. Wilkins, J. S. Moore, *J. Am. Chem. Soc.* **1995**, *117*, 2159; (d) F. Zeng, S. C. Zimmerman, *J. Am. Chem. Soc.* **1996**, *118*, 5326; (e) R. Klopsch, P. Frnake, A.-D. Schlüter, *Chem. Eur. J.* **1996**, *2*, 1330.

7 J. Ruiz, G. Lafuente, S. Marcen, C. Ornelas, S. Lazare, E. Cloutet, J.-C. Blais, D. Astruc, *J. Am. Chem. Soc.* **2003**, *125*, 7250.

8 N. Launay, A.-M. Caminade, J.-P. Majoral, *J. Am. Chem. Soc.* **1995**, *117*, 3282.

9 (a) E. C. Constable, *Chem. Comm.* **1997**, 1073; (b) C. Gorman, *Adv. Mater.* **1998**, *10*, 295; (c) G. R. Newkome, E. He, C. N. Moorefield, *Chem. Rev.* **1999**, *99*, 1689; (d) F. J. Stoddart, T. Welton, *Polyhedron* **1999**, *18*, 3575; (e) A.-M. Caminade, V. Maraval, Valerie; R. Laurent, J.-P. Majoral, *Curr. Org. Chem.* **2002**, *6*, 739; (f) Y. Tor, *Comp. Ren. Chim.* **2003**, *6*, 755; (g) K. Onitsuka, S. Takahashi, *Top. Curr. Chem.* **2003**, *228*, 39; (h) P. A. Chase, R. J. M. K. Gebbink, G. van Koten, *J. Organomet. Chem.* **2004**, *689*, 4016.

10 (a) F. Zeng, S. C. Zimmerman, *Chem. Rev.* **1997**, *97*, 1681; (b) H.-J. van Manen, F. C. J. M. van Veggel, D. N. Reinhoudt, *Top. Curr. Chem.* **2001**, *217*, 121; (c) D. K. Smith, *Chem. Comm.* **2006**, 34.

11 For example; (a) C. C. Mak, H.-F. Chow, *J. Chem. Soc., Chem. Commun.* **1996**, 1185; (b) S. A. Ponomarenko, N. I. Boiko, V. P. Shibaev, R. M. Richardson, I. J. Whitehouse, E. A. Rebrov, A. M. Muzafarov, *Macromolecules* **2000**, *33*, 5549; (c) A. Adronov, J. M. J. Fréchet, *Chem. Commun.* **2000**, 1701; (d) M. A. Carnagan, M. W. Grinstaff, *Macromolecules* **2001**, *34*, 7648; (e) D. Lagnoux, E. Delort, C. Douat-Casassus, A. Esposito, J.-L. Reymond, *Chem. Eur. J.* **2004**, *10*, 1215.

12 E. M. M. de Brabander-van den Berg, E. W. Meijer, *Angew. Chem. Int. Ed.* **1993**, *32*, 1308.

13 (a) G. R. Newkome, Z. Yao, G. R. Baker, V. K. Gupta, *J. Org. Chem.* **1985**, *50*, 2004; (b) G. R. Newkome, R. K. Behera, C. N. Moorefield, G. R. Baker, *J. Org. Chem.* **1991**, *56*, 7162.

14 (a) A. W. van der Made, P. W. N. M. van Leeuvan, *J. Chem. Soc., Chem. Commun.* **1992**, 1400; (b) J. W. J. Knapen, A. W. van der Made, J. C. de Wilde, P. W. N. M. van Leeuwen, P. Wijkens, D. M. Grove, G. van Koten, *Nature* **1994**, *372*, 659; (c) D. Seyferth, D. Y. Son, *Organometallics* **1994**, *13*, 2682; (d) A. M. Herring, B. D. Steffey, A. Miedaner, S. A. Wander, D. L. DuBois, *Inorg. Chem.* **1995**, *34*, 1100. (e) M. Petricci-Samija, V. Guillemette, M. Dasgupta, A. K. Kakkar, *J. Am. Chem. Soc.* **1999**, *121*, 1968.

15 K. Onitsuka, A. Shimizu, S. Takahashi, *Chem. Commun.* **2003**, 280.

16 V. Percec, W.-D. Cho, M. Möller, S. A. Prokhorova, G. Ungar, D. J. P. Yeardley, *J. Am. Chem. Soc.* **2000**, *122*, 4249.

17 (a) T. R. Krishna, N. Jayaraman, *J. Org. Chem.* **2003**, *68*, 9694; (b) G. Jayamurugan, N. Jayaraman, *Tetrahedron* **2006**, *62*, 9582.

18 W. T. S. Huck, F. C. J. M. van Veggel, D. N. Reinhoudt, *Angew. Chem. Int. Ed.* **1996**, *35*, 1213.

19 R. H. E. Hudson, M. J. Damha, *J. Am. Chem. Soc.* **1993**, *115*, 2119.

20 H. Meier, M. Lehmann, U. Kolb, *Chem. Eur. J.* **2000**, *6*, 2462.

21 K.-I. Sugiura, H. Tanaka, T. Matsumoto, T. Kawai, Y. Sakata, *Chem. Lett.* **1999**, 1193.

22 J. Nithyanandhan, N. Jayaraman, *J. Org. Chem.* **2002**, *67*, 6282.

23 P. R. Ashton, K. Shibata, A. N. Shipway, J. F. Stoddart, *Angew. Chem. Int. Ed.* **1997**, *36*, 2781.

24 M. Petrucci-Samija, V. Guillemette, M. Dasgupta, A. K. Kakkar, *J. Am. Chem. Soc.* **1999**, *121*, 1968.

25 H.-T. Chang, C.-T. Chen, T. Kondo, G. Siuzdak, K. B. Sharpless, *Angew. Chem. Int. Ed.* **1996**, *35*, 182.

26 O. Enoki, H. Katoh, K. Yamamoto, *Org. Lett.* **2006**, *8*, 569.
27 M. A. Carnagan, M. W. Grinstaff, *Macromolecules* **2001**, *34*, 7648.
28 (a) E. C. Constable, C. E. Housecroft, M. Cattalini, D. Phillips, *New. J. Chem.* **1998**, 193; (b) E. C. Constable, P. Harverson, *Inorg. Chim. Acta* **1996**, *252*, 9.
29 D. Seyferth, D. Y. Son, *Organometallics* **1994**, *13*, 2682.
30 S. Wang, R. C. Advincula, *Org. Lett.* **2001**, *3*, 3831.
31 A. Rajca, S. Utamapanya, *J. Am. Chem. Soc.* **1993**,*115*, 10688.
32 F. Morgenroth, E. Reuther, K. Müllen, *Angew. Chem. Int. Ed.* **1997**, *36*, 631.
33 H. Ihre, A. Hult, E. Söderlind, *J. Am. Chem. Soc.* **1996**, *118*, 6388.
34 Y. Ishida, M. Jikei, M.-a. Kakimoto, *Macromolecules* **2000**, *33*, 3202.
35 H.-T. Chen, F. Neerman, A. R. Parrish, E. E. Simanek, *J. Am. Chem. Soc.* **2004**, *126*, 10044.
36 P. Murer, D. Seebach, *Angew. Chem. Int. Ed.* **1995**, *34*, 2116.
37 W. J. Feast, S. P. Rannard, A. Stoddart, *Macromolecules* **2003**, *36*, 9704.
38 N. Vijayalakshmi, U. Maitra, *J. Org. Chem.* **2006**, *71*, 768.
39 H.-F. Chow, I. Y.-K. Chan, C. C. Mak, M.-K. Ng, *Tetrahedron* **1996**, *52*, 4277.
40 E. W. Kwock, T. X. Neenan, T. M. Miller, *Chem. Mater.* **1991**, *3*, 775.
41 G. R. Newkome, C. N. Moorefield, G. R. Baker, G. R. Saunders, S. H. Grossman, *Angew. Chem. Int. Ed.* **1991**, *30*, 1178.
42 C. Xia, X. Fan, J. Locklin, R. C. Advincula, *Org. Lett.* **2002**, *4*, 2067.
43 G. M. Dykes, L. J. Brierley, D. K. Smith, P. T. McGrail, G. J. Seeley, *Chem. Eur. J.* **2001**, *7*, 4730.
44 T. M. Miller, T. X. Neenan, R. Zayas, H. E. Bair, *J. Am. Chem. Soc.* **1992**, *114*, 1018.
45 F. M. H. de Groot, C. Albrecht, R. Koekkoek, P. H. Beusker, H. W. Scheeren, *Angew. Chem. Int. Ed.* **2003**, *42*, 4490.
46 J. Hu, D. Y. Son, *Macromolecules* **1998**, *31*, 8644.
47 A. Miedaner, C. J. Curtis, R. M. Barkley, D. L. DuBois, *Inorg. Chem.* **1994**, *33*, 5482.
48 J. Louie, J. F. Hartwig, *J. Am. Chem. Soc.* **1997**, *119*, 11695.
49 A. Sekiguchi, M. Nanjo, C. Kabuto, H. Sakurai, *J. Am. Chem. Soc.* **1995**, *117*, 4195.
50 C. Clausnitzer, B. Voit, H. Komber, D. Voigt, *Macromolecules* **2003**, *36*, 7065.
51 M. Nanjo, A. Sekiguchi, *Organometallics* **1998**, *17*, 492.
52 A. N. Ozerin, D. I. Svergun, V. V. Volkov, A. I. Kuklin, V. I. Gordelyi, A. Kh. Islamov, L. A. Ozerina, D. S. Zavorotnyuk, *J. Appl. Crystallogr.* **2005**, *38*, 996.
53 A. M. Naylor, W. A. Goddard III, G. E. Keiffer, D. A. Tomalia, *J. Am. Chem. Soc.* **1989**, *111*, 2339.
54 P. K. Maiti, T. Çağin, G. Wang, W. A. Goddard III, *Macromolecules* **2004**, *37*, 6236.
55 R. Scherrenberg, B. Coussens, P. van Vliet, G. Edouard, J. Brackman, E. De Brabander, *Macromolecules* **1998**, *31*, 456.
56 C. Jana, G. Jayamurugan, R. Ganapathy, P. K. Maiti, N. Jayaraman, A. K. Sood, *J. Chem. Phys.* **2006**, *124*, 204719.
57 (a) T. H. Mourey, S. R. Turner, M. Rubinstein, J. M. J. Fréchet, C. J. Hawker, K. L. Wooley, *Macromolecules* **1992**, *25*, 2401; (b) C. Cai, Z. Y. Chen, *Macromolecules* **1998**, *31*, 6393; (c) M. Jeong, M. E. Mackay, R. Vestberg, C. J. Hawker, *Macromolecules* **2001**, *34*, 4927; (d) P. M. Drew, D. B. Adolf, *Soft Matter* **2005**, *1*, 146.
58 C. L. Larson, S. A. Tucker, *Appl. Spectrosc.* **2001**, *55*, 679.
59 W. I. Lee, Y. Bae, A. J. Bard, *J. Am. Chem. Soc.* **2004**, *126*, 8358.
60 (a) D. Wang, T. Imae, *J. Am. Chem. Soc.* **2004**, *126*, 13204; (b) J.-F. Huang, H. Luo, C. Liang, I.-W. Sun, G. A. Baker, S. Dai, *J. Am. Chem. Soc.* **2005**, *127*, 12784.
61 G. Jayamurugan, N. Jayaraman, unpublished observation.
62 (a) D. K. Smith, F. Diederich, *Chem. Eur. J.* **1998**, *4*, 1353; (b) S. Hecht, J. M. J. Fréchet, *Angew. Chem. Int. Ed.*

2001, *40*, 74; (c) D. L. Richter-Egger, T. Tesfai, S. A. Tucker, *Anal. Chem.* **2001**, *73*, 5743; (d) S. Chen, Q. Yu, L. Li, C. L. Boozer, J. Homola, S. S. Yee, S. Jiang, *J. Am. Chem. Soc.* **2002**, *124*, 3395.

63 (a) J. F. G. A. Jansen, E. M. M. de Brabander-van den Berg, E. W. Meijer, *Science* **1994**, *266*, 1226; (b) J. F. G. A. Jansen, E. M. M. de Brabander-van den Berg, E. W. Meijer, *J. Am. Chem. Soc.* **1995**, *117*, 4417.

64 (a) L. S. Kaanumalle, J. Nithyanandhan, M. Pattabiraman, N. Jayaraman, V. Ramamurthy, *J. Am. Chem. Soc.* **2004**, *126*, 8999; (b) L. S. Kaanumalle, R. Ramesh, V. S. N. M. Maddipatla, J. Nithyanandhan, N. Jayaraman, V. Ramamurthy, *J. Org. Chem.* **2005**, *70*, 5062.

65 P. J. Dandliker, F. Diederich, M. Gross, C. B. Knober, A. Louati, E. M. Sanford, *Angew. Chem. Int. Ed.* **1994**, *33*, 1739.

66 (a) R. M. Crooks, M. Zhao, L. Sun, V. Chechik, L. K. Yeung, *Acc. Chem. Res.* **2001**, *34*, 181; (b) S. Deng, J. Locklin, D. Patton, A. Baba, R. C. Advincula, *J. Am. Chem. Soc.* **2005**, *127*, 1744; (c) Z. B. Shifrina, M. S. Rajadurai, N. V. Firsova, L. M. Bronstein, X. Huang, A. L. Rusanov, K. Müllen, *Macromolecules* **2005**, *38*, 9920; (d) T. Goodson III, O. Varnavski, Y. Wang, *Int. Rev. Phys. Chem.* **2004**, *23*, 109.

67 (a) D. C. Tully, J. M. J. Fréchet, *Chem. Commun.* **2001**, 1229; (b) F. Vögtle, S. Gestermann, R. Hesse, H. Schwierz, B. Windisch, *Prog. Polym. Sci.* **2000**, *25*, 987.

68 For example, see: (a) H.-F. Chow, C. C. Mak, Chi Ching, *J. Chem. Soc., Perkin Trans. 1*, **1997**, 91; (b) P. B. Rheiner, D. Seebach, *Chem. Eur. J.* **1999**, *5*, 3221; (c) C.-O. Turrin, J. Chiffre, J.-C. Daran, D. de Montauzon, A.-M. Caminade, E. Manoury, G. Balavoine, J.-P. Majoral, *Tetrahedron* **2001**, *57*, 2521. (d) J. Nithyanandhan, R. Davis, S. Das, N. Jayaraman, *Chem. Eur. J.* **2004**, *10*, 689.

69 For example, see: (a) R. Breinbauer, E. N. Jacobsen, *Angew. Chem. Int. Ed.* **2000**, *39*, 3604; (b) J. Ruiz, G. Lafuente, S. Marcen, C. Ornelas, S. Lazare, E. Cloutet, J.-C. Blais, D. Astruc, *J. Am. Chem. Soc.* **2003**, *125*, 7250; (c) S. Wang, B. S. Gaylord, G. C. Bazan, *Adv. Mater.* **2004**, *16*, 2127.

70 (a) D. Astruc, F. Chardac, *Chem. Rev.* **2001**, *101*, 2991; (b) R. van Heerbeek, P. C. J. Kamer, P. W. N. M. van Leeuwen, J. N. H. Reek, *Chem. Rev.* **2002**, *102*, 3717.

71 (a) M. Liu, J. M. J. Frechet, *Pharm. Sci. Technol. Today* **1999**, *2*, 393; (b) K. Kono, *Drug Delivery System* **2002**, *17*, 462; (c) T. D. McCarthy, P. Karellas, S. A. Henderson, M. Giannis, D. F. O'Keefe, G. Heery, J. R. A. Paull, B. R. Matthews, G. Holan, *Mol. Pharm.*, **2005**, *2*, 312.

72 T. Vossmeyer, B. Guse, I. Besnard, R. E. Bauer, K. Müllen, A. Yasuda, *Adv. Mater.* **2002**, *14*, 238.

73 (a) N. Krasteva, I. Besnard, R. E. Bauer, K. Müllen, A. Yasuda, T. Vossmeyer, *Nano Lett.* **2002**, *2*, 551; (b) N. Krasteva, B. Guse, I. Besnard, A. Yasuda, T. Vossmeyer, *Sens. Actuators B*, **2003**, *92*, 137; (c) Y. Josep, N. Krasteva, I. Besnard, B. Guse, M. Rosenberger, U. Wild, A. Knop-Gericke, R. Schlögl, R. Krustev, A. Yasuda, T. Vossmeyer, *Faraday Discuss.* **2004**, *125*, 77; (d) M. Schlupp, T. Weil, A. J. Berresheim, U. M. Wiesler, J. Bargon, K. Müllen, *Angew. Chem. Int. Ed.* **2001**, *40*, 4011.

74 For a recent report on assessing the role of nitrogen atoms in dendrimers to a chelate metal ion, see: M. L. Tran, L. R. Gahan, I. R. Gentle, *J. Phys. Chem. B* **2004**, *108*, 20130.

75 (a) M. Wells, R. M. Crooks, *J. Am. Chem. Soc.* **1996**, *118*, 3988; (b) H. Tokuhisa, R. M. Crooks, *Langmuir* **1997**, *13*, 5608; (c) R. M. Crooks, A. J. Ricco, *Acc. Chem. Res.* **1998**, *31*, 219.

76 (a) T. Gao, E. S. Tillman, N. S. Lewis, *Polym. Mater. Sci. Eng.* **2004**, *90*, 174; (b) T. Gao, E. S. Tillman, N. S. Lewis, *Chem. Mater.* **2005**, *17*, 2904.

77 J. Ledesma-García, J. Manríquez, S. Gutiérrez-Granados, L. A. Godínez, *Electroanalysis* **2003**, *15*, 659.

78 L. L. Miller, Y. Kunugi, A. Canavesi, S. Rigaut, C. N. Moorefield, G. R. Newkome, *Chem. Mater.* **1998**, *10*, 1751.

79 (a) B. W. Koo, C. K. Song, C. Kim, *Sens. Actuators B*, **2001**, *77*, 432; (b) C. Kim, E. Park, C. K. Song, B. W. Koo, *Synth. Met.* **2001**, *123*, 493; (c) C.-K. Song, B.-W. Koo, C.-K. Kim, *Jpn. J. Appl. Phys.* **2002**, *41*, 2735.

80 M. Sülü, A. Altindal, Ö. Bekaroğlu, *Synth. Met.* **2005**, *155*, 211.

81 (a) M. Albrecht, R. A. Gossage, A. L. Spek, G. van Koten, *Chem. Commun.* **1998**, 1003; (b) M. Albrecht, R. A. Gossage, M. Lutz, A. L. Spek, G. van Koten, *Chem. Eur. J.* **2000**, *6*, 1431; (c) M. Albrecht, N. J. Hovestad, J. Boersma, G. van Koten, *Chem. Eur. J.* **2001**, *7*, 1289.

82 (a) R. P. Briñas, T. Troxler, R. M. Hochstrasser, S. A. Vinogradov, *J. Am. Chem. Soc.* **1995**, *127*, 11851; (b) S. A. Vinogradov, L.-W. Lo, D. F. Wilson, *Chem. Eur. J.* **1999**, *5*, 1338; (c) S. A. Vinogradov, D. F. Wilson, *Chem. Eur. J.* **2000**, *6*, 2456; (d) S. A. Vinogradov, I. B. Rietveld, O. S. Finikova, R. P. Briñas, D. F. Wilson, *Polym. Mater. Sci. Eng.* **2004**, *91*, 203.

83 W.-S. Li, D.-L. Jiang, Y. Suna, T. Aida, *J. Am. Chem. Soc.* **2005**, *127*, 7700.

84 D. K. Smith, F. Diederich, *Chem. Commun.* **1998**, 2501.

85 (a) A. Juris, *Annu. Rep. Prog. Chem., Sect. C*, **2003**, *99*, 177; (b) P. Ceroni, V. Vicinelli, M. Maestri, V. Balzani, S. Lee, J. van Heyst, M. Gorka, F. Vögtle, *J. Organomet. Chem.* **2004**, *689*, 4375.

86 (a) V. Balzani, P. Ceroni, S. Gestermann, M. Gorka, C. Kauffmann, F. Vögtle, *J. Chem. Soc., Dalton Trans.* **2000**, 3765; (b) V. Balzani, P. Ceroni, M. Maestri, C. Saudan, V. Vicinelli, *Top. Curr. Chem.* **2003**, *228*, 159; (c) V. Balzani, F. Vögtle, *C. R. Chim.* **2003**, *6*, 867.

87 F. Vögtle, S. Gestermann, C. Kauffmann, P. Ceroni, V. Vicinelli, V. Balzani, *J. Am. Chem. Soc.* **2000**, *122*, 10398.

88 I. Grabchev, J.-P. Soumillion, B. Muls, G. Ivanova, *Photochem. Photobiol. Sci.* **2004**, *3*, 1032.

89 (a) I. Grabchev, X. Qian, V. Bojinov, Y. Xiao, W. Zhang, *Polymer* **2002**, 5731; (b) I. Grabchev, J.-M. Chovlon, X. Qian, *New. J. Chem.* **2003**, *27*, 337; (c) I. Grabchev, S. Guittonneau, *J. Photochem. Photobiol. A. Chem.* **2006**, *179*, 28.

90 Q.-Q. Chen, L. Lin, H.-M. Chen, S.-P. Yang, L.-Z. Yang, X.-B. Yu, *J. Photochem. Photobiol. A. Chem.* **2006**, *180*, 69.

91 (a) D. Astruc, J.-C. Blais, M.-C. Daniel, S. Gatard, S. Nlate, J. Ruiz, *C. R. Chim.* **2003**, *6*, 1117; (b) M.-C. Daniel, J. Ruiz, D. Astruc, *J. Am. Chem. Soc.* **2003**, *125*, 1150; (c) D. Astruc, M.-C. Daniel, J. Ruiz, *Chem. Commun.* **2004**, 2637; (d) M.-C. Daniel, J. R. Aranzaes, S. Nlate, D. Astruc, *J. Inorg. Organomet. Polym. Mater.* **2005**, *15*, 107.

92 C. Valerio, J.-L. Fillaut, J. Ruiz, J. Guittard, J.-C. Blais, D. Astruc, *J. Am. Chem. Soc.* **1997**, *119*, 2588.

93 J. R. Aranzaes, C. Belin, D. Astruc, *Angew. Chem. Int. Ed.* **2006**, *45*, 132.

94 C. Valério, J. Ruiz, J.-L. Fillaut, D. Astruc, *C. R. Acad. Sci. Paris* **1999**, 79.

95 C. M. Casado, I. Cuadrado, B. Alonso, M. Morán, J. Losada, *J. Electroanal. Chem.* **1999**, *463*, 87.

96 V. K. Gupta, S. Chandra, S. Agarwal, H. Lang, *Sens. Actuators B* **1995**, *107*, 762.

97 F. L. Derf, E. Levillain, G. Trippé, A. Gorgues, M. Sallé, R. M. Sebastían, A. M. Caminade, J.-P. Majoral, *Angew. Chem. Int. Ed..* **2001**, *40*, 224.

98 (a) C. C. Lee, J. A. MacKay, J. M. J. Fréchet, F. C. Szoka, *Nat. Biotechnol.* **2005**, *23*, 1517; (b) I. Amato, *Cell* **2005**, *123*, 967; (c) S. Sridhar, M. Amiji, D. Shenoy, D. Nagesha, V. Weissig, W. Fu, *Proc. SPIE* **2005**, *6008*, 600816; (d) N. G. Portney, M. Ozkan, *Anal. Bioanal. Chem.* **2006**, *384*, 620.

99 R. Malik, R. Wiwattanapatapee, R. Klopsch, K. Lorentz, H. Frey, J. W. Weener, E. W. Meijer, W. Paulus, R. Duncan, *J. Controlled Release* **2000**, *65*, 133.

100 T. R. Krishna, S. Jain, U. S. Tatu, N. Jayaraman, *Tetrahedron* **2005**, *61*, 4281.

101 M. Snejdarkova, L. Svobodova, D. P. Nikolelis, J. Wang, T. Hianik, *Electroanalysis* **2003**, *15*, 1185.

102 (a) M. Snejdarkova, L. Svobodova, G. Evtugyn, H. Budnikov, A. Karyakin, D. P. Nikolelis, T. Hianik, *Anal. Chim. Acta.* **2004**, *514*, 79; (b) L. Svobodová, M. Šnejdárková, K. Tóth, R. E. Gyurcsanyi, T. Hianik, *Bioelectrochemistry* **2004**, *63*, 285.

103 (a) H. Chuang, P. Macush, M. B. Tabacco, *Anal. Chem.* **2001**, *73*, 462; (b) A.-C. Chang, J. B. Gillespie, M. B. Tabacco, *Anal. Chem.* **2001**, *73*, 467; (c) J. Ji, A. Schanzle, M. B. Tabacco, *Anal. Chem.* **2004**, *76*, 1411.

104 S. S. Mark, N. Sandhyarani, C. Zhu, C. Campagnolo, C. A. Batt, *Langmuir* **2004**, *20*, 6808.

105 P. J. G. Goulet Jr., D. S. dos Santos, R. A. Alvarez-Puebla Jr., O. N. Oliveria, R. F. Aroca, *Langmuir* **2005**, *21*, 5576.

106 L. Shen, N. Hu, *Biomacromolecules* **2005**, *6*, 1475.

107 H. C. Yoon, H. Yang, Y. T. Kim, *Analyst* **2002**, *127*, 1082.

108 (a) H. C. Yoon, D. Lee, H.-S. Kim, *Anal. Chim. Acta.* **2002**, *456*, 209; (b) H. C. Yoon, M.-Y. Hong, H.-S. Kim, *Anal. Biochem.* **2000**, *282*, 121.

109 H. C. Yoon, M.-Y. Hong, H.-S. Kim, *Anal. Chem.* **2000**, *72*, 4420.

110 (a) E. Kim, K. Kim, H. Yang, Y. T. Kim, J. Kwak, *Anal. Chem.* **2003**, *75*, 5665; (b) H. C. Yoon, M.-Y. Hong, H.-S. Kim, *Langmuir* **2001**, *17*, 1234.

111 J. Wang, M. Jiang, T. W. Nilsen, R. C. Getts, *J. Am. Chem. Soc.* **1998**, *120*, 8281.

112 (a) A. Sarkar, P. S. Satoh, P. R. Dvornic, S. N. Kaganove, *Polym. Prepr.* **2003**, *44*, 1055; (b) A. Sarkar, S. N. Kaganove, P. R. Dvornic, P. S. Satoh, *Polym. News* **2005**, *30*, 370.

113 (a) B. L. Ibey, H. T. Beier, R. M. Rounds, G. L. Coté, V. K. Yadavalli, M. V. Pishko, *Anal. Chem.* **2005**, *77*, 7039; (b) B. L. Ibey, H. T. Beier, R. M. Rounds, M. V. Pishko, G. L. Coté, *Proc. SPIE*, **2006**, *6094*, 609401; (c) H. T. Beier, B. L. Ibey, R. M. Rounds, M. V. Pishko, G. L. Coté, *Proc. SPIE*, **2006**, *6094*, 609401.

114 M. Snejdarkova, L. Svobodova, V. Gajdos, T. Hianik, *J. Mater. Sci.: Mater. Medicine* **2001**, *12*, 1079.

115 L. Svobodova, M. Snejdarkova, T. Hianik, *Anal. Bioanal. Chem.* **2002**, *373*, 735.

116 L. Svobodová, M. Šnejdárková, K. Tóth, R. E. Gyurcsanyi, T. Hianik, *Bioelectrochemistry* **2004**, *63*, 285.

117 H. Zhu, Y. Zhu, X. Yang, C. Li, *Chem. Lett.* **2006**, *35*, 326.

118 B. Alonso, P. G. Armada, J. Losada, I. Cuadrado, B. González, C. M. Casado, *Biosens. Bioelectron.* **2004**, *19*, 1617.

119 J. Losada, I. Cuadrado, M. Moran, C. M. Casado, B. Alonso, M. Barranco, *Anal. Chim. Acta* **1997**, *338*, 191.

9
Molecular Approaches in Organic/Polymeric Field-effect Transistors

K.S. Narayan and S. Dutta

9.1
Introduction

The invention of organic field-effect transistors (OFETs) with reasonable efficiencies provided the basic impetus to the field of inexpensive macro-electronics. Polymer-based FETs, PFET devices hold enormous prospects as active elements in driver circuits to realize low-cost, large-area electronic structures [1–4]. During the past decade, intense research activity has taken place in the areas pertaining to molecular design and synthesis of organic semiconductors, purification procedures, solution processing methods, and engineering the metal/organic interfaces to realize optimum device structures [1–4]. A simple prototypical OFET/PFET device is typically comprised of three electrodes, namely, source, drain and gate, along with organic (polymer) layers. In principle, the operation of the metal–insulator–semiconductor FET structure (OFET/PFET) is based on the control of charge density within the semiconductor through the external gate voltage. The option of having all the components of the device structure, insulator, semiconductor and metal to be polymeric/organic in nature opens up a new paradigm for sensors and smart applications. These devices, which rely on low cost processing techniques, offer several possibilities to a wide range of applications. A variety of applications have been proposed and demonstrated, for organic-semiconductor circuits, including display backplanes, sensors, RFID (radio frequency identification) transponders and tags [1–5].

Even though the field-effect action was known for organic and polymer-based semiconductors twenty years ago, it was not possible to realize performance parameters which could be utilized. The earliest known approach to the fabrication of a FET using organic semiconductors was reported by Barbe and Westgate in 1970 [6]. The discovery of conducting polymers turned the future of electronics in a very fascinating direction due to the tractable mechanical and semiconducting properties of the materials, low-cost processing, the possibility of large area electronics etc. The polymer FET (PFET) based on polyacetylene was first reported in 1983 [7] and was followed by marginal, incremental improvements in the performance. Real breakthroughs in the performance parameters were achieved during the last decade when regioregular polyalkylthiophene-based FET device structures [9, 10] were introduced on appropriate compatible dielectric substrates. Organic small molecules and conjugated oligomers have also attracted considerable interest in the field of transistors [1–4].

In this chapter, we cover different aspects of field-effect transistors based on organic materials and issues related to improving the performance of the device. Since this interdisciplinary field has attracted numerous researchers, resulting in a large number of publications, we have included representative topics and highlights to provide a general summary rather than an exhaustive comprehensive report. The organization of the chapter includes a brief introduction to the device operation, electrical characterization, followed by device fabrication techniques, progress including the molecular approaches and prospects and problems related to the field.

9.2
Device Operations and Electrical Characterization

The operation of a metal–insulator–semiconductor FET structure (OFET/PFET) is based on the principle of modulating the current density in the semiconductor via the external gate voltage. The purpose of the source and drain is to serve as barrier-free charge-injecting electrodes to the semiconductor upon application of a certain external voltage; the gate electrode is isolated from the active semiconductor by a thin dielectric layer, controlling the field-induced charge distribution near the semiconductor/insulator interface. High electric field at a trap-free interface between the semiconductor and the insulator leads to bending of the energy band towards or away from the Fermi level, depending on the polarity of the applied gate voltage. For example, in case of a p-type semiconductor, the energy bands bend in an upward direction when the gate is subjected to a negative voltage. This results in accumulation of positive charges at the interface providing a highly conducting (on) state, called the *accumulation* or *enhancement* mode. On the other hand, upon applying positive gate bias, the positive charges are pushed away from the interface due to opposite band bending, which leads to depletion of mobile charges. Basically the *depletion* mode corresponds to the low conducting (off) state of the device. The electrical performance of the FET is characterized mainly in terms of two parameters *field-effect mobility*, representing the speed, and *on/off current ratio*, describing the switching performance of the device.

In order to derive the transistor characteristic of the OFET/PFET structure quantitatively, it is convenient to assume the idealized conditions: (i) the gradient of the transverse field that is directed perpendicular to the channel is much larger than that of the longitudinal field along the channel, (ii) there are no interfacial traps or difference in Fermi level between the gate and the semiconductor, (iii) carrier mobility is constant throughout the channel, and (iv) reverse leakage current is negligibly small. The first condition essentially means that the charge induced by the transverse field is much more effective than that induced by the longitudinal field. This eventually requires the channel length to be quite large compared to the insulator thickness. The condition termed *gradual channel approximation*, was first introduced by Shockley for modeling the traditional FET device characteristics [11].

The field-induced surface charge density is represented by the product of capacitance of the insulator per unit area (C_i) and the gate voltage V_g. When a negative drain-source voltage (V_d) is applied, a gradient of surface potential appears along the channel (say, in the *x*-direction). Under this circumstance, the conductivity is contributed by the following terms: (i) V_g-induced surface charge density and (ii) doping-induced charge density in the bulk of the semiconductor. The final expression for the drain-source current I_d can be written as:

$$I_d = \frac{W\mu C_i}{L}\left(V_g - V_0 - \frac{V_d}{2}\right)V_d + \frac{Ne\mu W d_s}{L}V_d$$

where, W and L are the channel width and length respectively, μ is the mobility, d_s the thickness of the semiconductor, N the doping concentration and V_0 the small offset voltage (often called switch-on voltage). The first term is the same as in the conventional *inversion mode* MISFET (metal insulator semiconductor FET) devices and the additional residual ohmic term arises primarily from bulk-doping. This sizable contribution can only be decreased by minimizing the unintentional doping [12].

The linear term predominates at lower drain voltage whereas the latter term takes over as the drain voltage approaches the saturation voltage $V_{d,sat} = V_g$ (assuming $V_g \gg V_0$). In the saturation regime, one may consider that the current–voltage (I–V) characteristic is independent of V_d. However, in reality, the saturation voltage exceeds V_g due to the presence of the residual charge in the bulk, and requires more voltage to be depleted at the drain region. The field-effect mobility of the transistor can be defined and extracted from the saturation or linear regime, depending on the operating region of interest. Alternatively, the field-effect mobility in the saturation regime is also estimated from the slope of the plot of $(I_d)^{1/2}$ against V_g for $|V_d| > |V_g|$. Typically, the mobility calculated in the saturation regime is much higher than in the linear regime, since the latter factor is more affected by deviations from linearity in the transistor characteristics (I_d versus V_d curves), at low V_d. The field-effect mobility is a very sensitive parameter and depends on several internal and external factors. It is controlled by the semiconductor/insulator interface, orientation of the polymers or organic molecules, morphology, grain size (for organic molecules and oligomers), carrier density, and impurity concentration.

9.3
Device Fabrication

The fabrication of silicon-based or traditional inorganic semiconductor-based MISFET structures involves either diffusion or implantation-based processes,

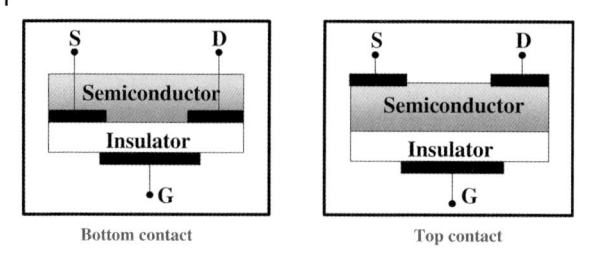

Fig. 9.1 Top contact and bottom contact FET structures.

where source and drain electrode deposition is followed by the growth of the insulating layer, normally thermally grown silicon oxide, and finally deposition of the metal electrodes. In OFETs (or PFETs), the semiconductor is not a bulk substrate but a thin film and so the device structure is fabricated in an *inverted* architecture like TFTs. Typically the OFET structures are designed in two different categories. In *coplanar* (or *bottom-contact*) geometry, all the layers are on the same side of the semiconductor, whereas in *staggered* (or *top-contact*) geometry, the gate and the source-drain electrodes are on opposite sides of the semiconductor layer (Fig. 9.1).

Coplanar configuration is the most adopted geometry for OFETs due to its significant advantages, especially in integrated circuit (IC) applications. It is easier to fabricate the bottom-contact devices on oxide-dielectric layers with miniaturized dimensions using different types of lithographic techniques without damaging the soft polymer films. In addition, the bulk-controlled current becomes less effective in this geometry. The main disadvantage of this device structure is the high contact resistance at the drain and source regions, which dominates in the case of a short-channel device and has injection limited nonlinear transport. The origin of the high resistance lies in: (i) under etching of the drain and source electrodes forming an air gap between the electrodes, semiconductor and the insulator, (ii) small contact area between metal and semiconductor ($\approx d_m \times W$, where d_m is the thickness of the metal deposited) [13, 14]. On the other hand, in the staggered geometry the contact resistance is less susceptible than that in the coplanar geometry as the contact area in the former (in particular, overlap of the source and gate contacts) is large enough. Moreover, there is a component of electric field pointing vertically up that can induce injection from the top of the source. However, there are severe limitations in this architecture, such as restricted channel length due to lack of lithographic technique and bulk dominated current depending on the thickness of the semiconductor film. The device geometry plays a major role in driving the performance as well as issues related to stability and degradation [11, 12].

Most of the high performance OFETs have been constructed on highly doped silicon wafers covered with a thermally grown silicon oxide layer. The highly doped substrate acts as the gate electrode. In the case of a coplanar structure, the source and drain electrodes are designed on top of the insulating layer using standard microlithography techniques, prior to the deposition of semiconductor.

In a staggered configuration, the semiconductor is deposited, followed by the thermal evaporation of the metal electrodes (drain and source) using a shadow-mask of desired dimension.

A crucial step in the fabrication process of FETs involves the deposition of semiconductor on the dielectric insulator. There are various deposition techniques that are adopted, depending on the nature of the semiconducting material. Thermal deposition is the conventional method to deposit the organic small molecule on the insulating surface by heating the material under high vacuum conditions. The base pressure of the deposition system is an important parameter to control the mean free path of the sublimed organic molecules and the presence of unwanted atoms and molecules, which results in contamination in the polymer film [15–17]. Substrate temperature and deposition rate also influence the thin film morphology which affects the transport characteristics of OFETs [15–17]. The advantages of vacuum deposition are (i) control of thickness and (ii) high purity films with the possibility of high order which can be controlled by the deposition rate and the temperature of the substrate. High performance devices can be achieved if the process parameters can be correlated to the morphology of the films. As intermolecular charge hopping across grain boundaries and disordered domains is not as efficient as transport in an ordered domain, increasing the grain size is a promising approach to increasing the charge carrier mobility.

Solution-processed deposition is an elegant, convenient and facile route to obtain high quality polymer film over a large area. This technique involves the homogeneous deposition of polymer film from its solution by using *spin-casting* [18], *drop-casting* [19], *printing technology* [20], *dip-coating* [21] or the *Langmuir–Blodgett* (LB) technique [22]. A basic requirement for this technique is good solubility of the polymer in common organic solvents. *Spin-casting* is one of the most widely used methods for solution-processed deposition, where the substrate is rotated at a speed of few hundred to few thousand rotations per minute after dropping an appropriate amount of solution on it. Uniform thickness can be maintained by optimizing the coating parameters such as viscosity of the solution and rotation speed. Typically the film thickness is proportional to the viscosity and inversely proportional to the rotation speed. *Drop-casting* is another deposition scheme in which the solution is allowed to dry slowly in inert conditions. This is a comparatively slow-growth process controlled by the natural evaporation of the solvent and is known to yield better performance parameters than spin-casting methods. A comprehensive study dealing with the comparison between spin- and drop-casting and their influence on the mobility and orientation of the film has been carried out recently [21]. *Printing technology* is the technique that has attracted contemporary interest for the development of organic electronics. This method involves printing the conducting or conjugated polymers, which can be used as a form of *suspension* or *ink* [23, 24]. The primary goal of this technique includes the fabrication of ICs at a far greater production speed, lower cost, and with less manufacturing complexity, incorporating the possibility of processing at room temperature in ambient atmosphere. *Dip-coating* is one of the efficient processing methods to achieve large area devices on both side of the substrate. In this process, the substrate is dipped into the polymer solution in a repeated and automated manner. The *Langmuir–Blodgett* (LB) technique has also been explored in polymer electronics. It allows fine control of both the structure and thickness of the film. This technique is limited to amphiphilic molecules that consist of a hydrophobic chain and a hydrophilic head group.

9.3.1
Substrate Treatment Methods

The major challenge in solution-processed techniques lies in the nature of the interface formed at the insulator and semiconductor boundary, the adhesive forces arise primarily from weak Van der Waal interactions. The interface quality can be improved by using a surface modification treatment prior to the deposition of the semiconductor [25]. In principle, silane-based compounds are used to develop the self-assembled monolayers on top of the dielectric. The surface treatment can substantially increase the mobility (by as much as two orders of magnitude) [25]. The basic purpose of surface treatment is to make the dielectric surface more hydrophobic in nature, which in turn attracts the side chain of the polymers forming a highly ordered interface. An alternative method to improve the interface is a surface-directed, self-organized bilayer of two organic bulk phases with spontaneous solvent evaporation and this recent work has been demonstrated by Sirringhaus and coworkers [25]. An enhancement of field-effect mobility due to surface-mediated molecular ordering has been widely reported in regioregular polyhexylthiophene, P3HT thin film transistors. Structural ordering at the interface of P3HT and the insulator, SiO_x, can be controlled by functionalized self-assembled monolayers SAMs ($-NH_2$, $-CH_3$). γ-aminopropyltriethoxysilane (γ-APS) and octyltrichlorosilane (OTS) have been used for $-NH_2$ and $-CH_3$, respectively, to fabricate the SAMs. Synchrotron grazing-incident X-ray diffraction (GI-XRD) and atomic force microscopy (AFM) are typically employed to determine the microstructure and ordering of P3HT films where two different chain orientations (edge-on for a $-NH_2$ functionalized surface and face-on for a $-CH_3$ functionalized surface) are observed. It is observed that the orientation is introduced by the unshared electron pairs (polar groups) and alkyl chains of SAMs. It is speculated that there is a repulsive force between the π-electron clouds of the thienyl backbone and the unshared electron pairs of the SAMs on insulator substrates with polar groups ($-NH_2$ and $-OH$). There are $\pi-H$ interactions between the thienyl backbone and H atoms of the SAM end groups in the π-conjugate systems, as depicted in Fig. 9.2. In systems that are devoid of the unshared electron pairs (P3HT-CH_3), the parallel orientation of P3HT chains to the insulator substrate turns out to be the most thermodynamically stable because of the $\pi-H$ interactions. However, in systems with unshared electron pairs (P3HT $-NH_2$ and $-OH$), perpendicular (edge-on) orientation is achieved because of the repulsive force between the π-electron clouds of the thienyl backbone and unshared electron pairs of the SAMs. AFM images for P3HT-NH_2 systems show worm-like lamellar crystallite structures that arise due to perpendicular orientations of the chains. The P3HT-CH_3 system indicated disk-type crystallites with large domains

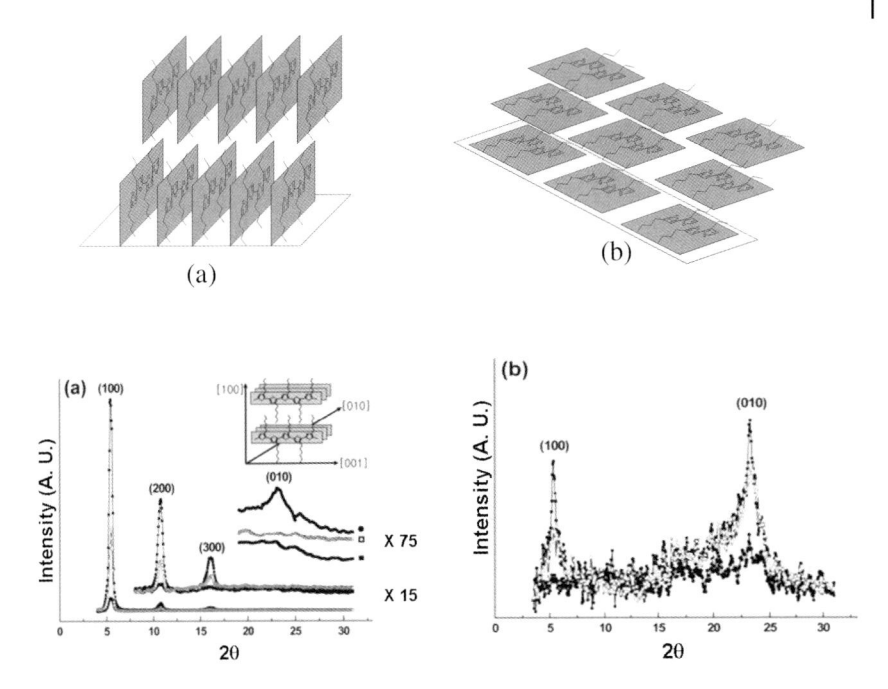

Fig. 9.2 Schematic representation of self-assembled structural ordering in regioregular P3HT providing 2-dimensional charge transport. Orientation of P3HT planes on functionalized surfaces. Top: (a) edge-on orientation, (b) face-on orientation. Bottom: XRD as a function of scattering angle for RRP3HT films (adapted from Ref. [28]).

because of the parallel orientation. The increase in mobility for the P3HT-NH$_2$ annealed sample was attributed to the perpendicular orientation with respect to the insulator that enabled the charge carrier to transport in the two-dimensional conjugation direction. These studies which correlate the morphology and FET performance suggest a wide scope for improving the device parameters [26–28].

9.3.2
Electrode Materials

Prior to (bottom contact) or after (top contact) the semiconductor deposition, metals such as gold (Au, $\phi_m = 5.1$ eV), aluminum (Al, $\phi_m = 4.2$ eV), magnesium (Mg, $\phi_m = 3.7$ eV), calcium (Ca, $\phi_m = 2.8$ eV), silver (Ag, $\phi_m = 4.6$ eV) are used as drain and source contacts. The choice of the metals, depending on their work function (ϕ_m), is a crucial factor in determining the efficient charge transport mechanism in OFETs/PFETs. In most cases, Au forms the drain and source contacts due to an optimum work function match with the HOMO level of the organic/conjugated polymers (~ 5 eV) for p-FETs. Aluminum is generally used as a gate electrode rather than drain and source electrodes, because it is unstable in air with a high tendency to be oxidized. Indium tin oxide (ITO)-coated glass is also used more often due to its transparent behavior coupled with metallic feature. Apart from the inorganic metals, the water-soluble conducting polymer poly(3,4-ethylenedioxythiophene)/poly(4-styrenesulfonate) (PEDOT/PSS) is widely used for its processing advantages to realize all-organic/polymeric FETs. The efficiency of injecting holes (or electrons) from the metal contact to the HOMO (LUMO) of the organic semiconductor depends on the energy barrier ϕ_{Bh} (ϕ_{Be}) for the holes (electrons) which has to be overcome. A finite shift of the vacuum level of the organic semiconductor has been observed at the metal/semiconductor interface, where the magnitude and sign depend on the specific metal/organic combination. This shift can be attributed to an ultrathin, interfacial electric dipole layer, that can be as high as 1.5 eV [29–33]. The issue, related to barrier formation, becomes more evident in the case of n-channel OFETs. In this context, it is essential to mention that despite the low work function of Al, it is not an ideal electron injecting electrode. This is likely due to (i) the oxidation of Al, which creates an insulating layer of aluminum oxide, and (ii) the formation of a dipole barrier. Most n-channel FETs reported have used Ca as the electron injecting electrode. These devices however have the limitation of operation in inert conditions or require a high degree of encapsulation.

9.4
Progress in Electrical Performance

The performance of OFET in terms of mobility, on/off ratio, stability etc. has been improving continuously over the last decade. Considerable research efforts have also resulted in an increase in field-effect mobility by five orders of magnitude over the past two decades to values in the range 0.1–1 cm^2 V^{-1} s^{-1}. The typical route of impressive improvements can be summarized by the following processes: (i) design and synthesis of organic semiconductor; (ii) optimization of the film deposition parameters for the active layer to obtain superior morphology and structural orientation; and (iii) optimization of injection from the source and drain contacts. OFETs can be divided into two categories, p- and n-channels, according to the nature of the charge carriers that form the conducting channel at the semiconductor/insulator interface. However, it is noteworthy to specify that p- or n-channel does not imply the type of doping, unlike for the inorganic counterparts, but it signifies the more mobile charge carriers, either holes or electrons, respectively. Recently, extensive research is being conducted on design and optimization of *ambipolar OFET*, consisting of both types of charge carriers with balanced mobility.

9.5
Progress in p-Channel OFETs

Most of the organic semiconductors that are used for OFET, display p-type behavior in the transistor characteristics. The unintentional doping during synthesis and the appearance of electron-trapping centres at the semiconductor/insulator interface during the fabrication process make them convenient p-type conductors. The most widely used p-type organic semiconductors are pentacene and the other acene-based molecules and thiophene oligomers as active materials.

Numerous efforts have been carried out on thiophene oligomers such as *sexithiophene* (6T) to improve and optimize the organic-based transistor characteristics. The first report on 6T-based FET, which was also the first report on a small conjugated molecule, showed the mobility, μ, to be of the order of 10^{-3} cm^2 V^{-1} s^{-1} [34]. The importance of the ordering of the film was first realized with a 10–100-fold increase of thiophene oligomers upon substitution of dimethyl at both ends of the oligomers [35]. The interpretation was confirmed using dihexyl-sexithiophene (DH6T) obtaining μ as high as 0.05 cm^2 V^{-1} s^{-1} due to the very regular microscopic arrangement [36]. On proper purification treatment of the source materials, a high μ of 0.03 cm^2 V^{-1} s^{-1} and on/off ratio $\sim 10^6$ were obtained, which are amongst the highest reported values for unsubstituted 6T-based OFETs [37]. The variation of field-effect mobility with different polymer dielectric layer has been studied comprehensively with DH6T on a polymethylmethacrylate (PMMA) layer ($\mu \sim$ 0.04–0.08 cm^2 V^{-1} s^{-1}), DH6T on a polyamide layer ($\mu \sim$ 0.09–0.10 cm^2 V^{-1} s^{-1}) and DH6T on perylene ($\mu \sim$ 0.095–0.13 cm^2 V^{-1} s^{-1}) [37]. Recently, values of $\mu \sim$ 1.1 cm^2 V^{-1} s^{-1} have been reported for alkyl-substituted oligothiophene [38]. Over the past decade, pentacene has proved to be a leading candidate for OFET due to its morphology and crystalline property. Pentacene can be deposited using the organic vapor phase deposition technique [39], the solution process method [40] and even the pulse-laser deposition technique [41]. The crystalline pentacene with controlled growth has a remarkably high value of field-effect mobility ranging from 0.038 [42] to 0.62 [43] to 1.5 cm^2 V^{-1} s^{-1} [44]. Recent progress has registered μ value of pentacene-based transistor up to 3 cm^2 V^{-1} s^{-1} using a polymer dielectric layer [45] and 6.4 cm^2 V^{-1} s^{-1} for a treated alumina dielectric layer [46].

The mobilities in the case of polymer FET are usually lower than the small molecule counterpart due to the poor molecular ordering and low crystallinity obtained by solution techniques. One of the first solution-processable organic semiconductors, used for an efficient FET structure, was (P3HT) [47, 48] where the mobility value of the OFET based on P3HT, spun from the chloroform solution, was reported to lie in the range 10^{-5}–10^{-4} cm^2 V^{-1} s^{-1}. Typical FET characteristics of a P3OT-based device are shown in Fig. 9.3. A comparative study of P3AT transistors as a function of the side chain showed that μ decreases from 1–2×10^{-4} cm^2 V^{-1} s^{-1} for poly(3-butylthiophene) and P3HT to 6×10^{-7} cm^2 V^{-1} s^{-1} for poly(3-decylthiophene) [49]. Regioregular P3HT-based FET device structures consisting of 98.5% or more head-to-tail linkages have shown a dramatic increase in mobility when compared to that of regiorandom P3AT-based FETs [19, 25]. With the drop-casting method, μ has been increased to 0.045 cm^2 V^{-1} s^{-1}, due to formation of a self-assembled lamella structure of RRP3HT film [19]. Studies on the correlation between the mobility and the deciding factors such as degree of regioregularity, deposition technique and orientation of polymer stacking atop the insulator surface have evoked tremendous interest [25]. It

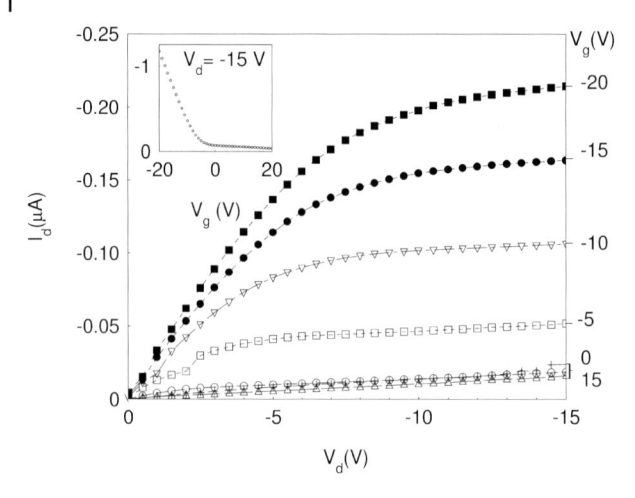

Fig. 9.3 Typical polymer FET characteristics, semiconductor: P3HT, dielectric: PVA. See Refs. [67–69] for details.

is observed that highly regioregular P3HT (>91% head-to-tail linkage) forms a lamellar structure with an edge-on orientation (π–π stacking in the plane of the substrate) when spun from chloroform. High field-effect mobilities of 0.05–0.1 cm^2 V^{-1} s^{-1} were obtained for 96% RRP3HT-based FET devices. In contrast, spin-coated films of P3HT with low regioregularity (81% head-to-tail linkage) give lamella with a face-on orientation (π–π stacking perpendicular to the substrate) resulting in a low field-effect mobility of the order of 2×10^{-4} cm^2 V^{-1} s^{-1}. However, drop-cast deposition of 81% regioregular P3HT reproduces the edge-on orientation of the lamella structure that increases μ by an order of magnitude ($\sim 10^{-3}$ cm^2 V^{-1} s^{-1}). This study states that, in addition to the degree of the order of the film, the method of deposition also influences the π–π stacking direction relative to the substrate, which in turn has direct impact on the mobility of the materials. The mobility of RRP3HT is also influenced by the solvents used and the surface modification of the insulator. μ was observed to vary by almost two orders of magnitude, depending on the solvents, and the highest mobility was obtained using chloroform as the solvent for RRP3HT. In the case of RRP3HT-based FET, it has been observed that treatment of silicon dioxide with hexamethyldisilazane (HMDS) or alkyltrichlorosilane replaces the hydroxy group at the silicon dioxide surface with methyl or alkyl groups. The apolar nature of these groups apparently attracts the hexyl side-chains of P3HT, favoring lamella with an edge-on orientation. Mobilities of 0.05–0.1 cm^2 V^{-1} s^{-1} have been achieved using HMDS treatment of the insulator surface. Recently, OFETs based on RRP3HT have been fabricated using a dip-coating technique. The highest field-effect mobility value of 0.18 cm^2 V^{-1} s^{-1} has been recently reported. The origin of the high μ was attributed to the formation of a rod-like morphology [50, 51]. However, exposure of RRP3AT to air causes an increase in conductivity, subsequent degradation of the transistor, and low on/off ratio that have been attributed to oxidation defects [19, 50].

Poly(9,9-dioctylfluorene-co-bithiophene) (F8T2) is another promising polymer semiconductor and is more environmentally stable than P3HT [51, 52]. F8T2 possesses a thermotropic, nematic liquid crystalline phase above 265 °C. It can be oriented using a rubbed polyamide alignment layer to form a macroscopic domain. The carrier mobility is higher in the case of oriented films along the rubbing direction rather than in isotropic films and can be as high as 0.02 cm^2 V^{-1} s^{-1} with an on/off ratio $\sim 10^5$ at room temperature [51, 52]. The promising potential of this polymer was demonstrated using device processing methods involving ink-jet printing technology. Design of an all-polymer transistor inverter-circuit with channel length down to 5 μm was implemented [52].

It has been shown that both molecular self-assembly and stability against oxidative doping by atmospheric oxygen are key to designing solution processable organic semiconductors for low-cost OTFTs. Oxidative doping stability and excellent TFT performance characteristics can be built into a polythiophene system via structural design [54]. It has been found that systems such as regioregular poly(3,3-dialkyl-quaterthiophene) **3** (PQT) [54] exhibit stable efficient properties. This is essentially a semiconducting polythiophene that is designed to assemble into large crystalline domains on crystallization from a liquid-crystal phase, and to possess an extended, planar π electron system that allows close intermolecular π–π distances, which facilitate high charge carrier mobility. In addition, rather than increasing the IP by sterically twisting the repeat units in the backbone, which intuitively should reduce the crystalline perfection, a linear conjugated comonomer, thieno[3,2-b]thiophene was incorporated. The delocalization of electrons from this fused aromatic unit into the backbone is less favorable than from a single thiophene ring, due to the larger resonance stabilization energy of the fused ring over the single thiophene ring. This reduced delocalization along the backbone results in a lowering of the polymer HOMO. Furthermore, the formation of highly ordered crystalline domains can be promoted due to rotational invariance of the linearly symmetrical thieno[3,2-b]thiophene in the backbone which facilitates the adoption of the low-energy backbone conformation. Tail-to-tail regiopositioning of the alkyl chains on the bithiophene monomer helps promote self-organization while minimizing any steric interactions between the neighbouring alkyl groups, thus preserving backbone planarity [54].

9.6
Progress in n-Channel OFET

The major problem in fabricating n-channel OFETs is the presence of traps. A low trace of oxygen is enough to degrade the device by introduction of deep electron trap sites. Protonic sites, contributed by moisture, can also attach to the functional groups by hydrogen bonding, and can serve as additional trap sites for electrons. However, the deciding factors for the device engineering are the electron affinity of the semiconductor and the energy level matching between the semiconductor and the metal electrodes (source and drain) to promote efficient electron injection. An early study using lutetium (Pc_2Lu) and thulium (Pc_2Tm) bisphthalocyanines-based n-channel OFET was reported, where the mobilities were between 2×10^{-4} and 1.4×10^{-3} cm^2 V^{-1} s^{-1} for both the materials [55]. C_{60} and C_{60}/C_{70} fullerenes were also used as the active layer with mobility 5×10^{-4} cm^2 V^{-1} s^{-1}. The mobility of C_{60} was raised to 0.08 cm^2 V^{-1} s^{-1} by modifying the deposition technique [56]. Further research has produced a mobility of 0.56 cm^2 V^{-1} s^{-1} in C_{60}-based FET, which is quite a reasonable value for a n-channel transistor system [57]. It has recently been shown that OFETs based on N,N'-dioctyl-3,4,9,10-perylenetetracarboxylicdiimide(PTCDI-C18H) as the organic semiconductor provide bottom contact devices with mobility as high as 0.6 cm^2 V^{-1} s^{-1} [58]. There are only a few reports available on n-type based PFET device structures. Some of these polymers are ladder type, rigid systems such as poly(benzobisimidazobenzophenanthroline)(BBL), poly(p-phenylene-2,6-benzobisthiazole) [59, 60]. Field-effect mobilities of $3–4 \times 10^{-3}$ cm^2 V^{-1} s^{-1} have been achieved in the encapsulated, bottom-gate FET configuration consisting of methanofullerene [6,6]-phenyl-C61-butyric acid methyl ester (PCBM) as active semiconductor and calcium as the drain-source contacts [61].

9.7
Progress in Ambipolar OFET

In addition to the development of p-channel and n-channel FETs, there are a few reports on ambipolar transistors, where both p- and n-type transport can be achieved in a single device with an interesting property of light emission from these devices. The advantages of such devices are: (i) the possibility of a complementary circuit with low power consumption analogous to CMOS technology; (ii) the possibility of having radiative recombination in a transistor by balancing the injection of both electrons and holes at the same place. Despite high purity with few acceptors ($<10^{13}$ cm^{-3}), unipolar transport characteristics are typically observed in organic semiconductors. This can be attributed to different injection efficiencies for electrons and holes and, to some extent, can be related to the effective mass of the charge species. An approach using heterostructure FET was reported to obtain ambipolar transport in a single device with bottom-contact architecture [36]. The device structure consisted of 6T and C_{60} as the two active layers for p- and n-channel operations, respectively. The active layers were sublimed on top of a silicon oxide layer, drain and source contact pads. The two materials were chosen because the HOMO level of 6T is energetically lower for holes than the HOMO of C_{60} while the LUMO of C_{60} is energetically lowered for electrons. The drawback of this structure was the use of similar metal electrodes, one of which forms a schottky barrier with C_{60}. Recently two different drain and source were adopted according to the semiconductors pentacene and N,N'-ditridecylperylene-3,4,9,10-tetracarboxylic diimide (PTCDI-$C_{13}H_{27}$) to obtain balanced mobility [62]. An alternative approach was adopted by making a film of heterogeneous polymer blends consisting of interpenetrating networks and also by using narrow-bandgap organic semiconductors [63]. A blend of hole-transporting poly(methoxy dimethyloctyloxy)-phenylene vinylene (OC1C10-PPV) with electron-transporting PCBM was used for ambipolar conduction. The electron mobility in such blends (7×10^{-4} cm^2 V^{-1} s^{-1}) was two orders of magnitude lower than the electron mobility of a pure film of PCBM, while the hole mobility was similar to that of single-component OC1C10-PPV (3×10^{-5} cm^2 V^{-1} s^{-1}). Ambipolar transport was also found in the OFET structure fabricated with a layer of PCBM [64]. Pentacene film together with calcium as source and drain were also used to extract ambipolar conduction incorporating hole mobility and electron mobility of 4.5×10^{-4} cm^2 V^{-1} s^{-1} and 2.7×10^{-5} cm^2 V^{-1} s^{-1} respectively [65]. A recent breakthrough in ambipolar transport in most of the common p-type semiconductors has received considerable attention in this remarkable research field. According to this report, the origin of unipolar transport in such materials is the presence of the hydroxy group (OH) which may act as a trap centre [67] and oxygen has high electron affinity, which replaces the hydrogen ion of OH with induced electrons, providing hole-only transport. However, the inclusion of hydroxy-free insulator, for example divinyltetramethylsiloxane-bis(benzocyclobutene) (BCB), has produced ambipolar transport in almost all the well-known p-type conjugated polymers [66].

9.8 PhotoPFETs

Interesting, unique phenomena in these device structures, arising from the combination of device physics and photophysics in these systems, were observed by Narayan et al. [67]. They demonstrated the utility of light as an additional controlling parameter of the transistor state. The transistor exhibits large photosensitivity, indicated by the sizable changes in the drain-source current, I_d, at low levels of light. The photo-induced response reported was considerably higher than that of efficient conventional 2-terminal organic/polymeric photodiodes due to a process of internal amplification. The light responsive polymer-FET opens up a new device architecture concept for polymer-based electronics. The transistor action is considerably modified with photoexcitation with large changes in the I_d, and the saturation value of drain-source current I_d, depending on both the intensity of the light and the gate voltage, as shown in Fig. 9.4. The large photo-induced I_d is a consequence of an internal amplification process resulting from the photogenerated carriers which is possible only in this transistor configuration. The studies using regioregular polythiophenes were then extended to other polymers and organic molecules [67–69].

Apart from these large photoinduced changes in the transport processes, interesting long-lived relaxation behavior is observed upon switching off the photo-

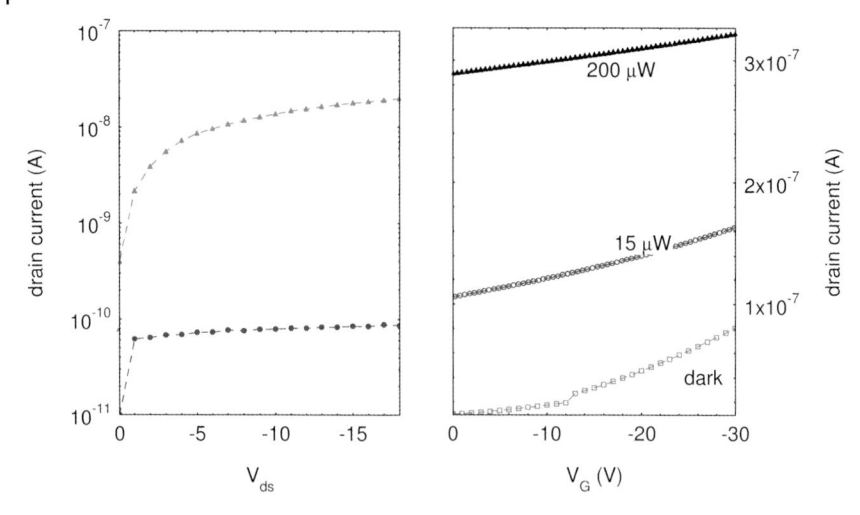

Fig. 9.4 Light induced photo-polymer FET based on P3OT as the semiconductor, and PVA characteristics: (i) Photo-induced amplification of source-drain current at −3 V gate bias. (ii) Transconductance characteristics, I_d vs. V_g in dark and light conditions (low intensity regime).

excitation. An illustrative experimental investigation with a suitable theoretical approach was accomplished to interpret the slow relaxation of the photoinduced current in organic FET. A V_g dependence of photocurrent relaxation in the intermediate stages of the decay in this three-terminal device was observed [68]. The asymmetrical charge distribution, which is inherent in this device geometry upon low-intensity photoexcitation can, to some extent, be manipulated by V_g. This feature was exploited to obtain the optical memory effect in a single device structure [69]. The optically generated charged regions which can be accessed by V_g were used to demonstrate memory effects comprising writing, storing and reading operations. The writing process involves optical excitation of the FET operated under a depletion mode, while the reading and erasing procedures involve appropriate gate voltage bias conditions as shown in Fig. 9.5. These results also provide a measure of the carrier distribution profile and the electric field prevalent in the semiconducting layer. The remarkably large photoresponsivity of these FETs has also attracted considerable attention subsequently [70, 71]. The ability to tailor photocarrier generation schemes in conjugated organic materials, both small molecules and polymers [67–71], and the demonstration of FET characteristics in the same media can form a set of potentially useful features to design optoelectronic elements. An important figure of merit for photodetectors is light-current/dark-current ratio. The ability to tune the two quantities independently compared to two-terminal photodiodes is one of the most obvious advantages of these photodetectors.

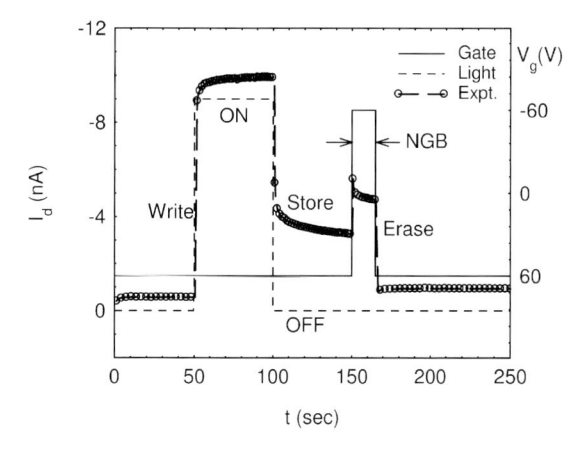

Fig. 9.5 Optoelectronic memory effect in the photo-polymer FET based on P3HT as the semiconductor, and PVA as the dielectric insulator at 20 K, using a periodic light pulse for writing (dashed line) and periodic gate bias (solid line, erasure is carried out by the negative gate bias NGB) as the two input parameters.

9.9
Photoeffects in Semiconducting Polymer Dispersed Single Wall Carbon Nanotube Transistors

Compared to the organic/polymer FETs, single wall carbon nanotube SWCNT-based FETs in terms of performance exhibit high drive currents, excellent transconductance, fast switching response and large on/off ratios [72]. A comparison between the two types of FETs, especially under photoexcitation, is instructive in understanding the fundamental process involved. Direct forms of specific chemical sensing have also been demonstrated using a variety of device structures. SWCNTFET are particularly appealing for use as optical detectors. However, methods based on optical excitation of isolated SWNT structures have been rare, possibly because of physical constraints such as the low photon-capture area of the isolated nanotubes, a large recombination probability and other intrinsic reasons leading to low internal photon to current conversion efficiencies. However, changes in the I_d current and threshold voltage shift of the SWCNTFET upon laser illumination have been reported recently [73]. The excitation spectrum of the photocurrent features was correlated to the optical absorption by the third interband gap of the van Hove singularity of the semiconducting SWCNT [74] and, in certain cases, was attributed to the schottky character of the electrode/semiconductor interface [75]. In absolute terms, for a given number of photons per unit area, the photoinduced changes in pristine SWCNTFETs are relatively small compared to semiconducting polymer-based FETs which offer a large area of absorption cross section for photocarrier generation and transport [62–65]. Under dark conditions, the polymer-based FETs have a much lower performance efficiency than SWNTFETs. A combination of the characteristics of these two classes of FETs in a single device is observed to offer interesting possibilities [76].

Recent methods for fabrication of such structures consisting of optically and electrically active semiconducting polymers on the nanotube have been reported [76, 77]. Upon photoexcitation, FET-characteristics display shifts in threshold voltage along with a presence of memory features [77]. The nanotubes have been suggested to act as conduits for hole-transport and studies of hybrid photovoltaic devices consisting of these set of materials exhibit increased efficiency of photo-induced charge separation at the active interface. Such hybrid systems offer the combination of efficient transport features of SWNT and the significant optical properties of the conjugated polymers. The percolation behavior in such hybrid systems has also been used to achieve an effective reduction in channel length, thus increasing transconductance [78].

Recent studies of conjugated polymer poly(3-hexylthiophene) P3HT and poly[2-methoxy-5-(2′-ethylhexyloxy]-1,4-phenylenevinyelene MEHPPV dispersed SWNTFET structures reveal interesting features resulting from a combination of a molecular process initiated by photoexcitation of the polymer and facilitated by device geometry conditions. The photoexcitation primarily affects the SWCNTFET characteristics in the depletion mode. The changes in the current in the depletion mode corresponding to the photoexcitation span several orders of magnitude, and the transistor I_d attains the saturation value corresponding to the on-state. It is to be noted that the current traverses only across the SWNT channel and the polymer network around the nanotube acts as the optically active antenna. The large change in I_d in the depletion mode is not instantaneous. Upon a close examination of experiments involving a periodic V_g, it is observed that the FET enters into the off-state (for $V_g > 0$) and under these conditions the photogenerated charge carriers induce the large increase in I_{ds} in the nanotube channel. The mechanism of this feature is speculated to be photocarrier generation in the polymer and transport to the charging centers around the SWCNT, facilitated by the difference in the energy levels in the depletion bias conditions. The changes in the current span several orders in magnitude, controlled by the initial on/off ratio of the SWCNT FET [76].

9.10
Recent Approaches in Assembling Devices

Apart from the approaches involving modifications in the device geometry and in the semiconductor/insulator interface an important research direction has been in the area of the molecular design of the active elements which permits self-assembled structures where transistor properties can be realized. An attractive paradigm is the self-organization of functional materials into molecularly ordered domains that are hierarchically organized across length scales.

In the process, it should be possible to achieve short-channel FET characteristics where one can study the scaling of mobility with respect to the molecular dimension. It may be possible to achieve a high degree of performance if the effect of the domain boundary is reduced. Apart from this it may be useful for organic chips and RFID tag application if low threshold voltages are obtained. The remarkable reduction of channel length was observed in the 30 nm channel FET based on pentacene, where electron beam lithography was used to pattern the platinum electrodes [79]. The field-effect mobility and the on/off ratio were observed to be of the order of 10^{-2} cm^2 V^{-1} s^{-1} and 10^2, respectively. This was attributed to the effect of bottom contact geometry, where pentacene tends to form smaller grains near the electrodes. The comparatively low on/off ratio was attributed to the large off current, which might be due to the large accumulation of charge at contact. Another alternative approach using a de-wetting technique was demonstrated where the top gate staggered geometry was maintained to obtain a channel length of 30 nm [80]. F8T2 was used as semiconducting material on top of PMMA dielectric to form an all-polymer short channel FET with mobility of 0.006 cm^2 V^{-1} s^{-1}.

The interface between the insulator and semiconductor is a very critical aspect in transistor performance. The modification of the interface property is a promising approach to the improvement of device performance. The influence of different deposition techniques and surface modification of the dielectric interface has been extensively reported [81]. A twenty-fold improvement over mobility on bare silicon oxide has been observed upon treating the dielectric surface with octadecyltrichlorosilane self-assembled monolayer. The surface treatment using a layer of ultrathin solution-based polymer after defining the drain-source electrodes has been reported more recently [82]. The self-organizing property selects the SAM to be within the channel region and thiol groups on the electrodes leading to better injection. Another promising method to improve the interface is the surface-directed, self-organized bilayer of two organic bulk phases with spontaneous solvent evaporation technique [81]. This technique completely avoids exposure of the interface to ambient contamination leading to the spontaneous formation of ultrathin and conformal semiconductor–insulator bilayers.

Microcontact printing (μCP), a parallel printing technique capable of structuring large areas with high-resolution chemical patterns, in contrast to nanoimprint lithography, can be readily applied to, e.g., flexible or structured substrates. It has been found that the process yields OTFTs with similar or better electrical performance than devices fabricated using conventional lithography [83]. The possibility of achieving molecular engineering and assembly via soft lithography methods has immense potential. A precisely defined array of nanostripes consisting of crystalline and highly ordered molecules obtained by a stamp-assisted deposition of the molecular semiconductors from a solution, has been utilized as a short-channel structure [84]. Interesting model systems such as rotaxane molecules have also shown self-organization and manipulation to offer a prospective route of improvement from the molecular level [85].

In summary, intense research activity over the last decade has been pursued in the areas pertaining to the molecular design and synthesis of organic semiconductors, purification procedures, solution processing methods, and engineering the metal/organic interfaces to realize optimum device structures. Some of these systems are also capable of exhibiting a variety of interesting phenomena such as optically induced current-enhancement, where the magnitude and relaxation processes can be controlled by V_g. Organic-polymeric materials offer a promising and commercially viable route for low-cost, large-area electronics. Prospects for large scale integration of the OFETs will offer tremendous alternatives in designing a variety of applications. The research breakthroughs and demonstration of new concepts in fabrication can be expected to be translated into large scale manufacturing in the coming decade.

Acknowledgments

We thank Dr. Vasuda Bhatia and Mr. Manohar Rao for assistance in preparing the manuscript.

References

1 C. D. Dimitrakaopoulos, D. J. Mascaro, *IBM J. Res. Dev.* **2001**, *45*, 11.
2 G. H. Gelinck et al., *Nat. Mater.* **2004**, *3*, 106.
3 H. Sirringhaus, *Adv. Mater.* **2005**, *17*, 2411.
4 G. Horowitz, *J. Mater. Res.* **2004**, *19*, 106.
5 R. Rotzoll et al., *Appl. Phys. Lett.* **2006**, *88*, 123502.
6 D. F. Barbe, C. R. Westgate, *J. Phys. Chem. Solids* **1970**, *31*, 2679.
7 F. Ebisawa, T. Kurokawa, S. Nara, *J. Appl. Phys.* **1983**, *54*, 3255.
8 H. Koezuka, A. Tsumura, T. Ando, *Synth. Met.* **1987**, *18*, 699.
9 T. A. Chen, X. M. Wu, R. D. Reike, *J. Am. Chem. Soc.* **1995**, *117*, 233.
10 A. Dodabalapur et al., *Appl. Phys. Lett.* **1998**, *73*, 142.
11 S. M. Sze, *Physics of Semiconductor Devices*, John Wiley, New York, **1981**.
12 F. Garnier, G. Horowitz, D. Fitchou, X. Peng, in *Science and Application of Conducting Polymers*, W. R. Salaneck, D. T. Clark, E. J. Samuelsen (Eds.), Adam Hilger, USA, **1990**.
13 M. Ahles, A. Hepp, R. Schmechel, H. von Seggern, *Appl. Phys. Lett.* **2004**, *84*, 428.
14 R. A. Street, A. Salleo, *Appl. Phys. Lett.* **2002**, *81*, 2887.
15 C. D. Dimitrakopoulos, B. K. Furman, T. Graham, S. Hegde, and S. Purushothaman, *Synth. Met.* **1998**, *92*, 47.
16 P. R. L. Malenfant et al., *Appl. Phys. Lett.* **2002**, *80*, 2517.
17 C. D. Dimitrakopoulos, A. R. Brown, A. Pomp, *J. Appl. Phys* **1996**, *80*, 2501.
18 H. Sirringhaus, N. Tessler, R. H. Friend, *Science* **1998**, *280*, 1741.
19 Z. Bao, A. Dodabalapur, A. J. Lovinger, *Appl. Phys. Lett.* **1996**, *69*, 4108.
20 Z. Bao, Y. Feng, A. Dodabalapur, V. R. Raju, A. J. Lovinger, *Chem. Mater.* **1997**, *9*, 1299.
21 G. M. Wang, J. Swensen, D. Moses, A. J. Heeger, *J. Appl. Phys.* **2003**, *93*, 6137.
22 G. Xu, Z. Bao, J. T. Groves, *Langmuir* **2000**, *16*, 1834.
23 *Printed Organic and Molecular Electronics*, D. Gamota, P. Brazis, K. Kalyanasundaram, J. Zhang (Eds.), Kluwer Academic, Dordrecht, **2004**.
24 H. Sirringhaus et al., *Science* **2000**, *290*, 2123.
25 H. Sirringhaus et al., *Nature* **1999**, *401*, 685.

26 R. J. Kline, M. D. Mcghee, M. F. Toney, *Nat. Mater.* **2006**, *5*, 222.
27 Y. Kim et al., *Nat. Mater.* **2006**, *5*, 197.
28 D. H. Kim et al., *Adv. Fun. Mater.* **2005**, *15*, 77.
29 S. Nariola et al., *Appl. Phys. Lett.* **1995**, *67*, 1899.
30 H. Ishii, K. Seki, *IEEE Electron. Devices* **1997**, *44*, 1295.
31 I. G. Hill, A. Rajagopal, A. Kahn, *Appl. Phys. Lett.* **1998**, *73*, 662.
32 I. H. Campbell et al., *Phys. Rev. B* **1996**, *54*, 14321.
33 I. H. Campbell et al., *Appl. Phys. Lett.* **1997**, *71*, 3528.
34 H. Akimichi, K. Waragai, S. Hotta, H. Kano, H. Sakati, *Appl. Phys. Lett.* **1991**, *58*, 1500.
35 F. Garnier et al., *J. Am. Chem. Soc.* **1993**, *115*, 8716.
36 A. Dodabalapur, L. Torsi, H. E. Katz, *Science* **1995**, *268*, 270.
37 C. D. Dimitrakopoulos, B. K. Furman, T. Graham, S. Hegde, S. Purushothaman, *Synth. Met.* **1998**, *92*, 47.
38 M. A. Loi et al., *Nat. Maer.* **2005**, *4*, 81.
39 G. Horowitz, X. Peng, D. Fichou, F. Garnier, *Synth. Met.* **1992**, *51*, 419.
40 A. R. Brown, A. Pomp, C. M. Hart, D. M. de Leeuw, *Science* **1995**, *270*, 972.
41 A. J. Salih, J. M. Marshall, J. M. Maud, *Philos. Mag. Lett.* **1997**, *75*, 169.
42 C. D. Dimitrakopoulos, A. R. Brown, A. Pomp, *J. Appl. Phys.* **1996**, *80*, 2501.
43 Y. Y. Lin, D. J. Gundlach, S. Nelson, T. N. Jackson, in *54th Annual Device Research Conference Digest* **1996**, p. 80.
44 Y. Y. Lin, D. J. Gundlach, S. Nelson, T. N. Jackson, *IEEE Electron. Device Lett.* **1997**, *18*, 606.
45 K. Shin, C. Yang, S. Y. Yang, H. Jeon, C. E. Park, *Appl. Phys. Lett.* **2006**, *88*, 072109.
46 D. K. Hwang et al., *Appl. Phys. Lett.* **2006**, *88*, 243513.
47 A. Assadi, C. Svensson, M. Willander, O. Inganäs, *Appl. Phys. Lett.* **1988**, *53*, 195.
48 A. Tsumura, H. Fuchigami, H. Koezuka, *Synth. Met.* **1991**, *41*, 1181.
49 J. Paloheimo, H. Stubb, P. Yli-Lahti, P. Kuivalainen, *Synth. Met.* **1991**, *41–43*, 563.
50 G. Wang, T. Hirasa, D. Moses, A. J. Heeger, *Synth. Met.* **2004**, *146*, 127.
51 H. Sirringhaus et al., *Appl. Phys. Lett.* **2000**, *77*, 406.
52 H. Sirringhaus et al., *Science* **2000**, *290*, 2123.
53 B. S. Ong, Y. Wu, P. Liu, *Proc. IEEE* **2005**, *93*, 1412.
54 I. Mcculloch et al., *Nat. Mater.* **2006**, *5*(4), 328.
55 G. Guillaud, M. A. Sadound, M. Maitrot, *Chem. Phys. Lett.* **1990**, *167*, 503.
56 R. C. Haddon et al., *Appl. Phys. Lett.* **1995**, *67*, 121.
57 S. Kobayashi, T. Takenobu, S. Mori, S. Fujiwara, Y. Iwasa, *Appl. Phys. Lett.* **2003**, *82*, 4581.
58 P. R. L. Malenfant et al., *Appl. Phys. Lett.* **2002**, *80*, 2517.
59 A. Babel, S. A. Jenekhe, *Adv. Mater.* **2002**, *14*, 371.
60 A. Babel, S. A. Jenekhe, *J. Am. Chem. Soc.* **2003**, *125*, 13656.
61 C. Waldauf, P. Schilinsky, M. Perisutti, J. Hauch, C. J. Brabec, *Adv. Mater.* **2003**, *15*, 2084.
62 C. Rost, D. J. Gundluch, S. Kars, W. Rieb, *J. Appl. Phys* **2004**, *95*, 5782.
63 E. J. Meijer et al., *Nat. Mater* **2003**, *2*, 678.
64 T. D. Anthopoulos et al., *Adv. Mater.* **2004**, *16*, 2174.
65 T. Yasuda, T. Goto, K. Fujita, T. Tsutsui, *Appl. Phys. Lett.* **2004**, *85*, 2098.
66 L. L. Chua et al., *Nature* **2005**, *434*, 194.
67 K. S. Narayan, N. Kumar, *Appl. Phys. Lett.* **2001**, *79*, 1891.
68 S. Dutta, K. S. Narayan, *Phys. Rev. B* 200, *68*, 125208.
69 S. Dutta, K. S. Narayan, *Adv. Mater.* **2004**, *16*, 2151.
70 V. Podzorov, V. M. Pudalov, M. E. Gershenson, *Appl. Phys. Lett.* **2004**, *85*, 6039.
71 Y. Noh et al., *Appl. Phys. Lett.* **2005**, *86*, 043501.
72 P. L. McEuen, J. Y. Park, *MRS Bull.* **2004**, *29*, 272.

73 K. Balasubramanian, Y. Fan, M. Burghad, K. Kern, M. Friedrich, U. Wannek, A. Mews, *Appl. Phys. Lett.* **2004**, *84*, 2400.

74 Y. Ohno, S. Kishimoto, T. Mizutani, *Jpn. J. Appl. Phys.* **2005**, *44*(4A), 1592.

75 M. Freitag, Y. Martin, J. A. Misewich, R. Martel, Ph. Avouris, *Nano Lett.* **2003**, *3*, 8.

76 K. S. Narayan, Manohar Rao, R. Zhang, P. Maniar, *Appl. Phys. Lett.* **2006**, *88*, 243507.

77 A. Star, Y. Lu, K. Bradley, G. Grüner, *Nano Lett.* **2004**, *4*(9), 1587.

78 X.-Z. Bo, C. Y. Lee, M. S. Strano, M. Goldfinger, C. Nuckolls, G. Blanchet, *Appl. Phys. Lett.* **2005**, *86*, 182102.

79 Y. Zhang, J. R. Petta, S. Ambily, Y. Shen, D. C. Ralph, G. G. Malliaras, *Adv. Mater.* **2003**, *15*, 1632.

80 N. Stutzmann, R. H. Friend, H. Sirringhaus, *Science* **2003**, *299*, 1881.

81 C. Arias, *J. Macromol. Sci., Part C: Polym. Rev.* **2006**, *46*, 103.

82 S. Kobayashi et al., *Nat. Mater.* **2004**, *3*, 317.

83 J. Z. Wang, J. F. Chang, H. Sirringhaus, *Appl. Phys. Lett.* **2005**, *87*, 083503.

84 M. Cavallini et al., *Nano Lett.* **2005**, *5*(12), 2422–2425.

85 A. H. Flood, J. F. Stoddart, D. W. Steuerman, J. R. Heath, *Science* **2004**, *306*, 2055.

10
Supramolecular Approaches to Molecular Machines

M.C. Grossel

10.1
Introduction

Since the early days of supramolecular chemistry, the creation of molecular assemblies having a function has been a key goal [1], arising to a considerable extent from an interest in the mimicking of biological function. However, a key aspect of chemistry is the creation of novel structures rather than just functional mimics, and as the field has grown so have the targets expanded, inspired for example by the earlier proposals of Drexler [2] and Feynman [3]. Key functions of supramolecular assemblies relevant to the construction of molecular machines [4] are recognition, transport, and modification of a substrate. In addition supramolecular assemblies are dynamic (as a result of internal motions within the complex) and switchable (mechanically, optically, electronically, etc.). They can change their size and shape, and can be chiral. Each of these properties has potential applications in materials science.

Stoddart and others [4a] have characterised a machine by a number of different attributes which include:
- the type(s) of energy required to make it work
- the type(s) of movements performed by its components
- the manner by which its operation can be monitored and controlled
- the possibility of repeating the operation at will and establishing a cyclic process
- the timescale needed to complete a cycle of operation
- the function(s) performed by the machine.

Photons and electrons are the best options for controlling molecular-scale machines since they can be involved in both the input and output operations. They produce no chemical by-products and circumvent the problem of the need for a continual supply of reagents. Processes can be readily monitored and controlled

Nanomaterials Chemistry. Edited by C.N.R. Rao, A. Müller, and A.K. Cheetham
Copyright © 2007 WILEY-VCH Verlag GmbH & Co. KGaA, Weinheim
ISBN: 978-3-527-31664-9

by optical spectroscopy or electrochemistry. However, as will be seen later, pH changes have also been successfully used to drive such systems.

Supramolecular assemblies are more likely to undergo larger amplitude variations in shape etc. in comparison with the internal conformational or configurational changes which can occur within a single molecule, though Feringa, Tour and others [5] have constructed elegant "molecular motors" and nanovehicles. To date one of the most promising approaches to "mechanical" molecular machines has been through the use of catenanes and rotaxanes.

10.2
Catenanes and Rotaxanes

Catenanes and rotaxanes are two types of structure which have only become readily accessible with the advent of supramolecular chemical techniques A *catenane* is a structure in which two or more rings are threaded through each other but are not joined covalently (i.e. they behave like links in a chain, see Fig. 10.1). An [n]-catenane refers to a catenane chain consisting of 'n' interlinked rings.

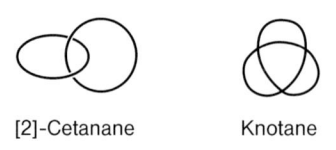

[2]-Cetanane   Knotane

Fig. 10.1 Catenanes and knotanes.

A further level of sophistication arises when the interlocked rings are unsymmetrical or are multiply entwined – the latter form molecular knots (also known as *knotanes*); such structures can be chiral (as in the case of the trefoil knot shown in Fig. 10.1).

A *rotaxane* consists of a ring surrounding a thread which is stoppered so that the ring cannot escape (Fig. 10.2), whereas the corresponding structure consisting of a ring around an unstoppered thread is called a *pseudorotaxane* (Fig. 10.2). An [n]-rotaxane has $n - 1$ rings surrounding the thread.

One key difference between the two is that pseudorotaxanes can thread and unthread, depending on their environment etc., whereas rotaxanes cannot. In addi-

Rotaxane   Pseudorotaxane

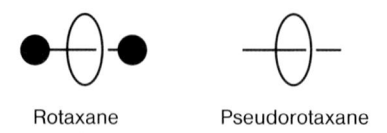

Fig. 10.2 Rotaxanes and pseudorotaxanes.

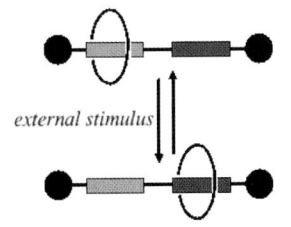

Fig. 10.3 Schematic of a rotaxane acting as a molecular shuttle.

tion both structures are dynamic through internal motion; the thread can rotate within the ring (or the ring about the thread) and, if the thread is long enough and contains several binding sites, the ring can shuttle between various "stations" on the thread, effectively behaving like a molecular-scale abacus (Fig. 10.3). Catenanes are also dynamic structures in which one ring can either hop around or rotate through another.

10.2.1
Synthetic Routes to Catenanes and Rotaxanes

Basic strategies for catenane and rotaxane synthesis are summarised in Fig. 10.4. In the case of catenanes one ring is preformed, a chain is threaded through it and the ends of this are then joined together. The two (or more) rings involved in the catenane need not, of course, be chemically identical. For rotaxanes and pseudorotaxanes, the ring can either be constructed around a stoppered rod or an unstoppered rod can be threaded through a ring to form a pseudorotaxane which can then be stoppered if required.

Several successful routes to catenanes have been developed using a variety of self-assembly techniques. These employ one of three basic approaches involving:

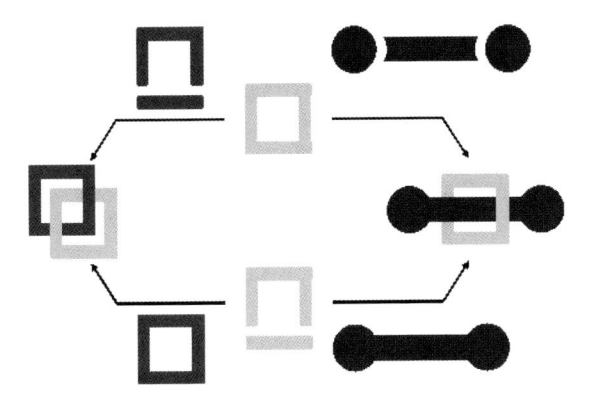

Fig. 10.4 Strategies for catenane and rotaxane synthesis.

(i) aromatic π–π association; (ii) hydrogen-bonded preorganisation; or (iii) ion templating.

10.2.2
Aromatic π–π Association Routes to Catenanes and Rotaxanes

10.2.2.1 Preparation and Properties of [2]-Catenanes

One of the leading pioneers in assembling structures of this type has been Fraser Stoddart. His observation [6] that the electron-rich aromatic rings in large crown ethers such as dibenzo-30-crown-10 form complexes with electron-deficient bipyridinium cations, led him to develop an efficient route to [2]- and higher order catenanes and rotaxanes. For example, when 4,4′-bipyridine is reacted with 1,4-bis-(bromomethyl)benzene in the presence of dibenzo-34-crown-10 the [2]-catenane **1** is formed in good yield (Scheme 10.1) as a result of the strong complexation of the dialkoxybenzene units by the cyclobis-(paraquat-p-phenylene) ring (PQT^{4+}) [7].

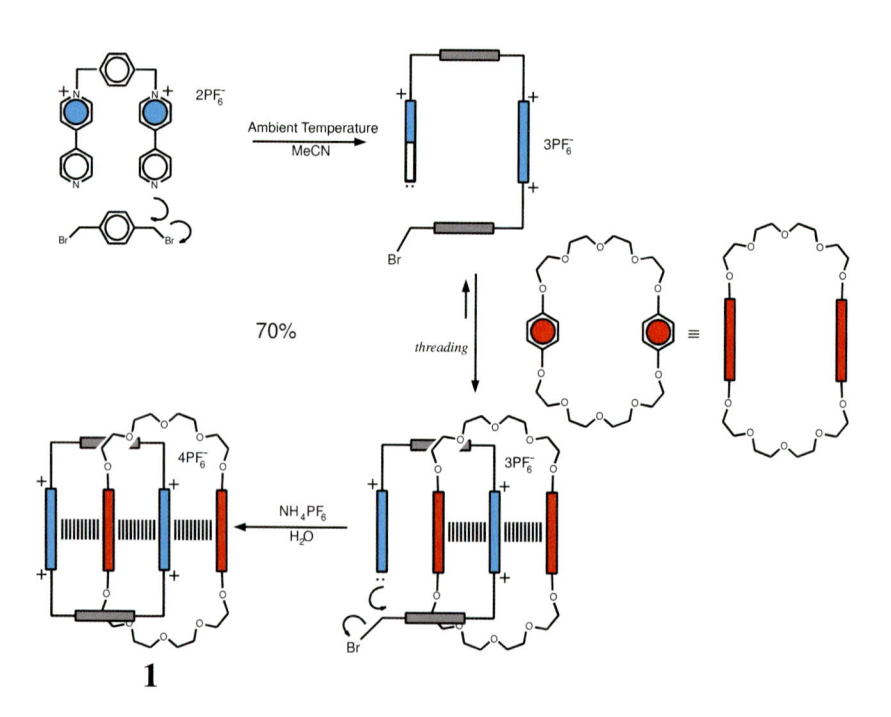

Scheme 10.1 Stoddart's route to a [2]-catenane [7].

As has already been noted, catenanes are potentially dynamic structures; indeed in the [2]-catenane **1** for example the crown ether rotates through the

PQT^{4+} ring at ca. 300 Hz, whilst precessing around this latter at ca. 2 kHz (at room temperature).

10.2.2.2 Multiple Catenanes

In the original synthesis outlined in Scheme 10.1 there is insufficient space within the structure for the threading of further rings. However, if the PQT^{4+} box is enlarged, higher order catenanes can be prepared (Scheme 10.2). These structures are also dynamic with the dibenzo-34-crown-10 rings rotating through the PQT^{4+} ring at ca. 8 kHz (at ca. 298 K).

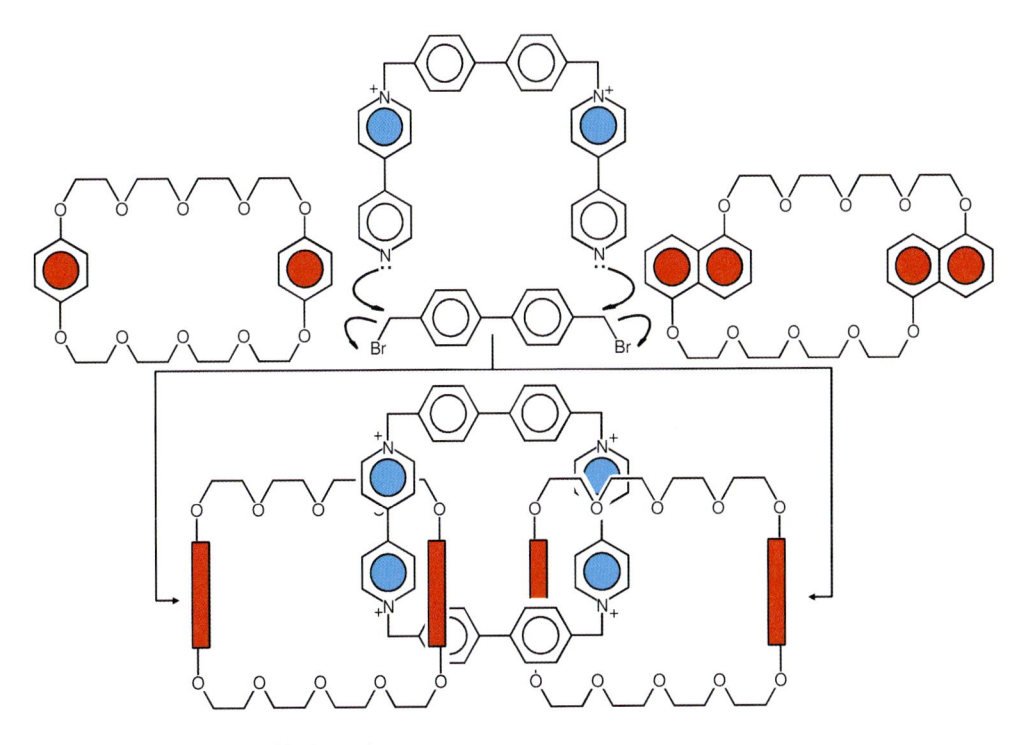

Scheme 10.2 Preparation of higher-order catenanes.

Alternatively expansion of the crown ether unit, thereby introducing additional dialkoxybenzene units, also provides a route to a [3]-catenane (Fig. 10.5). In this structure the two paraquat boxes roll through the crown ether at ca. 28 kHz at room temperature and chase each other around the crown ring at ca. 300 Hz behaving somewhat like a molecular train set, always keeping one "station" apart.

Further development of this synthetic approach with careful choice of the crown ethers and cationic boxes used has enabled Stoddart to prepare a variety of catenane and rotaxane-based polymers including the [5]-catenane known as olympiadane [8].

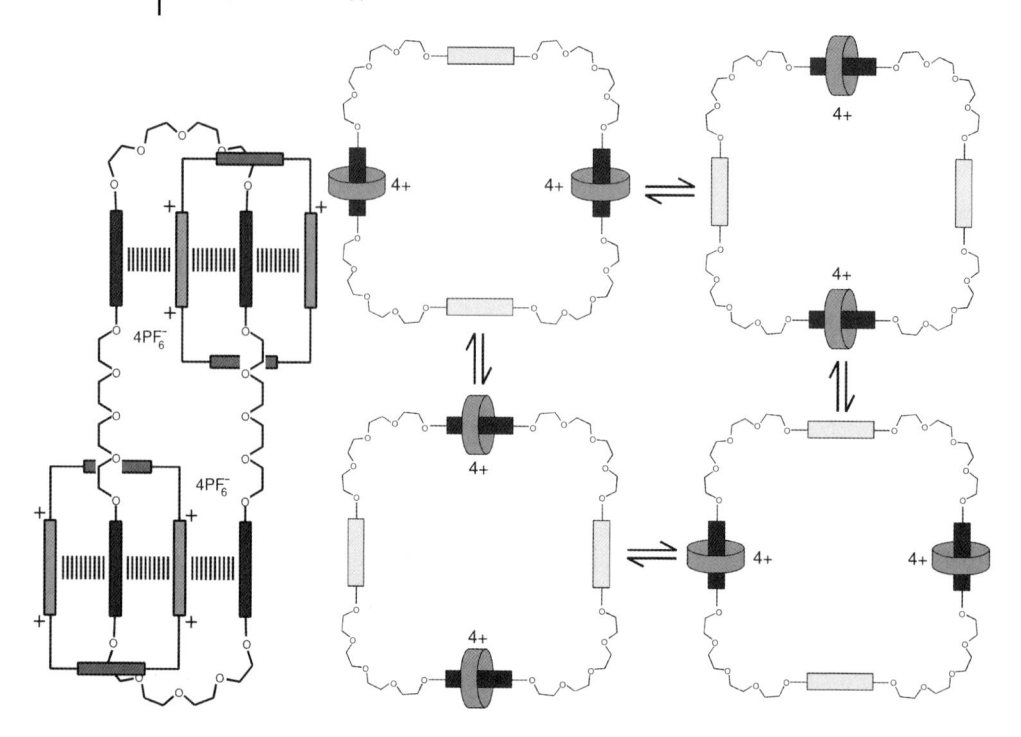

Fig. 10.5 The molecular train set: a [3]-catenane derived from tetrabenzo-68-crown-20 and two PQT^{4+} rings.

10.2.2.3 Switchable Catenanes

This approach to catenanes has proved quite versatile. By varying the nature and number of the electron-rich aromatic rings in the crown ether it is possible to create more complex catenanes and to fine-tune their dynamic behavior. Replacement of the 1,4-dialkoxybenzene moiety by a better electron donor such as 1,5-dialkoxynaphthalene or tetrathiafulvalene (TTF) results in stronger complexation with the PQT^{4+} box. A key advantage which arises from incorporation of the TTF unit into a catenane is that the structure can be switched between different states. TTF^0 is an electron donor which forms complexes with the PQT^{4+} box, whereas TTF$^+$ does not show donor/acceptor character and TTF^{2+} is an electron acceptor which forms charge transfer complexes with the 1,5-dialkoxynaphthalene moiety [9].

For example, in the TTF/1,5-dialkoxy-naphthalene-based catenane shown in Fig. 10.6 a neutral TTF ring is the favored guest within the PQT^{4+} box but oxidation of the TTF leads to switching of the location of the crown ether, placing the naphthalene ring inside the PQT^{4+} box cavity as the preferred guest.

Another key feature of such catenanes is that the complexes are colored, the PQT^{4+}/1,5-dialkoxynaphthalene complex being red (λ_{max} 473 nm) whereas the TTF complex is green (λ_{max} 854 nm) – see Table 10.1 [10]. Thus the catenane

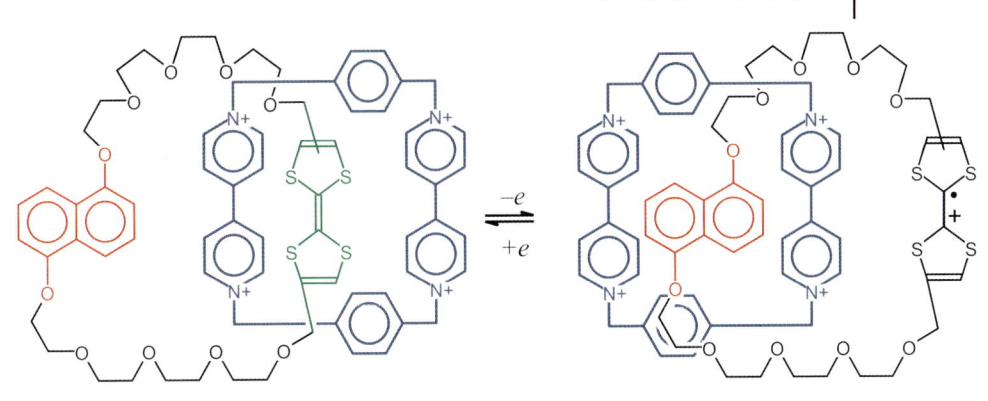

Fig. 10.6 A bistable color-switching [2]-catenane.

shown in Fig. 10.6 is a bistable, electrochemically switchable system, i.e. a color-switchable dye.

A further development of this concept is the introduction of a third "blue" station thereby affording a tristable structure (Fig. 10.7) which is switchable between three color states and is thus capable of acting as a pixel in a solid-state display device. This has been achieved through the addition of a substituted benzidine. Key variables in this system are the display colors, the binding energies of the individual complexes, and the oxidation potentials of the electron-donor guests. The 2,2'-difluorinated benzidine derivative has been found to give a particularly good blue (Table 10.1). Thus the neutral catenane is green but oxidation of the TTF unit pushes the PQT^{4+} ring onto the benzidine site changing the color to blue (Fig. 10.7). Subsequent oxidation occurs at the benzidine unit resulting in a shift of the PQT^{4+} ring onto the dialkoxynaphthalene position resulting in a change of color to red [11].

Table 10.1 Colors of various complexes with the PQT^{4+} box [10].

λmax (nm)					
333	473	494	601	677	854
PQT^{4+} complex Colour					
	red		blue		green

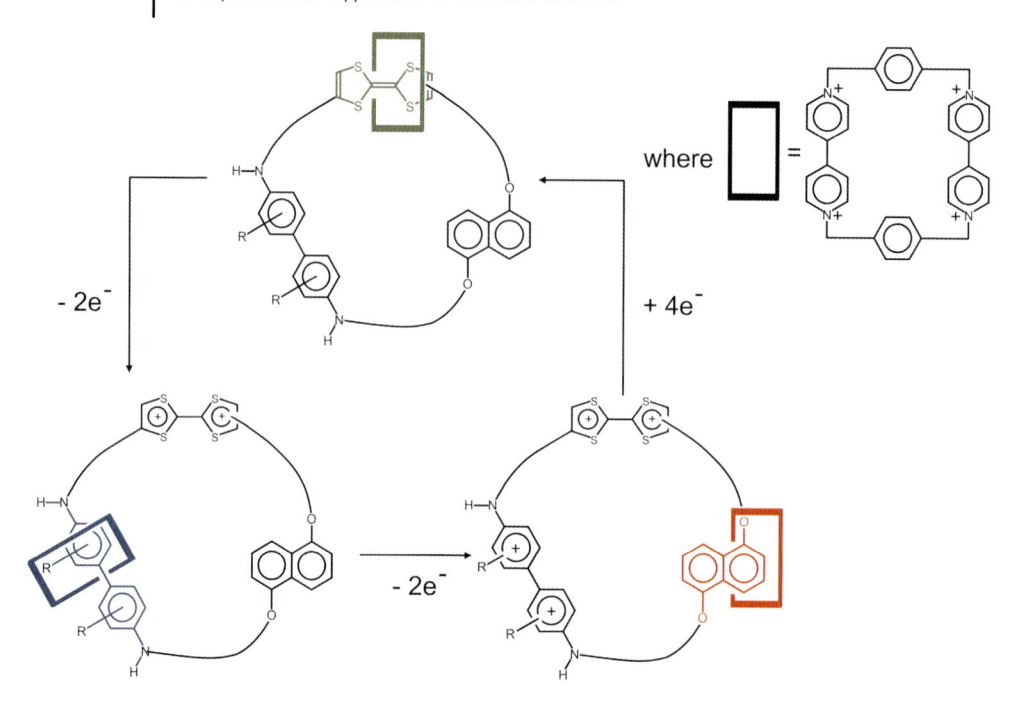

Fig. 10.7 A three-color switchable [2]-catenane [11].

10.2.2.4 Other Synthetic Routes to Paraquat-based Catenanes

A number of alternative approaches to paraquat-based catenanes have been reported recently. One method exploits intramolecular 1,3-dipolar cycloadditions involving an azide and a terminal alkyne (the latter is needed to avoid isomer issues), a process known as "Click" chemistry (Scheme 10.3) [12].

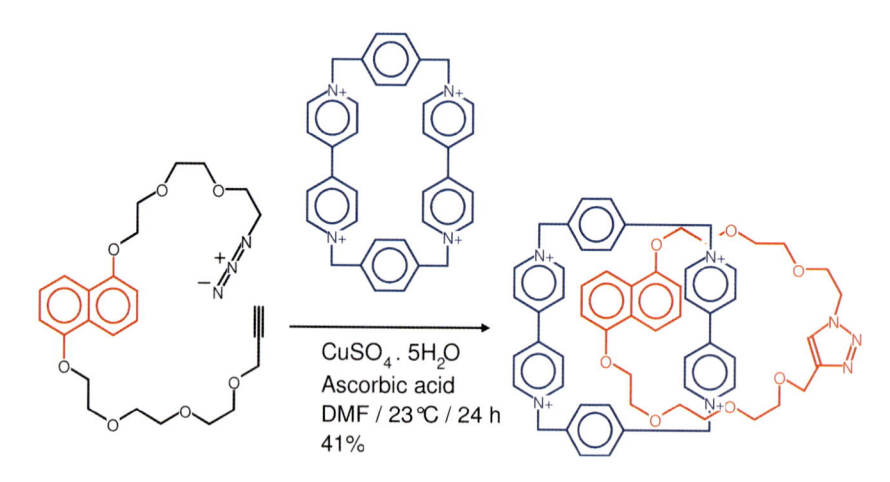

Scheme 10.3 Preparation of a [2]-catenane using "Click" chemistry [12].

10.2 Catenanes and Rotaxanes | 327

Another route uses the Eglington coupling of terminal alkynes which can be carried out as a one-pot catenane synthesis (Scheme 10.4) [13].

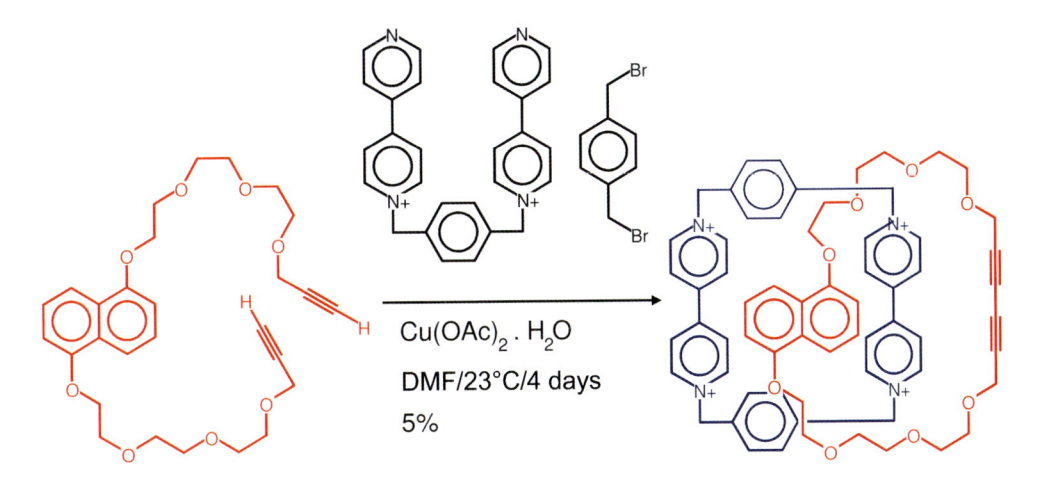

Scheme 10.4 A "one-pot" route to [2]-catenanes using Eglinton coupling [13].

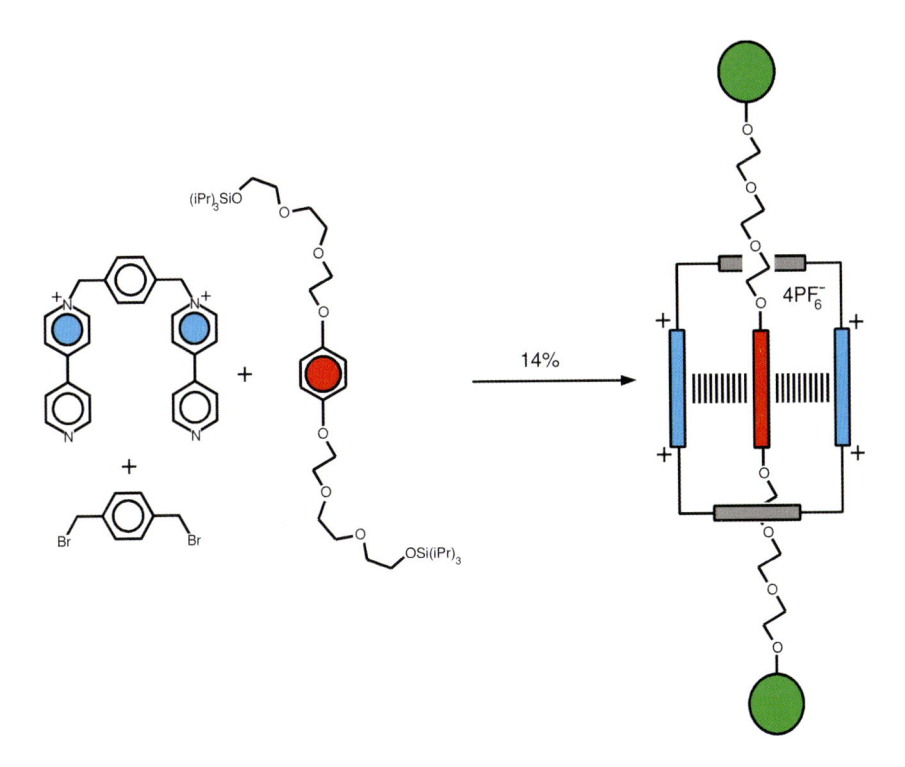

Scheme 10.5 Paraquat-based route to a [2]-rotaxane.

10.2.2.5 Rotaxane Synthesis

Rotaxanes can be prepared using the same chemical approaches as have been devised for the synthesis of catenanes. Thus formation of the paraquat box in the presence of a bulky stoppered or unstoppered 1,4-bis-poly(ethyleneoxy) benzene or similarly 1,5-disubstituted naphthalene core affords a [2]-rotaxane (Scheme 10.5).

If a sufficient number of electron-rich alkoxybenzene units are incorporated into a thread and the PQT^{4+} box is added, a pseudorotaxane will readily form (Scheme 10.6).

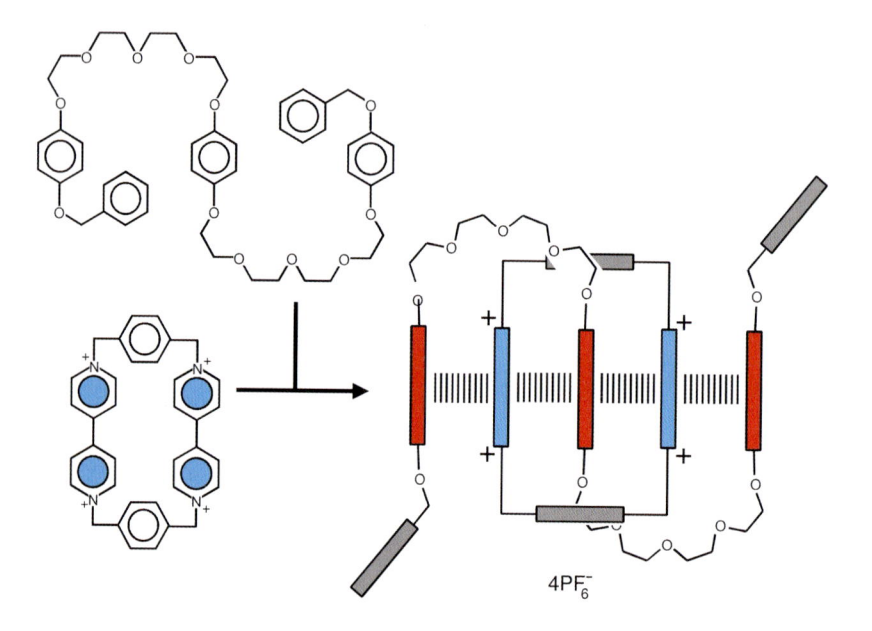

Scheme 10.6 A self-assembling PQT^{4+} box-based pseudorotaxane.

10.2.2.6 Switchable Catenanes

The TTF/Dialkoxynaphthalene/PQT^{4+} ring combination has also been used to create an electromechanically controllable device which might be regarded as an analog of a muscle [14].

The design uses a symmetrical [3]-rotaxane (Fig. 10.8) each side of which has two stations (a TTF and a dialkoxynaphthalene site, allowing redox switching of a PQT^{4+} box between them in the usual manner. However, in this case each PQT^{4+} box has been functionalised to allow it to be attached to one side of a gold-surfaced array of microcantilever arms. Chemical or electrochemical oxidation and reduction of the TTF units in such (non-aligned) coatings causes the PQT^{4+} boxes to move towards the middle of the rod and out again and, in order

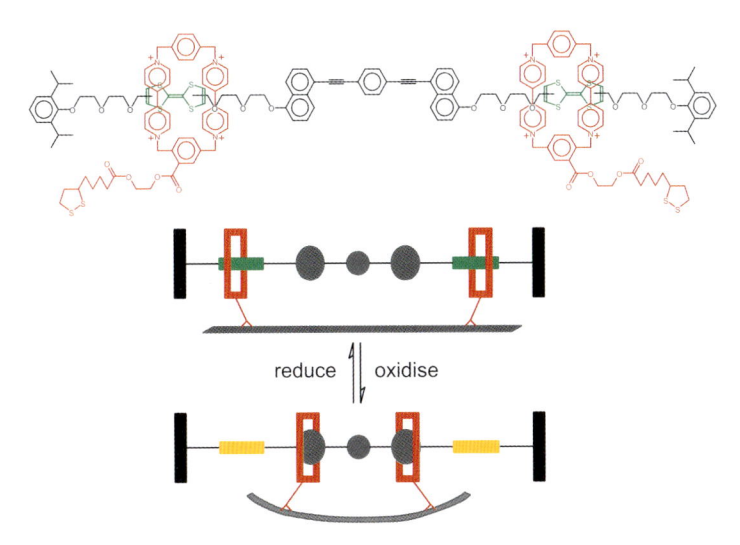

Fig. 10.8 An artificial muscle mimic [14].

to accommodate these movements, the surface flexes, behavior which can be detected optically. This system can by cycled about 25 times but slowly degrades.

10.2.2.7 Neutral Catenane Assembly

Sanders and coworkers [15] have exploited the complexation of pyromellitic di-imides by bis-1,5-(dinaphtho)-38-crown-10 as a means of templating the formation of a mixed [2]-catenane using acetylene coupling to create the second ring in good yield (Scheme 10.7). Similar results were obtained using the corresponding 1,4,5,8-naphthalenecarboxylic acid di-imide. The structures are dynamic with the pirouetting of the crown around the *bis*-di-imide ring being the lower energy process since it does not involve complete breaking of the donor–acceptor interactions present in these systems.

10.2.3
Ion Templating

10.2.3.1 Approaches to Redox-switchable Catenanes and Rotaxanes

The use of 2,2′-bipyridyl, phenanthroline and related ligands for catenane formation has been successfully exploited by Sauvage and others [16]. Scheme 10.8 outlines the basic principles of this approach.

The initial step involves formation of a tetrahedral complex of Cu(I) using two bipyridyl ligands, one of which is part of a crown ether. The subsequent linking of the ends of the acyclic ligand results in formation of a catenate (a metal com-

Scheme 10.7 Strong π-donor–π-acceptor interactions between neutral substrates facilitate catenane assembly [15].

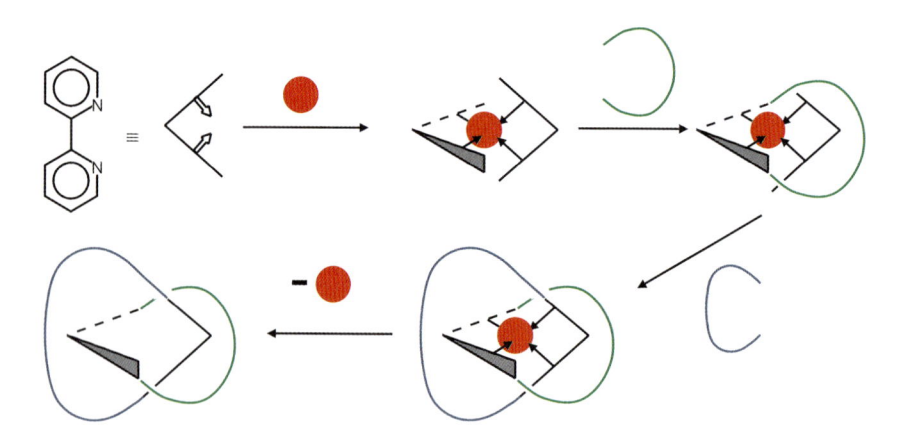

Scheme 10.8 Metal ion-templated catenane synthesis.

plex of a catenane) which can finally be demetallated using KCN (Scheme 10.9) [17]. Sauvage and coworkers [18] have also explored the use of other metal-ion templates, e.g. pentacoordinate Zn^{2+} complexes, for catenane synthesis.

Incorporation of additional coordination sites (e.g. by use of a terpyridine unit) within one crown opens up the possibility of producing a redox-switchable device. In the presence of Cu(I) the two bipyridyl units assemble in a tetrahedral manner

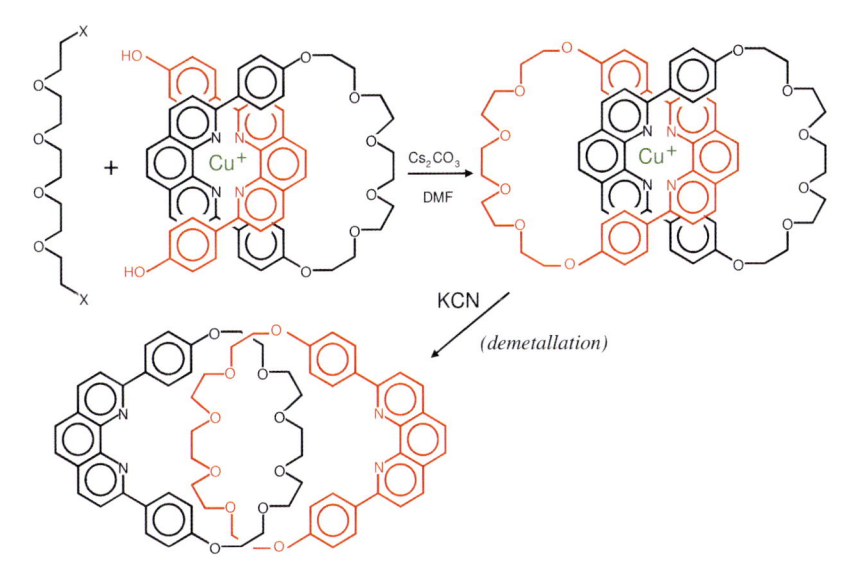

Scheme 10.9 Metal ion-templated catenane synthesis [16, 17].

around the metal ion, but oxidation to Cu(II) results in a switching of the catenane (see Fig. 10.9) so that the terpyridine coordinates with the metal ion. In this way one ring can be rotated about the other in a controlled manner by redox switching [19].

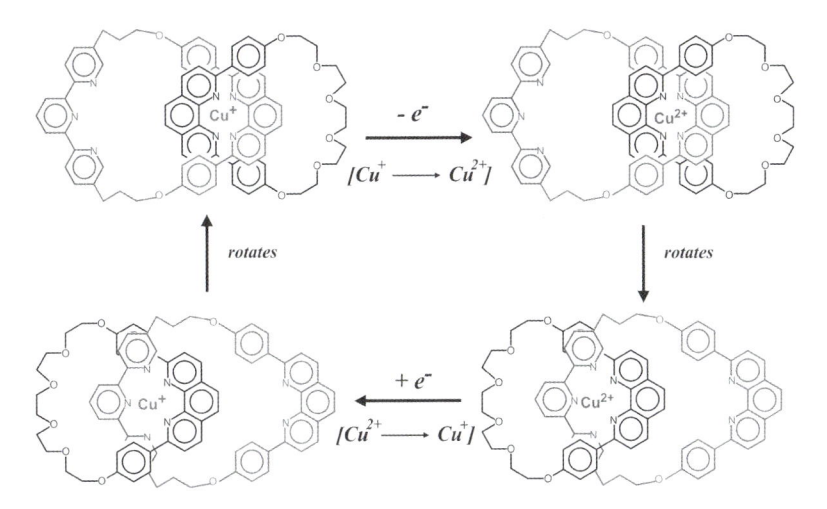

Fig. 10.9 A simple catenane-based rotary motor [18, 19].

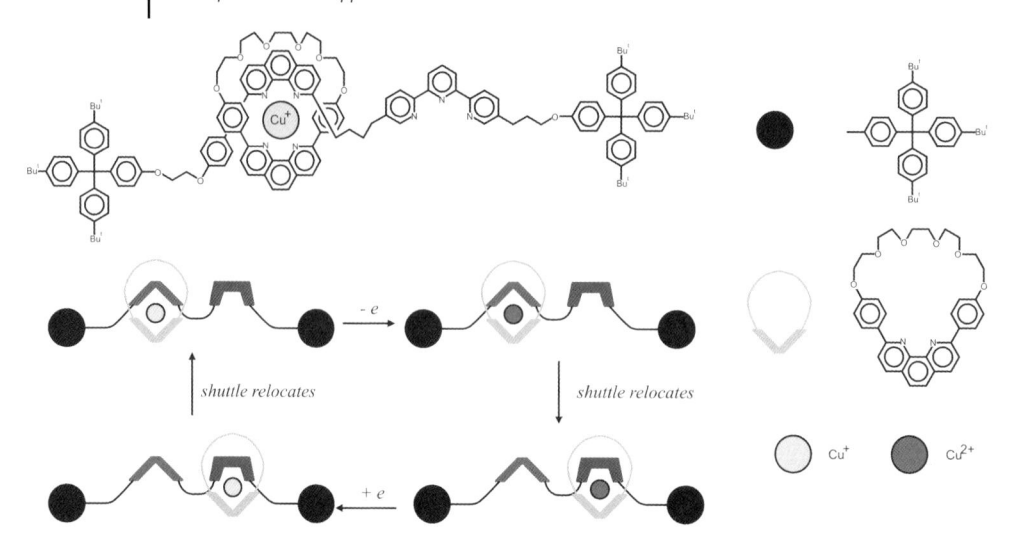

Fig. 10.10 A simple molecular abacus [20].

The same approach can be used in a rotaxane to produce a redox-switchable shuttle, providing a simple approach to a molecular abacus (Fig. 10.10) [20] and more recently has been used to develop muscle mimics [21].

10.2.3.2 Making More Complex Structures

Metal-ion templating also allows the multiple entwining of a poly-pyridyl-containing thread and a poly-pyridyl-containing ring to give a multiply entwined catenane which, upon demetallation, produces a knotane [see Fig. 10.11 and Scheme 10.10] [22].

This concept has recently been further developed by Stoddart in a thermodynamically controlled synthesis of a Borromean ring **4** (Fig. 10.12) [23]. In such a structure there are three interlocked but non-cocatenated rings such that if one is cut all the rings fall apart.

This synthesis exploits the Zn^{2+}-templated condensation of a bipyridyl-cored diamine ligand with pyridine 2,6-dicarboxaldehyde. Imine formation occurs, leading to a macrocycle having two exo- and two endo-receptor sites where zinc-complexation can occur (Fig. 10.13). The two other macrocycles assemble around this core in the required fashion to give a hexa-zinc complex. Careful borohydride reduction of the imines is accompanied by demetallation to give the Borromean ring. The potential use of Borromean rings remains unclear though it is suggested that the unique symmetrical architecture of the metal complex may lead to interesting electronic and optical properties, particularly since such structures can be made chiral [24].

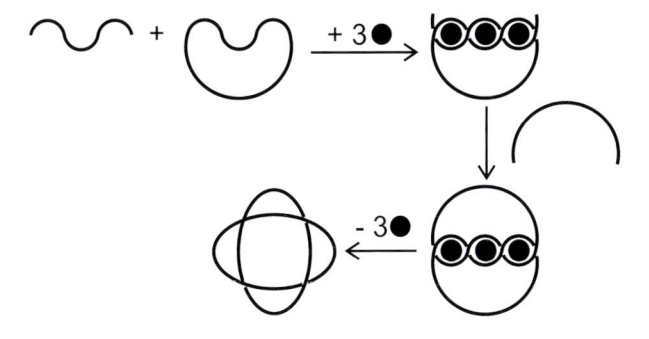

Fig. 10.11 Metal ion-templated knotane synthesis.

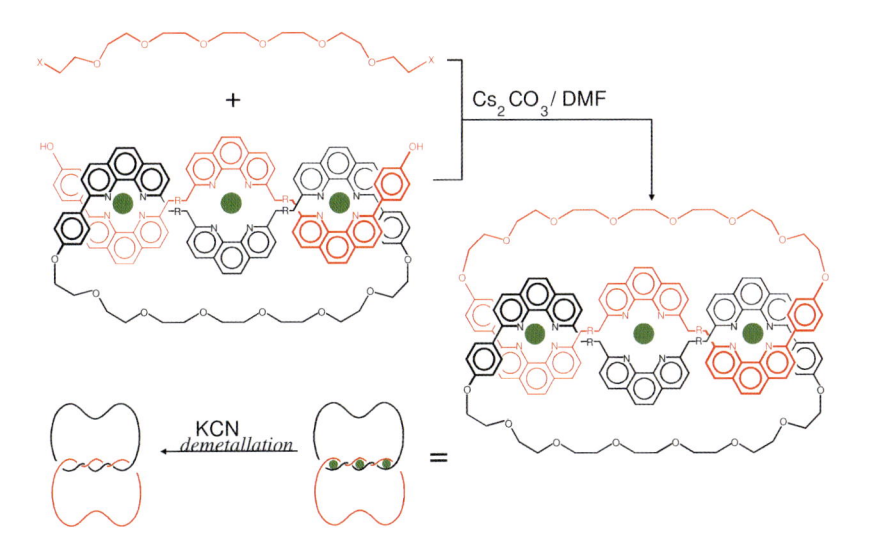

Scheme 10.10 Sauvage's knotane synthesis [22].

10.2.3.3 Routes to [n]-Rotaxanes using Olefin Metathesis – Molecular Barcoding

The strategies described so far for metal-ion templating have exploited tetrahedral or octahedral complexes formed from multidentate ligands to achieve interwoven/intertwined architectures. However, Sauvage [25] and Leigh [26] and coworkers have successfully demonstrated the application of square-planar metal complexes in catenane and rotaxane synthesis. The key here is to use one tridentate and one monodentate ligand in order to achieve the orthogonal ligand geometry required. The combination of an N,N'-disubstituted pyridine 2,6-dicarboxamide ring precursor (which affords a tridentate ligand when both amide

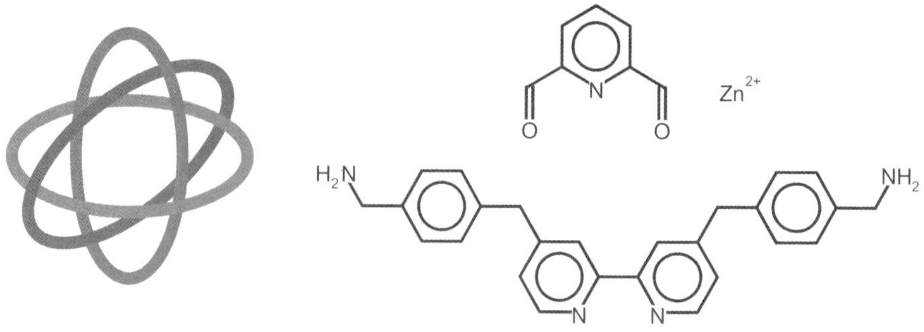

Fig. 10.12 A Borromean ring **4** and the components required for its synthesis [23].

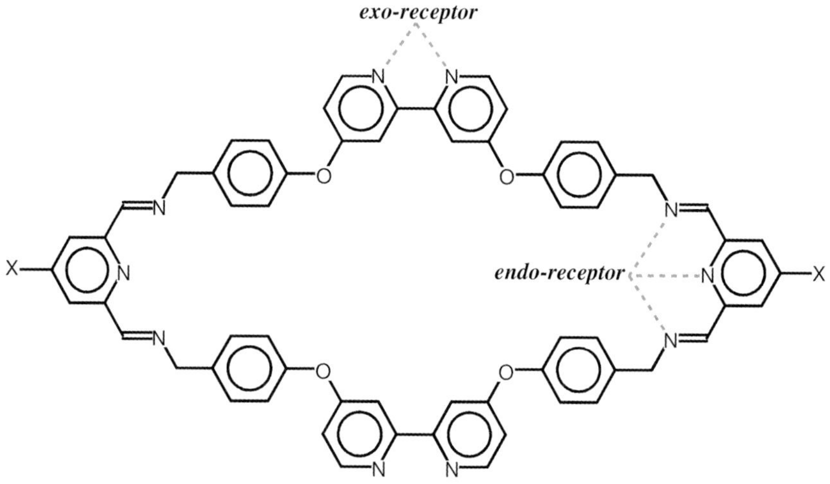

Fig. 10.13 The Borromean ring synthesis – key assembly interactions around one ring component.

groups are deprotonated [27]) and a stoppered thread containing a simple disubstituted pyridine centre have been docked around a Pd^{II} core (Scheme 10.11). Subsequent treatment of the complex with Grubb's catalyst followed by reduction and demetallation results in rotaxane formation in ca. 77% yield.

This strategy has subsequently been used to develop a controlled synthetic route to [*n*]-rotaxanes – structures which might be regarded as molecular scale barcodes since several different rings can be attached around the thread in a controlled sequence. The general concept is outline in Scheme 10.12. The use of a longer thread with just one pyridine complexation site encourages the first ring formed to move away from this point after demetallation. Addition of more

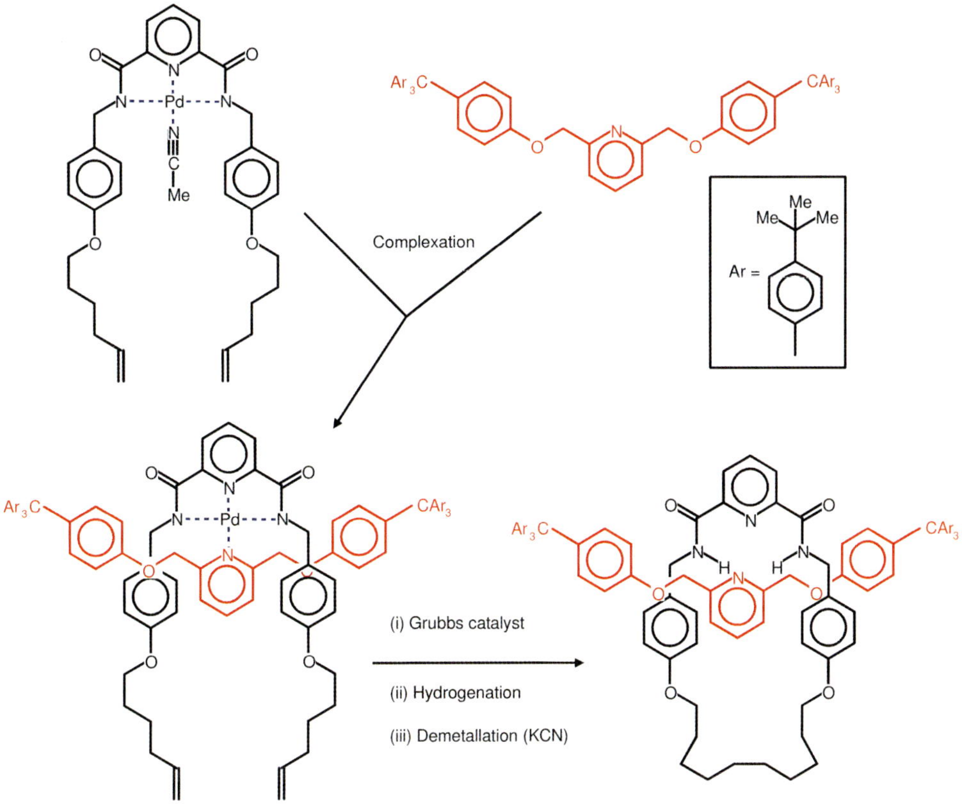

Scheme 10.11 Metal-ion templated rotaxane formation using olefin metathesis [26].

palladium-complexed ring precursor (bearing different substituent(s) on the pyridine dicarboxamide) results in further assembly around the thread. Alkene metathesis, followed by reduction and demetallation produces a [3]-rotaxane. This process can be repeated at will provided that there is sufficient room on the thread.

The same approach has also been applied successfully to [2]-catenane synthesis (Scheme 10.13) [29]. Various combinations of thread, preformed ring and preformed complex have been explored, the most successful involving use of a preformed acyclic pyridine dicarboxamide Pt^{II} complex with a pyridine-containing macrocycle.

10.2.3.4 Anion-templating

Another recent development in ion-templated catenane and rotaxane synthesis has been the use of anion templates. A pioneer of this approach has been Beer

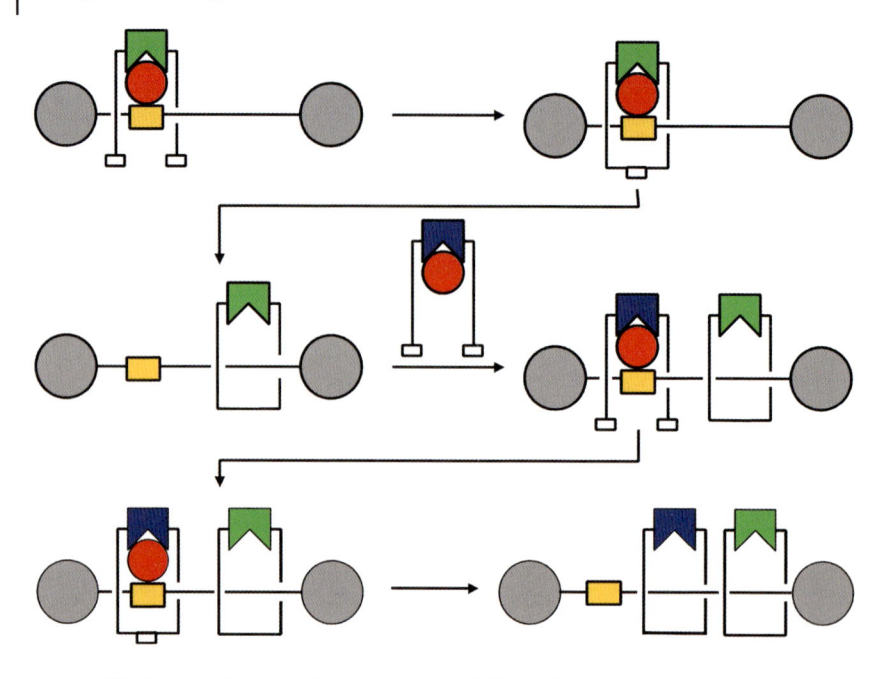

Scheme 10.12 A simple approach to molecular-scale bar-coding [28].

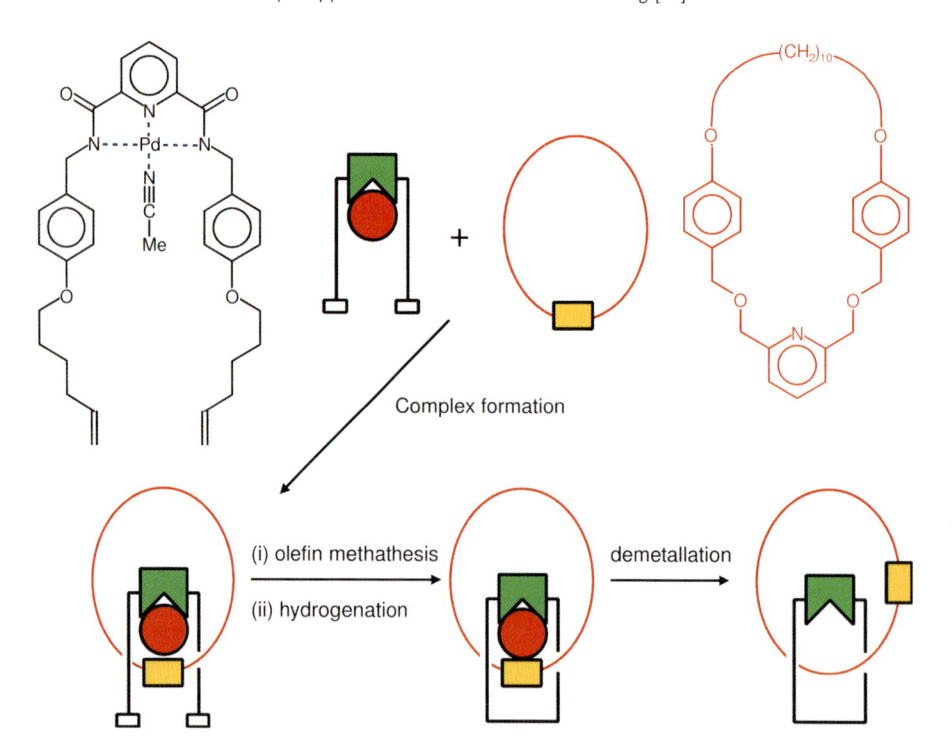

Scheme 10.13 Using olefin metathesis to make a [2]-catenane [29].

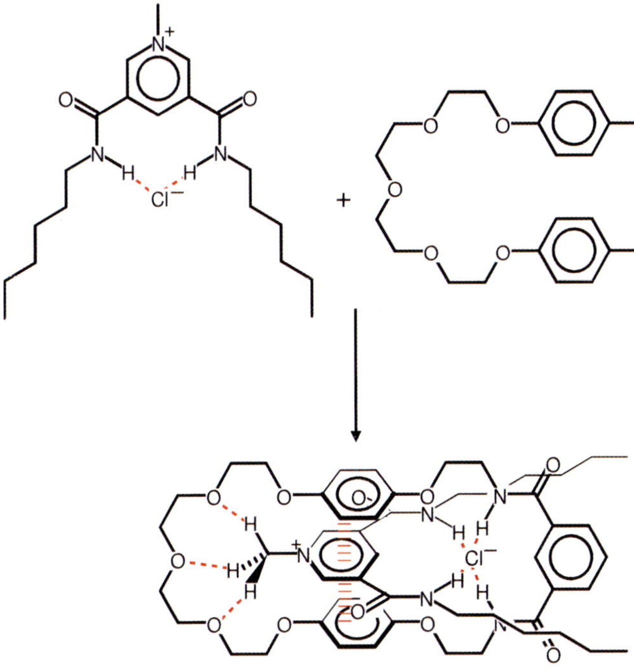

Scheme 10.14 Anion-templated pseudorotaxane assembly [30].

[30] who has used chloride ion as a template for pseudorotaxane threading (Scheme 10.14).

In a variation on this procedure a rotaxane has been formed by complexation of a 3,5-dialkyl N-methyl-pyridinium-stoppered thread with a bis-N,N'-(ω-alkenyl) isophthaloyl dicarboxamide in the presence of chloride ion and subsequent metathesis-based ring closure [31].

10.2.3.5 Other Approaches to Ion-templating

Fujita has prepared [2]-catenanes by a metal-ion mediated self-assembly process involving Pd^{II}-complexes and a variety of bis-pyridine ligands in water (Fig. 10.14) [32]. At low concentrations the equilibrium lies in favor of the simple macrocycle but at high concentrations the catenane is the dominant species present. The key to this process is the cavity size within the macrocycle, the van der Waals' separation of the aromatic walls being just right (at 3.5 Å) for inclusion of an aromatic guest.

Fujita has used this self-assembly process to generate a molecular lock [33]. In this case a stable macrocycle is formed irreversibly by reaction of a bis-pyridyl ligand with a Pt^{II}-complex in water but on addition of $NaNO_3$ and heating to 100 °C the ring unlocks and reassembles as a [2]-catenane. Cooling and removal of the salt "relocks" the catenane structure.

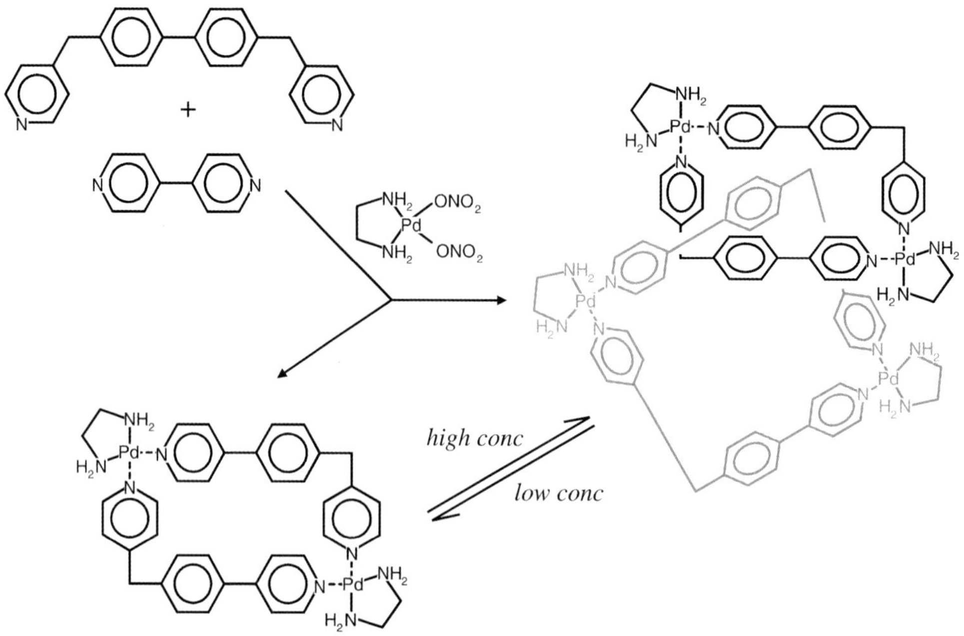

Fig. 10.14 An example of Fujita's metal-ion mediated [2]-catenane self-assembly [32].

Puddephat has produced catenanes using di-gold di-acetylide complexes [34]. Another interesting approach involves catenane self-assembly from preformed Cu^{II} dithiocarbamate-based macrocycles by controlled oxidation [35].

10.2.4
Hydrogen-bonded Assembly of Catenane, Rotaxanes, and Knots

A third approach to the construction of catenanes and rotaxanes exploits hydrogen-bonding interactions to control the assembly of such structures. The work of Hunter, Vögtle, Leigh and others has focussed on the use of neutral components whilst that of Stoddart has explored the behavior of complexes formed between dialkyl ammonium cations and crown ethers.

10.2.4.1 Catenane and Knotane Synthesis

The first successful preparation of a catenane using simple hydrogen-bonded self assembly was carried out by Hunter [36], who obtained a significant yield (ca. 34%) of a symmetrical [2]-catenane as an unexpected by-product from the synthesis of a macrocyclic benzoquinone receptor (Scheme 10.15).

In this synthesis the intertwining of the developing macrocyclic rings arises from the presence of a number of hydrogen-bonding contacts and several edge-to-face and face-to-face π–π interactions between the aromatic rings as the threaded structure assembles. This discovery opened an important new approach

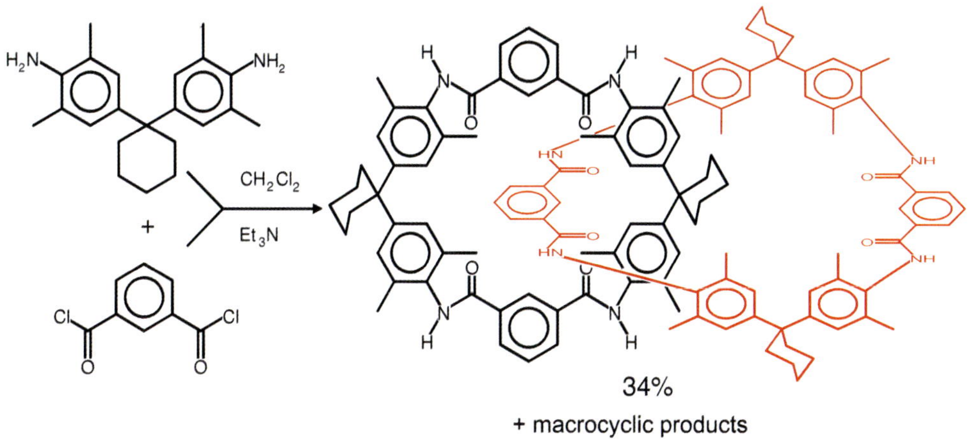

Scheme 10.15 Hunter's catenane synthesis [36].

to catenane and rotaxane synthesis, particularly from the groups of Vögtle and Leigh.

At the same time as Hunter reported his catenane synthesis, Vögtle was also exploring this one-pot route to catenanes and the effect of incorporating a variety of aromatic and heterocyclic groups into the structure. In the course of this work it was discovered that multiply intertwined structures (molecular knots or knotanes) are also formed if pyridine 2,6-dicarbonyl chloride is used as one of the components [37], and that the sequence in which the components are reacted with each other is critical in determining the outcome of this process (Scheme 10.16).

Subsequently Vögtle's group demonstrated [38] that a knotane (incorporating three diamine and three 2,6-dicarbonyl pyridine units) can be formed in a one-pot synthesis. Some knotanes are chiral.

10.2.4.2 Routes to Functional Catenanes and Rotaxanes

Just as Hunter had discovered catenanes present while trying to synthesise a benzophenone receptor, Leigh obtained a significant quantity of [2]-catenane while trying to synthesise a macrocycle capable of binding CO_2 (Scheme 10.17) [39]. Subsequently Leigh and coworkers have demonstrated that this approach provides a versatile route to catenanes and rotaxanes.

A particularly interesting feature of the isophthaloyl diamide route to catenane formation is that it occurs under thermodynamic control. This has been elegantly demonstrated using olefin metathesis reaction conditions to open and reclose an isophthaloyl diamide-based macrocycle containing an alkene unit (Scheme 10.18), as a result of which catenane is formed in >95% yield [40].

Leigh has also shown that the bis-isophthaloyl-based macrocycle can be formed around other hydrogen-bonding cores such as simple diamides [41]. This has al-

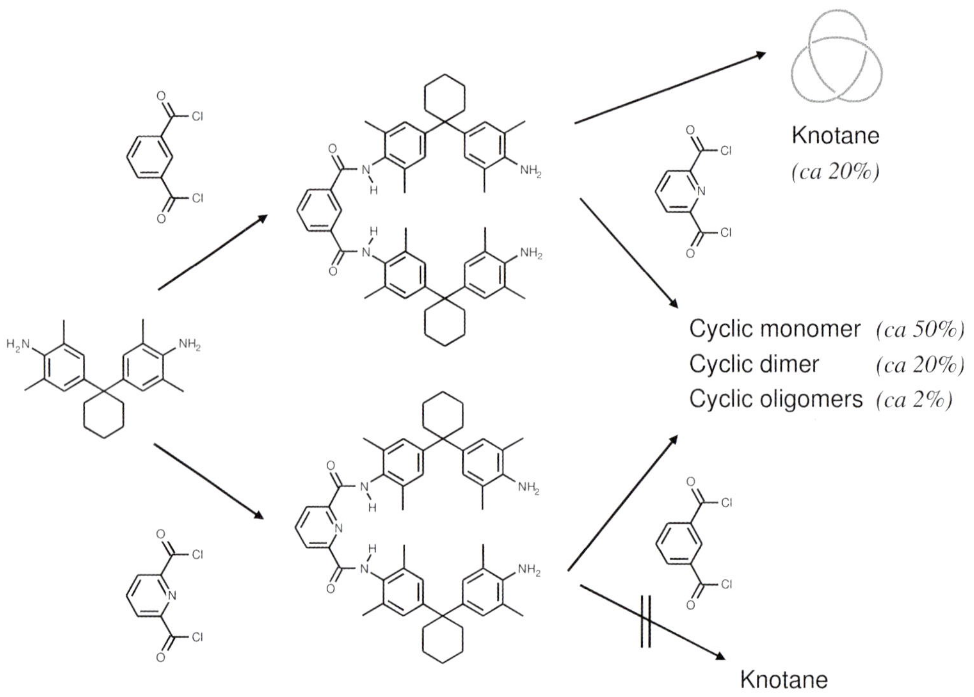

Scheme 10.16 Vögtle's preparative route to knotanes [37].

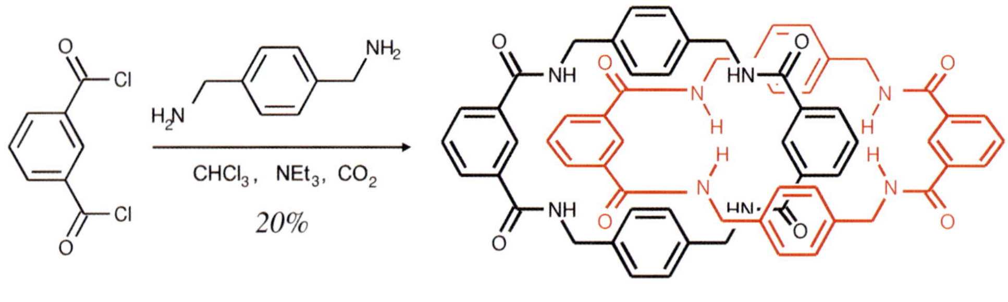

Scheme 10.17 Leigh's catenane synthesis [39].

lowed him to explore a number of different ways of controlling the position of the shuttle along the thread. For example, the use of a thread incorporating a succinamide station connected by a C_{12} linker to an N-alkyl naphthalimide stopper produces a photochemically driven molecular piston-like system capable of rapid cycling [42].

Another example of particular interest is the use of a fumaramide station to create a photo-switchable rotaxane [43]. The fumaramide group forms four hydrogen bonds to an isophthaloyl dicarboxamide-based macrocycle (see Fig. 10.15) but

Scheme 10.18 Hydrogen-bonded isophthaolyl diamide catenane assembly occurs under thermodynamic control as demonstrated under olefin metathesis reactions [40].

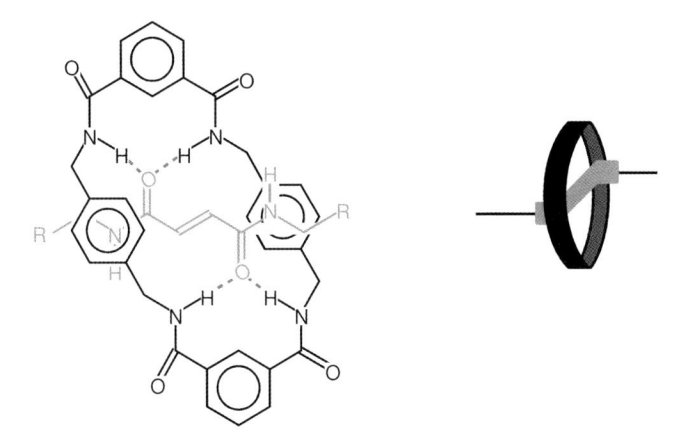

Fig. 10.15 Hydrogen-bonding contacts in Leigh's isophthaloyl dicarboxamide-based macrocycle/fumaramide complex (together with a schematic representation of it) [41].

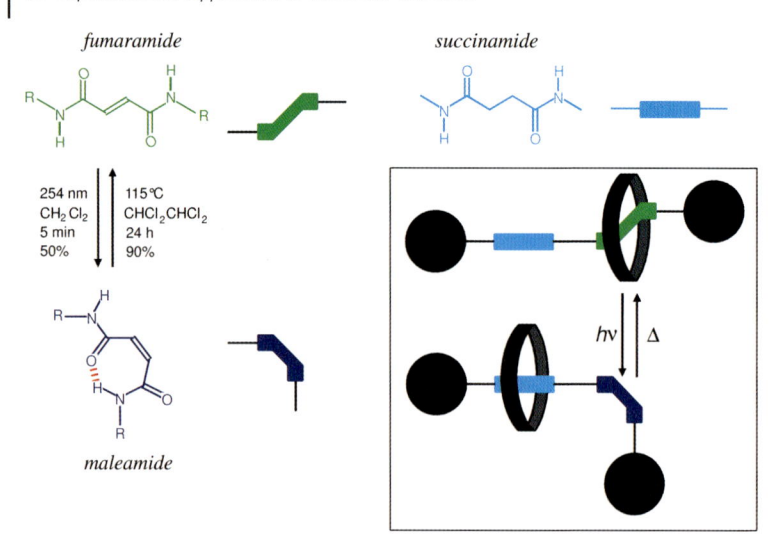

Fig. 10.16 The components of Leigh's hydrogen-bonded switchable rotaxane (plus schematic representations) [43].

photoisomerisation to the maleamide stereoisomer results in a significant weakening of the complex as a result of there being fewer hydrogen-bonding contacts between this stereoisomer and the macrocycle's amide substituents.

If another hydrogen bonding site is introduced onto the thread having a complexation energy lying between those of the two stereoisomers (but with $\Delta G_{complexation}$ at least 2 kcal mol^{-1} less than that for the fumaramide complex) it is possible to achieve >95% shift of the macrocylic shuttle from the fumaramide station to a second station upon photoisomerisation of the fumaramide to its maleamide form (at 298 K). In practice a succinamide group has proved to be the ideal partner for achieving good light-induced switching and the process can be readily reversed thermally, as summarised in Fig. 10.16.

In an elegant extension of this concept Leigh's group have created a rotaxane in which the succinamide station is tetrafluorinated (Fig. 10.17) [44]. In the fumaramide form the polarophobic fluorinated succinamide region is exposed whereas, when the alkene is isomerised, the shuttle moves to cover the fluorinated region, thereby increasing the polar nature of the rotaxane. Using a pyridinium 3,5-dicarboxamide-based shuttle (Fig. 10.17) the rotaxane has been coated onto a self-assembled monolayer obtained by treating a gold surface with 11-mercaptoundecanoic acid. The resulting surface has been shown to change its wetability on exposure to light (Fig. 10.17), the contact angle of CH_2I_2 for example being reduced by ca. 22° as the surface becomes less polarophobic upon irradiation. Consequently when a drop of CH_2I_2 is placed on the surface and one end of the drop is irradiated, it moves across the surface (Fig. 10.17 inset), driven by the pressure gradient arising from differences in contact angles between the opposite sides of the drop. Motion ceases when the all of the surface has reached photo-

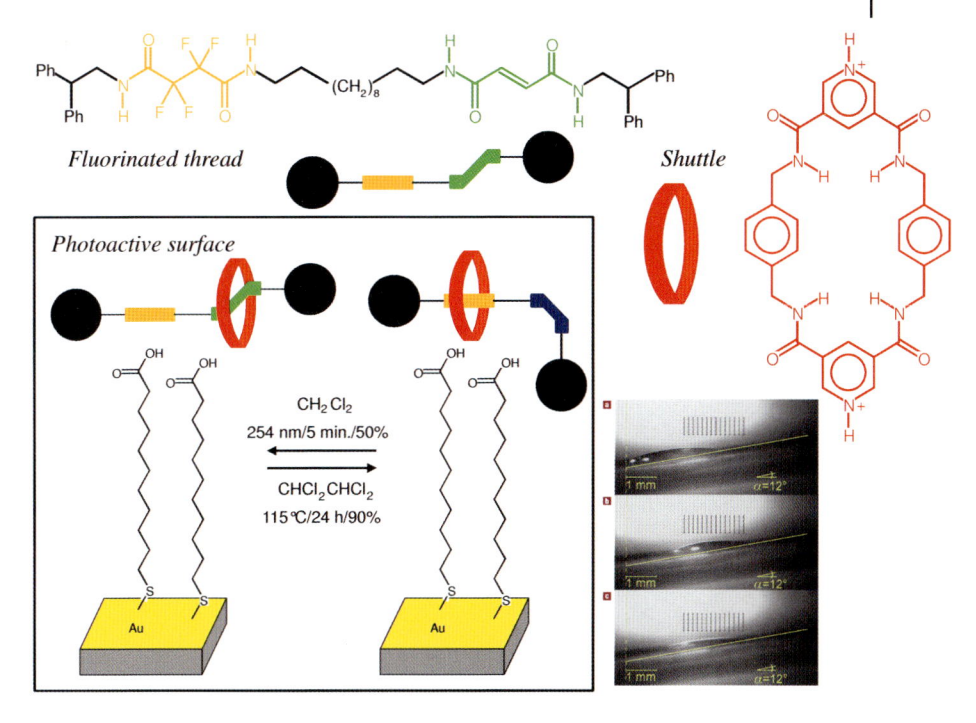

Fig. 10.17 A photoswitchable fluorinated rotaxane coating which can cause a droplet of CH_2I_2 to move up a slope (inset photograph [4b]) upon irradiation [44].

equilibrium (n.b. under these conditions there is a ca. 50:50 mixture of fumaramide:maleamide present but this produces sufficient change in contact angle to be useful).

It has already been noted that a variety of catenanes and rotaxanes can be readily made from these components. The incorporation of the fumaramide-based photoswitchable station introduces the possibility of controlling the position of the aromatic dicarboxamide shuttle around a ring containing multiple binding sites/stations, thus providing another approach to catenane-based molecular motors but one in which there is the opportunity for control of the direction of motion of the shuttle, a key problem in molecular-scale motors. Two examples are of particular interest.

A simple 3-station [2]-catenane has been assembled [45] in which each of the stations has a different binding affinity for the bis-isophthaloyl dicarboxamide-based shuttle. One station is a secondary fumaramide which binds the shuttle well (see Fig. 10.15), the second is a tertiary bis-N,N'-dimethyl fumaramide (which binds the shuttle less effectively because of the steric effect of the methyl groups), and the third is a succinic half amide half ester in which only the amide group can hydrogen bond with the shuttle (Fig. 10.18). In addition the design also includes a benzophenone group which is used to photosensitise the alkene

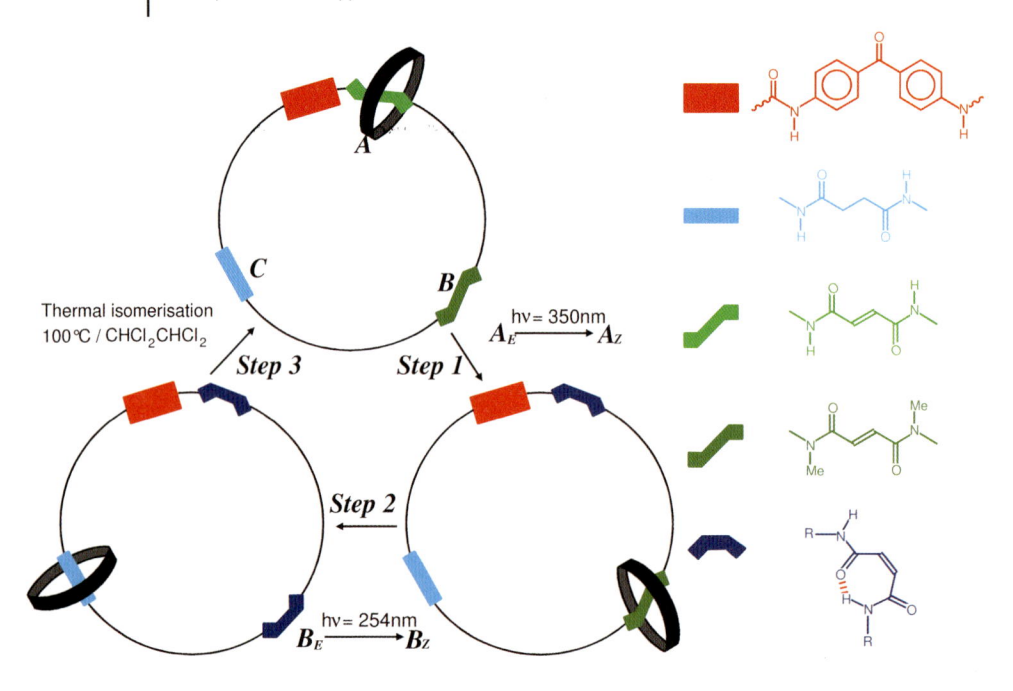

Fig. 10.18 A switchable 3-station [2]-catenane-based unidirectional motor [45].

isomerisation. Sequential photoisomerisation of the fumaramide groups followed by thermal re-equilibration does indeed move the shuttle around the ring in the required sequence but in this system there is no control over the actual direction in which the shuttle moves between the different stations. One way of overcoming this is to introduce a second shuttle onto the ring, taking advantage of the fact that the benzophenone amide linker is also a potential, albeit rather weaker, hydrogen-bonding site. In the resulting [3]-catenane the presence of the second shuttle blocks motion of the first in one direction and leads to a system in which there is much greater control of ring movement.

Another approach to this problem is to use a catenane which incorporates switchable "gates" to control the direction of shuttle movement thereby producing a ratcheted system [46]. This has been achieved through the incorporation of two removable stoppers into a switchable [2]-catenane through the use of two orthogonal protecting groups (a trityl and a bulky silyl group) as shown in Fig. 10.19. In the initial state the bis-isophthaloyl dicarboxamide shuttle (Fig. 10.15) is located on the fumaramide station but photo-switching of the alkene destabilises this complex and removal of the silyl protecting group (stopper 1) allows the shuttle to migrate to the succinamide station. Reintroduction of stopper 1 prevents further return via this route. Subsequent chemical re-isomerisation of the maleamide station back to fumaramide followed by removal of stopper 2 (the trityl protecting group), thermal equilibration, and final reintroduction of the trityl

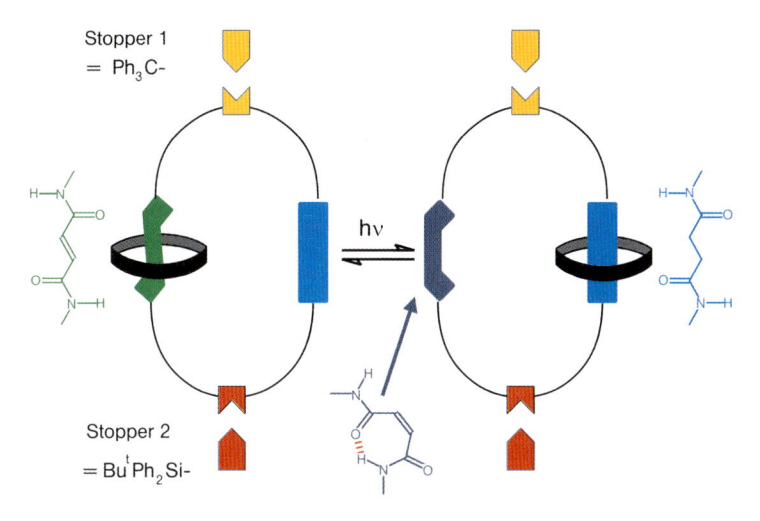

Fig. 10.19 Leigh's approach to a ratcheted-motor [46].

stopper returns the catenane to its starting point having undergone a controlled, unidirectional rotation cycle. This system is robust but its chemical complexity makes it impractical for general use. None-the-less it demonstrates a viable approach to achieving unidirectional motion of a motor-like structure through the use of ratcheting groups.

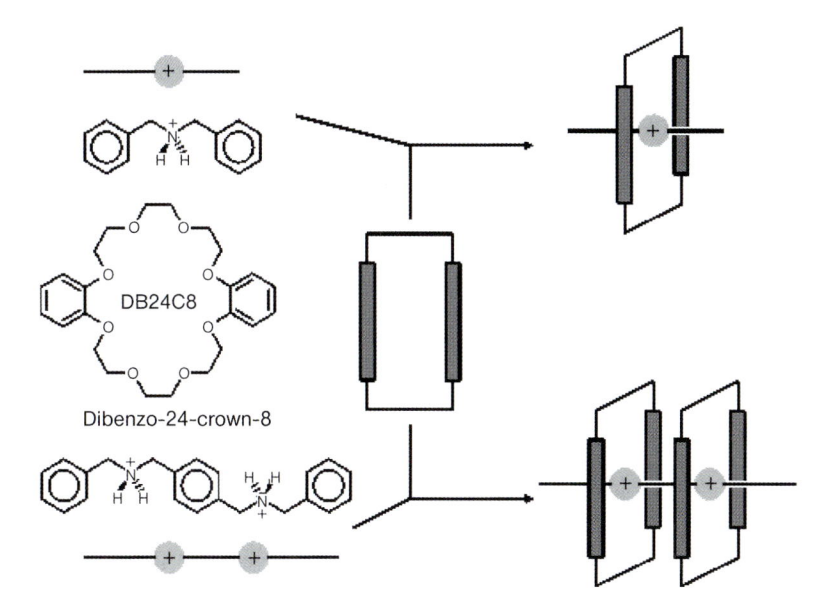

Fig. 10.20 Dialkyl ammonium salt complexes with crown ethers [47].

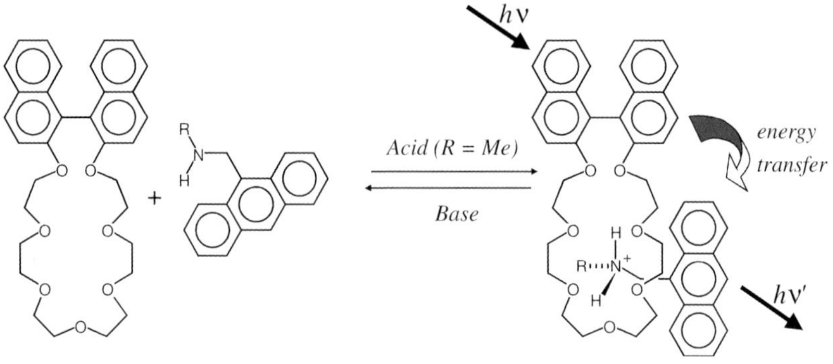

Fig. 10.21 A pH-switchable pseudorotaxane-based energy transfer system [48].

10.2.4.3 Catenanes and Rotaxanes Derived from Dialkyl Ammonium Salts

Stoddart and others [47] have also explored the formation of rotaxanes and pseudorotaxanes in non-polar solvents by crown ether complexation of rod-like dialkylammonium salts, principally through $N^+\text{–}H\cdots OR_2$ hydrogen bonding

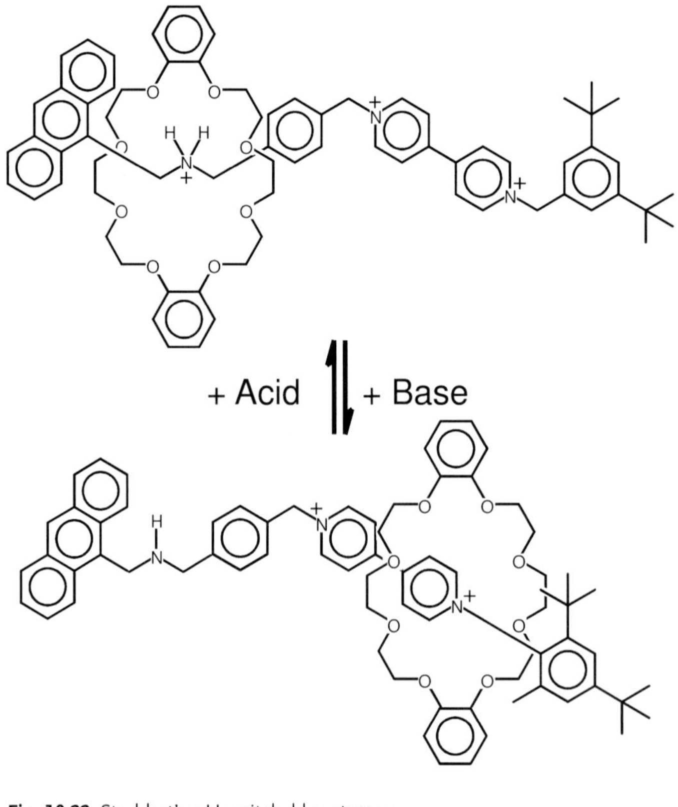

Fig. 10.22 Stoddart's pH-switchable rotaxane.

(Fig. 10.20). Such structures have the advantage of being pH-controllable since deprotonation of the ammonium ion results in unthreading of the complex.

An elegant demonstration of an optical "plug and socket" system using such ammonium ion complexation is shown in Fig. 10.21. When the binaphthyl crown and 9-(N-methyl)aminomethyl anthracene are mixed in acid solution and illuminated with ultraviolet radiation, anthracene luminescence is seen, but when base is added this light switches off. Within the complex, energy transfer occurs leading to visible light emission. The system is size selective, no luminescence being observed for the N-benzyl analog which is too large to form a pseudorotaxane. Complex formation is also anion-sensitive, threading occurring in the presence of PF_6^- but when chloride ion is added the rotaxane falls apart [48].

A further development of this concept is outlined in Fig. 10.22. Here the rotaxane thread incorporates two stations, one a paraquat dication unit and the other a dialkyl ammonium center attached to an anthracene stopper. In acid conditions, when the latter is protonated, it complexes with the crown ether, but in base deprotonation occurs and the shuttle moves to the paraquat dication station.

Elements of these two experiments have been combined to produce a *"molecular elevator"*. If three dibenzo-24-crown-8 units are linked via a triphenylene core, they form a tripod-like complex with a suitably matched 1,3,5-trisubstituted benzene core connected to three dibenzylamine-containing side arms in acid conditions (Fig. 10.23). Attachment of paraquat^{2+} units to the ends of each of the amine side arms (as in Fig. 10.22) produces a switchable complex in which the position of the tris-crown platform can be shunted up and down, much like an elevator, by changes in pH. The platform moves about 0.7 nm, generating a force

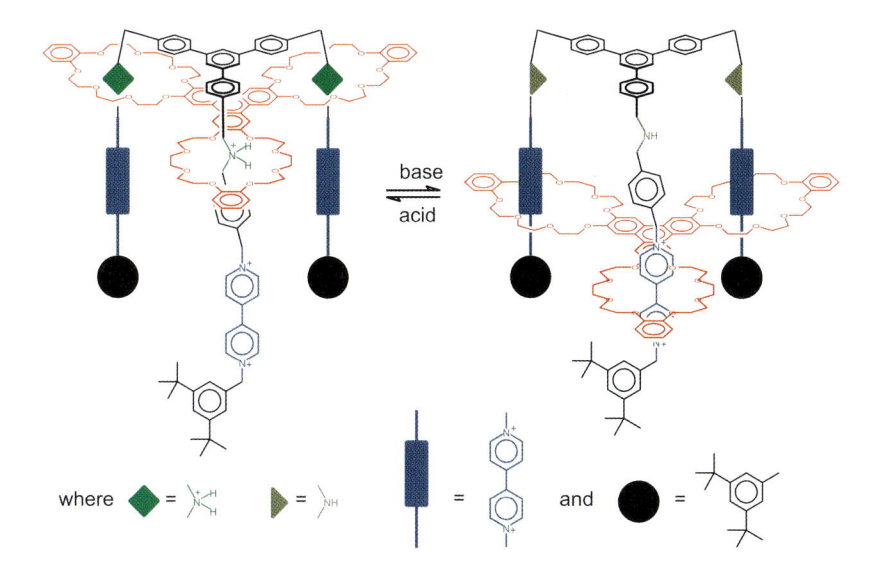

Fig. 10.23 The molecular elevator [49].

348 *10 Supramolecular Approaches to Molecular Machines*

of ca. 200 pN, more than an order of magnitude larger than that generated by natural linear motors such as kinesin and myosin [49].

10.2.5
Cyclodextrin-based Rotaxanes

Anderson and coworkers have shown that cyclodextrins can be used as rings for the formation of rotaxanes. Suitable threads include azo-dyes [50] and stilbenes [51]. Threading is mainly driven by the hydrophobic effect and so such rotaxanes

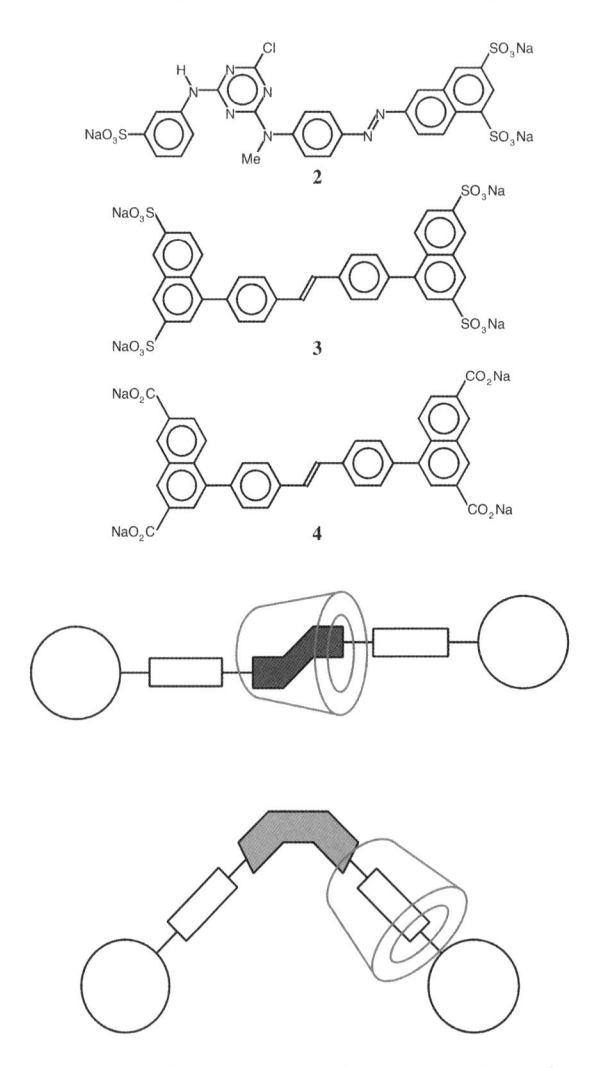

Fig. 10.24 Switching the geometry of the unsaturated-core (**2**, **3** or **4**) of a cyclodextrin-threaded rotaxane forces the shuttle to move, a process which occurs preferentially in one direction [51, 52].

are generally synthesised in water. Cyclodextrin-encapsulation has been found to increase both the photostability and the fluorescence behaviour of solutions of azo dyes thereby making them potentially more suitable for use in luminescent display devices [52]. Such systems can show different photoisomerisation behaviour in comparison with the free dyes.

10.2.5.1 Controlling Motion

Anderson has used steric constraints to control the photo-behavior of such complexes. For example whereas *E*-azobenzene **2** undergoes photoisomerisation affording a photostationary state in which both *E* and *Z* isomers are present, the corresponding α-cyclodextrin rotaxane complex remains unchanged. Whilst other azobenzene/α-cyclodextrin rotaxane complexes do isomerise [53], there is considerable steric constraint in [**2**.α-cyclodextrin] preventing the *Z*-isomer from being formed here. The stilbene-based rotaxane [**3**.α-cyclodextrin] also proves to be rigid; the corresponding β-cyclodextrin complex does photoisomerise as do both the α and β-cyclodextrin complexes of **4**, showing that the smaller end-groups present in **4** allow the cyclodextrin shuttle to move away from the double bond sufficiently to accomodate its isomerisation (Fig. 10.24). NOE studies indicate that in the (**4**.α-cyclodextrin) complex the shuttling *only* occurs in the direction which places the wide rim close to the rotaxane stoppers, thereby introducing some directional control into the direction of shuttling. (i.e. the ring would be shuttled only in one direction along a thread.) Anderson has also recently used such shuttling to prevent enzyme-catalysed hydrolysis of a peptide-linked azobenzene chain [54].

10.3
Molecular Logic Gates

Molecular scale information processing will only become possible when molecular systems capable of performing logic functions become available. Consequently molecular scale logic gates which can simultaneously treat multiple inputs are potentially very exciting intelligent "bottom-up" components for molecular-scale computing. Particularly attractive targets are systems which produce a fluorescence signal, since this can be detected even in one molecule, thereby addressing the inherent problem of input and output present for many rotaxane-based devices.

YES and NOT single input gates are the simplest logical devices. In the former the input is passed to the output unchanged whereas in the latter the input is reversed (Table 10.2).

In molecular terms systems of this sort can be achieved very simply. For example, a molecule which fluoresces only when protonated (whilst under suitable irradiation) can be regarded as a YES gate whereas a system in which fluorescence is quenched upon protonation is a simple NOT gate.

Table 10.2 Logical operations associated with an YES/NOT logic gate.

Logic	Input	Output
YES	0	0
YES	1	1
NOT	0	1
NOT	1	0

de Silva has been particularly active in this field using systems based on PET (photo-induced electron transfer) [55, 56]. The anthryl crown ether (Fig. 10.25) was the first efficient example of an AND gate [57]. In this case there are two ionic inputs and a fluorescence output. This system consists of an anthracene fluorophore attached to two PET-active and ion-selective receptors: the amine which can be protonated; and the benzo-15-crown-5 which can complex with a sodium ion. The "*off*"-states which are essentially non-fluorescent arise when there is unprotonated amine or uncomplexed crown present. In such situations fluorescence is quenched by electron transfer, either from the free nitrogen atom or from the uncomplexed catechol unit. If, however, the amine is protonated *and* the crown is complexed to a sodium ion, no PET can occur and anthracene fluorescence is observed, i.e. there is output only when both cations are present.

More recently de Silva has reported a fluorescent polymeric AND logic gate with temperature and pH as inputs [58].

In an elaboration of the plug and socket systems described earlier, Balzani has reported a pseudorotaxane-based XOR (eXclusive OR) logic gate (Figs. 10.26 and 10.27) [59].

Whilst the dinaphtho-30-crown-10 and the dibenzyl diazapyrenium dication both strongly fluoresce, the charge-transfer complexed pseudorotaxane which is

Fig. 10.25 de Silva's PET logic gate [57].

$h\nu' = 432$ nm

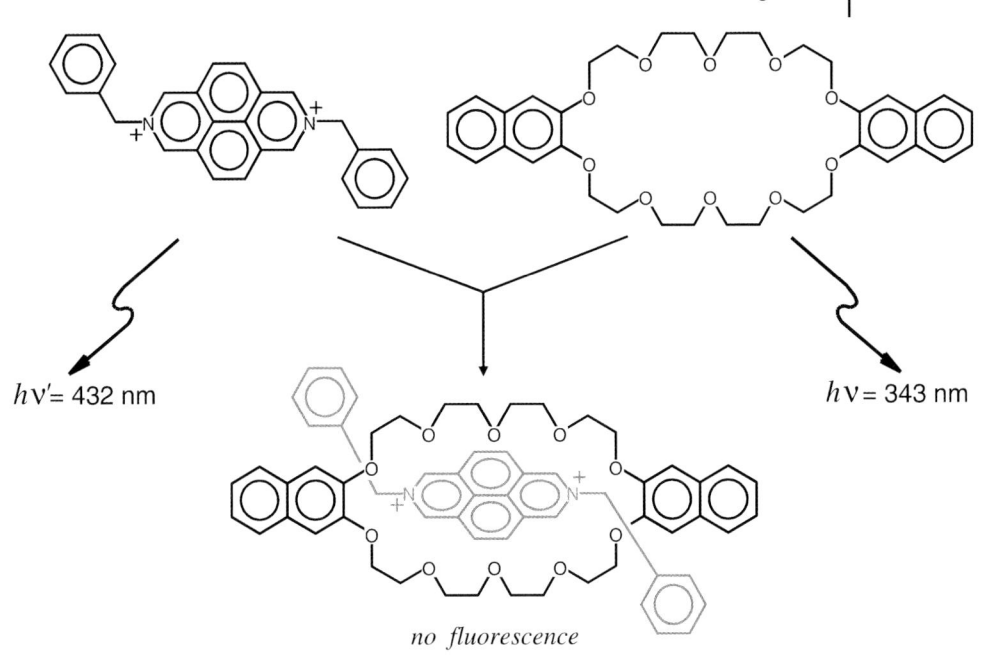

$h\nu = 343$ nm

no fluorescence

Fig. 10.26 A pseudorotaxane capable of acting as an XOR logic gate [59].

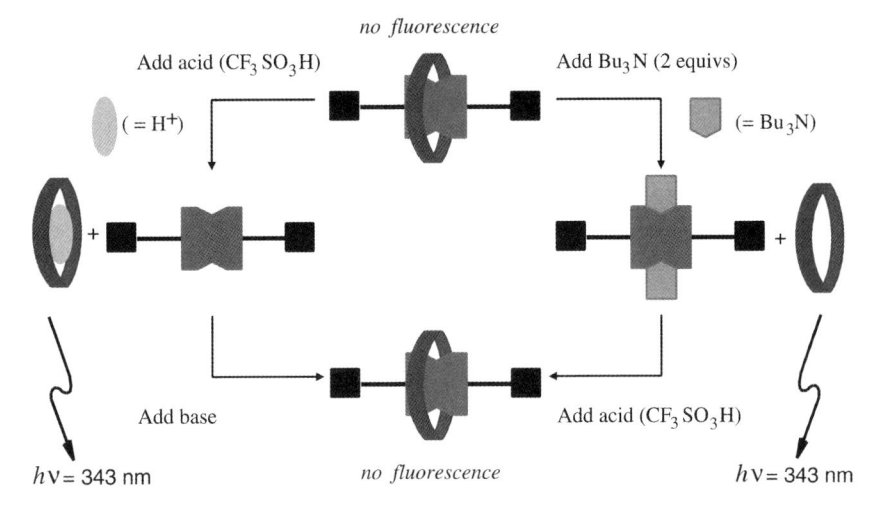

no fluorescence

Add acid (CF_3SO_3H)

Add Bu_3N (2 equivs)

(= H^+)

(= Bu_3N)

Add base

Add acid (CF_3SO_3H)

$h\nu = 343$ nm

no fluorescence

$h\nu = 343$ nm

Fig. 10.27 The operation of Balzani's catenane-based XOR logic gate [59].

formed when they are mixed does not (Fig. 10.26). However, if tributylamine is added, it reacts with the diazapyrenium salt breaking down the pseudorotaxane and liberating free crown which does then fluoresce. Subsequent addition of acid breaks down the amine/diazapyrenium complex, the pseudorotaxane reforms and fluorescence ceases (Fig. 10.27).

If the proton source is added first, this also breaks up the pseudorotaxane by forming a complex with the crown (which again fluoresces). Subsequent addition of a stoichiometric amount of amine (acting as a base), leads to reformation of the pseudorotaxane and loss of the fluorescence once again. This system is an XOR gate because fluorescence at 343 nm is *only* seen when either a proton source *or* tributylamine are present (Fig. 10.27).

10.4
Conclusions

It should be evident from the work described in this chapter that a wealth of complex, dynamic and controllable supramolecular assemblies having a wide range of optical and electronic behaviour are readily accessible. In 1960 in his now famous lecture "There is plenty of room at the bottom" [3] Richard Feynman considered the possibility of molecular-scale machines:

> "What would be the utility of such machines? Who knows? I cannot see exactly what would happen, but I can hardly doubt that when we have some control of the arrangements of things on a molecular scale we will get an enormously greater range of possible properties that substances can have, and of different things that we can do."

In recent years chemists have created a considerable variety of supramolecular structures showing a range of controllable and monitorable functions. Some of these have already been shown to produce macroscopic effects which are potentially useful. Despite their inherent thermal and oxidative instability, organic-based structures offer unlimited and exciting opportunities for creating complex and polyfunctional "designer" materials constrained only by the imagination of the scientists involved. As Feynman implies in his lecture, we should not assume that nanoscale machines should be direct mimics of those in the macroscopic world but rather we need to be imaginative in developing new ways of exploiting the behavior of these molecular-scale systems.

References

1 J.-M. Lehn, *Supramolecular Chemistry. Concepts and Perspectives*, VCH, Weinheim, 1995.
2 K. E. Drexler, Nanosystems: Molecular Machinery, Manufacturing, and Computation, Wiley Interscience, New York, 1992.
3 R. Feynman, There is plenty of room at the bottom, *Eng. Sci.*, **1960**, *23*, 22–36; *Saturday Rev.*, **1960**, *43*, 45–47;

R. P. Feynman, *The Pleasure of Finding Things Out: The Best Short Works of Richard Feynman*, Penguin Books, London, 2001, Ch. 5.

4 (a) V. Balzani, A. Credi, F. M. Raymo, J. F. Stoddart, *Angew Chem. Int. Ed.*, **2000**, *39*, 3348–3391. (b) E. R. Kay, D. A. Leigh, F. Zerbetto, *Angew Chem. Int. Ed.*, **2007**, *46*, 72–191.

5 B. L. Feringa, R. A. Van Delden, N. Koumura, E. M. Geertsema, *Chem. Rev.*, **2000**, *100*, 1789–1816. Y. Shirai, J.-F. Morin, T. Sasaki, J. M. Guerrero, J. M. Tour, *Chem. Soc. Rev.*, **2006**, *35*, 1043–1055.

6 B. L. Allwood, N. Spencer, H. Shahriari-Zawareh, J. F. Stoddart, D. J. Williams, *J. Chem. Soc., Chem. Commun.*, **1987**, 1064–1066; P. R. Ashton, A. M. Z. Slawin, N. Spencer, J. F. Stoddart, D. J. Williams, *J. Chem. Soc., Chem. Commun.*, **1987**, 1066–1069; P. R. Ashton, E. J. T. Chrystal, J. F. Mathias, K. P. Parry, A. M. Z. Slawin, N. Spencer, J. F. Stoddart, D. J. Williams, *Tetrahedron Lett.*, **1987**, *28*, 6367–6370.

7 D. B. Amabilino, P. R. Ashton, C. L. Brown, E. Córdova, L. A. Godínez, T. T. Goodnow, A. E. Kaifer, S. P. Newton, M. Pietraszkiewicz, D. Philp, F. M. Raymo, A. S. Reder, M. T. Rutland, A. M. Z. Slawin, J. F. Stoddart, D. J. Williams, *J. Am. Chem. Soc.*, **1995**, *117*, 1271–1293.

8 D. B. Amabilino, P. R. Ashton, A. S. Reder, N. Spencer, J. F. Stoddart, *Angew. Chem. Int. Edn.*, **1994**, *33*, 433–436.

9 P. R. Ashton, V. Balzani, J. Becher, A. Credi, M. C. T. Fyfe, G. Mattersteig, S. Menzer, M. B. Nielsen, F. M. Raymo, J. F. Stoddart, M. Venturi, D. J. Williams, *J. Am. Chem. Soc.*, **1999**, *121*, 3951–3857.

10 C. P. Collier, G. Mattersteig, E. W. Wong, Y. Luo, K. Beverly, J. Sampaio, F. M. Raymo, J. F. Stoddart, J. R. Heath, *Science*, **2000**, *289*, 1172.

11 W.-Q. Deng, A. H. Flood, J. F. Stoddart, W. A. Goddard III, *J. Am. Chem. Soc.*, **2005**, *127*, 15994–15995.

12 O. Š. Miljanić, W. R. Dichtel, J. F. Stoddart, Poster 188, International Symposium on Macrocyclic and Supramolecular Chemistry, Victoria, Canada, June 25–30, 2006; see also: W. R. Dichtel, O. Š. Miljanić, J. M. Spruell, J. R. Heath, J. F. Stoddart, *J. Amer. Chem. Soc.*, **2006**, *128*, 10388–10390.

13 O. Š. Miljanić, W. R. Dichtel, S. Mortezaei, J. F. Stoddart, *Org. Lett.*, **2006**, *8*, 4835–4838; see also M. J. Gunter, S. M. Farquhar, *Org. Biomol. Chem.*, **2003**, *1*, 3450.

14 T. J. Huang, B. Brough, C.-M. Ho, Y. Liu, A. H. Flood, P. A. Bonvallet, H.-R. Tseng, J. F. Stoddart, M. Baller, S. Magonov, *App. Phys. Lett.*, **2004**, *85*, 5391–5393; Y. Liu, A. H. Flood, P. A. Bonvallet, S. A. Vignon, B. H. Northrop, H.-R. Tseng, J. O. Jespersen, T. J. Huang, B. Brough, M. Baller, S. Magnov, S. D. Solares, W. A. Goddard, C.-M. Ho, J. F. Stoddart, *J. Am. Chem. Soc.*, **2005**, *127*, 9745–9759.

15 D. G. Hamilton, J. E. Davies, L. Prodi, J. K. M. Sanders, *Chem. Eur. J.*, **1998**, *4*, 608–620.

16 For an introduction see: J.-P. Sauvage, *Acc. Chem. Res.*, **1998**, *31*, 611–619; M.-J. Blanco, M. C. Jiménez, J.-C. Chambron, V. Heitz, M. Linke, J.-P. Sauvage, *Chem. Soc. Rev.*, **1999**, *28*, 293–305.

17 D. J. Cardenas, A. Livoreil, J.-P. Sauvage, *J. Am. Chem. Soc.*, **1996**, *118*, 11980–11981.

18 C. Hamann, J.-M. Kern, J.-P. Sauvage, *Inorg. Chem.*, **2003**, *42*, 1877–1883.

19 A. Livoreil, C. O. Dietrich-Buchecker, J.-P. Sauvage, *J. Am. Chem. Soc.*, **1994**, *116*, 9399; See also D. J. Cardenas, A. Livoreil, J.-P. Sauvage, *J. Am. Chem. Soc.*, **1996**, *118*, 11980–11981 for a three-state system.

20 J.-P. Collin, P. Gavana, J.-P. Sauvage, *New J. Chem.*, **1997**, *21*, 525.

21 M. C. Jimenez-Molero, C. Dietrich-Buchecker, J.-P. Sauvage, *Chem. Eur. J.*, **2002**, *8*, 1456–1466.

22 J. J. Nierengarten, C. O. Dietrich-Buchecker, J.-P. Sauvage, *J. Am. Chem. Soc.*, **1994**, *116*, 375–376.

23 A. J. Peters et al., *Chem. Commun.*, **2005**, 3394–3396.

24 C. D. Pentecost, A. J. Peters, K. S. Chichak, G. W. V. Cave, J. F. Stoddart, *Angew. Chem. Int Edn.*, **2006**, *45*, 4099–4104.
25 C. Hamann, J.-M. Kern, J.-P. Sauvage, *Dalton Trans.*, **2004**, 3770–3775.
26 A.-M. L. Fuller, D. A. Leigh, P. J. Lusby, I. D. H. Oswald, S. Parsons, D. B. Walker, *Angew. Chem Int. Ed.*, **2004**, *43*, 3914–3918.
27 A. N. Dwyer, M. C. Grossel, P. N. Horton. *Supramolecular Chem.*, **2004**, *16*, 405–410.
28 A.-M. L. Fuller, D. A. Leigh, P. J. Lusby, Poster 72, International Symposium on Macrocyclic and Supramolecular Chemistry, Victoria, Canada, June 25–30, 2006.
29 A.-M. L. Fuller, D. A. Leigh, P. J. Lusby, A. M. Z. Slawin, D. B. Walker, *J. Am. Chem. Soc.*, **2005**, *127*, 12612–12619.
30 J. A. Wisner, P. D. Beer, M. G. B. Drew, *Angew. Chem. Int. Ed.*, 2001, **40**, 3606.
31 J. A. Wisner, P. D. Beer, M. G. B. Drew, M. R. Sambrook., *J. Am. Chem. Soc.*, 2002, **124**, 12469.
32 M. Fujita, *Nature*, **1994**, *367*, 720–723; M. Fujita, M. Aoyagi, F. Ibukuro, K. Ogura, K. Yamaguchi, *J. Am. Chem. Soc.*, **1998**, *120*, 611–612; M. Fujita, *Acc. Chem. Res.*, **1999**, *32*, 53–61.
33 M. Fujita, F. Ibukuro, K. Yamaguchi, K. Ogura, *J. Am. Chem. Soc.*, **1995**, *117*, 4175–4176.
34 C. P. McArdle, M. J. Irwan, M. C. Jennings, R. J. Puddephat, *Angew. Chem. Int. Ed.*, **1999**, *38*, 3376–3378.
35 M. E. Padilla-Tosta, O. D. Fox, M. G. B. Drew, P. D. Beer, *Angew. Chem. Int. Ed.*, **2001**, *40*, 4235–4239.
36 C. A. Hunter, *Chem. Commun*, **1991**, 741; C. A. Hunter, *J. Am. Chem. Soc.*, **1992**, *114*, 5303–5311; C. A. Hunter, D. H. Purvis, *Angew. Chem. Int. Ed.*, **1992**, *31*, 792–795; C. A. Hunter, *Chem Soc. Rev.*, **1994**, 101–109.
37 F. Vögtle, S. Meier, R. Hoss, *Angew. Chem. Int. Ed.*, **1992**, *31*, 1619; R. Jäger, F. Vögtle, *Angew. Chem. Int. Ed.*, **1997**, *36*, 931–944.
38 O. Safarowsky, M. Nieger, R. Fröhlich, F. Vögtle, *Angew. Chem. Int. Ed.*, **2000**, *39*, 1616–1618; O. Lukin, F. Vögtle, *Angew. Chem. Int. Ed.*, **2005**, *44*, 1456–1477.
39 A. J. Johnston, D. A. Leigh, R. J. Prichard, M. D. Deegan, *Angew. Chem. Int. Ed.*, **1995**, *34*, 1209–1212; A. J. Johnston, D. A. Leigh, L. Nezhat, J. P. Smart, M. D. Deegan, *Angew. Chem. Int. Ed.*, **1995**, *34*, 1212–1216.
40 T. J. Kidd, D. A. Leigh, A. J. Wilson, *J. Am. Chem. Soc.*, **1999**, *121*, 1599–1600.
41 W. Clegg, C. Gimenez-Saiz, D. A. Leigh, A. Murphy, A. M. Z. Slawin, S. J. Teat, *J. Am. Chem. Soc.*, **1999**, *121*, 4124–4129.
42 A. M. Brouwer, C. Fronchet, F. G. Gatti, D. A. Leigh, L. Mottier, F. Paolucci, S. Roffia, G. W. H. Wurpel, *Science*, **2001**, *291*, 2124–2128.
43 A. Altieri, G. Bottari, F. Dehez, D. A. Leigh, J. K. Y. Wong, F. Zerbetto, *Angew. Chem. Int. Ed.*, **2003**, *42*, 2296–2300.
44 J. Berná, D. A. Leigh, M. Lubomska, S. M. Mendoza, P. Rudolf, G. Teobaldi, F. Zerbetto, *Nature Mater.*, **2005**, *4*, 704–710.
45 D. A. Leigh, J. K. Y. Wong, F. Dehez, F. Zerbetto, *Nature*, **2003**, *424*, 174–179.
46 J. V. Hernández, E. R. Kay, D. A. Leigh, *Science*, **2004**, *306*, 1532–1537.
47 P. R. Ashton, P. T. Glink, M.-V. Martinez-Diaz, J. F. Stoddart, A. J. White, D. J. Williams, *Angew. Chem. Int. Ed.*, **1996**, *35*, 1930–1933.
48 E. Ishow, A. Credi, V. Balzani, F. Spadola, L. Mandolini, *Chem. Eur. J.*, **1999**. *5*, 984–989; see also S. J. Cantrell, M. C. T. Fyfe, A. M. Heiss, J. F. Stoddart, A. J. P. White, D. J. Williams, *Chem Commun.*, **1999**. 1251–1252. M. Montalti, L. Prodi, *Chem. Commun.*, **1998**, 1461.
49 J. D. Badjic, V. Balzani, A. Credi, S. Silvi, J. F. Stoddart, *Science*, **2004**, *303*, 1845–1849.
50 M. R. Craig, M. G. Hutchings, T. D. W. Claridge, H. L. Anderson, *Angew. Chem. Int. Ed.*, **2001**, *40*, 1071–1074.

51 C. A. Stanier, S. J. Alderman, T. D. W. Claridge, H. L. Anderson, *Angew. Chem. Int. Ed.*, **2002**, *41*, 1769–1772.

52 J. E. H. Buston, J. R. Young, H. L. Anderson, *Chem. Commun.*, **2000**, 905–906.

53 See for example: T. Fujimoto, A. Nakamura, Y. Inoue, Y. Sakata, T. Kaneda, *Tetrahedron Lett.*, **2001**, *42*, 7987–7989.

54 C. A. Stanier, S. J. Alderman, T. D. W. Claridge, H. L. Anderson, *Angew. Chem. Int. Edn.*, **2002**, *41*, 1769–1772; A. G. Cheetham, M. G. Hutchings, T. D. W. Claridge, H. L. Anderson, *Angew. Chem. Int. Ed.*, **2006**, *45*, 1596–1599.

55 A. P. de Silva, H. Q. N. Gunaratne, C. P. McCoy, *Nature*, **1993**, *364*, 42–44.

56 A. P. de Silva, H. Q. N. Gunaratne, T. Gunnlaugsson, A. J. M. Huxley, C. P. McCoy, J. T. Rademacher, and T. E. Rice., *Chem. Rev.*, **1997**, *34*, 963–972.

57 A. P. de Silva, H. Q. N. Gunaratne, C. P. McCoy, *J. Am. Chem. Soc.*, **1997**, *119*, 7891–7891.

58 S. Uchiyama, N. Kawai, P. de Silva, K. Iwai, *J. Am. Chem. Soc.*, **2004**, *126*, 3032–3033.

59 A. Credi, V. Balzani, S. J. Langford, J. F. Stoddart, *J. Am. Chem. Soc.*, **1997**, *119*, 2679–2681.

11
Nanoscale Electronic Inhomogeneities in Complex Oxides

V. B. Shenoy, H. R. Krishnamurthy, and T. V. Ramakrishnan

11.1
Introduction

The study of complex oxide materials has been a great source of stimulation for quantum condensed matter physics. Many fundamental concepts such as that of a Mott insulator, spin frustration, exotic superconductivity etc., have been motivated by and are realized in these systems. The fascinating and bewildering plethora of phenomena seen in them have thrown up many surprises which continue to challenge condensed matter physicists, and perhaps offer great future technological opportunities for the imaginative. One of these surprises is the phenomenon of electronic inhomogeneity.

A standard paradigm of condensed matter theory is the notion of homogeneous systems (suitably enlarged to include periodic arrangements of "broken symmetry" variables) with periodic boundary conditions. However, there is strong experimental evidence suggesting that a new paradigm may be necessary to describe many novel complex oxides. Most of the doped oxides (including manganites, cuprates and cobaltates) are found to be inhomogeneous on scales ranging from nanometers to microns [1–3]. These inhomogeneities are "electronic" in nature in the sense that different regions of the material have very different electronic properties, e.g., they can be metallic or insulating, have differing magnetic order, etc. In addition to the large range of length scales over which these inhomogeneities are found, they also exhibit a large range of time scales – in some systems these are found to be static while in others they are dynamic and even evolve under external influences such as magnetic/electric fields. This phenomenon observed in oxides is often referred to as "phase separation" and is the subject of this chapter.

Several obvious and fundamental questions arise with regard to the origin of these inhomogeneities. Are they "intrinsic" or "extrinsic"? That is, are they *intrinsic* to the system in that they correspond to the low energy configurations of the Hamiltonian of the system, or, are they *extrinsic* in that their presence is contingent on the existence of external influences, such as surfaces, stresses, etc.?

Nanomaterials Chemistry. Edited by C. N. R. Rao, A. Müller, and A. K. Cheetham
Copyright © 2007 WILEY-VCH Verlag GmbH & Co. KGaA, Weinheim
ISBN: 978-3-527-31664-9

What is the origin of the wide range of length and times scales seen? Furthermore, as we discuss below, there are suggestions that the presence of such inhomogeneities is the key to understanding many of the interesting properties exhibited by oxides. For example, in the context of doped manganites, one of the important questions that arises is whether these inhomogeneities are essential for the material to show such responses as colossal magnetoresistance. In addition to these fundamental issues, there have been suggestions in the literature that transition metal oxides are characterized by "electronic softness" [3, 4] and this property can be used to generate spatially modulated electronic properties, not unlike the spatial tuning of the properties of a liquid crystal. This provides an important research direction in tailoring atomic arrangements at nanometric scales using ideas of solid state chemistry and materials science to design "electronically patterned" materials.

In this chapter, we discuss electronic inhomogeneities in oxides. In Section 11.2 we discuss the experimental work which shows conclusive evidence for the presence of electronic inhomogeneities. Following this, we present a brief discussion (Section 11.3) of the past theoretical work aimed at understanding the experiments. We then discuss (Section 11.4) a strong-correlation model recently introduced by two of us and collaborators [5–8], called the ℓb model, especially appropriate for the doped manganites which, as mentioned above, prominently show electronic inhomogeneities. The ℓb model, when treated in a *homogeneous* dynamical mean field theory framework, has been shown to successfully explain the colossal magneto-resistance and many other hitherto poorly understood features of doped manganites seen in experiments. However, this model, which has only short-range interactions, has the limitation that it leads to "macroscopic phase separation" induced by strong local electronic correlations. Since this macroscopic phase separation involves 'phases' *with different charge densities*, it is rendered unfavorable by the ubiquitous long-ranged Coulomb interaction. Hence, in Section 11.5, we discuss very recent work by us [9] on an 'extended ℓb model' which includes the long-range Coulomb interaction, showing that its presence frustrates the macroscopic phase separation resulting in an electronic inhomogeneity with a scale of nanometers. Based on these points we argue (Section 11.6) that the *nanoscale electronic inhomogeneities in manganites arise from long-range Coulomb interaction frustrating macroscopic phase separation induced by strong local electronic correlations*, whereas the large length scale inhomogeneities seen in experiments owe their origin to unscreened long-range elastic interactions, and we suggest an experiment to test our hypothesis.

11.2
Electronic Inhomogeneities – Experimental Evidence

Electronic inhomogeneities manifest themselves in many ways, so that their presence can be discerned from structural, magnetic and transport properties, and

11.2 Electronic Inhomogeneities – Experimental Evidence

also by direct observation. In cases where these inhomogeneities are of a large length scale (exceeding about one hundred nanometers), probes such as X-ray and neutron diffraction show their presence; where the length scale is small (a few nanometers), local probes such as NMR and Mössbauer spectroscopies have proved to be useful tools. Thermodynamic and transport properties also contain signatures of electronic inhomogeneities. The most convincing proof for the inhomogeneities is provided by direct observational evidence such as electron microscopy, scanning tunneling microscopy, photo-emission spectroscopy etc. In this section, we briefly review the experimental work on electronic inhomogeneities, focusing on doped perovskite manganites with the formula $Re_{1-x}Ak_xMnO_3$, where Re is a rare-earth ion (such as La, Pr, Nd, etc.) and Ak is an alkaline earth ion (such as Ca, Sr, Ba, etc.).

Some of the first studies on manganites actually contained evidence for electronic inhomogeneities although they were not recognized as such. Wollan and Koehler [10] studied the magnetic structure of $La_{1-x}Ca_xMnO_3$ (LCMO) by neutron diffraction. They observed both ferromagnetic and antiferromagnetic peaks being simultaneously present in the doping range of $x \lesssim 0.3$, as shown in Fig. 11.1. More recently, Woodward et al. [11] studied the magnetic structure of $Nd_{1/2}Sr_{1/2}MnO_3$ by neutron scattering. This compound becomes ferromagnetic at 250 K, and upon further cooling to 220 K seems to become a mixture of ferromagnetic and A-type antiferromagnetic phases. At an even lower temperatures of about 150 K, a third, CE-type, antiferromagnetic phase appears, as seen in the neutron data. Clearly the size scales (greater than a hundred nanometers) of the electronic inhomogeneities are large enough to produce well defined neutron peaks. Figure 11.2(A) shows the evolution of the volume fraction of the phases with temperature. The influence of external perturbations on the electronic inhomogeneities in the same compound was investigated by Ritter et al. [12]. Figure 11.2(B) shows the transformation of the antiferromagnetic regions into ferromagnetic regions on application of an external magnetic field. These experiments provide evidence for mesoscale phase separation in this manganite, their temperature dependence and the influence of external stimuli such as a magnetic field. A further example [13] is $Pr_{0.7}Ca_{0.3}MnO_3$ which shows the presence of two distinct phases below the charge ordering transition at 80 K. Neutron diffraction shows a charge-ordered AFM phase and a charge-delocalized phase.

Although the experimental works cited above clearly indicate the presence of different electronic and magnetic 'phases' within a structurally and chemically pure sample, they do not provide precise information on the scale and nature of these inhomogeneities. The only inference to be made from X-ray and neutron data is that the inhomogeneities are of mesoscale (few hundred nanometers).

There are, however, several experiments that provide direct visual evidence of electronic inhomogeneities including the length scales associated with them. Possibly the first of these was by Mori et al. [15], who observed charge stripes in high resolution transmission electron microscopy of thin films of $La_{1-x}Ca_xMnO_3$ (LCMO) with $0.5 \leq x \leq 0.75$. The spatial structure consisted of paired Jahn–

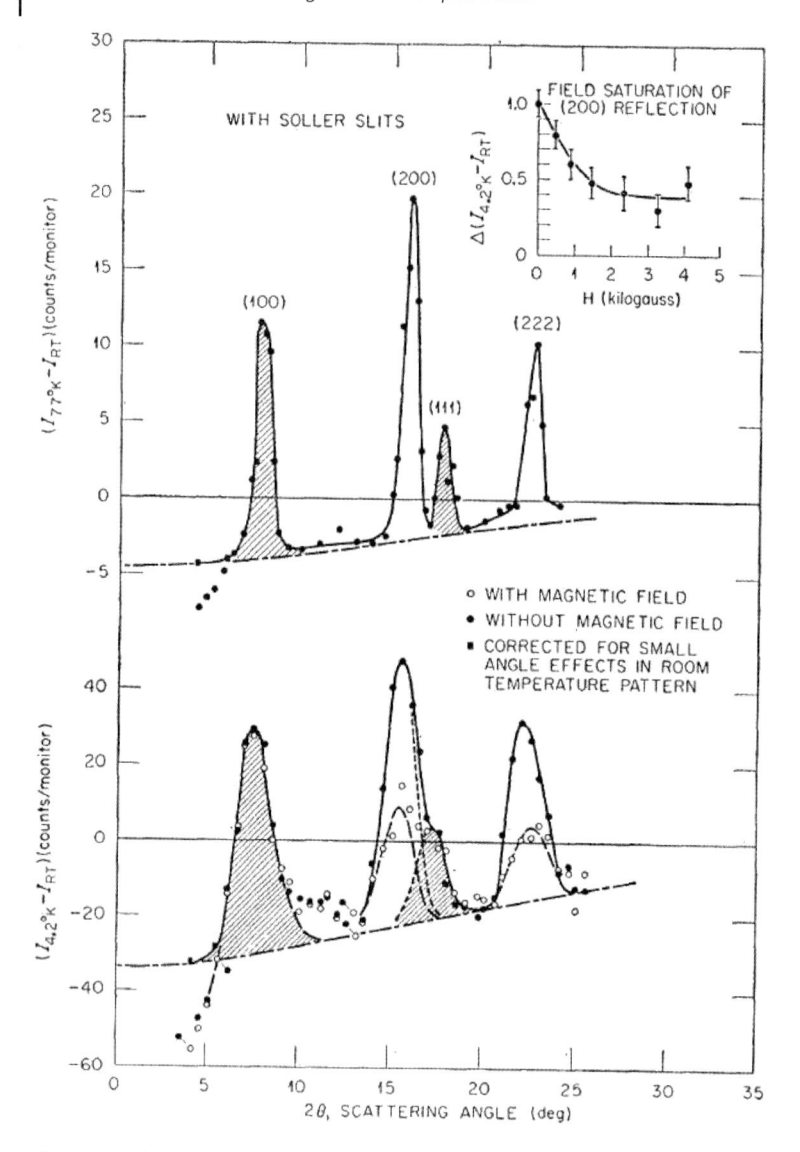

Fig. 11.1 Earliest evidence of electronic inhomogeneites in LCMO. Hatched regions show antiferromagnetic peaks while other peaks are ferromagnetic. After Wollan and Koehler [10].

Teller distorted oxygen octahedra surrounding the Mn^{3+} ions, separated by stripes of $Mn^{4+}-O_6$ octahedra. The stripes were spaced at about five to ten lattice spacings, the spacing being doping dependent, providing clear evidence of electronic inhomogeneity on the nanoscale. There are other direct observations of

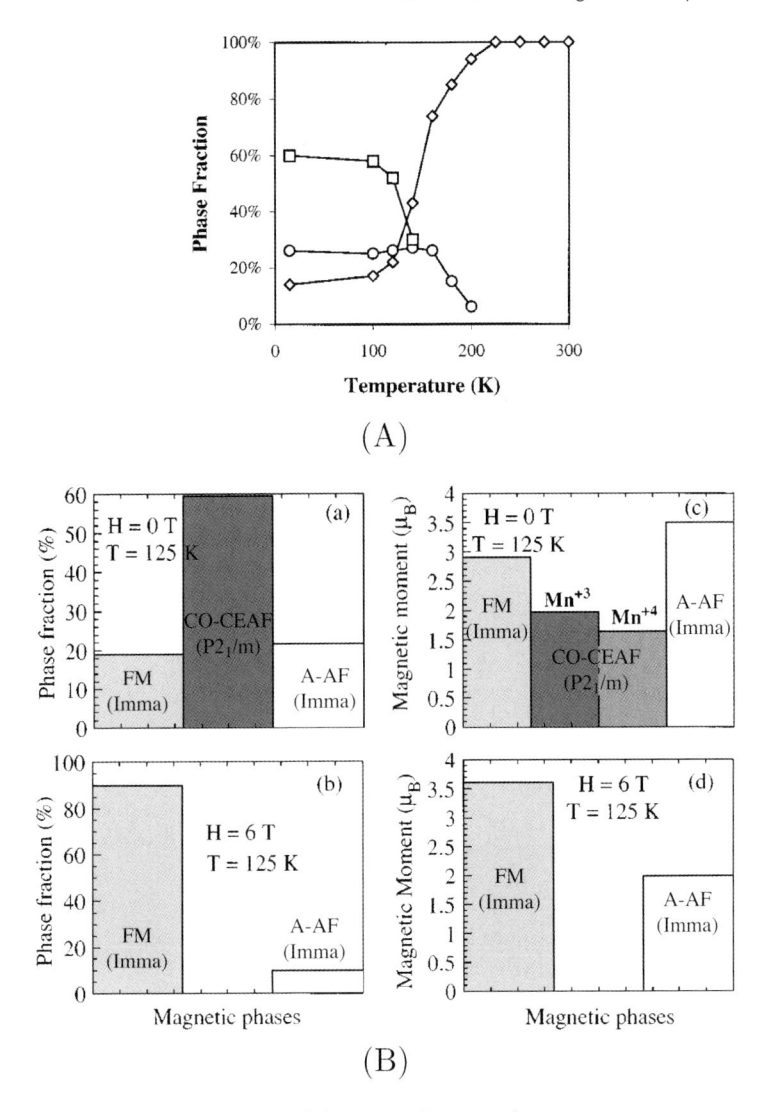

Fig. 11.2 (A) Volume fraction of phases as a function of temperature in $Nd_{1/2}Sr_{1/2}MnO_3$. Diamonds, circles and squares respectively represent ferromagnet (FM), antiferromagnet (A-type, A-AF) and antiferromagnet (CE-type, CO-CEAF) respectively. (After Woodward et al. [11]) (B) Phase fractions at 125 K at different magnetic fields (a) 0 T and (b) 6 T; (c) and (d) show the magnetic moment in each of the phases. After Ritter et al. [12].

nanoscale electronic inhomogeneities; for example, in a scanning tunneling microscopy study of $Bi_{1-x}Ca_xMnO_3$ ($x \approx 0.75$), Renner et al. [16] found nanoscopic charge-ordered and metallic domains correlated with structural distortions.

Possibly the first direct evidence of *mesoscale* inhomogeneities in manganites was provided by Uehara et al. [17]. Their transmission electron microscopy study of $La_{5/8-y}Pr_yCa_{3/8}MnO_3$ demonstrated the coexistence of insulating charge-ordered regions with interspersed metallic ferromagnetic regions with a size scale of about 0.2 µm. Based on these observations they suggested that these inhomogeneities are key to understanding the colossal magneto-response of manganites. Their argument is that the spin alignment direction in different ferromagnetic regions is different, and hence conduction electrons which are spin polarized cannot easily hop to neighboring domains. On application of a magnetic field, the moments in each of the ferromagnetic domains aligns with the magnetic field and this allows conduction electrons to freely hop from one domain to the other, drastically reducing the resistance and producing the colossal magnetoresistance effect.

Another interesting direct observation of mesoscale inhomogeneities was reported by Fäth et al. [18]. They studied the metal–insulator transition in $La_{0.7}Ca_{0.3}MnO_3$ by scanning tunneling microscopy and found electronic inhomogeneities below the ferromagnetic transition temperature with a length scale of about 0.2 µm. These samples also showed metallic ferromagnetic regions interspersed within insulating regions similar to the observations of Uehara et al. [17]. Fäth et al. [18] also studied the evolution of the inhomogeneities in an applied magnetic field. They found that the volume fraction of the metallic ferromagnetic domains increased at the expense of the insulating parts with increasing magnetic field, closely correlating with the decrease in the resistivity with the magnetic field. Again, this work suggested a possible mechanism for colossal magnetoresistance (this may be contrasted with the suggestion of Uehara et al. [17]) in which the electronic inhomogeneity is the key player. Zhang et al. [19] reported a magnetic force microscopy study of $La_{0.33}Pr_{0.34}Ca_{0.33}MnO_3$. They found that the magnetic domains on the mesoscopic scale evolved with temperature showing magnetic hysteresis which coincided with the resistivity hysteresis. This study again reinforced the idea that the electronic inhomogeneities are the crucial players in producing the colossal responses in manganites.

A transmission electron microscopy and electron holography study of $La_{0.5}Ca_{0.5}MnO_3$ by Loudon et al. [14] found mesoscopic domains of ferromagnetic regions interspersed in insulating regions (see Fig. 11.3). Most interestingly they found that some of the ferromagnetic regions were charge ordered! This experiment suggests that even the mesoscopic ferromagnetic region is inhomogeneous on the nanoscale with coexisting metallic and charge-ordered nanodomains.

Our last example of mesoscale electronic inhomogeneities is from a novel photoemission spectro-microscopy study of $La_{1/4}Pr_{3/8}Ca_{3/8}MnO_3$ reported by Sarma et al. [20]. This technique has the advantage of spatially resolving the local

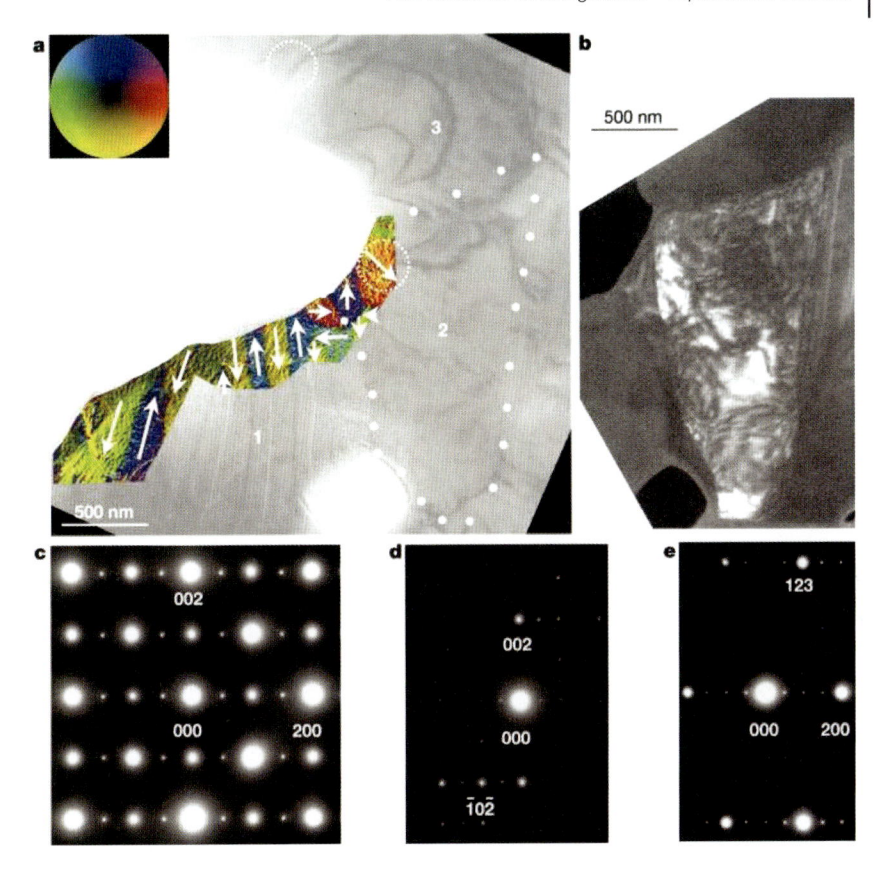

Fig. 11.3 Transmission electron micrography and electron holography showing ferromagnetic domains with charge order. After Loudon et al. [14].

metallic/non-metallic nature along with a simultaneous determination of the local chemical composition with a spatial resolution of about 0.5 μm. They found very large (15 μm × 5 μm) domains of insulating patches surrounded by a metallic background (see Fig. 11.4). Moreover, with the increase in temperature, the regions evolved with the metallic regions undergoing a metal–insulator transition at higher temperatures. A key finding of the experiment was that the insulating regions appeared essentially at their initial location on cooling the sample, indicating a novel memory effect associated with the electronic inhomogeneities.

Electronic inhomogeneities have also been observed in other oxides, notably cuprates [21–24] and cobaltates [25–32]. But in this chapter we confine our attention mostly to manganites, although some of our conclusions may have implications for these other contexts as well.

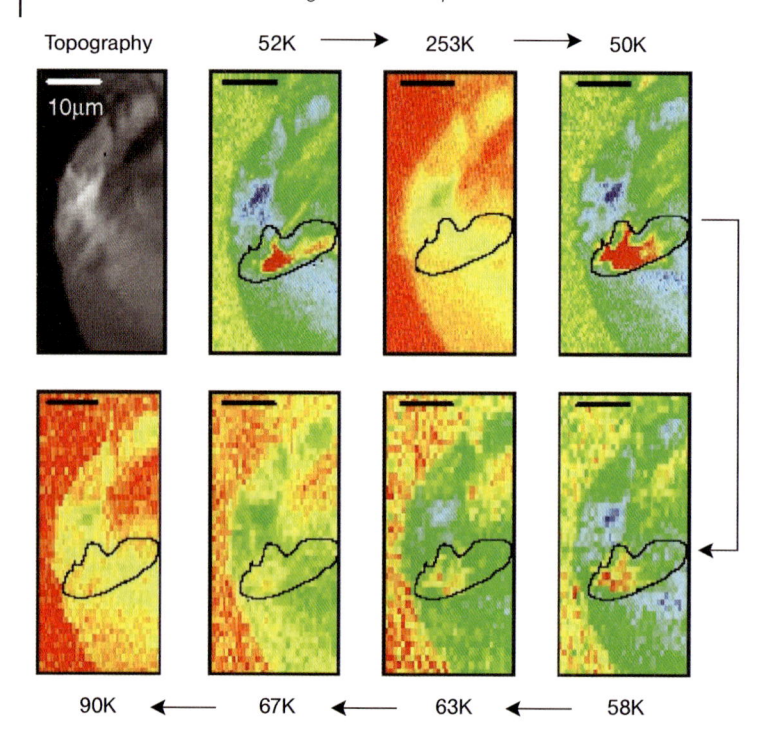

Fig. 11.4 Microscopic electronic inhomogeneities in $La_{1/4}Pr_{3/8}Ca_{3/8}MaO_3$. The indicated region represents a microscopic insulating patch of 15 μm × 5 μm. On warming the sample, the surrounding metallic regions become insulating. On cooling the sample, the insulating patch appears in essentially the same locations (where it was present before the warming process). Thus the electronic inhomogeneities exhibit memory effects. On the length scale of 0.5 μm (the spatial resolution) the material is chemically homogeneous. Colour key: Blue/green – metallic, Yellow-red – insulating. After Sarma et al. [20].

11.3
Theoretical Approaches to Electronic Inhomogeneities

We now turn to a discussion of theoretical ideas that have been proposed to explain the origin of the electronic inhomogeneities, with specific focus on manganites.

Manganites are known to have many competing energy scales which give rise to a fascinating range of phenomea in them. The simplest possible description of manganites at a microscopic level involves the Mn d-orbitals and the lattice distortion of the oxygen octahedron surrounding the Mn ion [33]. The octahedral crystal field splits the degeneracy of the five Mn d-orbitals into three degenerate t_{2g} orbitals and two degenerate e_g orbitals. In the doped manganites, both Mn^{3+} and Mn^{4+} configurations are present; both have three electrons in the t_{2g} orbitals, and Mn^{3+} has, in addition, a lone electron in one of the e_g orbitals. The spins of the t_{2g} electrons are aligned parallel, due to strong Hund coupling, and it is only the resulting "core spin" ($S = 3/2$) of the t_{2g} electrons that affects the low energy physics of manganites. Thus, the relevant degrees of freedom at each manganese site are: an average of $(1 - x)$ electrons per site populating the two e_g orbitals, the t_{2g} core spins, and lattice (phonon) degrees of freedom corresponding to the distortion of the oxygen octahedra surrounding the manganese ion. The e_g electrons hop from a Mn site to a neighboring Mn site with an amplitude t (~ 0.2 eV). The spin of the e_g electron has a strong ferromagnetic Hund coupling J_H (~ 2.0 eV) with the local t_{2g} core spin. Another important energy scale in the electronic sector is the on-site Mott–Hubbard repulsion U (~ 5.0 eV) which forbids double occupancy of the local e_g sector. Neighboring t_{2g} spins interact with each other via an antiferromagnetic superexchange coupling J_{SE} (~ 0.02 eV). Finally, the energy gained by the Jahn–Teller distortion of the oxygen octahedron is given by E_{JT} (~ 0.5 eV). As discussed elsewhere [6, 34–37], the competition between these different interactions, leading to a variety of states very close in energy, is responsible for the complex phase diagram of manganites, and their extreme sensitivity to external perturbations such as temperature, magnetic field and strain.

Because of the difficulties of dealing with all the degrees of freedom and competing interactions mentioned above, much of the early work aimed at understanding colossal responses and electronic inhomogeneities in manganites has been based on simplified models which neglect one or more of them. A prominent example is the work of Dagotto and coworkers [33] who studied simple *magnetic Hamiltonians* with competing phases that are separated by a first order transition. Based on these studies, they suggested that the system is prone to macroscopic phase separation, which is frustrated by disorder leading to the electronic inhomogeneities at various scales, the magnitude of the disorder determining the scale. From the real space structure obtained from simulations they constructed a random resistor network to explain the colossal responses by a mechanism similar to that proposed by Fäth et al. [18], again suggesting that the phase separation is key to colossal responses. Other simplified models have been studied, and alternate scenarios have been proposed. There are also suggestions that manganites close to half doping are near a multicritical point with competing phases affected by disorder [38, 39]. Ahn et al. [40] consider a model Hamiltonian including electron–phonon interaction, and long-range elastic coupling between local lattice distortions. They present a scenario for mesoscopic/microscopic inhomogeneities, and suggest that these are responsible for the colossal responses.

Next, we turn to our study of electronic inhomogeneities in manganites [9] which attempts to take into account all the degrees of freedom and their interactions. It is based on the ℓb model developed by Ramakrishnan et al. [7], of which we present a brief review in the next section. A more detailed discussion of this model and earlier work on it may be found in the paper by Ramakrishnan et al. [6].

11.4
The ℓ/b Model for Manganites

The key physical idea behind the ℓb model is that under the conditions prevailing in the doped manganites, namely interactions that inhibit the kinetic energy of the e_g electrons and the strong e_g electron – JT phonon coupling, the e_g electrons spontaneously reorganize themselves into two types of fluids, labeled ℓ and b, with very different quantum dynamics.

The majority of the e_g electrons go into *polaronic* ℓ states, each of which is a specific linear combination of the e_g orbitals accompanied by a strong local JT distortion (of magnitude $Q_0 \simeq g/K$ where K is the force constant for the JT phonons and g is the e_g electron – JT phonon coupling strength), thereby gaining local JT energy E_{JT} ($\simeq g^2/(2K)$). The quantum coherent hopping of an ℓ polaron involves the transfer of the electron from one site to the next, accompanied by relaxation of the lattice distortion at the first site and the formation of lattice distortion at the second site. Its amplitude is therefore much reduced, by the Huang–Rhys or JT-distortion wavefunction overlap factor (given by $\eta \sim \exp\{-(E_{JT}/2\hbar\omega_0)\} \ll 1$ where $\hbar\omega_0 \simeq 0.05$ eV is the JT optical phonon energy). Thus at temperature scales larger than ηt the ℓ polarons do not hop coherently and can be considered static. The direction of the JT distortion Θ_i, which determines the orbital state of the ℓ polaron, can vary from site to site, and Θ_i as well as the sites where the polarons are present can also change on long timescales determined by ηt. In the metallic regime for $x < 0.5$, the orbital angles typically have only short-range correlations, corresponding to an *orbital liquid* state. States with long-range orbital order are also known, e.g., at $x = 0.5$.

However, some kinetic energy gain is still possible by having a small number of e_g electrons in *broad-band, non-polaronic, mobile b* states. They have large amplitudes on *undistorted or weakly distorted sites*, and hence have zero site energy, and nearly undiminished inter-site hopping amplitude $\sim t$. There is a strong repulsion between the two fluids, as double occupancy on a polaronic site costs a large extra energy $\bar{U} = (U + 2E_{JT})$. So the b electrons mostly run around on the fraction [$\sim x$] of sites which are typically unoccupied and undistorted. Both types of electrons have a strong Hund's coupling J_H with the t_{2g} core spins. Over long timescales, sites can exchange roles between being hosts for ℓ polarons and being empty and hosts for b electrons.

Finally, the presence of the localized polaronic ℓ states, large U and J_H gives rise to a new, occupancy-dependent, ferromagnetic, "virtual double exchange" coupling between nearest neighbour t_{2g} core spins. This comes about due to *virtual, fast (adiabatic) hopping* of an ℓ electron to neighboring sites and back, i.e., leaving the local lattice distortion unrelaxed, by paying an energy cost of $2E_{JT}$ in the intermediate state, provided the neighboring site is empty and the t_{2g} spins on the two sites are parallel. (Otherwise the energy of the intermediate state increases by U and J_H respectively.) For $E_{JT} \gg t$, from second-order perturbation theory the interaction can be seen to be of the form

$$\left(\frac{t^2}{2E_{JT}S^2}\right)\frac{1}{2}(\mathbf{S}_i \cdot \mathbf{S}_j + S^2)[n_{\ell i}(1-n_j) + n_{\ell j}(1-n_i)] \tag{11.1}$$

Here, for simplicity, the dependence on the angles Θ_i etc., has been ignored (extensions in contexts with orbital ordering effects where one must include these are discussed in Refs. [37, 41], and the t_{2g} spins have been approximated as classical spins of fixed length $S = \frac{3}{2}$. The $\frac{1}{2S^2}(\mathbf{S}_i \cdot \mathbf{S}_j + S^2)$ factor above comes from large J_H, and the occupancy-dependent terms from large $\bar{U}$. This term (and longer range extensions which become important if E_{JT} is not very large compared to t) is crucially responsible for many features of doped manganites, in particular the ubiquitous appearance of ferromagnetic correlations upon doping.

The above considerations lead to the following ℓb model Hamiltonian:

$$\mathcal{H}_{\ell b} = -E_{JT}\sum_{i,\sigma}\ell_{i\sigma}^\dagger \ell_{i\sigma} - t\sum_{\langle ij \rangle}b_{i\sigma}^\dagger b_{j\sigma} - J_H\sum_i (\mathbf{s}_{i\ell} + \mathbf{s}_{ib})\cdot \mathbf{S}_i + U\sum_i n_{i\ell}n_{ib}$$
$$- \mu\sum_i (n_{il} + n_{ib}) - (J_{VDE} - J_{SE})\sum_{\langle ij \rangle}\mathbf{S}_i \cdot \mathbf{S}_j \tag{11.2}$$

where $\mathbf{s}_{i\ell}(n_{i\ell})$ and $\mathbf{s}_{ib}(n_{ib})$ are spin (number) operators respective to the ℓ and b states at site i, and μ is the chemical potential determined from the condition that $\langle n_l + n_b \rangle = 1 - x$. The last term contains the virtual double exchange process discussed above approximated as a homogenized, average ferromagnetic interaction, of strength $J_{VDE} \sim x(1-x)\frac{t^2}{E_{JT}}$, which dominates J_{SE}, the antiferromagnetic superexchange interaction once x is substantial. Two other simplifying approximations have been made in writing the Hamiltonian (11.2). The first is that orbital quantum numbers of the ℓ and b states have been ignored, and the kinetic energy in the Hamiltonian is 'orbitally averaged', i.e., the hopping amplitude t represents an average over the possible orbital configurations at the two pertinent sites. This is a reasonable approximation for the parameter ranges where the manganite is an 'orbital liquid' (i.e., when there is no long-range orbital order, and orbital correlations are short ranged). Second, there are no cooperative/long-range lattice effects, i.e., no intersite polaron correlations. Despite all these simplifications, the ℓ-b Hamiltonian has the merit that it includes *all* the main energy scales that govern manganite physics, whereas much of the previous work on manganites is based on simplified models [33, 42, 43] that neglect one or more of the main energy scales.

The Hamiltonian (11.2) closely resembles the Falicov–Kimball model. The model was solved [5–7] using the dynamical mean field theory [44] (DMFT). It is successful in capturing the colossal magnetoresistance effect, the role of ion radius, the ferromagnetic insulating state and the ferro-insulator to ferrometal transition found at low doping in low band width manganites, etc., in the orbital fluid regime. The key result is that the on-site Coulomb interaction U (the largest energy scale in the problem) effectively renormalizes the effective half bandwidth of the b states to a smaller value D_{eff} which is doping (x) dependent; with increase in x, the effective bandwidth increases. At zero temperature [5] with $U, J_H \to \infty$, and a ferromagnetic ordering of the core spins, the effective half bandwidth (in terms of the bare half bandwidth D_0) is given by

$$D_{\mathrm{eff}} = \sqrt{x} D_0 \qquad (11.3)$$

At low doping, the effective band bottom is above the polaronic energy level $-E_{\mathrm{JT}}$. Hence only the polaronic ℓ states are occupied, the chemical potential μ is pinned at $-E_{\mathrm{JT}}$, and the system is an insulator. A critical doping x_c is necessary for b-states to become occupied, with

$$x_c = \left(\frac{E_{\mathrm{JT}}}{D_0}\right)^2 \qquad (11.4)$$

signaling an insulator to metal transition. For a given doping, the effective bandwidth decreases from its zero temperature value as the temperature increases (see Fig. 11.5), because the hopping of the b electrons, which are strongly Hund's rule coupled to the core spins, becomes greatly inhibited when the spins get thermally disordered. This leads to the thermal ferro-metal to para-insulator transition. Application of an external magnetic field causes the spins to align and thus increases the effective bandwidth and consequently decreases the insulating gap,

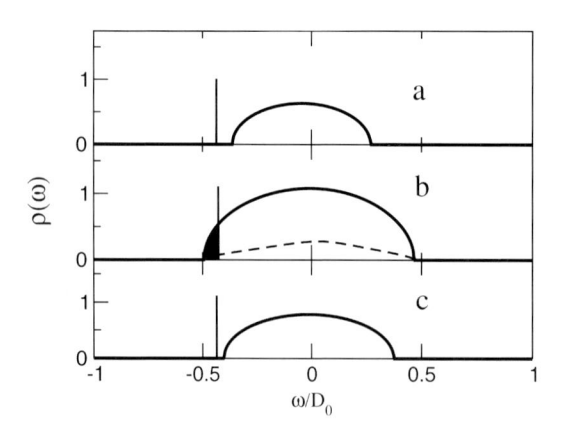

Fig. 11.5 Spectral density of b-electrons, for various doping x and temperature T. The polaron level is indicated by the vertical line. The results are for the parameters $E_{\mathrm{JT}} = 0.5$ eV, $D_0 = 1.2$ eV, $U = 5.0$ eV. The effective virtual double exchange is 2.23 meV. (a) $x = 0.1$, $T = 0$ K; ferromagnetic insulator. (b) $x = 0.3$, $T = 180$ K; ferromagnetic metal. (c) $x = 0.3$, $T = 350$ K, paramagnetic insulator. After Ramarkishnan et al. [7].

11.4 The ℓb Model for Manganites

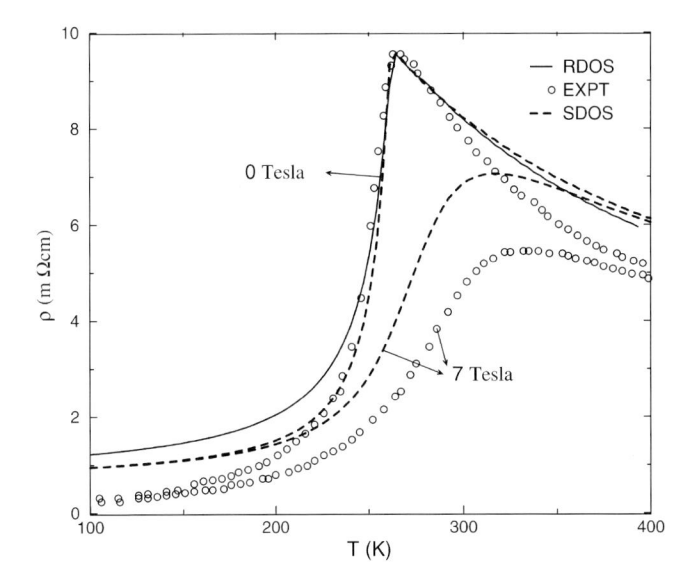

Fig. 11.6 Resistivity of $La_{1-x}Ca_xMnO_3$ as a function of temperature. Circles – experimental data [45], solid lines – calculation with realistic density of states, dashed lines – calculation with semi-circular density of states. After Ranmakrishnan et al. [7].

leading to a dramatic reduction in the resistivity. This causes the colossal magnetoresistance. The theory is able to reproduce several observed experimental trends in the resistivity (see Fig. 11.6) as well as in other propeties [5–7].

The work of Ramakrishnan et al. [7] did not address the issue of electronic inhomogeneities. In fact, an effective homogeneous state is *imposed* in the DMFT treatment. This is a drastic assumption, for, from previous work on the Falicov–Kimball model [46] the ground state of the ℓb Hamiltonian (11.2) (under the parameter ranges discussed above) is *known to be a macroscopically phase separated state* due to the strong on-site Coulomb correlation U, which forbids the simultaneous site occupancy of ℓ and b states at a site. All the holes in the system move to one side of the box; this allows the band states (b) to gain (negative) kinetic energy by hopping amongst these sites freely, without being scattered from repulsive ℓ sites. We emphasize that this phase separation is induced by strong electronic correlation rather than by phase competition. However, as is well known from earlier work on other such cases [47–49], since the two portions have different electron densities, such a phase separation is, of course, rendered unfavorable by the long-range Coulomb interactions. Indeed, as we have shown elsewhere [9], in an ℓb model extended to include the long-range Coulomb interactions, these interactions suppress phase separation and lead to nanoscale inhomogeneities with 'puddles' of b and ℓ regions. We describe this work in the next section.

11.5
The Extended ℓb Model and Effects of Long-range Coulomb Interactions

The model we consider assumes that the Mn ions occupy the sites of a cubic lattice (taken to be of unit lattice parameters), while the dopant Ak ions occupy an x fraction of the 'body center' sites of each unit cube formed by the Mn ions. Since our aim is to study the effect of long-range Coulomb interactions on phase separation, we make further simplifying assumptions. We assume that the t_{2g} core spins are aligned ferromagnetically, and assume that $J_H \to \infty$, and this effectively projects out ℓ or b electron spin opposite to that of the t_{2g} core spins – we obtain an effectively spinless model. The above considerations lead us naturally to the following *extended ℓb Hamiltonian*

$$H = H_{\ell b} + H_C \qquad (11.5)$$

with

$$H_{\ell b} = -E_{JT}\sum_i n_{\ell i} - t\sum_{\langle ij \rangle}(b_i^\dagger b_j + \text{h. c.}) + U\sum_i n_{\ell i} n_{bi} \qquad (11.6)$$

and

$$H_C = \sum_i \Phi_i q_i + \frac{V_0}{2}\sum_{i \neq j}\frac{q_i q_j}{r_{ij}} \qquad (11.7)$$

Here $\ell_i^\dagger$ and $b_i^\dagger$ create ℓ and b electrons respectively at site i, and $n_{\ell i} \equiv \ell_i^\dagger \ell_i$ and $n_{bi} \equiv b_i^\dagger b_i$ are the corresponding number operators. In terms of the hole operator ($h_i^\dagger \equiv \ell_i$ which removes an ℓ polaron at site i) the electron charge operator $q_i \equiv h_i^\dagger h_i - b_i^\dagger b_i$, and has the average value x per site because of overall charge neutrality. The Coulomb term H_C has two parts; the charge at site i has energy $q_i \Phi_i$, where Φ_i is the electrostatic potential there due to Ak^{2+} ions, and the interaction between the charges at site i and j leads to an energy $V_0 \frac{q_i q_j}{r_{ij}}$. The model in Eq. (11.5) has several ingredients which are generic to strongly correlated systems: two different quantum states locally, of comparable energy; the competition strongly influenced by phase coherence (kinetic energy gain due to hopping) and strong coupling (e.g. to the lattice, leading to a localized polaronic state); tendency of these states to phase separate due to short-range electron correlation, and the inevitable long-range Coulomb interactions. In what follows, we take the short-range Coulomb correlation U to be large (∞).

We first briefly describe a simple approximate analytical treatment [9] of the extended ℓb Hamiltonian (11.5) by considering the Ak ions to be distributed uniformly in space. We then try a variational ground state with a periodic arrangement of clusters of ℓ polarons (the 'bulk'), and of vacant sites (referred to as 'clumps') as sketched in Fig. 11.7 which has parameters n_b (the b electron density

11.5 The Extended ℓ/b Model and Effects of Long-range Coulomb Interactions

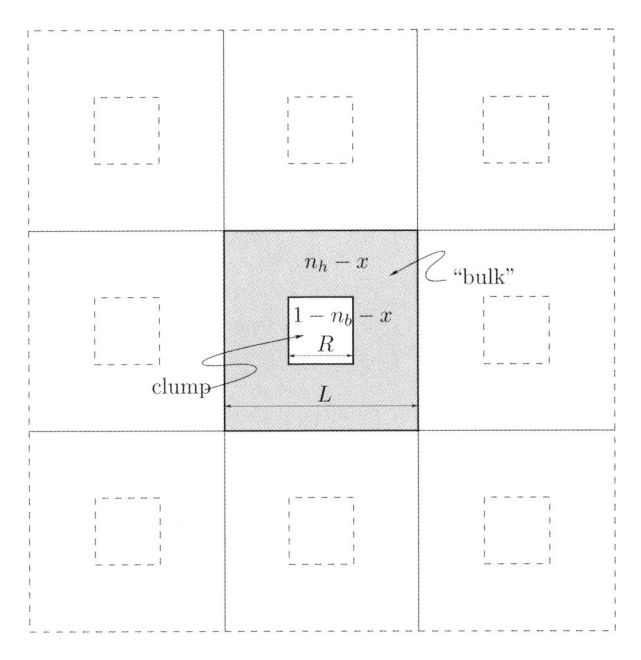

Fig. 11.7 Schematic of the ground state used in the approximate analytical calculation. "Phase separation" takes the system to two distinct types of regions called the 'bulk' (cluster of ℓ polarons) and the 'clump' (cluster of hole sites). The dashed lines indicate the assumed periodic nature of the clump distribution – clumps of size R (volume R^3) are assumed to be arranged in a periodic fashion (period L) with intervening 'bulk' regions. Delocalized b electronic states are found in the clumps. The initial hole density is x (also equal to the background negative charge density). The charge density in the clump is $(1 - n_b - x)$ and that in the 'bulk' is $n_h - x$. n_b is the fraction of electrons that are promoted to delocalized states, and n_h is the fraction of holes created in the 'bulk'. The fractions x, n_b and n_h are related via charge balance. After Shenoy et al. [9].

in the clump), the size R of the clump, and the parameter $\alpha = L/R$ representing the spacing of the clumps (see Fig. 11.7). The hole density n_h (see Fig. 11.7) and the b electron density n_b are related via the conservation condition. There are three competing energies: the electrostatic energy, the polaronic energy of the ℓ states, and the kinetic energy of the b states. Estimating these three energies using a series of physically motivated approximations, we arrive at (details may be found in Ref. [9]) the following expression for the clump size

$$R^2 = \frac{9n_b(6t(1 - n_b) - E_{JT})}{4\pi K V_0 (1 - n_b - x)^2} \tag{11.8}$$

where $n_b = \frac{1}{2}\left(1 - \frac{E_{JT}}{6t}\right)$, and K is a number of order 1. Note that the clump size varies as $1/\sqrt{V_0}$. Clump sizes for $E_{JT} = 2.5t$ and $x = 0.3$ are hence roughly between 10 and 5 lattice spacings for V_0 between 0.01 and 0.1. Thus, for realistic values of parameters, even this simple analytical treatment predicts electronic inhomogeneities of nanometric scale.

We now discuss results of full scale numerical simulations of the Hamiltonian (11.5) on finite 3d periodic lattices, allowing for a random distribution of the Ak ions. In our discussion below all energy scales are normalized by the bare intersite hopping amplitude t. We have considered systems as large as $20 \times 20 \times 20$ (more details may be found in the paper by Shenoy et al. [9]). The numerical determination of the ground state of the Hamiltonian (11.5) requires further simplifying approximations. The most important approximation is the Hartree approximation, i.e., the charge operator q_i is replaced by its expectation value in the ground state $\langle q_i \rangle = h_i^\dagger h_i - \langle b_i^\dagger b_i \rangle$. Since we have assumed that $U \to \infty$, the b electrons do not hop to sites where an ℓ polaron is present. This leads to the segregation of the two types of electrons into disjoint clusters. In a cluster of hole sites (which has at least two nearest neighbor hole sites) which we call a 'clump', the b electron states are determined by solving the quantum kinetic energy Hamiltonian exactly. The determination of the ground state begins with a random charge neutral configuration on which successive electron moves are performed. An electron move entails removing an electron from an ℓ or a b state and placing it in another ℓ or b state (implying that there are four distinct types of electron moves possible). At each step we calculate the energies of possible electron moves and adopt the move that most reduces the energy. The ground state is the configuration at which there are no electron moves that can further reduce the energy. *This is a new generalization of the common Coulomb glass simulation [50–52] which includes the quantum mechanically obtained b states within their clump or puddle.* The electrostatic energy is calculated accurately using the Ewald technique and fast Fourier transform routines.

In the absence of the long-range Coulomb interaction ($V_0 = 0$) this procedure leads to macroscopically phase separated state (see Fig. 11.8). The holes aggregate to one side of the simulation box and a fraction of the electrons occupying the ℓ states is promoted to the b states in the one large hole clump – with their concentrations determined by the equality of the chemical potential in the two regions (i.e., the highest occupied b level equals $-E_{JT}$). This phase separation is due to the strong coulomb repulsion between the two types of electron fluids and is in agreement with known results on the Falicov–Kimball model [46, 53].

Long-range Coulomb interaction "frustrates" this phase separation, i.e., macroscopic phase separation costs prohibitive energy in the presence of long-range Coulomb interactions. The precise nature of the resulting ground state electronic configuration depends on the Jahn–Teller energy E_{JT} and the doping level x. There are two critical doping levels x_{c1} and x_{c2} for any given Jahn–Teller energy E_{JT} (see Fig. 11.9). When the doping level is less than x_{c1}, i.e., $x < x_{c1}$, no b states are occupied – the holes (and hence also the ℓ polarons) form a Coulomb glass [54]. When the doping exceeds x_{c1}, some of the hole clumps are occupied and the ground state consists of isolated b-electron puddles dispersed in a polaronic background (see Fig. 11.10(a)). There is a second critical doping level x_{c2} (see Fig. 11.9); when $x \gtrsim x_{c2}$ the occupied b-electron puddles percolate and the system

11.5 The Extended ℓb Model and Effects of Long-range Coulomb Interactions

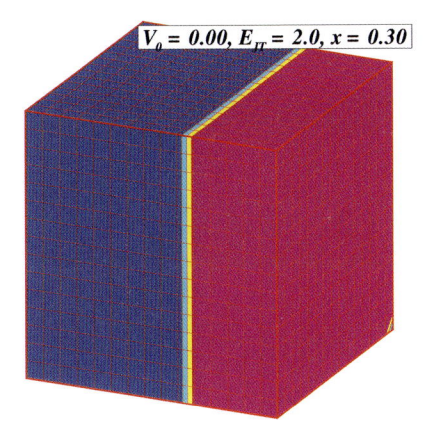

$V_0 = 0.00$, $E_{JT} = 2.0$, $x = 0.30$

Fig. 11.8 "Macroscopic phase separation" in the ℓb model in the absence of long-range Coulomb interaction. The lighter regions correspond to holes that form a large clump (with occupied b states), the darker regions are occupied by ℓ polarons. This simulation is performed with a cube of size $16 \times 16 \times 16$. After Shenoy et al. [9].

attains metallicity (see Fig. 11.10(b)). Interestingly, the critical doping levels x_{c1} and x_{c2} are insensitive to V_0 for small values of V_0 ($V_0 \leq 0.5$; typical values of V_0 in manganites are in the range 0.01–0.1). For larger values of V_0, the electrostatic energy cost forbids large clumps. This renders b-electron puddles to be energetically unfavorable and the system is an insulator.

The underlying reasons for the results stated above can be understood from a study of the density of states of the ℓ-polarons and b-electrons as shown in Fig. 11.11. The results indicate that the chemical potential μ (as noted in Fig. 11.11) essentially follows $\mu = -E_{JT} + V_0$. As is evident from Fig. 11.11, the polarons form a Coulomb glass with a soft gap [54]. Note also that the polarons have an energy spread of the order of V_0. The b electron density of states is band-like, with a width that is determined by the doping x. The simulation results show that the b band center (marked by E_b in the figure) is *not* affected by doping, and scales as $E_b \approx V_0$ (for small $V_0 \leq 0.5$). Interestingly, our results show that the effective band width of the b electrons obtained from the simulations scales as $D_{\text{eff}} = D_0 \sqrt{x}$, in remarkable agreement with the DMFT result (11.4).

The random distribution of the Ak ions is crucial to the observed behavior of the band center and the chemical potential. From the positional correlation function between two holes, and a hole and an Ak ion, shown in Fig. 11.12, it is evident that the holes do cluster around the Ak ions, as one might expect. However, we find that the effective electrostatic screening is quite strong in this system, as is evident from the fact that the correlation functions reach a plateau within a few lattice spacings. Due to the clustering of the holes around the Ak ions, the ℓ polarons have an increased number of ℓ polaron as neighbors, whence the average energy of the ℓ polarons is increased by V_0, i.e., the average energy becomes

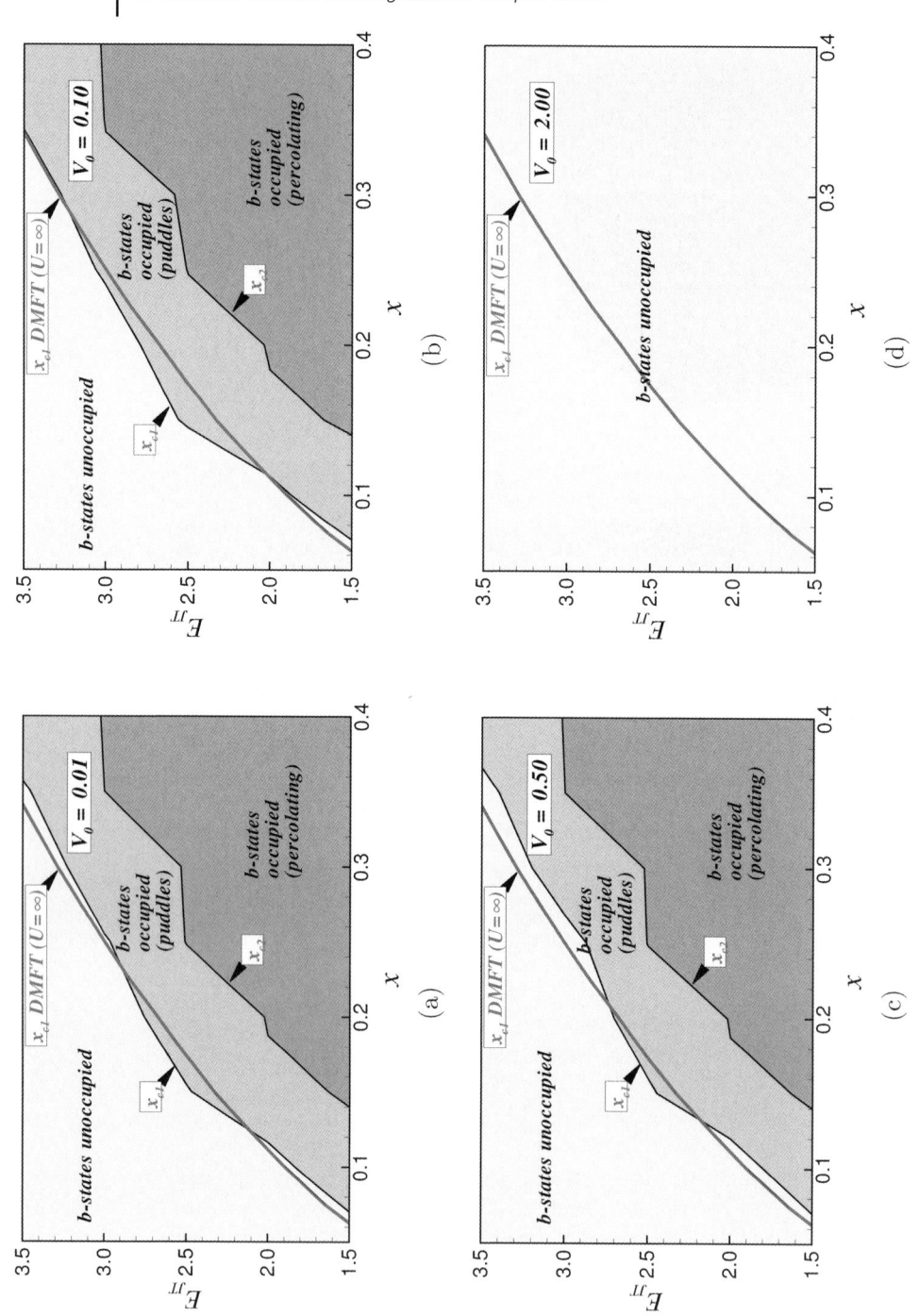

11 Nanoscale Electronic Inhomogeneities in Complex Oxides

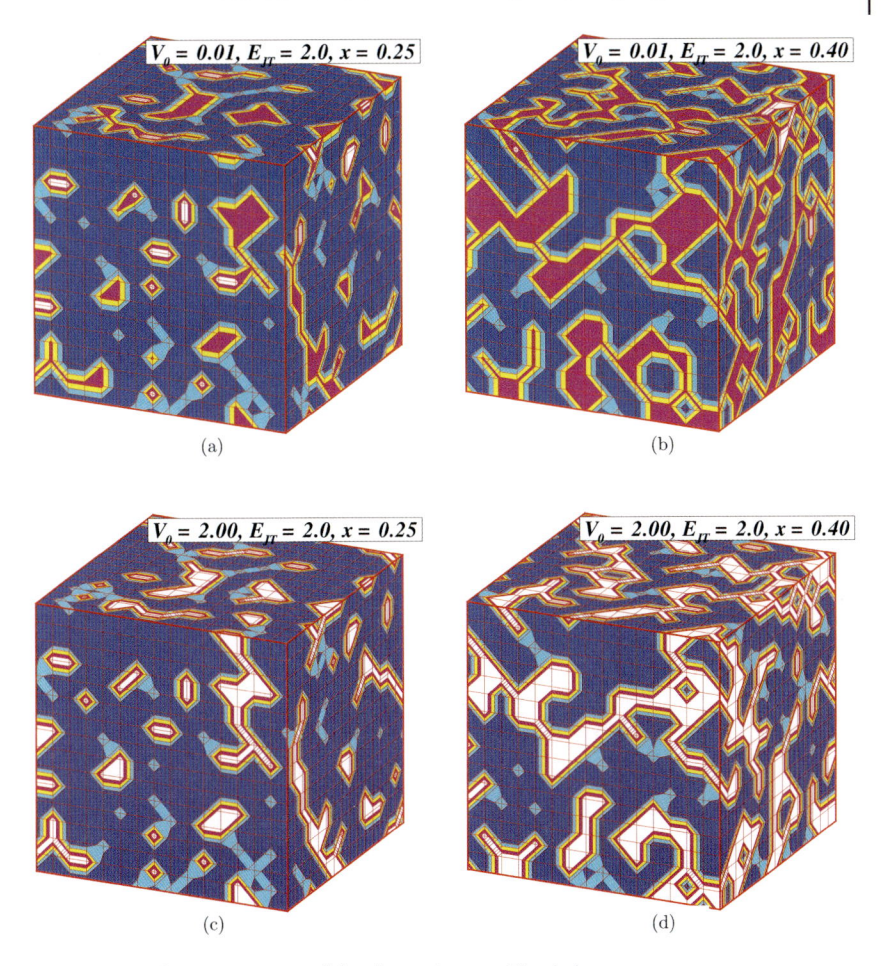

Fig. 11.10 Real space structure of the electronic state. The darkest regions (magenta) denote hole clumps with occupied b electrons, the lightest (white) denote hole clumps with no b electrons, the second lighter shade (cyan) denotes singleton holes, and the second darkest shade (light blue) represents regions with ℓ polarons. The simulations for each V_0 are for the same realization of the random distribution of Ak ions. The cell size is $16 \times 16 \times 16$. After Shenoy et al. [9].

Fig. 11.9 Critical doping levels x_{c1} and x_{c2} obtained from simulations. The lightest region in the contour plot contains no b electrons, the intermediate shade has "trapped" b states occupied, and the darkest regions correspond to b states that percolate through the simulation box. The solid line corresponds to the analytical DMFT result for x_{c1}. After Shenoy et al. [9].

(a) $\mu = -2.4$, $E_b = 0.0$, $V_0 = 0.10$, $E_{JT} = 2.50$, $x = 0.30$

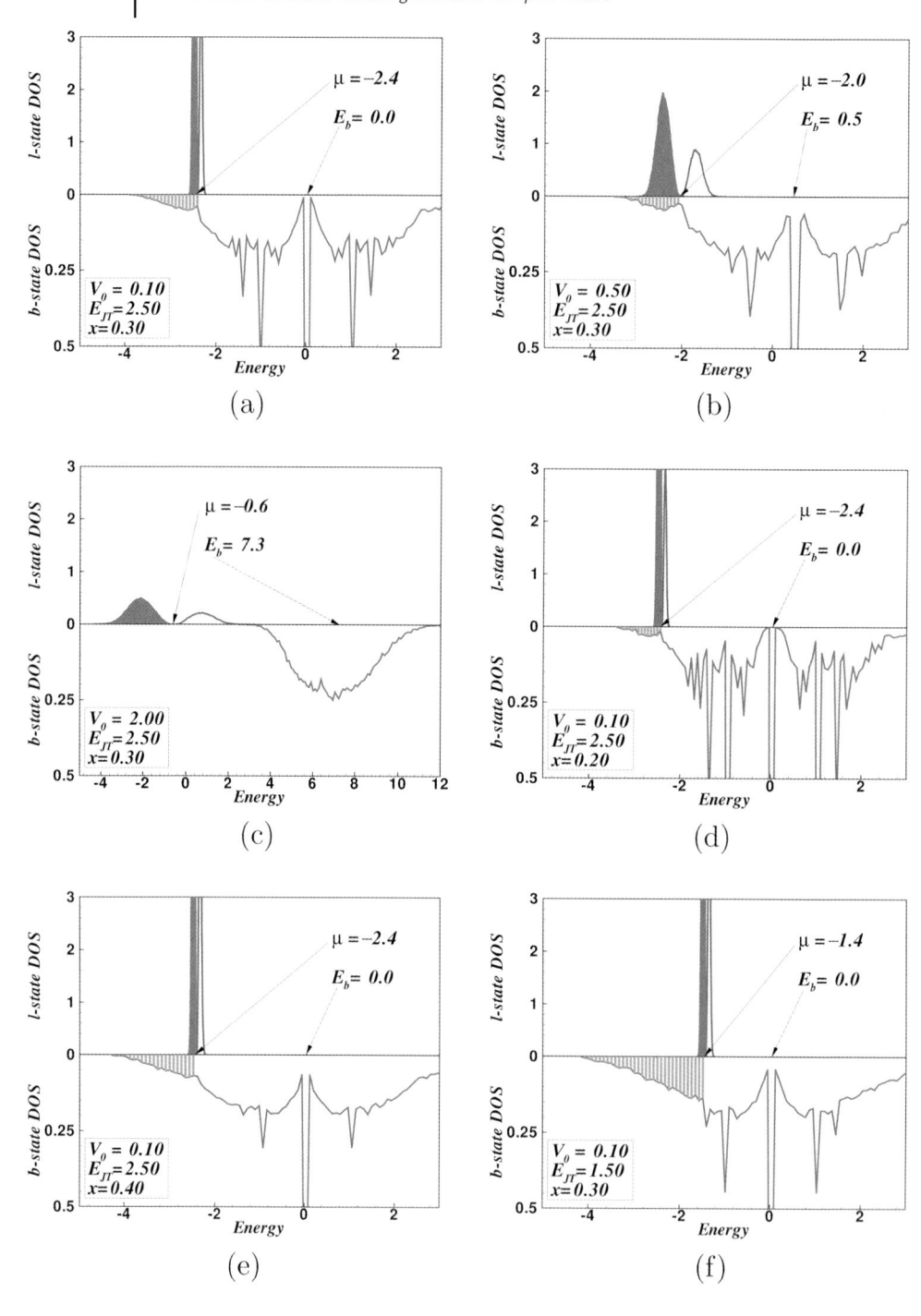

(b) $\mu = -2.0$, $E_b = 0.5$, $V_0 = 0.50$, $E_{JT} = 2.50$, $x = 0.30$

(c) $\mu = -0.6$, $E_b = 7.3$, $V_0 = 2.00$, $E_{JT} = 2.50$, $x = 0.30$

(d) $\mu = -2.4$, $E_b = 0.0$, $V_0 = 0.10$, $E_{JT} = 2.50$, $x = 0.20$

(e) $\mu = -2.4$, $E_b = 0.0$, $V_0 = 0.10$, $E_{JT} = 2.50$, $x = 0.40$

(f) $\mu = -1.4$, $E_b = 0.0$, $V_0 = 0.10$, $E_{JT} = 1.50$, $x = 0.30$

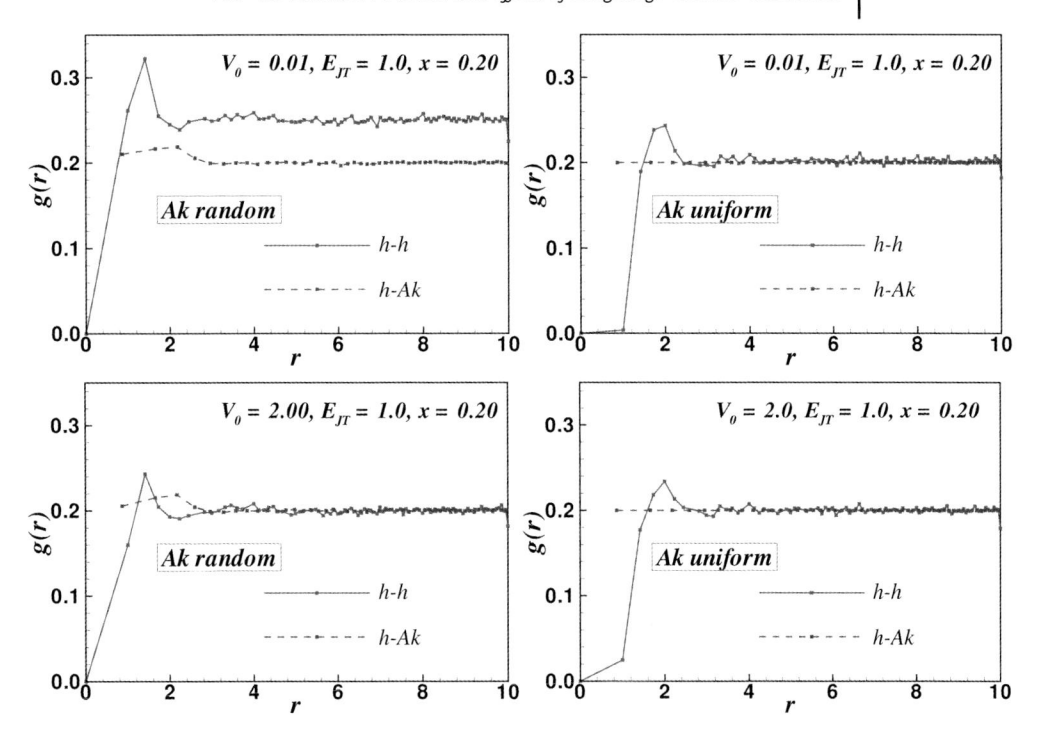

11.5 The Extended ℓb Model and Effects of Long-range Coulomb Interactions | 377

Fig. 11.12 Position correlation function for $h - h$ and $h - Ak$. The figures in the left column are for random distribution of Ak ions, while those in the right column are for uniform distribution of Ak ions (uniform distribution means that the total charge of the Ak ions is distributed equally among the Ak sites). The result is from simulations with a $20 \times 20 \times 20$ cube (for a single realization of the random distribution of the Ak ions in the left-side graphs). After Shenoy et al. [9].

$-E_{JT} + V_0$. The chemical potential is now given by $-E_{JT} + V_0$ (since the average energy of the states is the chemical potential in a Coulomb glass [54]). Similarly, every b-electron sees an effective repulsive potential of V_0 due to the Ak ions (since the hole sites cluster around the Ak ions), and the b band center is shifted from zero to V_0. Noting that the b band width scales as $\sqrt{x}$ as noted above, we conclude that the minimum doping for the b-band bottom to touch the chemical

Fig. 11.11 Density of states (DOS) of ℓ polarons and b electrons obtained from simulations with $10 \times 10 \times 10$ cubes averaged over one hundred random initial conditions. The occupied states are shaded. The chemical potential μ and the b band center E_b are marked. In all cases the polarons have a soft Coulomb gap at the chemical potential. Panels (a), (b) and (c) show the effect of the long-range Coulomb parameter V_0. Comparison of (a), (d) and (e) shows how the band width increases with doping. The effect of E_{JT} on the chemical potential and the bandwidth can be seen from a comparison of (a) and (f). After Shenoy et al. [9].

potential (allowing for b state occupancy) is $x_{c1} = \left(\frac{E_{JT}}{D_0}\right)^2$ which is precisely the DMFT prediction [5].

It should be noted that the full scale simulations presented here provide further insights into the physics of the system beyond the single site DMFT. As indicated above, an important feature seen here, distinct from that obtained in DMFT, is that the b site occupancy *per se* does not make the system metallic. The occupied b-electron puddles must percolate through the sample, and furthermore, extended states in the percolating puddles must get occupied, to obtain metallicity, and this cannot be captured in a single site DMFT. We are able to obtain reasonable estimates for the second critical doping x_{c2} based on calculations of inverse participation ratio and Kubo conductivity (see Fig. 11.9). We are able to make further inferences and predictions based on our simulations, especially useful in the study of low bandwidth manganites (such as Pr-based manganites) which have a large ferro-insulating regime. For very low doping ($x < x_{c1}$), the conductivity will be due to the thermally activated motion of the ℓ polarons which now form a Coulomb glass. In this case the low temperature conductivity is expected to be that of the Coulomb glass i.e., $\sigma(T) \sim e^{-A/\sqrt{T}}$ where A is a constant. In the doping range $x > x_{c2}$ we expect a "bad metal" – with small metallic conductivity that decreases with temperature. In the doping regime $x_{c1} \leq x \leq x_{c2}$, many different types of excitations contribute to the conductivity. These will include polaron hopping, variable range hopping of b electrons from one puddle to another etc. We note here that the transport measurements on doped manganites reported by Rao [55] do show such features.

We next address the issue of the size scale of the electronic inhomogeneities. The b-electron puddle sizes obtained from our simulations are shown in Fig. 11.13. We note that the size scale of the electronic inhomogeneity is nanometric, as expected from the analytical calculations. However, its dependence on V_0 differs fundamentally from the analytical result presented before, which assumed a

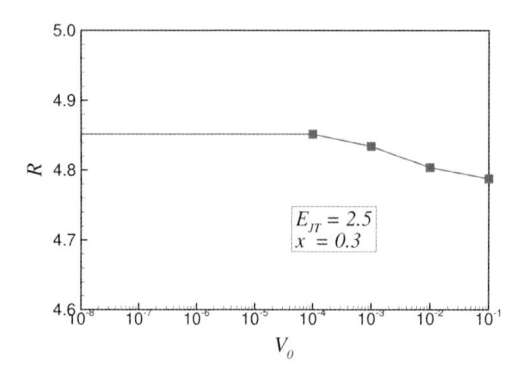

Fig. 11.13 Clump size as a function of the Coulomb interaction parameter V_0 (normalized by t) for $E_{JT} = 2.5t$, $x = 0.3$ obtained from simulations. After Shenoy et al. [9].

homogeneous distribution (jellium) of the dopant Ak ions. For the more realistic, random distribution of the Ak ions, the size of the electronic inhomogeneity is almost independent of the long-range Coulomb parameter V_0, in that even an extremely small V_0 produces clump sizes that are of the size scale of a few lattice spacings. The dependence of the average size of the electronic inhomogeneity is thus due to the long-range Coulomb interaction which, *along with* the random distribution of Ak ion, acts as a 'singular perturbation' that frustrates macroscopic phase separation. However, the sizes and the distribution of the clumps are determined by the random distribution of the Ak ions, and thus we conclude that doped manganites (and similarly, possibly many other correlated oxides) are necessarily and intrinsically electronically inhomogeneous, on a *nanometric scale*.

The results of our extended ℓb model with realistic energy parameters presented in this chapter provide several new insights into the complex electronic inhomogeneities seen in correlated oxides, and in particular the low bandwidth manganites with a large ferromagnetic region in their phase diagram. Our arguments may be sharply contrasted with earlier work. In particular our work suggests that the nanoscale inhomogeneities in manganites arise not out of 'phase competition' induced phase separation frustrated by disorder as argued from the studies of model spin Hamiltonians [33], but from short-range Coulomb correlation induced phase separation frustrated by long-range Coulomb interaction, similar to what has been suggested in cuprates [47, 48]. Most importantly, nanoscale electronic inhomogeneities are present in *both* the insulating and metallic phases of doped manganites. As is evident from Fig. 11.10, each of these constitute a single thermodynamic phase that is homogeneous at mesoscales. One has a metal insulator transition between two such nanoscopically inhomogeneous phases at x_{c2} as a function of doping. These observations are in agreement with experiments; indeed the electron holography results of Loudon et al. [14] show that even the ferrometallic regions have interspersed in them charge-ordered insulating regions which can be interpreted as the cluster of ℓ states. We note here that these experiments are at half filling and the orbital degrees of freedom are important (our model here is an orbitally averaged one appropriate for describing an "orbital liquid"). Modifications to the $\ell - b$ Hamiltonian to account for these are available in the literature [37, 41]. However we believe that the key features of the electronic inhomogeneities discussed above would remain even after the inclusion of these modifications.

Are the mesoscale electronic inhomogeneities essential for colossal responses? We infer from our work, in contrast to previous work [33, 56], that *there is no need to appeal to mesoscale phase separation to explain CMR in manganites*. The full scale simulations with a random distribution of dopant ions including the long-range Coulomb interactions reproduce the results of a homogenized DMFT calculation (which does not consider mesoscale phase separation). Colossal responses in the ℓb model are obtained by changes in the b electron energetics brought about by the external field, that causes exponential changes in their occupancy. As is well known experimentally, there are many manganites that are mesoscopically homogeneous and that show colossal responses [57]. We also point out that our arguments are substantially different from those put forth by Ahn et al. [40], who again argue for the necessity of mesoscale phase separation for colossal responses. Their model considers electron lattice coupling with two *localized states* at every site with an additional long-range *elastic* interaction between the sites. They, however, neglect the local Coulomb correlation, long-range Coulomb interactions and doping disorder. Furthermore, the relationship between the parameter values in their model and the microscopic parameters, and hence the relevance of their conclusions in the context of manganites, is unclear without further investigations.

Finally, we turn to the issue of mesoscale patterns clearly seen in experiments [17, 19, 20]. They could have two possible origins. First, due to 'incomplete transformations', i.e., they are metastable configurations in which the system gets trapped due to 'kinetic arrest' [58]. A second, in our opinion, more likely, cause is the effect of unscreened long-range elastic inhomogeneities [1, 20, 59, 60]. It is well established that uniform pressure has a strong influence on the stability of different manganite phases; pressure is indeed known to drive metal–insulator and magnetic transitions [61]. Moreover, there is solid experimental evidence [1, 20, 59, 60] for *preexisting* strain sources (such as cracks etc.) seen in surface probes. These could lead to the coexisting metal/insulator regions. It is not hard to think of experiments that could corroborate this point further. One proposal we make is that a pre-cracked single crystal manganite sample on which a load is applied should be studied (Fig. 11.14); the load-dependent singular stress fields near the crack tip should drive metal insulator/magnetic transitions which can be observed using a scanning probe microscope.

There remains an intriguing question. Can a chemically homogeneous, defect-free, large single crystal of doped manganite spontaneously generate *mesoscale* electronic inhomogeneities? It is known that strain strongly affects the microscopic energy scales in manganites [62, 63]. The key question is whether the system will spontaneously generate spatial patterns in the elastic strain that will reduce the electronic energy. The length scale of the pattern chosen will depend on the competition between the electronic energies (which would, in general, depend on the local strains and their spatial gradients) and the elastic energy.

11.6
Conclusion

It is evident from our survey of the experiments on oxides that electronic inhomogeneities are very common in them. Their scales can vary from nanoscopic to mesoscopic. We have presented a theoretical model that throws light on the origins of these inhomogeneities. Our model has two fluids, one that is localized and polaronic, the other delocalized and band-like. Strong on-site Couloumb correlation between these fluids induces their phase separation into two regions with

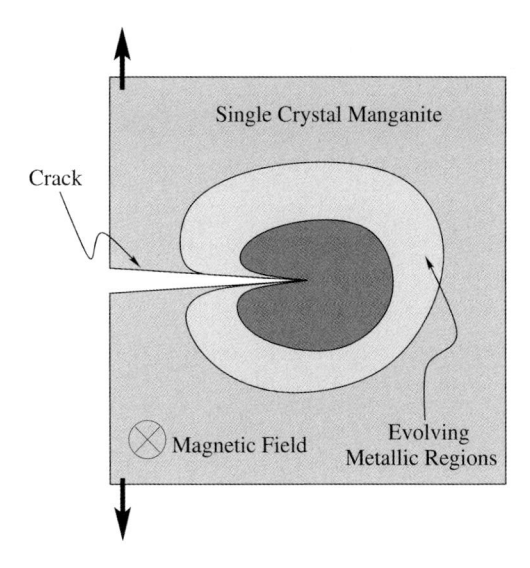

Fig. 11.14 A possible experiment to study the effect of long-range strain fields on the evolution of mesoscale inhomogeneities. On application of the load to the sample (indicated by bold arrows), the singular strain fields near the crack tip drive metal–insulator/magnetic transitions which can be studied with a scanning probe microscope. The experiments can also be conducted in the presence of a magnetic field. After Shenoy et al. [9].

different charge densities. This phase separation is frustrated by the ubiquitous long-range Coulomb interactions giving rise to a nanoscale structure with puddles of band-like electrons dispersed in a polaronic background. The puddles are *nanometric* in size, which is determined by the distribution of dopant ions. Nanoscale electronic inhomogeneities in manganites are, therefore, a result of strong correlation physics (as in cuprates [47, 48]). Further, we infer that the mesoscale inhomogeneities seen in experiments are likely to owe their origin to long-range unscreened elastic strain fields, possibly due to defects. Our work also suggests the possibility of controlling and tailoring these nanoscale inhomogeneites by means of materials chemistry to make "electronically patterned" materials on the nanoscale.

Acknowledgment

VBS wishes to thank the Indian National Science Academy for financial support of this work. HRK thanks the Department of Science and Technology and the Indo-French Centre for the Promotion of Advanced Research for support. TVR acknowledges support from the Department of Atomic Energy.

References

1. N. D. Mathur, P. B. Littlewood, *Phys. Today*, 25–30 (January 2003).
2. C. N. R. Rao, A. K. Kundu, M. M. Seikh, L. Sudheendra, *Dalton Trans.* 3003–3011 (2004).
3. E. Dagotto, *Science* **309**, 257 (2005).
4. G. C. Milward, M. J. Calderón, P. B. Littlewood, *Nature* **433**, 607 (2005).
5. G. V. Pai, S. R. Hassan, H. R. Krishnamurthy, T. V. Ramakrishnan, *Europhys. Lett.* **64**, 696 (2003).
6. T. V. Ramakrishnan, H. R. Krishnamuthy, S. R. Hassan, G. V. Pai, Theory of Manganites Exhibiting Colossal Magnetoresistance, cond-mat/0308396.
7. T. V. Ramakrishnan, H. R. Krishnamurthy, S. R. Hassan, G. V. Pai, *Phys. Rev. Lett.* **92**, 157203 (2004).
8. H. R. Krishnamurthy, *Pramana* **64**, 1063 (2005), Invited talk published in the Proceedings of the 22nd IUPAP International Conference on Statistical Physics, S. Dattagupta, H. R. Krishnmurthy, R. Pandit, T. V. Ramakrishanan, D. Sen (Eds.).
9. V. B. Shenoy, T. Gupta, H. R. Krishnamurthy, T. V. Ramakrishnan, Coulomb Interactions and Nanoscale Electronic Inhomogeneities, to be published.
10. E. O. Wollan, W. C. Koehler, *Phys. Rev.* **100**, 545 (1955).
11. P. M. Woodward, D. E. Cox, T. Vogt, C. N. R. Rao, A. K. Cheetham, *Chem. Mater.* **11**, 3528 (1999).
12. C. Ritter, R. Mahendiran, M. R. Ibarra, L. Morellon, A. Maignan, B. Raveau, C. N. R. Rao, *Phys. Rev. B* **61**, R9229 (2000).
13. P. G. Radaelli, R. M. Ibberson, D. N. Argyriou, H. Casalta, K. H. Andersen, S.-W. Cheong, J. F. Mitchell, *Phys. Rev. B* 172419 (2001).
14. J. C. Loudon, N. D. Mathur, P. A. Midgley, *Nature* **420**, 797 (2002).
15. S. Mori, C. H. Chen, S.-W. Cheong, *Nature* **392**, 473 (1998).
16. C. Renner, G. Aeppli, B. G. Kim, Y.-A. Soh, S.-W. Cheong, *Nature* **416**, 518 (2002).
17. M. Uehara, S. Mori, C. H. Chen, S. W. Cheong, *Nature* **399**, 560 (1999).
18. M. Fäth, S. Freisem, A. A. Menovsky, Y. Tomioka, J. Aarts, J. A. Mydosh, *Science* **285**, 1540 (1999).
19. L. Zhang, C. Israel, A. Biswas, R. L. Greene, A. de Lozanne, *Science* **298**, 805 (2002).
20. D. D. Sarma, D. Topwal, U. Manju, S. R. Krishnakumar, M. Bertolo, S. La Rosa, G. Cautero, T. Y. Koo, P. A. Sharma, S.-W. Cheong, A. Fujimori, *Phys. Rev. Lett.* **93**, 097202 (2004).
21. E. Sigmund, K. A. Müller (Eds.), *Phase Separation in Cuprate Superconductors* (Springer, Heidelberg, 1994).
22. D. M. Broun, P. J. Turner, W. A. Huttema, S. Ozcan, B. Morgan, R. Liang, W. N. Hardy, D. A. Bonn, In-plane Superfluid Density of Highly Underdoped $YBa_2Cu_3O_{6+x}$, arxiv.org:cond-mat/0509223.
23. J. M. Tranquada, H. Woo, T. G. Perring, H. Goka, G. D. Gu, G. Xu, M. Fujita, K. Yamada, *Nature* **429**, 534 (2004).
24. K. McElroy, J. Lee, J. A. Slezak, D.-H. Lee, H. Eisaki, S. Uchida, J. Davis, *Science* **309**, 1048 (2005).
25. C. N. R. Rao, O. Parkash, D. Bahadur, P. Ganguly, S. Nagabhushana, *J. Solid State Chem.* **22**, 353 (1977).
26. M. A. Señaris Rodriguez, J. B. Goodenough, *J. Solid State Chem.* **118**, 323 (1995).
27. J. Wu, J. W. Lynn, C. J. Glinka, J. Burley, H. Zheng, J. F. Mitchell, C. Leighton, *Phys. Rev. Lett.* **94**, 037201 (2005).
28. J. C. Burley, J. F. Mitchell, S. Short, *Phys. Rev. B* **69**, 054401 (2004).
29. A. K. Kundu, P. Nordblad, C. N. R. Rao, *Phys. Rev. B* **72** (2005).
30. A. K. Kundu, P. Nordblad, C. N. R. Rao, *J. Solid State Chem.* **179**, 923 (2006).
31. V. Bhide, D. Rajoria, C. N. R. Rao, G. Rama Rao, V. Jadhao, *Phys. Rev. B* **12**, 2832 (1975).

32. P. L. Kuhns, M. J. R. Hoch, W. G. Moulton, A. P. Reyes, J. Wu, C. Leighton, *Phys. Rev. Lett.* **91**, 127202 (2003).
33. E. Dagotto, *Nanoscale Phase Separation and Colossal Magnetoresistance* (Springer-Verlag, Berlin, 2003).
34. D. D. Sarma, N. Shanthi, S. R. Barman, N. Hamada, H. Sawada, K. Terakura, *Phys. Rev. Lett.* **75**, 1126 (1995).
35. S. Satpathy, Z. S. Popović, Vukajlović, *Phys. Rev. Lett.* **76**, 960 (1996).
36. A. J. Millis, *Philos. Trans. R. Soc. London, Ser. A* **356**, 1473 (1998).
37. O. Cepas, H. R. Krishnamurthy, T. V. Ramakrishnan, *Phys. Rev. B* **73**, 035218 (2006).
38. Y. Motome, N. Furukawa, N. Nagaosa, *Phys. Rev. Lett.* **91**, 167204 (2003).
39. Y. Tokura, *Rep. Progr. Phys.* **69**, 707 (2006).
40. K. H. Ahn, T. Lookman, A. R. Bishop, *Nature* **428**, 401 (2004).
41. O. Cepas, H. R. Krishnamurthy, T. V. Ramakrishnan, *Phys. Rev. Lett.* **94**, 247207 (2005).
42. N. Furukawa, *J. Phys. Soc. Jpn.* **64**, 2754 (1995).
43. A. J. Millis, R. Mueller, B. I. Shraiman, *Phys. Rev. B* **54**, 5405 (1996).
44. A. Georges, G. Kotliar, W. Krauth, M. J. Rozenberg, *Rev. Mod. Phys.* **68**, 13 (1996).
45. G. J. Snyder, R. Hiskes, S. DiCarolis, M. R. Beasley, T. H. Geballe, *Phys. Rev. B* **53**, 14434 (1995).
46. J. K. Freericks, E. H. Leib, D. Ueltschi, *Phys. Rev. Lett.* **88**, 106401 (2002).
47. V. J. Emery, S. A. Kivelson, H. Q. Lin, *Phys. Rev. Lett.* **64**, 475 (1990).
48. V. J. Emery, S. A. Kivelson, *Physica C* **209**, 597 (1993).
49. R. Jamei, S. Kivelson, B. Spivak, *Phys. Rev. Lett.* **94**, 056805 (2005).
50. S. D. Baranovskii, A. L. Efros, B. L. Gelmont, B. I. Shklovskii, *J. Phys. C: Solid State Phys.* **12**, 1023 (1979).
51. J. H. Davies, P. A. Lee, T. M. Rice, *Phys. Rev. B* **29**, 4260 (1984).
52. T. Vojta, M. Schreiber, *Philos. Mag. B* **81**, 1117 (2001).
53. J. K. Freericks, V. Zlatić, *Rev. Mod. Phys.* **75**, 1333 (2003).
54. A. L. Efros, B. I. Shklovskii, *J. Phys. C: Solid State Phys.* **8**, L49 (1975).
55. L. Sudheendra, C. N. R. Rao, *J. Physics: Condens. Matter* **18**, 3029 (2003).
56. J. Burgy, M. Mayr, V. Martin-Mayor, A. Moreo, E. Dagotto, *Phys. Rev. Lett.* **87**, 277202 (2001).
57. R. Mathieu, D. Akahoshi, A. Asamitsu, Y. Tomioka, Y. Tokura, *Phys. Rev. Lett.* **93**, 227202 (2004).
58. M. K. Chattopadhyay, S. B. Roy, P. Chaddah, *Phys. Rev. B* **72**, 180401 (2005).
59. P. Soh, Y.-A. Evans, Z. Cai, B. Lai, C.-Y. Kim, G. Aeppli, N. Mathur, M. Blamire, E. Isaacs, *J. Appl. Phys.* **91**, 7742 (2002).
60. M. Paranjape, A. K. Raychaudhuri, N. D. Mathur, M. G. Blamire, *Phys. Rev. B* **67**, 214415 (2003).
61. P. Postorino, A. Congeduti, P. Dore, A. Sacchetti, F. Gorelli, L. Ulivi, A. Kumar, D. D. Sarma, *Phys. Rev. Lett.* **91**, 175501 (2003).
62. A. J. Millis, T. Darling, A. Migliori, *J. Appl. Phys.* **83**, 1588 (1998).
63. C. A. Perroni, V. Cataudella, G. D. Filippis, G. Iadonisi, V. M. Ramaglia, F. Ventriglia, *Phys. Rev. B* **68**, 224424 (2003).

Index

a

abacus, catenanes 332
absorption
  – anisotropic noble metal nanoparticles 202
  – growth kinetics 155
  – plasmon resonances 200–204
absorption cross section, single wall carbon nanotube FETs 313
absorption edge shift
  – capping agents 161
  – growth kinetics 155
acetylacetonates, sol–gel routes 132–133
acetylcholinesterase (AChE), biosensors 281–282
acids, dendrimers 270
activated carbon-ruthenium oxide nanoparticles 240
additive, flame-retardant 72
adenovirus, amyloid fibrils 177
ADNT see aromatic dipeptide nanotubes
adsorption, effect on growth kinetics 166–167
Ag nanoparticles, extinction spectra 199
aggregation, growth process 160
alcohols, sol–gel routes 127–131
aldehydes, sol–gel routes 131–132
aldol-like condensation 133
alignment of gold nanorods 90
aliphatic linkers, molecular self-assembly 171
alkoxides, sol–gel routes 130–132
alkyl halide elimination, nonaqueous sol–gel routes 133
aluminum, electrode materials 305
ambipolar OFETs, progress 310–311
ammonium salts, molecular machines 346–348
amphiphilic peptides, molecular self-assembly 175

amyloid fibrils 177–178
  – molecular self-assembly 172
  – pathologically associated proteins 178
amyloid structures 178–180
anion sensing, dendrimers 279–281
anion-templating, molecular machines 335–337
aniosotropic surface binding, assembly of nanorods 211
anisotropic growth
  – metal nanocrystals 14–17
  – nanocrystals 7–14
  – semiconductor nanocrystals 7–14
anisotropic noble metal nanoparticles
  – optical response 201–202
  – phase retardation 202
antenna-prophyrin metal complex 274
anti-epidermal growth factor receptor 104
antibacterial agents, peptide nanostructures 173
antigen-antibody decomplexation 286
aqueous electrolytes, potential 244
aqueous medium, ordered assemblies 24
arc method, purification 51
aromatic dipeptide nanotubes (ADNT) 179
aromatic homodipeptides 179
aromatic interactions, amyloid fibrils 178
aromatic p–p association routes, rotaxanes 322–329
artificial muscle, catenanes 329
assemblies
  – dendrimers 283–284
  – dimerization 214
  – nanocrystals 22–30
  – see also self-assembly
assembling devices, field-effect transistors 314–316
asymmetric capacitors
  – conducting polymers 237
  – hybrid systems 239

asymmetric molecules, chiral sensing 275–277
atmospheric pressure CVD 86
Au nanocrystals
– growth 153–154
– lattice 25
Au nanoprisms 199
Au nanorods 92
Au nanowires 101

b

bacterial contamination, dendrimers 282
bandgap
– inorganic nanowires 96
– particle size measurement 152
– quantum dots 19
– tunable 76
bandgap variation, nanocrystals 18–19
bar-coding, rotaxanes 336
barium titanate nanoparticles 130
bathochromic shift, surface plasmons 190
BCB see divinyltetramethylsiloxane-bis(benzocyclobutene)
benzyl alcohol, solvent-controlled synthesis 127, 131
Bi nanotubes, electronic transport properties 79
bifunctional molecules, assembly of nanorods 208
binary lattices 25
biochemistry, carbon nanotubes 69–71
biocompatibility
– carbon nanotubes 70
– dendrimers 281
– peptide building blocks 171
biodiagnostics, surface plasmons 200
bioinspired material
– amyloid fibrils 177–178
– peptide-based composite 180
biosensors
– acetylcholinesterase 281–282
– bone growth 69
– calorimetric 288–289
– dendrimer-based 281–292
– inhibitory effect 282
– PAMAM dendrimers 281
– SAM–dendrimer conjugates 284–288
biotin-streptavidin binding detection 32
biotin-streptavidine connectors 208
bistable color-switching [2]-catenane 325
block copolypeptides 180
bolaamphiphiles 174
bone growth, biomedical sensors 69

boron nitride (BN)
– cathodoluminescence properties 94
– solubilization 77
Borromean ring synthesis 334
branched Au nanocrystals 17
bulk purity 52

c

CaC bond formation, nonaqueous sol–gel routes 133
calorimetric biosensors 288–289
cancer treatment
– inorganic nanowires 103
– laser photothermal destruction 197
– metal nanocrystals 34
capacitance 227
– carbononaceous materials 232
– conducting polymers 230, 237
– hybrid electrodes 235
– hybrid materials 226–229
capacitors
– conventional 225–226
– electrochemical 223–229
– electrochemical double layer 225–226
capacity, carbon nanotubes 61
capacity fading, conductive polymers 245
capping agents
– effect on growth kinetics 160–161
– monomer concentration 162
– NaOH concentration 164
– oleic acid 161
– PVP 163–166
– rate constant 162
– size distribution 166
– thiols 165
– ZnO nanocrystals 163–166
carbon dioxide detection 271
carbon monoxide detection 271
carbon nanotube FETs 73
carbon nanotube-molecule-silicon junctions 74
carbon nanotubes (CNT) 45–75
– biochemical and biomedical aspects 69–71
– biocompatibility 70
– catalyst precursor films 46
– chemical sensors 68–69
– electrical conductivity 71
– electrical properties 60–66
– electrochemical functionalization 56, 68
– electrode materials 229
– electrodeposition 58
– electron emission 72

- energy storage and conversion 68
- etching 47
- exciton-phonon bound states 61
- films 46
- flame-retardant additive 72
- floating-potential dielectrophoresis 73
- fluid mechanics 66–68
- functionalization 54–60
- hybrid materials 71, 231–234
- ion transport 69
- isotropic-nematic phase transition 66
- Kondo effect 66
- lattice defects 58
- luminescence 61
- mechanical properties 66–68
- metal-semiconductor transition 64
- microcatheter 72
- microwave plasma CVD 48
- near-infrared photovoltaic devices 72
- nebulized spray pyrolysis 47
- noncovalent functionalization 56
- optical properties 60–66
- phase transitions 66–68
- photoconductivity excitation spectra 64
- photovoltaic devices 72
- p-n junction diodes 73
- polymer composites 71
- precursors 46
- properties and applications 60–69
- PSS 46
- purification 50–54
- pyrolysis 47
- Raman spectra 63
- resistance 61
- sidewall functionalization 54
- solubilization 54–60
- supercapacitors 68
- superhydrophobic films 68
- superplastic deformation 66
- surfaces 58
- switches 66
- synthesis 45–50
- thermal conductivity 64
- transistors 72–75
- transport regimes 64
- walls 60
- water plasma CVD 48
- water-soluble graft copolymers 71
carbonaceous materials
    - capacitance 232
    - hybrid materials 231–234
    - power densities 231
    - specific capacitance 234

cascade molecules 249
catalyst precursor films 46
catalytic combustion, MWNT synthesis 45
catenanes
    - abacus 332
    - aromatic p–p association routes 322–329
    - artificial muscle 329
    - click chemistry 327
    - coordination sites 330
    - crown ethers 323
    - dialkyl ammonium salts 346–348
    - Eglington coupling 327
    - hydrogen-bonded assembly 338–348
    - metal-ion templating 332
    - molecular machines 320–348
    - molecular train 324
    - redox-switchable 329–332
    - rotary motor 331
    - self-assembly 321
    - Stoddart's route 322
    - synthetic routes 321–322
    - tetrathiafulvalene (TTF) 324
    - thermodynamic control 339
cathodoluminescence (CL)
    - boron nitride nanorods 94
    - inorganic nanowires 98
cation radicals, dendrimers 271
cations, detection 277–279
CdS nanowires 95
CdSe nanocrystals 2
    - capping agents 161
    - oleic acid effect 161–163
    - spectra 2
CdSe tetrapods 8
cell capture agents 282–283
cell surface, molecular self-assembly 172
centrifugation, in purification 52
ceramic coating 92
    - carbon nanotubes 58
    - metal oxide nanowires 59
characteristic length scales, surface plasmons 188
charge density, OFET 301
charge separation
    - single wall carbon nanotube FETs 314
    - supercapacitors 225
charge transport, barriers 97
chemical sensors
    - carbon nanotubes 68–69
    - dendrimers 267–281
chemical vapor deposition (CVD) 48
chiral indices 62
chiral sensing 275–277

CL see cathodoluminescence
click chemistry, catenanes 327
closed-caged nanospheres 179
cluster growth, ZnO nanocrystals 156
clusters 160
– supercapacitors 221
– theory of nucleation 141
CMOS technology, OFET 310
CO and CO_2, dendrimers 271–272
Co-CoO nanocrystals, ferromagnetic 21
coatings
– carbon nanotubes 71
– ceramic oxide 92
– conducting polymers 238
– nanowires 92
– rotaxanes 343
– SWNTs 49
coaxial nanowires 92
cobaltocene 291
collective oscillation, conduction electrons 185
colloid synthesis, theory of nucleation 140
colloidal crystals 29–30, 158
colloidal gold 189
color-switching [2]-catenane 325–326
combustion, catalytic 45
complementary circuit, OFET 310
complex oxides 357–385
complexing agents 162
composite detectors 271
composite electrolytes 242
composite films
– carbon nanotubes 71
– dendrimers 267
composites
– bioinspired peptide-based 180
– optical devices 30
concentration gradients, theory of nucleation 142
conducting films 271
conducting polymer-carbon nanotube hybrids 237–238
conducting polymer-transition metal oxide nanohybrids 235–237
conducting polymers 235
– capacitance 230, 237
– capacity fading 245
– coatings 238
– hybrid materials 230–231
– multi-walled carbon nanotubes (MWNT) 237
– polypyrrole (PPy) 237
conduction electrons, collective oscillation 185

conductivity
– OFET 301
– thermal 64
cone-like particle morphology, ZnO nanocrystals 124
confinement, electron plasma 189
continuous monitoring, dendrimers 289
continuously spun fibers, mechanical properties 67
contrast enhancing, metal nanocrystals 34
controlled oversaturation, ordered assemblies 29
controlling motion, molecular machines 349
coplanar fuel cells, inorganic nanowires 103
coplanar geometry, OFET 302
copper, nanocrystals 4
core-multishell heterostructure nanowire arrays 92
coulomb interactions, electronic inhomogeneities 369–381
critical doping levels 374
critical supersaturation level 141
cross-sections, calculation 201
crown ethers, catenanes 323
crystallinity, nonaqueous sol–gel routes 120
cubic nanorods, lattice 9
CVD see chemical vapor deposition
cyclic D,L-peptides 173
cyclic peptide-based nanostructures 172–174
cyclic peptides, functionalization 56
cyclodextrin-based rotaxanes 348–349
cysteine-lysine block copolypeptides 180
cytotoxicity, carbon nanotubes 69

d

DDA see discrete dipole approximation
decomposition, solvothermal 7
dendrimer-based biosensors 281–292
dendrimer-based glucose sensors 289–292
dendrimer-carbon composite detectors
dendrimer peripheries 266
dendrimers 249–298
– acids 270
– anion sensing 279–281
– antenna-prophyrin metal complex 274
– antigen-antibody decomplexation 286
– assemblies 283–284
– bacterial contamination 282
– biocompatibility 281
– calorimetric biosensors 288–289
– carbon dioxide detection 271
– carbon monoxide detection 271
– cascade molecules 249

- cation radicals 271
- cell capture agents 282–283
- chemical sensors 267–281
- chiral sensing 275–277
- CO and CO_2 271–272
- cobaltocene 291
- composite detectors 271
- composite films 267
- conducting films 271
- conformational features 268
- constitution 252–261
- continuous monitoring 289
- core 250
- covalent bond formation 262
- covalent coupling 289
- DNA sensing 287
- electrical response 267
- electroactive signal tracer 285
- encapsulation 265
- endo-receptor properties 265–267
- exo-receptor properties 265–267
- ferrocene 280, 291
- ferrocene methanol 285
- films 271
- fluorescence 264–265, 277–279, 289
- FRET 289
- functional groups 250, 266
- functionalization 267, 271
- gas sensing 212–275
- generation 251
- glucose monitoring 289
- growth 251
- guest binding 276
- intrinsic viscosity 262–264
- LBL assembly 267, 283–284
- linkers 250
- luminescence 264
- macromolecular properties 262–267
- metal-ligand coordination bond formation 262
- molecular modeling 262–264
- molecular recognition 279
- nanoscale sensors 249–298
- non-covalent bond formation 262
- organic amines 270
- oxyen quenching constant 274
- PAMAM 281
- phthalocyanins 271
- platinum^{II} complex 272
- radicals 271
- radius of gyration 263
- SAM-dendrimer conjugates 284–288
- sensors see sensors
- silicon linkages 281
- site-selective positioning 275
- structural features 250
- sulfur dioxide 272
- surface plasmon resonance sensor 283
- synthesis 250
- synthetic methods 250–562
- vapoconductivity 270–271
- vapor sensing 267–270, 272–275
- viscosity studies 262–264
- visual color detector 288
- VOC detection 269
- water-soluble 281
dendrites, ordered assemblies 24
density of states, electronic inhomogeneities 376
deposition
    - carbon nanotube surfaces 58
    - OFET 302
    - ordered assemblies 26
dialkyl ammonium salts 346–348
dielectric core 196
dielectrophoresis, SWNT transistors 73
diffusion, theory of nucleation 142
diffusion layer thicknesses 143
diffusion limited growth 143–147
diffusion-reaction control, nanocrystal growth 148–151
dilute magnetic semiconductors 22
dimerization, assembly of nanorods 214
diodes, carbon nanotubes 73
dip-coating, OFET 303
dipolar surface plasmon band 197
discrete dipole approximation (DDA) 201
divinyltetramethylsiloxane-bis(benzocyclobutene) (BCB) 311
DLCS see double-layer carbon supercapacitors
DNA
    - biomedical sensors 69
    - functionalization 57–58
    - optical devices 31
    - ordered assemblies 22–23
DNA-directed self-assembling, carbon nanotubes 60
DNA oligonucleotides, self assembly 90
DNA sensing, dendrimers 287
domain-wall propagation, inorganic nanowires 99
doped perovskite manganites 359
doped SWNTs 64
doping level improvement 74
double layer capacitor, hybrid materials 225–226

double-layer carbon supercapacitors (DLCS) 241
double-layered CNT films 48
double-walled carbon nanotubes (DWNTs) 49
drain, electrode materials 305
drop-casting, OFET 303
Drude-Lorentz model, surface plasmons 185
DWNT *see* double-walled carbon nanotubes

e

EDLC *see* electrochemical double layer capacitor
effective half bandwidth 368
effective mass approximation 18
Eglington coupling, catenanes 327
electrical double-layer, supercapacitors 226
electrically conducting polymers 237
– *see also* conducting polymers
electrocatalysis, layer-by-layer assembly 283–284
electrochemical capacitors, hybrid materials 223–229, 235
electrochemical deposition, gold nanorods 198
electrochemical double layer capacitor (EDLC), hybrid materials 225–226, 234
electrochemical power sources, Ragone plot 224
electrode materials 229–234
– field-effect transistors 305–306
– hydrous ruthenium 229
electrodeposition
– carbon nanotubes 58
– inorganic nanowires 81
electrodynamic modeling calculations 205
electroluminescence 94
– optical devices 31
electrolytes
– hybrid materials 241–243
– ionic liquids 242–243
– polymer composites 242
electrolytic capacitors 225
electron beam lithography, surface plasmon 204
electron clouds, collective oscillation 185
electron oscillations 198
electron plasma, localized surface plasmons 189
electronic absorption bands, purification 53
electronic inhomogeneities
– Coulomb interactions 369–381
– critical doping levels 374
– density of states 376
– doped perovskite manganites 359
– effective half bandwidth 368
– electron charge operator 370
– experimental evidence 358–363
– extended LB model 370–381
– Falicov-Kimball model 369
– insulating regions 363
– kinetic energy gain 366
– LB model 366–369
– long-range coulomb interactions 370–381
– magnetic Hamiltonians 365
– memory effect 363
– nanoscale 357–385
– orbital liquid state 366
– theoretical approaches 364–365
– weakly distorted sites 366
electronic state structure 375
enediol-functionalized titania nanoparticles 129
energy storage, electrochemical 222–223
energy storage and conversion, carbon nanotubes 68
enhanced capacitance, origin 226–229
enhanced light transmission, surface plasmons 186
enhancement mode, OFET 300
ester elimination, nonaqueous sol–gel routes 133
etching, carbon nanotubes 47
ethanol sensing, inorganic nanowires 101
ether elimination, nonaqueous sol–gel routes 133
exciton generation, nanocrystals 18
exciton-phonon bound states, carbon nanotubes 61
extended LB model 370–381
extinction
– anisotropic noble metal nanoparticles 202
– gold nanorods 193
– localized surface plasmons 189
– surface plasmons 197

f

F8T2 *see* poly(9,9-dioctylfluorene-co-bithiophene)
Falicov-Kimball model, electronic inhomogeneities 369
far-field dipolar coupling 204
Faradaic processes, supercapacitors 228, 244
FDTD *see* finite different time domain
Fe-Pt nanocrystals 30

Fe^{II}-porphyrin-cored dendrimer   266
ferrocene   280, 291
ferrocene methanol   285
ferrocenylated PAMAM dendrimer-enzyme conjugate   287
ferromagnetic Co-CoO nanocrystals   21
ferromagnetism, inorganic nanowires   99
FET   see field-effect transistors
field-effect mobility, OFET   300, 307
field-effect transistors (FET)
   – device fabrication   300–306
   – inorganic nanowires   98
   – substrate treatment methods   304–305
   – thin films   73
   – see also organic field-effect transistors
field-emission, inorganic nanowires   98
films
   – carbon nanotubes   46
   – composite   71
   – dendrimers   271
   – LBL assembly   283–284
   – ordered assemblies   26
   – superhydrophobic   68
   – see also thin films
finite different time domain (FDTD)   201
flame-retardant additives   72
floating-potential dielectrophoresis   73
fluid mechanics, carbon nanotubes   66–68
fluorescence
   – Au nanorods   92
   – dendrimers   264–265, 277, 289
   – quantum dots   7
fluorescence labeled dendrimers   277–279
fluorescence output, molecular logic gates   350
fluorescence resonance energy transfer (FRET)   32, 289
fluorescent tags, nanocrystals   33
free-space photon, surface plasmons   186
FRET   see fluorescence resonance energy transfer
fuel cells, inorganic nanowires   103
fumaramide, rotaxanes   340
functionalization
   – carbon nanotubes   54–60
   – catenanes   339–346
   – cyclic peptides   56
   – dendrimers   267, 271
   – DNA   57–58
   – electrochemical   56
   – inorganic nanotubes   77–79
   – inorganic nanowires   90–92
   – liquid-liquid extraction   55
   – noncovalent   56

– peptides   56
– piranha solutions   55
– rotaxanes   339–346
– sidewall   54
– solvent-controlled synthesis   129

g
GaN nanoparticles   3
GaN nanowires   89
gas detection
   – dendrimers   271, 272–275
   – inorganic nanotubes   79
gate, molecular   349–351
gate voltage, photoPFETs   311
gene expression, inorganic nanowires   104
generalized multipole technique (GMT)   201
GeNW, Langmuir–Blodgett film   91
glucose biosensors   289–292
glutathione, assembly of nanorods   208–212
GMT   see generalized multipole technique
gold
   – electrode materials   305
   – growth kinetics   153
   – see also Au
gold nanocrystals   see Au nanocrystals
gold nanorods   192
   – absorbance of light by   192
   – electrochemical deposition   198
   – extinction coefficient   193
   – fluorescence intensity   92
   – surface plasmon absorption   186, 193
gradual channel approximation, OFET   301
growth
   – anisotropic   7–17
   – Au nanocrystals   153–154
   – dendrimers   251
   – diffusion limited   143
   – GaN nanowire   88–89
   – in situ   151
   – inorganic nanowires   80
   – layer-by layer   28
   – LSW theory   143–147
   – nanocrystals   141–143
   – seed-mediated   15
   – selective   17–18
   – solid phase   139
   – in solution   139–170
   – SWNTs   48–49
   – TiO$_2$ nanorods   84
   – ZnO nanocrystals   154–160
   – ZnO nanowires   82
growth kinetics
   – absorption   155
   – adsorption   166–167

growth kinetics (*cont.*)
- capping agents 160–161
- gold 153
- mixed diffusion–reaction 150

guest binding, dendrimers 276

h

halides, sol–gel routes 127–130

Hamiltonians
- extended 370
- magnetic 365

heterostructure FETs 310
heterostructure nanowire arrays 92
heterostructured nanoparticles 196
higher order nanostructures, plasmon coupling 204–215
hole-transporting, OFET 311
hollow nanocrystals 5
HOMO, supercapacitors 228
Hunter's catenane synthesis 338
hybrid electric vehicles, supercapacitors 225
hybrid electrodes 234
- capacitance 235

hybrid materials 220
- carbon nanotubes 71, 231–234
- conducting polymers 230–231
- double layer capacitor 225–226
- electrochemical capacitors 223–229
- electrolytes 241–243
- limitations 243–244
- metal oxides 229–230
- nanostructured 219–248
- polymer–carbon nanotube 237–238
- polymer–transition metal oxide 235–237
- polymers 235
- polypyrrole (PPy) 235
- salts 241
- solid state electrochemical capacitor 235
- supercapacitors 241–243
- transition metal oxides 229–230
- transition metal oxides–carbon nanotube 238–241

hybrid surface waves 185

hybrid systems 239
- redox reaction 240
- specific capacitance 240

hydrogels, bioinspired composite nanomaterials 180
hydrogen-bonded assembly 338–348
hydrophobic tails 175
hydrous ruthenium, electrode materials 229

i

iced lipid nanotubes 76
ideally reversible electrode, supercapacitors 227
impurities, solvent-controlled synthesis 130
InAs quantum dots, spectra 20
indium tin oxide, electrode materials 305
inhomogeneities, electronic *see* electronic inhomogeneities
inorganic nanotubes 75–79
- covalent functionalizaion 77
- precursors 76
- properties and applications 79
- solubilization 77–79
- synthesis 75–77
- tunable bandgap 76

inorganic nanowires 79–104
- anti-epidermal growth factor receptor 104
- bandgap 96
- cancer treatment 103
- carbothermal synthesis 82
- cathodoluminescence 98
- coplanar fuel cells 103
- crystal structure 83
- CVD 86
- domain-wall propagation 99
- elastic modulus 102
- electrodeposition 81
- ethanol sensing 101
- ferromagnetism 99
- field effect transistors 98
- fuel cells 103
- functionalization 90–92
- gene expression 104
- growth 80
- interconnects 97
- nitridation 89
- optical trapping 96
- oriented attachment 86
- photothermal therapy 103
- piezoelectric generators 103
- planes 85
- rare-earth chlorides 89
- ring resonator lasers 96
- rotors 96
- self-assembly 90–92
- sensing characteristics 100
- SET transistors 103
- superconductivity 98
- surfactant 81
- synthesis 79–90
- thermal decomposition 87

– transistors 103
– waveguide behavior 95
– Young's modulus 101
inorganic–organic nanocomposite 221
InS nanorods 11
insulating regions, electronic inhomogeneities 363
intercalation 100
interconnects, inorganic nanowires 97
interfaces, surface plasmons 185
interparticle distance, surface plasmons 206
intrinsic viscosity, dendrimers 262–264
IO material *see* inorganic–organic nanocomposite
ion templating, molecular machines 329–338
ionic diffusion 159
ionic liquids 242–243
iron-platinum alloy, nanocrystals 5
isosbestic point, assembly of nanorods 213
isotropic–nematic phase transition 66

j
junctions, carbon nanotube–molecule–silicon 74

k
ketones 134
  – sol–gel routes 131–132
kinetic energy gain, electronic inhomogeneities 366
knotanes 333
  – molecular machines 320, 338–339
knots, hydrogen-bonded assembly 338–348
Kondo effect, carbon nanotubes 66

l
Langmuir-Blodgett (LB) films, self-assembly 91
Langmuir-Blodgett (LB) model
  – electronic inhomogeneities 366–369
  – extended 370–381
Langmuir-Blodgett (LB) technique, OFET 304
Lanreotide 173
laser photothermal destruction, localized surface plasmons 197
lasers, optical devices 31
lattice defects 58
lattices
  – binary 25
  – ordered assemblies 24
  – two-dimensional 4

layer-by-layer (LBL) assembly
  – dendrimers 267, 283–284
  – nanorods 214
  – ordered 28
  – superlattices 26
layered double hydroxides (LDH) 242
LCMO 359
LDH *see* layered double hydroxides
Leigh's catenane synthesis 340
ligand exchange reactions 127
ligands, nonaqueous sol–gel routes 121
light confinement 185
light conversion, localized surface plasmons 196
light-current/dark-current ratio, photoPFETs 312
light emitting diodes 31
light transmission, enhanced 186
linear *bis*-conjugated peptides, amyloid fibrils 178
linear oligopeptides 176
linear peptide-based nanostructures 174–177
linkers, dendrimers 250
lipid nanotubes 76
liquid-liquid extraction 55
lithium ion conductors 242
Llifshitz-Slyozov-Wagner (LSW) theory 143–147
localized surface plasmons (LSP) 185, 189–190
  – colloidal gold 189
  – excitation 196–204
  – laser photothermal destruction 197
  – light conversion 196
  – Mie theory 189
logic functions, molecular machines 349
logic gates, molecular 349–351
long-range Coulomb interactions 370–381
LSP *see* localized surface plasmons
LSW *see* Llifshitz-Slyozov-Wagner (LSW) theory
luminescence
  – carbon nanotubes 61
  – dendrimers 264
LUMO, supercapacitors 228
lymphocyte, fluorescent tags 33

m
M13 bacteriophage, self-assembly 91
machine, characterization 319
macrophage cells, fluorescent tags 33
magnetic filtration, purification 52
magnetic Hamiltonians 365

magnetic modes, surface plasmon   197
magnetic nanocrystals   7
magnetic particles, biochemical application   33
magnetic quantum dots   22
magnetic resonance imaging   34
magnetization, temperature dependence   21
magnetron-sputtering techniques   76
manganese oxide, electrode materials   229
manganites   366–369
Medin, amyloid fibrils   177
memory devices, magnetic nanocrystals   35
memory effect, electronic inhomogeneities   363
3-mercaptoproponic acid (MPA)   208–210
11-mercaptoundecanoic acid (MUA)   208–210
mesoscale inhomogeneities   362
metal acetylacetonates   132–133
metal alkoxides   130–132
metal cations, detection   277–279
metal/dielectric interface, surface plasmons   185–186
metal halide-benzyl alcohol system   127
metal halides   127–130
metal ion-templated synthesis
   – catenanes   330–332
   – knotanes   333
   – rotaxanes   335
metal-ligand coordination bond formation   262
metal nanocrystals
   – anisotropic growth   14–17
   – cancer treatment   34
   – contrast enhancing   34
   – magnetic resonance imaging   34
   – optical anisotropy   32
   – optical response   31
   – recent developments   4–6
   – toxicity   33
metal oleylamine complexes   1
metal-organic vapor phase epitaxy (MOVPE)   92
metal oxide nanocrystals   119–138
   – recent developments   6–7
metal oxide nanoparticles
   – nonaqueous sol–gel routes to   121–127
   – process routes   126
   – solvent-controlled synthesis   127–133
   – surfactant-controlled synthesis   121–127
metal oxide nanowires, ceramic coating   59

metal oxides
   – hybrid materials   229–230, 238–241
   – precursors   132
   – synthesis   119
metal salts, reduction   4
metal–semiconductor transition   64
MgO nanowires, synthesis   82
micellar droplets, nanocrystals synthesis   13
microcatheters   72
microcrystals   29
microwave plasma CVD   48
Mie theory, localized surface plasmons   189
mineral acids, purification   51
MISFET   302
mixed diffusion-reaction control
   – Monte-Carlo simulations   150
   – nanocrystal growth   148–151
molecular abacus, catenanes   332
molecular barcoding   333–335
molecular design   314
molecular elevator   347–348
molecular lock   337
molecular logic gates   349–351
   – input operation   350
   – pseudorotaxane   351
molecular machines   319–356
   – ammonium salts   346–348
   – anion-templating   335–337
   – aromatic p-p association routes   322–329
   – catenanes see catenanes
   – complex structures   332–333
   – controlling motion   349
   – cyclodextrin-based rotaxanes   348–349
   – dialkyl ammonium salts   346–348
   – hydrogen-bonded assembly   338–348
   – ion templating   329–338
   – knotanes   338–339
   – light emission   346
   – logic functions   349
   – molecular logic gates   349–351
   – motion control   343–344, 349
   – multiple catenanes   323–324
   – olefin metathesis   333–335
   – photo-induced electron transfer   350
   – p-p association routes   322–329
   – ratcheted-motor   345
   – rotaxanes see rotaxanes
   – steric constraints   349
   – switchable catenanes   324–326
   – synthetic routes   321–322, 326–328
molecular modeling   262–264
molecular recognition   279
molecular-scale barcoding   336

molecular-scale motors 343
molecular self-assembly see self-assembly
molecular shuttle 321
molecular train 324
monodisperse magnetic nanoparticles 122
monodisperse oxide nanocrystals 6
monodisperse semiconductor nanocrystals 1
monomer concentration
– capping agents 162
– Monte-Carlo simulations 150
monomer diffusion 158
monomer flux, theory of nucleation 142
Monte-Carlo simulations, mixed diffusion-reaction control 150
motors
– catalytic nano- 100
– molecular 320, 343, 348
MOVPE see metal-organic vapor phase epitaxy
MPA see 3-mercaptoproponic acid
MUA see 11-mercaptoundecanoic acid
multi-walled carbon nanotubes (MWNT) 45
– conducting polymers 237
– see also carbon nanotubes; single-wall carbon nanotubes
multilayer deposition
– electronic devices 34
– ordered assemblies 27
multilayered ceramic capacitors 225
multiple catenanes 323–324
multiple exciton generation 18
multipolar excitation 197
multipole resonances 197–200
muscle analog, catenanes 328
MWNT see multi-walled carbon nanotubes

n

n-channel OFETs, progress 309–310
n-type carbon nanotube FETs 73
NADH see β-nicotinamide adenine dinucleotide
nano-switches 34
nanochain formation 212
nanocomposites 71–72
nanocrystals
– applications 30–34
– arrangements 22–29
– bandgap variation 18–19
– colloidal 29–30
– copper 4
– devices 30–35
– different shapes 7–17
– electronic properties 18–21
– electro-optical devices 30–31
– exciton generation 18
– fluorescent tags 33
– iron-platinum alloy 5
– low-dimensional arrangements 22–24
– magnetic properties 7, 21–22
– metal see metal nanocrystals
– metal oxide 6–7, 119–138
– nickel-iron alloy 5
– oleylamine complexes 1
– one-dimensional arrangements 22–24
– optical devices 30–31
– optical properties 18–21
– ordered assemblies 22–30
– organometallic precursors 5
– oxide 7–14
– precursors 2, 5
– properties 18–22
– quantum confinement 18
– quantum dots 7
– quantum yields 18
– reaction-limited growth 147–148
– recent developments 1–44
– selective growth 17–18
– semiconductor see semiconductor nanocrystals
– shapes 7–17
– spherical 1–7
– superparamagnetism 21
– surface atoms 10
– thermolysis 1
– three-dimensional superlattices 26–29
– two-dimensional arrays 24–26
– two-dimensional lattice 4
– water solubility 3
– ZnO see ZnO nanocrystals
nanocrystals growth
– anisotropic 7–17
– diffusion-reaction control 148–151
– in solution 139–170
– LSW theory 143–147
– mechanism 141–143
– oleic acid 161–163
– solution 139–170
– theory of nucleation 140–141
– thiols adsorption 166–167
nanocrystals synthesis
– micellar droplets 13
– solvothermal decomposition 7
– structure-directing agent 14
– thermolysis 4, 6
nanodiamond coated SWNTs 49
nanoelectronics 34–35
nanolasers, optical properties 95

nanoparticles
- shape 191–194
- size 190–191
nanoprisms 13, 16
- surface plasmon 200
nanorods 191
- assembly 208–215
- gold 192
- isosbestic point 213
- modal overlap 199
- MPA 208–210
- MUA 208–210
- synthesis 8–11
nanoscalar electronic devices 34–35
nanoscale electronic inhomogeneities 357–385
nanoscale sensors 249–298
nanosphere lithography 32
nanospheres, assembly 204–208
nanostructured materials
- hybrid 219–248
- multipole resonances 197–200
- peptide-based 172–180
- plasmon coupling 204–215
- surface plasmon resonances 185–218
nanotubes
- molecular self-assembly 175
- recent developments 45–118
nanovesicles formation 175
nanowires
- biological aspects 103–104
- electrical and magnetic properties 97–100
- recent developments 45–118
- sensor applications 100–101
- transistors and devices 102–103
NaOH concentration, capping agents 164
natural amyloid 177–178
near-field coupling 204
near-infrared photovoltaic devices 72
nebulized spray pyrolysis 47
neutral catenane assembly 329
nickel-iron alloy 5
β-nicotinamide adenine dinucleotide (NADH) 68
NiO nanocrystals 6
NiS nanoprism 13
nitridation, inorganic nanowires 89
noble metal nanoparticles, extinction spectrum 198
non-conjugated peptides, molecular self-assembly 176
non-covalent bond formation 262
nonaqueous sol–gel routes 119–138
- aldol-like condensation 133
- alkyl halide elimination 133
- CaC bond formation 133
- crystallinity 120
- ester elimination 133
- ligands 121
- metal oxide nanoparticles 121–127
- monodisperse magnetic nanoparticles 122
- particle morphology 122
- precipitation 121
- precursors 120
- reaction pathways 133
- solvents 121
- stabilizing ligands 121
- steric repulsion 122
- superlattices 122–123
- surface capping agents 122
noncovalent functionalization, carbon nanotubes 56
nonspherical nanostructures 191
nucleation 154
- capping agents 161
- La Mer's condition 141
- theory see theory of nucleation
numerical approximate methods 201

o

OFET see organic field-effect transistors (OFETs)
OI material see organic–inorganic nanocomposite
olefin metathesis 335–336
- molecular machines 333–335
oleic acid, nanocrystal growth 161–163
oleylamine complexes 1
oligo(phenylene ethynylene) (OPE) 74
oligopeptides 176
on/off current ratio, OFET 300
one-dimensional arrangements, nanocrystals 22–24
OPE see oligo(phenylene ethynylene)
optical anisotropy, metal nanocrystals 32
optical cross sections, size-normalized 93
optical devices
- biotin-streptavidin binding detection 32
- light emitting diodes 31
- nanocrystals 30–33
- nanosphere lithography 32
- solar cells 30
optical response
- metal nanocrystals 31
- surface plasmon 201

optical trapping
- inorganic nanowires 96
- plasmon coupling 207
optoelectronic memory effect, photoPFETs 313
optoelectronics, surface plasmons 204
ordered assemblies
- II-VI semiconductors 23
- dendrites 24
- DNA 22–23
- electrostatic self-assembly 29
- films 26
- LBL growth 28
- multilayer deposition 27
- nanocrystals 22–30
- PAH 27
- rings 23–24
- semiconductors 23
- superlattices 25
- templates 22
- weak interactions 28
organic amines, dendrimers 270
organic field-effect transistors (OFETs) 299
- accumulation mode 300
- BCB 311
- carbon nanotube 73
- charge density 301
- CMOS technology 310
- coplanar geometry 302
- dip-coating 303
- field-effect mobility 307
- heterostructure FET 310
- LB technique 304
- molecular approaches 299–318
- polymer FET 307
- polymer semiconductor 309
- printing technology 303
- progress 306–311
- self-organization 314
- staggered geometry 302
- traps 309
- unipolar transport 310
- see also field-effect transistors
organic-inorganic nanocomposite (OI materials) 221
organic solar cells, optical devices 30
organic solvents
- sol–gel routes 132–133
- solvent-controlled synthesis 127
osteoblast proliferation, biomedical sensors 69
Ostwald ripening 151
- theory of nucleation 142
oxide nanocrystals, anisotropic growth 7–14

oxide nanoparticles, biomedical applications 33–34
oxyen quenching constant 274

p

p-channel oFETs, progress 306–309
p-doped SWNTs 64
p-n junction diodes 73
p-p association routes 322–329
PAH see poly(allylamine hydrochloride)
PAMAM see poly(amido amine)
PANI see polyaniline
paraquat-based synthetic routes
- catenanes 326–328
- rotaxanes 327
particle size measurement
- bandgap 152
- electromagnetic waves 151
- SAXS 151
- UV absorption 152
pathologically associated proteins 178
PbS nanowires, synthesis 86
PDDA see poly(diallyldimethylammonium chloride)
peak shifts, distance dependent 206
PEDOT see poly (3,4-ethylenedioxythiophene)
PEDOT-MoO$_3$ nanohybrid electrode 244
PEI see polyethyleneimine
peptide-based nanostructures 178–180
- composites 180
- cyclic 172–174
- linear 174–177
peptide building blocks, molecular self-assembly 171
peptide fragments 177–178
peptide nanomaterials 171–184
peptides, functionalization 56
performance improvement, OFET 315
perovskite manganites, electronic inhomogeneities 359
pesticides, biosensors 281–282
pH-switchable rotaxane 346
photo-switchable rotaxane 340
photochemically prepared gold nanorods solution 192
photoconversion, carbon nanotubes 72
photodetector, inorganic nanowires 103
photoexcitation 314
photonics, surface plasmon-based 188
photons, surface plasmons 186
photopFETs 311–313
photoswitchable fluorinated rotaxane coating 340

photothermal therapy, inorganic nanowires   103
photovoltaic devices, carbon nanotubes   72
phthalocyanins   271
piezoelectric generators, inorganic nanowires   103
piranha solutions, functionalization   55
planes
   – inorganic nanowires   85
   – superlattices   26
plasma, resonance frequency   185
plasmon bands, particle size measurement   152
plasmon coupling   204–215
   – near-field   204
   – optical trapping   207
plasmon oscillations, tuning   190–196
plasmon peak   195
plasmon resonances   190
   – absorption   200–204
   – dielectric environment   194–196
   – heterostructured nanoparticles   196
   – nanostructured materials   185–218
   – scattering   200–204
   – *see also* surface plasmon resonances
platinum^{II} complex   272
polaritons   186
polarized light, surface plasmons   193
polaronic l states   366
poly(allylamine hydrochloride) (PAH)   27
poly(amido amine) (PAMAM)   249, 281
poly-amino acids   171
polyaniline (PANI)   237
polyaniline-coated carbon nanofiber   238
poly(diallyldimethylammonium chloride) (PDDA)   27
poly(9,9-dioctylfluorene-co-bithiophene) (F8T2)   309
polyelectrolytes
   – hybrid materials   242
   – layer-by-layer (LBL) approach   214
   – optical devices   30
   – structures   27
polyethyleneimine (PEI), ordered assemblies   27
poly(3,4-ethylenedioxythiophene) (PEDOT)   236
   – conducting polymers   237
polymer-carbon nanotube hybrids   237–238
polymer composites
   – alignment of gold nanorods   90
   – carbon nanotubes   71
polymer dispersed single wall carbon nanotube transistors   313–314
polymer FET, OFET   307
polymer films, OFET   303
polymer-fullerene photovoltaic devices   72
polymer ligand   3
polymer matrix, composite electrolytes   242
polymer semiconductors, OFET   309
polymer-transition metal oxide nanohybrids   235–237
polymeric field-effect transistors   *see* organic field-effect transistors
polymers
   – hybrid materials   230–231
   – optical devices   30
polypyrrole (PPy)
   – conducting polymers   237
   – hybrid polymer   235
poly(sodium-4-styrenesulfonate) (PSS)   46
poly(styrene sulfonate)sodium salt (PSS)   27
polyvinyl pyrollidone (PVP)   16, 163–166
post-LSW theories, nanocrystal growth   143–147
power densities
   – carbononaceous materials   231
   – supercapacitors   223
PPI dendrimers   281
PPy   *see* polypyrrole
precursors
   – carbon nanotubes   46
   – inorganic nanotubes   76
   – nanocrystals   2, 5
   – nonaqueous sol–gel routes   120
   – solvent-controlled synthesis   132
propagating surface plasmons (PSP)   185
prostate-specific antigen, detection   101
protein immunosensors   69
protein-nanotube conjugates   70
proton conducting properties   241
pseudocapacitance   227
pseudorotaxane
   – assembly   337
   – molecular logic gates   351
PSP   *see* propagating surface plasmons
PSS   *see* poly(sodium-4-styrenesulfonate); poly(styrene sulfonate)sodium salt
purification
   – carbon nanotubes   50–54
   – electronic absorption bands   53
   – magnetic filtration   52
   – mineral acids   51
PVP   *see* polyvinyl pyrollidone

q

QCM   *see* quartz crystal microbalance
quadrupole surface plasmon absorption   198

quantum confinement
— nanocrystals 18
— spatial 220
quantum dots 33
— bandgap 19
— magnetic 22
— nanocrystals 7
— spectra 20
— supercapacitors 220
quantum efficiency 95
quantum yields, nanocrystals 18
quartz crystal microbalance (QCM) 268
quasi-zero dimensional systems 220

r
RADA16-I hydrogel 176
radicals, dendrimers 271
radio frequency identification (RFID) 299, 315
Ragone plot, electrochemical power sources 224
Raman spectra, carbon nanotubes 52, 63
random telegraph signal, SWNT-FETs 75
rare-earth chlorides 89
ratcheted-motors 345
reaction-limited growth, nanocrystals 147–148
reaction pathways, nonaqueous sol–gel routes 133
redox-switchable catenanes 329–332
redox-switchable rotaxanes 329–332
resonance energy transfer (FRET) 289
resonance frequency, surface plasmons 185
RFID *see* radio frequency identification
ring resonator lasers, inorganic nanowires 96
rodlike micelles 193
rotary motors, catenanes 331
rotaxanes
— aromatic p–p association routes 322–329
— bar-coding 336
— coatings 343
— cyclodextrin-based 348–349
— dialkyl ammonium salts 346–348
— fumaramide 340
— hydrogen-bonded assembly 338–348
— molecular machines 320–348
— paraquat-based route 327
— redox-switchable 329–332
— routes 333–335
— synthetic routes 321–322, 328

rotors
— inorganic nanowires 96
— self-powered 99
RRP3HT, OFET 307
RTS *see* random telegraph signal

s
S-layers, molecular self-assembly 172
salts
— hybrid nanostructured materials 241
— reduction 4
SAM *see* self-assembled monolayers
SAM-dendrimer conjugates, dendrimers 284–288
SAXS *see* small angle X-ray scattering
scattering
— anisotropic noble metal nanoparticles 202
— plasmon resonances 200–204
seed-mediated growth 15
self-assembled monolayers (SAMs) 174
self-assembly
— catenanes 321
— DNA-directed 60
— DNA oligonucleotides 90
— inorganic nanowires 90–92
— Lanreotide 173
— LB film 91
— M13 bacteriophage 91
— naturally occurring 171–172
— peptides 171–184
— RADA16-I hydrogel 176
— silica nanowires 90
— S-layers 172
— viruses 172
— *see also* assemblies
self-organization, OFET 314
self-powered synthetic nanorotors 99
semiconducting polymer dispersed SWNT transistors 313–314
semiconductor nanocrystals
— anisotropic growth 7–14
— ordered assemblies 23
— recent developments 1–4
sensing
— anion 279–281
— biomolecular 284–288
— chiral 275–277
— CO and CO_2 271–272
— in solution 272–275
— organic amines and acids 270
— vapor 34, 267–270
sensing characteristics, inorganic nanowires 100

sensors 100–101
- biomedical 69
- chemical see chemical sensors
- dendrimers 249–298
- glucose 289–292
- β-nicotinamide adenine dinucleotide (NADH) 68
- surface plasmon resonance 283
SET see single electron tunneling transistors
shape-controlled synthesis, nanocrystals 13
shape dependence, surface plasmon resonance 191
shapes, nanocrystals 7–17
sidewall functionalization, carbon nanotubes 54
silicon linkages, dendrimers 281
silicon nanowires (SiNWs)
- self-assembly 90
- synthesis 79
silver, surface plasmons 186
silver nanoparticles, extinction spectra 199
single crystalline nanowires 81
single-electron tunneling (SET) transistors
- electronic devices 34
- inorganic nanowires 103
single-wall carbon nanotubes (SWNT)
- charge separation 314
- electronic spectrum 53
- FETs 75, 313
- semiconducting 313–314
- nanodiamond coated 49
- p-doped 64
- synthesis 48
- transistors 73
- see also carbon nanotubes; multi-walled carbon nanotubes
SiNW see silicon nanowires
site-selective positioning, dendrimers 275
size distribution
- capping agents 166
- mixed diffusion–reaction 150
- reaction-limited growth 147–148
small angle X-ray scattering (SAXS), particle size measurement 151
SNAP see superlattice nanowire pattern transfer
SnO_2 nanorods, ethanol sensing 101
sol–gel chemistry, aqueous and nonaqueous 120–121
sol–gel routes
- acetylacetonates 132–133
- alcohols 127–131
- aldehydes 131–132
- alkoxides 130–132

- chemical reactions 127–134
- halides 127–130
- ketones 131–132
- metal acetylacetonates 132–133
- metal alkoxides 130–132
- nanocrystalline metal oxides 119–138
- nonaqueous see nonaqueous sol–gel routes
- organic solvents 132–133
- selected reaction mechanisms 133–134
solar cells 30
solid phase, growth 139
solid polymer electrolytes (SPEs) 242
solid state electrochemical capacitor 235
solubilization
- carbon nanotubes 54–60
- inorganic nanotubes 77–79
solution-phase Raman spectroscopy, purification 52
solvent-controlled synthesis
- impurities 130
- metal oxide nanoparticles 119, 127–133
- organic solvents 127
- precursors 132
- surface functionalization 129
- Ti nanoparticles 129
solvents
- effect on kinetics 160
- nonaqueous sol–gel routes 121
solvothermal decomposition, nanocrystals synthesis 7
source, electrode materials 305
spatial quantum confinement, supercapacitors 220
SPE see solid polymer electrolytes
specific capacitance
- carbononaceous materials 234
- hybrid systems 240
spectra
- CdSe nanocrystals 2
- InAs quantum dots 20
spherical clusters, supercapacitors 221
spherical nanocrystals, recent developments 1–7
spin-casting, OFET 303
stabilizing ligands 121
staggered geometry, OFET 302
Stoddart's pH-switchable rotaxane 346
Stoddart's route, catenanes 322
Stokes-Einstein diffusion model 159
structure-directing agent 14

substrate treatment methods, field-effect transistors   304–305
subwavelength holes   188
subwavelength waveguide components   186
sulfur dioxide, dendrimers   272
supercapacitors
– carbon nanotubes   68
– charge separation   225
– clusters   221
– costs   243–244
– double-layer carbon   241
– electrical double-layer   226
– electrical storage   225
– electrochemical energy   222–223
– electrode materials   229–234
– electrolytes   241–243
– electrons   228
– Faradaic processes   228, 244
– HOMO   228
– hybrid electric vehicles   225
– hybrid materials   219–248
– ideally reversible electrode   227
– inorganic-organic nanocomposite   221
– ionic liquids   242–243
– long term stability   243
– LUMO   228
– organic-inorganic interface   221
– PEDOT   236, 244
– possible limitations of hybrid materials   243–244
– potential   244
– power densities   223
– quantum dots   220
– spatial quantum confinement   220
– spherical clusters   221
superconductivity, inorganic nanowires   98
supercrystals   122
superhydrophobic films   68
superlattice nanowire pattern transfer (SNAP)   97
superlattices
– nonaqueous sol–gel routes   122–123
– ordered assemblies   25
– planes   26
superparamagnetism, nanocrystals   21
superplastic deformation   66
superstructures   208
surface atoms
– nanocrystals   10
– percentage   12
surface capping agents   127, 161
– nonaqueous sol–gel routes   122
surface charge density, OFET   301
surface functionalization   129

surface plasmon absorption
– gold nanorods   193
– quadrupole   198
surface plasmon-based photonics   188
surface plasmon oscillations   190–196
surface plasmon polaritons   186
– *see also* propagating surface plasmons (PSP)
surface plasmon resonance   185–218
– dendrimers   283
– shape dependence   191
surface plasmons   185, 187
– bathochromic shift   190
– biodiagnostics   200
– characteristic length scales   188
– Drude-Lorentz model   185
– electron beam lithography   204
– enhanced light transmission   186
– excitation   196–204
– extinction modes   197
– free-space photon   186
– frequency of oscillation   190
– introduction   185–190
– localized   *see* localized surface plasmons
– photons   186
– polarized light   193
– propagating   186–189
– silver   186
– thin films   186
– TOAB   195
– waveguide components   186
surfactant-controlled synthesis   119, 121–127
surfactant-like peptides   175
surfactants, inorganic nanowires   81
SWCNTFET   *see* single wall carbon nanotube FETs
switchable catenanes   324–326, 328–329
switches, carbon nanotubes   66
SWNT   *see* single-wall carbon nanotubes
synthesis
– carbon nanotubes   45–50
– catenanes   329
– core-multishell heterostructure nanowire arrays   92
– dendrimers   *see* dendrimers
– inorganic nanotubes   75–77
– inorganic nanowires   79–90
– MgO nanowires   82
– molecular machines   321–322, 326–328
– nanocrystals   *see* nanocrystals synthesis
– nanorods   8–11

synthesis (*cont.*)
– PbS nanowires 86
– shape-controlled 13
– silicon nanowires 79
– surfactant-controlled 119, 121–127
– SWNTs 48
– VO$_2$ nanowires 84
synthetic rotors, self-powered 99

t

T-matrix method   *see* transition matrix method
TEM   *see* transmission electron micrograph
templates
– ordered assemblies 22
– self-assembly 91
tetraoctylammonium bromide (TOAB) 195
tetrathiafulvalene (TTF) 324
theory of nucleation
– concentration gradients 142
– diffusion 142
– diffusion limited growth 143–147
– nanocrystal growth 140–141
– particle growth 147
– particle radii 145
– supersaturation level 141
– surface reaction 147
thermolysis
– nanocrystals 1
– nanocrystals synthesis 4, 6
thin films
– field-effect transistors 73
– OFET 302
– optical devices 30
– ordered assemblies 26
– surface plasmons 186
– *see also* films
thiols, capping agents 165
thiols adsorption 166–167
thiophene oligomers, OFET 307
three-dimensional binary lattices 25
three-dimensional superlattices 26–29
Ti nanoparticles 129
tin oxide
– electrode materials 229
– nanorods growth 84
tissue engineering, linear peptides 176
TOAB   *see* tetraoctylammonium bromide
touch sensor 32
toxicity, metal nanocrystals 33
transistors
– carbon nanotubes 72–75
– inorganic nanowires 103
– SWNT 73, 313–314
– *see also* field-effect transistors

transition matrix (T-matrix) method
– assembly of nanospheres 204
– optical response 201
transition metal oxides
– electrode materials 229
– hybrid materials 229–230
– nanohybrids 235–237
– nanotube hybrids 238–241
transmission electron micrograph (TEM), particle size measurement 151
traps, OFET 309
TTF   *see* tetrathiafulvalene
tubular assemblies 173
tubulin, molecular self-assembly 172
tunable bandgap, inorganic nanotubes 76
tuning
– extinction cross-section 203
– optical properties of metal nanoparticles 190
– plasmon oscillations 190–196
two-dimensional arrays, nanocrystals 24–26
two-dimensional lattices 4
two-dimensional layers 221

u

ultra-thin films
– ordered assemblies 26
– *see also* thin films
ultracapacitors 223

v

van der Waals interactions, coalescence of nuclei 161
vapoconductivity 270–271
vapor pressure, ionic liquid 243
vapor sensing
– dendrimers 267–270, 272–275
– electronic devices 34
viruses, molecular self-assembly 172
viscosity studies, dendrimers 262–264
visual color detectors 288
VO$_2$ nanowires, synthesis 84
VOC   *see* volatile organic compounds
Vögtle's route to knotanes 340
volatile organic compounds (VOC) detection 269

w

walls, carbon nanotubes 60
water, dissociation constant 157
water concentration, ZnO nanocrystal size 158
water plasma chemical vapor deposition 48
water-soluble dendrimers 281
water-soluble graft copolymers 71

waveguide behavior, inorganic nanowires   95
waveguide components, surface plasmons   186
weakly distorted sites   366

y
Young's modulus, inorganic nanowires   101

z
ZnO nanocrystals
 – capping agents   163–166
 – cluster growth   156
 – cone-like particle morphology   124
 – growth   154–160
 – particle morphology   122
 – size   158, 163

ZnO nanowires
 – field-effect transistors   98
 – growth   82
 – optical properties   94
 – photodetector   103

Further Reading

S. A. Edwards

The Nanotech Pioneers

Where Are They Taking Us?

2006

ISBN: 978-3-527-31290-0

G. Schmid (Ed.)

Nanoparticles

From Theory to Application

2004

ISBN: 978-3-527-30507-0

P. M. Ajayan, L. S. Schadler, P. V. Braun

Nanocomposite Science and Technology

2003

ISBN: 978-3-527-30359-5

C. S. S. R. Kumar, J. Hormes, C. Leuschner (Eds.)

Nanofabrication Towards Biomedical Applications

Techniques, Tools, Applications, and Impact

2005

ISBN: 978-3-527-31115-6

P. Samori (Ed.)

Scanning Probe Microscopies Beyond Imaging

Manipulation of Molecules and Nanostructures

2006

ISBN: 978-3-527-31269-6